14th International Conference on Development and Modernization of Manufacturing (RIM 2023)

IOP Conference Series: Materials Science and Engineering
Volume 1298

Sarajevo, Bosnia and Herzegovina
20 - 22 September 2023

Editors:

Zijah Burzic
Budimir Mijovic
Fadil Islamovic
Atif Hodzic

ISBN: 978-1-7138-8735-5
ISSN: 1757-8981

Printed from e-media with permission by:

Curran Associates, Inc.
57 Morehouse Lane
Red Hook, NY 12571

Some format issues inherent in the e-media version may also appear in this print version.

Institute of Physics
Dirac House, Temple Back
Bristol BS1 6BE UK

Phone: 44 1 17 929 7481
Fax: 44 1 17 920 0979

techtracking@iop.org

Additional copies of this publication are available from:

Curran Associates, Inc.
57 Morehouse Lane
Red Hook, NY 12571 USA
Phone: 845-758-0400
Fax: 845-758-2633
Email: curran@proceedings.com
Web: www.proceedings.com

TABLE OF CONTENTS

IOP Conf. Series: Materials Science and Engineering 1298 (2023) 011001 doi:10.1088/1757-899X/1298/1/011001

Preface

This volume of "IOP Conference Series: Material Science and Engineering" contains the selected papers presented at the 14th International Conference on Development and Modernization of Manufacturing (RIM 2023), which was held in Sarajevo (Bosnia and Herzegovina) from 20th to 22nd September 2023.

Since 1997, the Faculty of Technical Engineering of the University of Bihać has successfully organized and held 14 international scientific conferences RIM on a bi-annual basis in 4 different countries: Bosnia and Herzegovina, Croatia, Egypt, and Montenegro. The main goal of the Conference is to bring together international scientists, engineers, industrialists, and experts providing them a unique platform for sharing new ideas and the recent developments in various areas of engineering.

The papers of this volume are related to the following thematic areas:

- Mechanical Engineering,
- Wood Products Engineering and Design,
- Electrical Engineering and Computer Science,
- Technologies and Structures in Civil Engineering.

The Organization Committee of the RIM 2023 conference would like to thank all the authors who contributed to this volume as well as to the reviewers, invited speakers, and all conference participants.

List of Editors:

- Prof. Zijah Burzić, PhD
- Prof. Budimir Mijović, PhD
- Prof. Fadil Islamović, PhD
- Prof. Atif Hodžić, PhD

List names of Organization Committee:

Atif Hodžić, PhD, president; Damir Hodžić, PhD, vice-president; Fadil Islamović PhD; Aladin Crnkić, PhD; MSc Mustafa Ružnić; MSc Edvin Alijanović; Fatka Kulenović, PhD; Jasna Hamzabegović, PhD; Edis Softić, PhD; Šefik Behrem, PhD; Redžo Hasanagić, PhD; Samir Vojić, PhD; Aida Husetić, PhD; Dženana Gačo, PhD; Minka Ćehić, PhD; Husein Rošić, PhD; Ermin Bajramović, PhD; Ratko Knežević, PhD; Blaženka Brlobašić Šajatović, PhD; Esad Bajramović, PhD; MSc Dejla Ramić, MSc Amel Džanić; MSc Amel Toroman; MSc Zinaid Kapić; BSc Irma Pašić; Nevzeta Ćehajić.

List names of Scientific Committee:

Salah Eldien Omer, PhD, president (CRO); Dr. Fadil Islamović, PhD (BIH); Atif Hodžić, PhD (BIH); Ismet Bišćević, PhD (BIH); Budimir Mijović, PhD (CRO); Thomas Stautmeister, PhD (GER); Milan Jurković, PhD (CRO); Refik Šahinović, PhD (BIH); Fadi Al-Turjman, PhD (TUR); Anatoly Chubinski, PhD (RUS); Necati Agiralioglu, PhD (TUR); Ali Danadeh Mehr, PhD (TUR); Zijah Burzić, PhD (SRB); Milan Šernek, PhD (SLO); Radivoje Topić, PhD (SRB); Branko Božić, PhD (CRO); Hamid Farrokh Ghatte, PhD (TUR); Zoran Trposki, PhD (MKD); Munir Jahić, PhD (BIH); Darko Ujević, PhD (CRO); Slobodan Bojanić, PhD (ESP); Siniša Ilić, PhD (SRB); Smail Klarić, PhD (BIH); Fatka Kulenović, PhD (BIH); Vencislav Grabulov, PhD (SRB); Taško Maneski, PhD (SRB); Sanin Džidić, PhD (BIH); Galina Varankina, PhD (RUS); Amir Čaušević, PhD (BIH); Dženana Gačo, PhD (BIH); Mersida Manjgo, PhD (BIH); Jovan Vučinić, PhD (CRO); Damir Hodžić, PhD (BIH); Endre Romhanji, PhD (SRB); Marko Petrič, PhD (SLO); Šefik Behrem, PhD (BIH); Meri Burzić, PhD

IOP Conf. Series: Materials Science and Engineering 1298 (2023) 011001 doi:10.1088/1757-899X/1298/1/011001

List names of reviewers:

Peer Review Statement

All papers published in this volume have been reviewed through processes administered by the Editors. Reviews were conducted by expert referees to the professional and scientific standards expected of a proceedings journal published by IOP Publishing.

- **Type of peer review:** Double anonymous
- **Conference submission management system:** Morressier
- **Number of submissions received:** 56
- **Number of submissions sent for review:** 51
- **Number of submissions accepted:** 34
- **Acceptance Rate (Number of Submissions Accepted / Number of Submissions Received X 100):** 60.7 %
- **Average number of reviews per paper:** 2
- **Total number of reviewers involved:** 57
- **Contact person for queries:**

Aladin Crnkić, PhD; Faculty of Technical Engineering, University of Bihać; Irfana Ljubijankića bb, 77000 Bihać, Bosnia and Herzegovina; aladin.crnkic@unbi.ba

Comparative Analysis of the Mechanical Exploitation Properties of the Steel Sheet Welded Joint of New and Old Materials

Fadil Islamović[1], Ermin Bajramović[1], Belma Islamović[1] and Esad Bajramović[1]
[1] Faculty of Technical Engineering, University of Bihać, Bihać, Bosnia and Herzegovina

E-mail: fadil.islamovic@unbi.ba

Abstract. The rehabilitation process of the war-damaged vertical cylindrical tank with a volume of 5.000 m^3 located at the oil terminal for liquid cargo TTT S-105 Ribić in Bihać, included a preliminary experimental examination of the mechanical exploitation properties of the steel sheet welded joints. A new steel sheet Č.0361 was acquired for this purpose. In addition, a sample of an old sheet of the same quality was taken from the construction of the devastated tank, which had been in use for more than 35 years. The experimental testing should determine the mechanical exploitation properties of the new material, examine the same properties of the old used material, compare the obtained results and confirm their mutual acceptability in the joint construction. The tank in question was damaged as a result of direct or indirect missile hits during the war period. Following the more economically acceptable rehabilitation, compared to demolition and complete reconstruction, the tank should be returned to a safe and usable condition.

1. Introduction

Liquid cargo terminal S-105 Ribić in Bihać is one of four oil terminals in the territory of the Federation of Bosnia and Herzegovina managed by the Operator - Terminali Federacije, d.o.o. Sarajevo. The total storage capacity of the five vertical cylindrical above-ground tanks of this terminal is 18.100 m^3. Technical data on capacities, diameters and type of derivatives for which individual tanks are intended are given in Table 1.

Table 1. Vertical above-ground cylindrical tanks for the storage of petroleum derivatives.

Tank	Capacity (m^3)	Diameter (m)	Type of derivatives
R-01	2.700	20	Diesel
R-02	2.700	20	Diesel
R-03	5.000	24	Petrol
R-04	5.000	24	Diesel
R-05	2.700	20	LUEL

One of these five tanks with 5.000 m^3 volume, intended for the storage of diesel fuel, was devastated during the war by direct and indirect missile hits, as shown in Figure 1. The picture shows three places of damage to the cylinder mantel of the tank, as well as the place where the old material was cut for the purpose of test tubes construction that will be used in experimental research. The experiment should determine the mechanical exploitation properties of the new material procured for this purpose, examine the same properties of the old, already used, material, compare the obtained results and confirm their mutual acceptability in the joint construction.

Figure 1. Devastated vertical cylindrical tank with 5.000 m^3 volume [1].

Comprehensive experimental tests cover a much wider area of research than the part of the results that will be presented in this paper due to the limited number of pages of the publication. The influence of the welding process, i.e., the selected technology in the plate welding process, on the mechanical and exploitation properties, was evaluated in detail based on several aspects that fully describe the behavior of the base material, the welded joint and the components of the welded joint, namely:

- tensile properties testing of test tubes of welded joint and weld metal according to EN 895 [2];
- hardness measurement by section (BM-HAZ-WM-HAZ-BM) according to EN 1043 [3];
- bending testing according to EN 910 [4];

In order to evaluate the condition of the old material from which the subject vertical cylindrical tanks with a flat bottom with 5.000 m^3 volume were made, certain tests were conducted on the old material, primarily tensile tests, measurement of cross-sectional hardness, impact tests, as well as determination of fracture toughness at straight deformation, K_{Ic}. Considering the limited amount of old material, it was estimated based on the standard EN 288-3 [6] that these tests can provide enough data to assess the quality of old material.

Generally speaking, these tests should answer the question of whether a partial, more economically acceptable, rehabilitation of existing tanks is possible in comparison to demolition and complete reconstruction. Namely, through experimental tests, it is necessary to determine the mechanical exploitation properties of the new purchased material, examine the same properties of the old, already used material from the sample taken from the tank, compare the obtained results and confirm their mutual acceptability in the joint construction.

2. Base material and welding technologies

The base material from which the above-ground vertical cylindrical tanks with a flat bottom are made is S.0361 steel, as well as the 5.000 m^3 volume tank located at the terminal in Bihać. A new material was used to analyze the influence of the choice of welding technology on the mechanical and exploitation properties. The material was delivered in the form of panels with dimensions of 1000x1000x12mm, which were cut to dimensions of 500x200x12mm and then welded. The mechanical properties of the delivered steel S.0361 are given in table 2.

Table 2. Mechanical properties of steel S.0361 in delivery condition.

Batch	Yield stress $R_{p0,2}$ (MPa)	Tensile strength, R_m (MPa)	Elongation A (%)	Impact energy A_{uk} (J)
234-349	247	398	24	55

With regard to the type of material and the technical-technological regulations that must be met, and bearing in mind that it is structural steel intended for pressure vessels, four technologies were used for welding the test samples [7]:

- technology A - REL procedure with electrode EVB 50 without preheating,
- technology B - REL procedure with electrode EVB 50 with preheating,
- technology C - MAG procedure with wire VAC 60 without preheating
- technology D - MAG procedure with wire VAC 60 with preheating.

After preparation, the plates are heated to the preheating temperature, which in this case is around 50°C (Technology B and D). Preheating is a thermal treatment prior to welding that aims to raise the initial temperature of the workpiece above the ambient temperature. Preheating achieves a reduction in the cooling rate, by which the critical cooling rate is avoided, i.e., the unfavorable structure of martensite and bainite is avoided. This means that the correct choice of preheating temperature is a very important step that ensures a more stable structure. The main reason for preheating is the increase of the general quality of welded joints, which is achieved through the following: reducing the cooling rate of the welded joint, reducing the residual stresses in the welded joint, reducing the hardness in the HAZ, increasing the impact toughness in the welded joint, and reducing the tendency to the formation of cracks [8]. A schematic representation of the preparation of the welded joint shape is given in Figure 2, whereas the welding sequence is given in Figure 3.

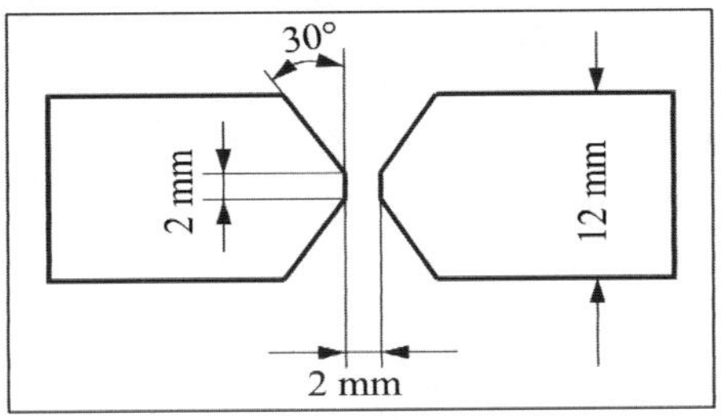

Figure 2. Schematic representation of the weld shape.

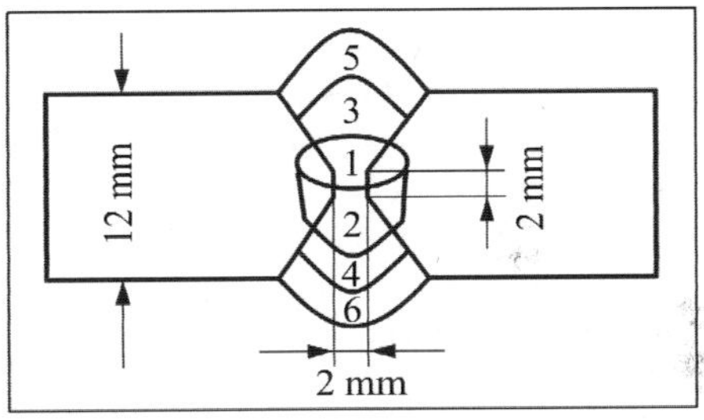

Figure 3. Schematic representation of the welding sequence.

The mechanical properties of the additional material according to the manufacturer's catalog are given in Table 3 for the EVB 50 electrode and in Table 4 for the VAC 60 wire. Figure 4 shows the cutting plan of the test tubes.

Table 3. Mechanical properties of additional material EVB 50.

Sample	Yield stress $R_{p0,2}$ (MPa) min.	Tensile strength, R_m (MPa)	Elongation A (%) min.	Impact energy A_{uk} (J) -20°C min.
EVB 50	450	500-550	26-30	50

Table 4. Mechanical properties of additional material VAC 60.

Sample	Yield stress $R_{p0,2}$ (MPa) min.	Tensile strength, R_m (MPa)	Elongation A (%) min.	Impact energy A_{uk} (J) -20°C min.
VAC 60	400	500-600	22-30	45

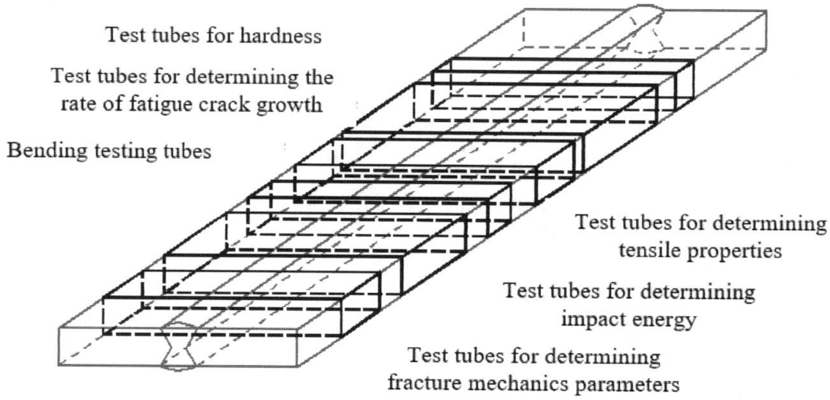

Figure 4. Plan of cutting test tubes [7].

The scheme of removing test tubes from welded plates is defined on the basis of the program of experimental tests, as shown in Figure 4. Test tubes for tensile and bending testing were extracted so that only the middle of the V-weld is taken. Test tubes for fracture mechanics were extracted so that the thickness of the test tube was the entire thickness of the welded plates with a welded joint in the middle. Charpy test tubes were extracted according to the thickness of the welded joint, depending on the placement of V-notch. Test tubes for testing the tensile properties of the weld metal (WM) and the heat affected zone (HAZ) were extracted longitudinally from the WM and HAZ.

3. Testing of tensile properties

The tensile testing of butt-welded joints, including the shape and dimensions of the test tubes, as well as the test procedure itself, are defined by the standard EN 895 Part I - Butt-welded joints on metallic materials - transverse tensile test [2]. This standard primarily defines transverse tension, that is, the introduction of a load transversely to the welded joint. In case of transverse tensioning of a butt-welded joint only the tensile strength of the tested sample is determined, as a rule, which should not be lower than the tensile strength of the base material.

Standard EN 895 Part II [9] provides for the testing of test tubes made of base metal, weld metal and HAZ, i.e., determination of tensile properties of base metal, weld metal and HAZ. As a rule, these tests determine the properties defined by the standard EN 10002-1 [10], as well as ASTM E8 [11]. These are yield stress, tensile strength, percentage elongation, contraction, stress-strain diagram, as well as analysis of the location and type of sample fracture. Depending on the size of the weld metal and the HAZ, macro or micro test tubes are used to determine the tensile properties.

The testing of the test tubes of the welded joint, the base metal and the weld metal was performed at room temperature on an electromechanical testing machine SCHENCK-TREBEL RM 100, in deformation control. The loading speed was 5 mm/min. Elongation was recorded using a double extensometer HOTTINGER DD. The accuracy of extensometer measurement was ±0.001 mm.

A typical appearance of the prepared test tubes of the welded joint for tensile tests is given in Figure 5. Figure 6 shows the tensile tests of the base metal and weld metal, while the micro test tube for testing the tensile properties of HAZ is shown in Figure 7.

IOP Conf. Series: Materials Science and Engineering 1298 (2023) 012001 doi:10.1088/1757-899X/1298/1/012001

Figure 5. Test tubes for determining tensile properties of a welded joint made of new and old material.

Figure 6. Test tubes for determining tensile properties of base metal and weld metal.

Figure 7. Micro test tubes for determining tensile properties of HAZ.

All of the above tests were performed on test tubes made of newly acquired steel material S.0361, but also on test tubes made of old material sampled from the tank structure amortized for more than 35 years.

3.1. The results of determining tensile properties of the welded joint, BM, WM and HAZ of the new material

The results of determining the tensile properties of butt-welded test tubes by transverse tension are given in Table 5. The results of determining the tensile properties of test tubes of the base steel material S.0361 are given in Table 6. The results of determining the tensile properties of test tubes of pure weld metal EVB 50 are given in Table 7, and the results of determining the tensile properties of test tubes of pure metal weld VAC 60 are given in Table 8.

Table 5. Results of tensile tests of butt-welded test tubes.

Sample mark	Welding technology	Yield stress $R_{p0,2}$ (MPa)	Tensile strength R_m (MPa)	Elongation A (%)
TF-1-1		249	415	25.8
TF-1-2	A	250	418	25.1
TF-1-3		247	415	25.4
TF-2-1		242	395	26.5
TF-2-2	B	249	405	27.2
TF-2-3		251	411	26.8
FT-3-1		253	417	25.8
FT-3-2	C	251	413	26.1
FT-3-3		255	421	25.5

FT-4-1		259	409	28.1
FT-4-2	D	257	411	28.6
FT-4-3		260	408	27.5
Mean value		252	412	26.5
Standard deviation		5.195	6.895	1.122

Table 6. Results of tensile tests of base material test tubes.

Sample mark	Upper yield stress R_{eH} (MPa)	Lower yield stress R_{eL} (MPa)	Tensile strength R_m (MPa)	Elongation A (%)
BM - 1	282	272	418	27.8
BM - 2	280	271	419	28.1
BM - 3	284	275	422	27.6
Mean value	282	273	419.7	27.8
Stand. deviation	2.000	2.082	2.082	0.252

Table 7. Results of tensile tests of test tubes of pure weld metal EVB 50.

Sample mark	Yield stress $R_{p0,2}$ (MPa)	Tensile strength R_m (MPa)	Elongation A (%)
EVB 50 - 1	455	538	26.9
EVB 50 - 2	462	543	26.2
EVB 50 - 3	467	548	26.5
Mean value	461	543	26.5
Stand. deviation	6.028	5.000	0.351

Table 8. Results of tensile tests of test tubes of pure weld metal VAC 60.

Sample mark	Yield stress $R_{p0,2}$ (MPa)	Tensile strength R_m (MPa)	Elongation A (%)
VAC 60 - 1	405	558	25.8
VAC 60 - 2	409	562	26.3
VAC 60 - 3	411	567	25.5
Mean value	408	562	25.9
Stand. deviation	3.055	4.509	0.404

Testing of HAZ micro tubes was performed at room temperature on an INSTRON 1122 electromechanical testing machine, with deformation (elongation) control. The loading speed was 1 mm/min. Elongation was recorded using an inductive sensor GS1. The results of testing the tensile properties of HAZ are given in Table 9.

Table 9. Results of tensile tests of micro tubes obtained in HAZ.

Sample mark	Yield stress $R_{p0,2}$ (MPa)	Tensile strength R_m (MPa)	Elongation A (%)
HAZ - A - 1	409	504	23.9
HAZ - A - 2	406	501	22.4
HAZ - A - 3	412	508	24.3
HAZ - B - 1	422	526	25.6
HAZ - B - 2	427	533	24.8
HAZ - B - 3	419	523	26.1
HAZ - C - 1	363	479	23.6
HAZ - C - 2	360	475	24.1

HAZ - C - 3	369	481	24.7
HAZ - D - 1	375	491	24.4
HAZ - D - 2	383	496	24.2
HAZ - D - 3	380	495	23.8
Mean value	394	501	24.3
Standard deviation	24.514	18.858	0.949

3.2. Results of determining the tensile properties of the old material

The tensile testing of the butt-welded joints of the test tubes obtained from the devastated tank with 5.000 m^3 volume at the terminal in Bihać, was conducted according to the Standard EN 895 Part I. Since the Standard EN 895 Part II [9] provides for the testing of test tubes made of base metal, weld metal and HAZ, that is, determining the tensile properties of the base metal, weld metal and HAZ of the new material, these properties were determined here, as well. The results of determining the tensile properties of butt-welded test tubes by transverse tension are given in Table 10, test tubes of basic old steel material S.0361 are given in Table 11, test tubes of pure weld metal in Table 12, and the test tube of HAZ in Table 13.

Table 10. Results of tensile testing of the butt-welded joint of old material.

Sample mark	Upper yield stress R_{eH} (MPa)	Lower yield stress R_{eL} (MPa)	Tensile strength R_m (MPa)	Elongation, A (%)
FS – 1	273	268	407	26.6
FS – 2	275	270	409	28.3
FS – 3	271	264	406	27.5
Mean value	273	267	407.3	27.5
Stand. dev.	2.000	3.055	1.528	0.850

Table 11. Results of tensile testing of test tubes of base old material.

Sample mark	Upper yield stress R_{eH} (MPa)	Lower yield stress R_{eL} (MPa)	Tensile strength R_m (MPa)	Elongation A (%)
BM – S – 1	276	273	415	26.7
BM – S – 2	278	272	420	25.9
BM – S – 3	272	267	413	26.9
Mean value	275	271	416.0	26.5
Stand. dev.	3.055	3.215	3.606	0.529

Table 12. Results of tensile tests of test tubes of pure weld metal.

Sample mark	Yield stress $R_{p0,2}$ (MPa)	Tensile strength R_m (MPa)	Elongation A (%)
WM – S – 1	419	520	26.1
WM – S – 2	416	515	26.8
WM – S – 3	422	523	25.3
Mean value	419	519	26.1
Stand. dev.	3.000	4.041	0.751

Table 13. Results of tensile tests of HAZ test tubes.

Sample mark	Yield stress $R_{p0,2}$ (MPa)	Tensile strength R_m (MPa)	Elongation A (%)
HAZ – S – 1	413	510	22.4
HAZ – S – 2	409	507	23.5
HAZ – S – 3	417	516	22.3
Mean value	413	511	22.7
Stand. dev.	4.000	4.583	0.666

The aim of the tensile testing of test tubes of welded joint, base material and welded joint components (WM and HAZ) of S.0361 steel was primarily:

- determining the tensile properties (yield stress, tensile strength, elongation) of the base material,
- determining the tensile properties of the welded joint by introducing a load transverse to the welded joint,
- determining the tensile properties of the components of the welded joint of WM and HAZ.

These tests provided the necessary data on how the selected welding technology affects the strength of the welded joint, as well as the components of the welded joint. Namely, the chosen technology has no effect on the strength of the welded joint, as shown in Table 5. All test tubes, regardless of the procedure used, i.e., REL procedure (electrode EVB 50 was used) without preheating (technology A) or with preheating (technology B), or MAG procedure (wire VAC 60 was used) without preheating (technology C) or with preheating (technology D), have cracked in the base material, which clearly indicates the character of the welded joint.

Based on the test results obtained by tensioning of the test tubes extracted from the base metal given in Table 6, it can be stated that the results are within the limits of the values prescribed by the standard for that material, that is, the values given by the manufacturer in the certification documentation.

The obtained values of the tensile properties of the test tubes of the welded joint and the base material are quite similar, which corresponds to the conclusion that the characteristics of the base material were actually obtained by testing the test tubes of the welded joint.

By testing the tensile properties of the weld metal, it is evident that the obtained values of the tensile strength of the VAC 60 wire are slightly better than the values obtained by testing the pure weld metal of the EVB 50 electrode, as shown in Table 7 and Table 8. However, the yield stress values obtained during the test of the weld metal EVB 50 electrode are about 10% better than those of the weld metal VAC 60 wire. The plasticity properties, i.e., the elongation, are approximately the same for both additional materials.

The obtained better yield stress values of the pure weld metal of EVB 50 electrode resulted in the strength properties (yield stress and tensile strength) of HAZ being about 10 to 15% better than the properties obtained from the HAZ testing of the welded joint where the VAC 60 wire was used as an additional material 60, as shown in Table 9. This improvement is even more pronounced in the case when preheating is used, which has a very positive effect on the improvement of the HAZ properties as the most critical zone of the welded joint.

As for the testing of the welded joint of the old material, i.e., the samples taken from the 5.000 m3 volume vertical cylindrical tank with a flat bottom, located at the terminal in Bihać, almost identical dependencies are obtained. This information is very important from the aspect of evaluating the usability and remaining service life of the existing working tanks, as well as the possible rehabilitation of damaged tanks.

4. Measurement of welded joint hardness

The hardness test of the welded joint was performed according to the Standard EN 1043 [3]. The Vickers method was used, and the applied load was HV 10. The indentations were made in the base material

BM, HAZ and WM, so that the areas on the welded joint were obtained and registered. The test was performed on a microhardness tester ''WOLPERT'' model V-Testor 2, as shown in Figure 8. The magnification at which the indentation was measured was 400x. The scheme of measurement locations is given in Figure 9.

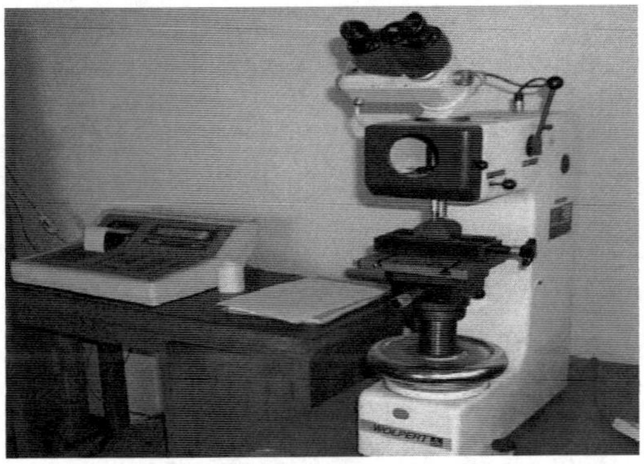

Figure 8. Microhardness tester "WOLPERT".

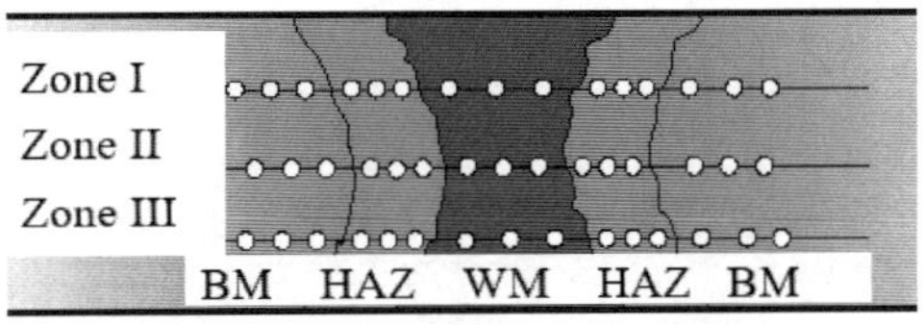

Figure 9. Scheme of hardness measurement on the welded joint.

The results of the hardness measurement on the new and old welded joint are also shown graphically in Figure 10.

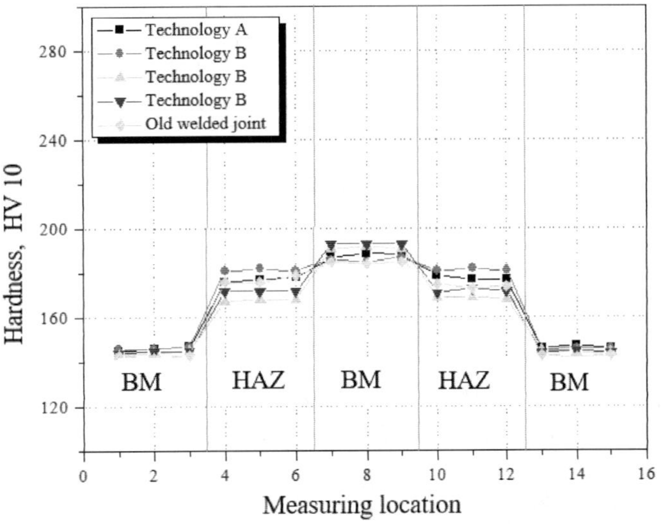

Figure 10. Graphic display of the results of the hardness measurement of the new and old welded joint.

The obtained results of the cross-sectional hardness measurement are in agreement with the values of tensile properties of BM and welded joint components of WZ and HAZ. Namely, the characteristics of BM, WZ and HAZ are clearly different. Since this is a question of ''over-matching'', which means that the strength of the weld metal is higher than the strength of BM, WM has the highest hardness value, which corresponds to the selected additional materials. The hardness of HAZ is slightly lower than that of WM, whereas BM has the lowest measured hardness value.

5. Bending testing

The testing of technological test tubes by bending the butt-welded joint of steel S.0361 was carried out with the aim of determining the ability of the steel to be deformed by bending around the welded joint. The test procedure, as well as the test tubes used for bending testing of butt-welded joints, are defined by the Standard EN 910 (Welded butt joints in metallic materials - Bend test) [4]. The test tubes, the geometry of which is given in Figure 11, were bent around the welded joint face and root using a pin with a diameter of 4a with a bending angle of up to 120°. During the test on the test tubes, errors with a length greater than 3 mm must not occur in any direction.

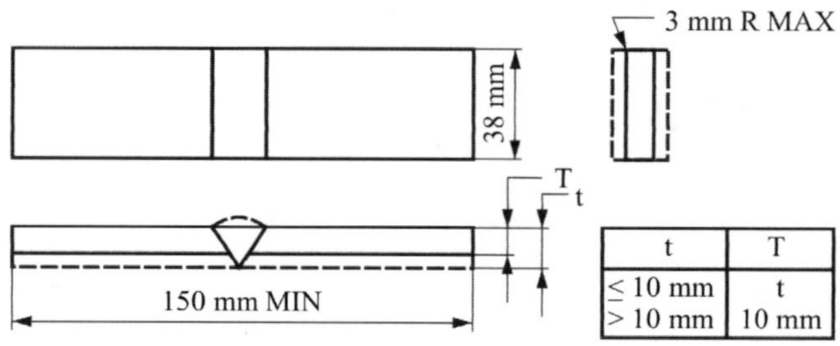

Figure 11. Test tube for transverse bending of a butt-welded joint of sheet metal.

The scheme of the three-point bending test procedure is given in Figure 12. The test was performed on the AMSLER mechanical testing machine in motion control.

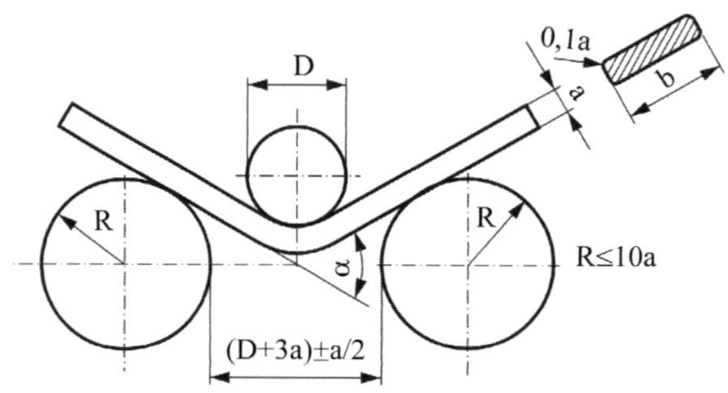

Figure 12. Bending testing scheme.

Geometric characteristics significant for bending testing are:

- L_0 = 300 mm, test tube length;
- a = 12 mm, test tube thickness;
- b = 30 mm, test tube width;
- D = 20 mm, pin diameter;
- R = 15 mm, radius of rollers-supports;
- $D + 3a = 20 + 3 \cdot 12 = 56$ mm, roller distance.

5.1. Bending testing of a butt-welded joint of a new material

The bending testing of the butt-welded joint of the new material was performed around the weld face and the weld root. Test tubes were tested for all four defined welding technologies. The design of test tubes for bending testing around the welded joint face is given in Figure 13 for the group of test tubes S-1, whereas Figure 14 shows the test tubes for bending testing around the welded joint root. The results of bending testsing of test tubes obtained from butt-welded sheets of new material are given in Table 14.

Figure 13. Test tube group S_L-1 for bending around the weld face.

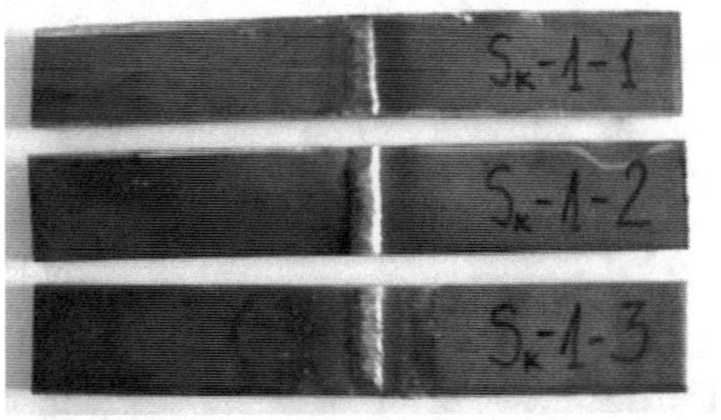

Figure 14. Test tube group S_K-1 for bending around the weld root.

Table 14. Test results of bending testing of test tubes of the new steel material S.0361.

Sample mark	Bending angle around the weld face (°)	Sample mark	Bending angle around the weld root (°)	Radiogram mark
		Technology A		
S_L - 1 - 1	120 (no cracks)	S_K - 1 - 1	120 (no cracks)	2
S_L - 1 - 2	120 (no cracks)	S_K - 1 - 2	120 (no cracks)	2
S_L - 1 - 3	120 (no cracks)	S_K - 1 - 3	120 (no cracks)	2
		Technology B		
S_L - 2 - 1	120 (no cracks)	S_K - 2 - 1	120 (no cracks)	2
S_L - 2 - 2	120 (no cracks)	S_K - 2 - 2	120 (no cracks)	2
S_L - 2 - 3	120 (no cracks)	S_K - 1 - 3	120 (no cracks)	2
		Technology C		
S_L - 3 - 1	120 (no cracks)	S_K - 3 - 1	120 (no cracks)	2
S_L - 3 - 2	120 (no cracks)	S_K - 3 - 2	120 (no cracks)	2
S_L - 3 - 3	120 (no cracks)	S_K - 3 - 3	120 (no cracks)	2
		Technology D		
S_L - 4 - 1	120 (no cracks)	S_K - 4 - 1	120 (no cracks)	2
S_L - 4 - 2	120 (no cracks)	S_K - 4 - 2	120 (no cracks)	2
S_L - 4 - 3	120 (no cracks)	S_K - 4 - 3	120 (no cracks)	2

5.2. Bending testing of a butt-welded joint of old material

The bending testing of the butt-welded joint of the old material was also performed around the weld face and the weld root. The layout of test tubes for bending testing around the weld face is given in Figure 15, whereas Figure 16 shows the test tubes for bending testing around the weld root. The results of the bending testing of the butt-welded test tubes of the old material are given in Table 15.

Figure 15. Test tube group S - S_L - 1 for bending around the weld face.

Figure 16. Test tube group S - S_K - 1 for bending around the weld root.

Table 15. Results of bending testing of old material test tubes of steel S. 0361.

Sample mark	Bending angle around the weld face (°)	Sample mark	Bending angle around the weld root (°)	Radiogram mark
S - S_L - 1	120 (no cracks)	S - S_K - 1	120 (no cracks)	2
S - S_L - 2	120 (no cracks)	S - S_K - 2	120 (no cracks)	2
S - S_L - 3	120 (no cracks)	S - S_K - 3	120 (no cracks)	2

Test results of test tubes extracted from welded sheets by bending testing, as shown in Table 14 and Table 15, indicate high-quality welding, because no cracks appeared when the bending angle of 120° was reached. Based on the test results and their analysis, it is possible to conclude that the selected welding technology is good, and that all the test results of the submitted steel plate samples of quality S.0361 are satisfactory.

6. Testing the correctness and safety of use of the tank

Since the experimental tests confirmed the acceptable mechanical exploitation properties of the new steel material S.0361, the same properties of the old, already used, material were examined by tension testing, hardness measurement and bending testing of welded joints and welded joint components. By comparing the obtained test results on test tubes made of new and old material, small acceptable deviations were observed, thus confirming their mutual acceptability in the common construction.

All of the above was a good basis to use the selected welding technology to partially rehabilitate the devastated tank with the aim of bringing it to a usable condition. Prior to pouring diesel fuel, it was necessary to fill the tank with water in order to determine the tightness of the welded joints and the entire structure. As an additional safety measure, it was important to have a proper bund wall, i.e., a concrete leak-proof tank that can receive fuel from the tank in the event of an accident. The form of the renovated tank with the associated bund wall is shown in Figure 17.

Figure 17. Restored vertical cylindrical tank with 5.000 m^3 volume and corresponding bund wall for receiving fuel spills [12].

Each such tank has security in the form of a concrete bund wall, with a waterproof PEHD film under the concrete, that acts as a barrier between the ground and above-ground products in the event of a spill. The bund walls are designed in such a way that they can accept the complete spilling of fuel from the tank and its retention until it is evacuated [11].

There are a total of five bund walls of different dimensions at the location of the Bihać terminal, namely:

- Bund wall 1 (for R-01) = 38 x 38 m
- Bund wall 2 (for R-02) = 38 x 38 m
- Bund wall 3 (for R-03) = 73 x 37 m
- Bund wall 4 (for R-04) = 62 x 41 m
- Bund wall 5 (for R-05) = 38 x 38 m

In addition to the above, and as a measure to prevent environmental incidents, this type of oil terminal must also have a separator for settling, purification and separation of oily waste and sediment, which is connected to the bund walls by pipelines in the event of an accident and fuel spillage from the tank [13]. Figure 18 shows the API oil separator at the terminal in Bihać.

Figure 18. API separators for oily water [11].

The concrete API separator with grate and tank/reservoir for collecting and storing waste oil (Figure 18) is located at the lowest altitude on the northwestern side of the Bihać terminal. Sedimentation, purification and separation of oily waste and sediment are performed in the separator. The dimensions of this separator are 25 x 5 m and it consists of two concrete chambers.

7. Concluding remarks

Part of the conducted research presented in this paper aimed to adequately select the optimal welding procedure and welding technology on the defined structure of the devastated cylindrical tank, with the best mechanical exploitation properties and the least introduced residual stresses in the welding process, i.e., to determine the impact of the heterogeneity of the material structure on the behavior of the welded joint. Based on the performed experimental tests, it is possible to conclude the following:

- The welding procedure has no effect on the tensile strength of the welded joint, because all tested test tubes, regardless of whether the REL (EVB 50) procedure or the MAG (VAC 60) procedure was used, cracked in the base material, which clearly indicates the character of the welded joint. This is ''over-matching'', which means that the strength of the weld metal is higher than the strength of the base material. However, the welding procedure and the selected welding technology affect the values of the tensile properties of HAZ, and the selected technology B shows the best tensile properties.
- The obtained results of cross-sectional hardness measurements are in agreement with the values of tensile properties of BM and welded joint components of WM and HAZ. The characteristics of BM, WM and HAZ are clearly different. The hardness of HAZ is slightly lower than that of WM, and BM has the lowest measured hardness value.
- The test results of test tubes extracted from welded sheets by bending indicate high-quality welding, because no cracks appeared when the bending angle of 120° was reached.

As for the examination of the welded joint of the old material, i.e., the samples obtained from the vertical cylindrical tank with 5.000 m^3 capacity, located at the terminal in Bihać, almost identical dependencies are obtained as in the examination of the new material, which provides us with the conclusion that the exploitation period of approximately 35 years did not have a significant influence on the properties of the material and the welded joint.

References

[1] Islamović F 2019 *Provjera deformacijskog i naponskog stanja na vertikalnim cilindričnim rezervoarima velike zapremine nakon izvršene sanacije* 12th International Scientific Conference on Production Engineering RIM 2019 - Sarajevo
[2] EN 895 1995 *Sučeono zavareni spojevi na metalnim materijalima - ispitivanje poprečnim zatezanjem* (Welded butt joints in metallic materials - Transverse tensile test)
[3] EN 1043 1996 *Destructive tests on welds in metallic materials-Hardness test - Part 2: Microhardness testing on welded joints*
[4] EN 910 1993 *Welded butt joints in metallic materials - Bend test*
[5] EN 875 1995 *Welded butt joints in metallic materials - Specimen, location and notch orientation for impact test*
[6] EN 288-3 1996 1996 *Obezbjeđenje kvaliteta u zavarivanju* Zbirka standarda DUZS Savezni zavod za standardizaciju Beograd
[7] Islamović F, Gačo Dž, Muratović P and Manjgo M 2009 *Eksperimentalna ispitivanja vlačnih svojstava zavarenog spoja čelika Č.0361 pri različito izabranim tehnologijama zavarivanja* 38. Međunarodna konferencija Zavarivanje u pomorstvu – Bol
[8] Kopas P, Sága M, Jambor M, Nový F, Trško L and Jakubovičová L 2018 Comparison of the mechanical properties and microstructural evolution in the HAZ of HSLA DOMEX 700MC welded by gas metal arc welding and electron beam welding *MATEC Web Conf.,* **244** (2018) **01009** doi.org/10.1051/matecconf/201824401009
[9] EN 895 Part II 1995 *Sučeono zavareni spojevi na metalnim materijalima - ispitivanje poprečnim zatezanjem* (Welded butt joints in metallic materials - Transverse tensile test)
[10] EN 10002-1 1999 *Mehanička ispitivanja metala Zatezna ispitivanja Dio 1: Termini i definicije*
[11] *Zahtjev za izdavanje okolinske dozvole za terminal tečnih naftnih derivata Ribić u Bihaću 2020* Institut za hidrotehniku d.d. Sarajevo
[12] Islamović F 2006 *Poboljšanje integriteta stacionarnih višekomornih tankostijenih posuda za tečna goriva* Doktorska disertacija Univerzitet u Tuzli - Mašinski fakultet Tuzla

[13] Amgad Fathey Fayek Tadros 2020 Environmental aspects of petroleum storage in above ground tank 2020 *E3S Web of Conferences* **166:01006** doi: 10.1051/e3sconf/202016601006

Numerical analysis of the construction integrity on the example of a prosthetic device

Fuad Hadžikadunić, Ahmed Mujkanović, Nedeljko Vukojević, Amna Bajtarević-Jeleč

Faculty of mechanical engineering of University of Zenica, Fakultetska 1, 72000 Zenica, Bosnia and Herzegovina

fuad.hadzikadunic@unze.ba

Abstract. The lower limb prosthesis, as the specific object of this biomechanical research, has the task of restoring the normal function of the missing part of the human body. Creating a prosthesis exclusively for the end user is a very complex process. The subject of analysis in this work is the analysis of forces, deformations and stress state on the example of structural elements of the lower prosthetic device with the aim of understanding the integrity of the structure under the action of variable load. This analysis is one of the bases for the overall approach to the design of prosthetic aids for the end user, and especially the creation of the socket, as the most important element of a prosthesis in contact with man's residuum. The main goal of this work is to estimate the real value of the load of a complex prosthesis, to understand the effect of the load on the elements of the observed example of construction, and to understand the load transfer on the connection of the prosthesis structure with the socket. After defining the real load values, the performed numerical analysis indicated the critical points of the construction, the need to use more accurate material properties in the "foot" element, and in the "socket" zone, the possibility for corrections of the shape and thickness of the wall.

1. Introduction

Human being represents a perfect complex structure of very complex construction, functionality and integrity. The structural integrity of a person can be damaged by various causes. In this case, maintaining the structural integrity and functionality of the human structure belongs to a very important area of biomechanics and is an integral part of the contemporary needs of society. Mobility is a key indicator of the quality of life of people with lower limb amputations. Critical aspects of mobility, such as stability and energy expenditure in walking, are greatly influenced by the mechanical behavior of leg prostheses. Technologies for the development and production of prosthetic devices have changed significantly over time, and contemporary needs include the collaboration of several scientific disciplines, the application of modern diagnostic and prosthetic techniques, the application of modern design technologies with the help of CAD technologies, the use of modern 'light-weight' materials, etc.

2. Literature overview and previous research

By researching previous works and scientific studies dealing with the creation and analysis of lower limb prosthetic devices with an emphasis on the creation of sockets, it can be concluded that a large number of research deals with the application and analysis of spatial scanning technologies, the creation of 3D models using CAD technologies, the application of finite element methods for analysis stress-deformation states of parts of prosthetic devices due to complex loads using different software, the relationship between the residue and the socket of the lower limb prosthesis, etc. In a certain number of papers, the focus is on the method of making a socket model using modern technologies, while in another group of papers, virtual methods of obtaining CAD are processed model and all the factors that influence it. A significant area of research is the impact of load on the relationship between the residual and the prosthesis socket as the most important area. Some of the researches are presented below.

In the work of Al-Khazraji K. at al., five laminated composite materials used for manufacturing lower-limb prosthetic sockets were fabricated using vacuum molding technique. The finite element technique (ANSYS-11) is used to analyze and evaluate fatigue characteristics. The results show that changing the type of reinforcement has a great influence on the measured properties, [1]. In the study of Jweeg J. M. at al., two laminated composite materials were used to manufacture residual limb prosthetic sockets using the vacuum moulding method. The ANSYS program was used to calculate the deformation, maximum principle stress, and safety factors. The results showed that laminations laidup from eight layers of perlon plus four layers of carbon gave optimum mechanical properties, [2]. In the paper of Colombo G. at al. a digital model is proposed of the patient around which the prosthetic devices, in particular for lower limb, can be designed. The procedure is identified to characterize a patient's avatar wearing the prosthesis and to simulate the prosthesis behavior by means of a virtual patient, [3]. The paper of Vitali A at al. concerns the design and manufacture of medical devices that require a high level of customization and the attention is appointed on lower limb prosthesis and in particular on the prosthetic socket, [4]. The paper of Colombo G. at al. concerns the design of lower limb prosthesis, both below and above knee. It describes a new computer-based design framework and a digital model of the patient around which the prosthesis is designed and tested in a completely virtual environment. The software platform adopts computer-aided and knowledge-guided approaches with the goal of replacing the current development process, mainly hand made, with a virtual one, [5]. In the study of Mohd S. J. at al. the pressure distribution profile between the residuum and two types of prosthetic sockets for transfemoral amputees by utilizing a finite element analysis is estimated and validated. The pressure distribution profile provides an understanding of the relationship between the socket design and the level of subject comfortability, [6]. In the paper of Henao C. S. at al. a FEA was performed to evaluate the interaction between residual limb and socket when considering the dynamic loads of the gait cycle. Fourteen transfemoral amputees participated in this study, where their residual limbs (i.e., soft tissues and bone), and their sockets were reconstructed. The socket and the femur were defned as elastic materials, while the bulk soft tissues were defned as a hyperelastic material. The infuence of adding the dynamic loads related to the gait cycle were compared against the modelling of the static load equivalent to the standing position resulting in changes of 23% ± 19% in the maximum values and in an increase in the size of the regions where they were located, [7]. The paper of Colombo G. describes a new design process based on the use of virtual humans and specifically conceived for a prosthesis component, the socket. The proposed design paradigm has been experimented both for trans-tibial and trans-femoral amputees and main problems related to the implementation of each step have been generally discussed, [8]. The paper of M. Buzzi at al. presents an innovative framework centred on virtual models of the patient's body, to design and configure lower limb prosthesis, both transfemoral and transtibial. The framework integrates virtual prototyping and knowledge-based tools to support the orthopaedic technician during all the steps of the lower limb prosthesis design, suggesting rules and procedures for each task, [9]. Authors Ali I. at al. used the FEM with the use of a number of different composite materials, in order to determine the distribution of stresses and deformations during loading with the use of Ansys software, [10]. In the research of authors Colombo G. at al. described the process of creating a 3D model of a socket, using new different manufacturing methods such as reversible engineering, virtual prototyping and rapid prototyping, [11]. In the work of authors Oldfrey B. at al. the process of obtaining a CAD model of the socket and its production using modern technologies is shown [12]. Krajňáková, V. and a group of

IOP Conf. Series: Materials Science and Engineering 1298 (2023) 012002 doi:10.1088/1757-899X/1298/1/012002

authors described one of the currently most reliable methods - scanning for obtaining CAD models, [13]. Sanders, J.E. with a group of authors conducted research on a socket made using CAD technologies, [14].

3. Prosthetic devices

Prostheses belong to orthopedic aids, with which the amputated limb or part of the limb is functionally and aesthetically compensated, Fig.1, [15]. The basic modules of the lower leg prosthesis are: 'socket', knee, lower leg part or pylon and foot with ankle, Fig.2. Each component of the prosthesis has an extremely important role, considering that it should imitate the extremely complex mechanisms of the natural leg. The 'socket' of the prosthesis is its most important part, because its correct construction is responsible for the good functionality of the entire prosthesis. The 'socket' must correspond to the shape of the remaining part of the limb, and must be made of a material that the body feels comfortable, that does not cause a feeling of cold, that does not lead to skin irritation or pain, that tolerates sweat well and is easy to work with. The prosthesis must have a certain weight so that too light prostheses do not cause sudden movements, which creates a feeling of instability. The ‘socket’ must be completely adapted to each individual anatomy of the remaining part of the limb and is realized based on the characteristics of the user. The 'socket' represents the connection between the rest of the amputated leg and the prosthesis. At the bottom of the 'socket' there is an adapter plate that connects the 'socket' to the knee module (if it is a transfemoral type of prosthesis). The shape of the 'socket' is very important for achieving functionality and comfort. It is important that the 'socket' fits tightly, but that it does not cut off the circulation in the rest of the leg, that it does not cause skin irritation or discomfort during wear. 'Sockets' are usually made of rigid materials: thermoplastics or composites with carbon fibers, glass or graphite to be thin, light and provide solid support, [16, 17].

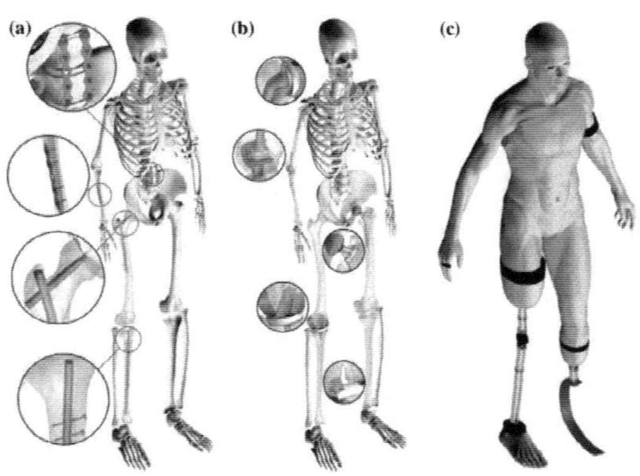

Figure 1. Examples of orthopedic aids: a) fracture fixators; b) articular prostheses; c) external prostheses, [1].

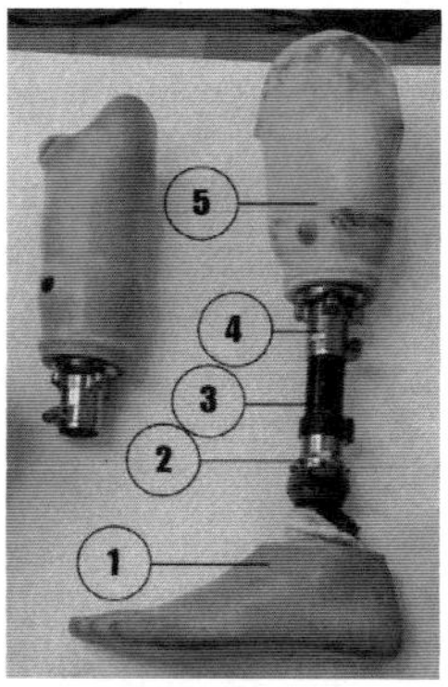

Figure 2. Lower limb prosthetic device; 1-Foot; 2-Lower pyramid adapter; 3-Pipe; 4-Upper pyramidal adapter; 5-"Socket".

4. Analysis of the human movement cycle

Walking is the basic and most important phenomenon in space and time, which reflects the characteristics of an individual's movement. It is characterized by orthogonal control of the body, with simultaneous bending of the body, head and upper limbs, and the use of the lower limbs. The movement cycle is defined by the time interval from the contact between the heel of one leg and the base until the next contact between the base and the heel of the same leg. Conventionally, the cycle is measured in percentages, not seconds. Initial heel strike is 0% cycle and subsequent heel strike is 100% cycle. This allows the cycle of different people to be normalized. The cycle is divided into 2 basic phases: the support phase, which includes the time period when the leg is on the ground, and the swing phase when the leg is in the air, Fig. 3. The support phase begins with the impact of the heel of the foot on the ground (0%), and ends with the separation of the toes from the ground (62%).

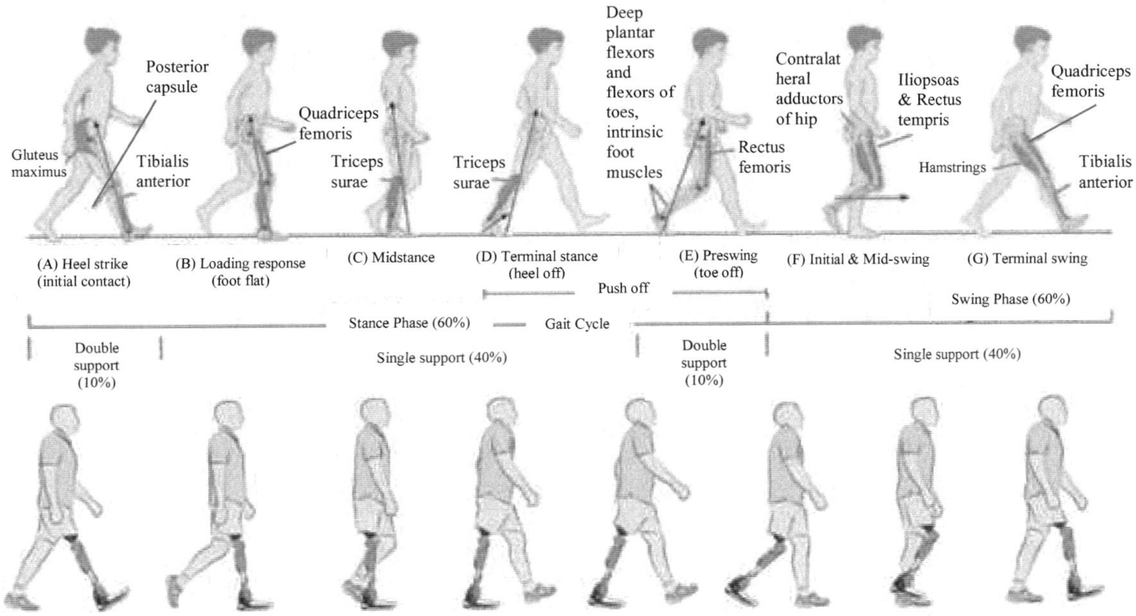

Figure 3. Normal walking cycle and walking cycle with prosthesis, [6].

The support phase can be double (at the beginning and end of the cycle), when support is provided by both limbs, and single, when only one limb touches the ground. In the stance phase, four leg positions can be distinguished: initial contact, loading, mid-stance, and preswing. The pre-swing begins with the contact of the opposite leg on the ground, and ends with the separation of the toes from the ground. The swing phase makes up about 38% of the gait cycle. It starts with the separation of the toes from the ground and ends with the impact of the heel on the ground. During the swing, the ankle is flexed dorsally by concentric contractions of the anterior tibialis muscle, and all other muscles are at rest. In this phase, three leg positions are distinguished: the beginning, middle and end of the swing, [18-20].

5. Creation of a CAD model of the prosthesis

Creating a total CAD model of a lower leg prosthesis involves creating each part individually, and then creating the overall assembly. Special importance is given to the creation of the "socket" of the prosthesis as the most complex, as it consists of a very complex series of spatial curved surfaces. Modeling of parts and assembly of the prosthesis was done in SolidWorks and Meshmixer software.

5.1. Creation of the 'socket'

The first step in creating the 'socket' of the prosthesis is 3D SCAN laser scanning and taking the precise geometry of the rest of the human extremity, i.e. the 'residue'. Each amputee has a unique remaining part of the limb, and thus each 'socket' is completely unique and special. The importance of the precise manufacturing process of the 'socket' lies in the fact that the 'socket' of the prosthesis represents the connection between the rest of the amputated leg and the prosthesis, that is, between the human and the prosthesis. The production of this part of the prosthesis is responsible for the comfort and functionality of the entire prosthesis. Fig. 4 shows the process of creating a CAD model of the 'socket', which begins with scanning the remaining part of the limb, and ends with the creation of an accurate model of the socket on the scanned surface.

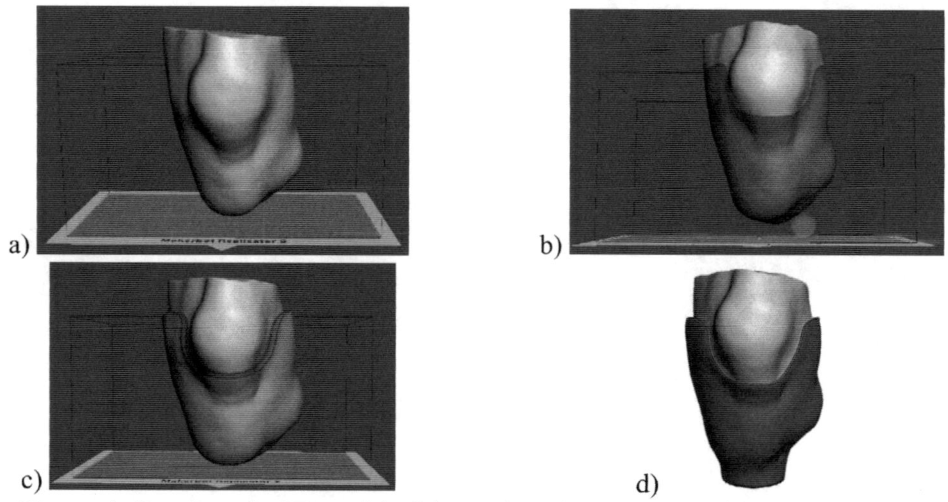

Figure 4. Creation of a 3D model of the socket: a) model of the scanned residuum; b) marking the surface of the socket; c) use of the Offset tool; d) 'socket' model according to the scanned residuum

5.2. Creation of a complete model of a lower limb prosthesis

Regardless of whether it is a lower leg or an upper leg prosthesis, the prosthetic foot has a very important functional role when making the prosthesis. In this case, as part of the student diploma research, the appearance of the prosthetic foot 1C30 Trias manufactured by Ottobock, as well as other connecting elements, was used only as a sample for assembling the 3D model of the prosthetic device, in order to obtain a more realistic overall structure. The foot, by the way, is made of carbon and provides the user with controlled support on uneven terrain. The model was created in SolidWorks and the desired shape was obtained. The essence of the consideration is actually the creation of an original 'socket' model based on a scan of the rest of the limb and its stress-deformation state, as a result of the load in different phases of movement, which is transmitted through elements of the prosthesis, Fig. 5.

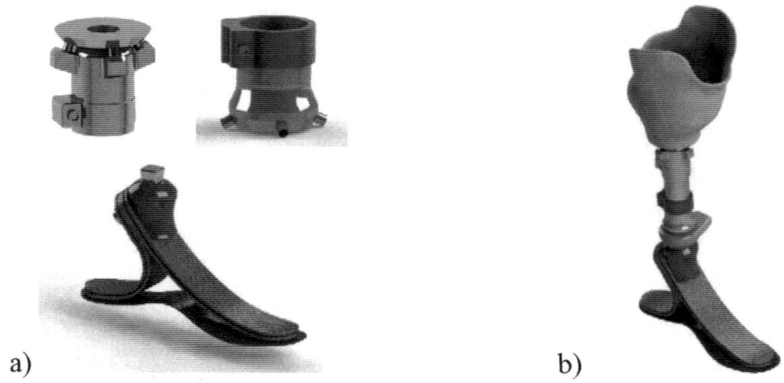

Figure 5. Creation of the overall 3D model of the prosthesis: a) element models; b) prosthesis model with socket and elements

6. Analysis of the construction of the lower leg prosthesis

The analysis of the entire complex structure of the lower leg prosthesis will be presented in several stages: experimental determination of the size and nature of the force during certain phases of movement; loading into the numerical model of the prosthesis, and numerical analysis of the stress-deformation state of the prosthesis elements with an emphasis on the "socket" of the prosthesis.

6.1. Experimental determination of the load force on the prosthesis

Fig. 6 shows the platform used to measure forces and acceleration during the human walking cycle (930 N). The procedure of socket and lower leg prosthetic device creation is oriented to 'custom made' for a specific person, and a measured force (presented in Fig. 7. to Fig. 11.) was used here as a result of his specific dynamic movement. The acquisition measurement system Spider 8, force transducer U9B 50kN and acceleration transducer SA62008-01-010 100mV/g METRIX, produced by HBM Germany, were used.

Figure 6. A measuring system with sensors specially created for this type of measurement

Fig. 7 shows the diagram of force and acceleration for 3 characteristic contacts: foot contact, heel contact and toe contact. The measurement was carried out by the subject leaning on the platform with different parts of the foot at a movement speed of 0.62 m/s in three passes. Fig. 8 shows the same diagrams with the difference in speed, which was 1.11 m/s during that measurement. If diagrams are compared, i.e. forces and accelerations at speed 1 and speed 2, it can be seen that when the speed increases (increase by approximately 44%), there is a slight increase in force (increase by approximately 10%), but there is a large increase in acceleration (an increase of approximately even 300%). Fig. 9 and Fig. 10 show the force and acceleration diagrams for the aforementioned contacts, with a difference in foot positioning.

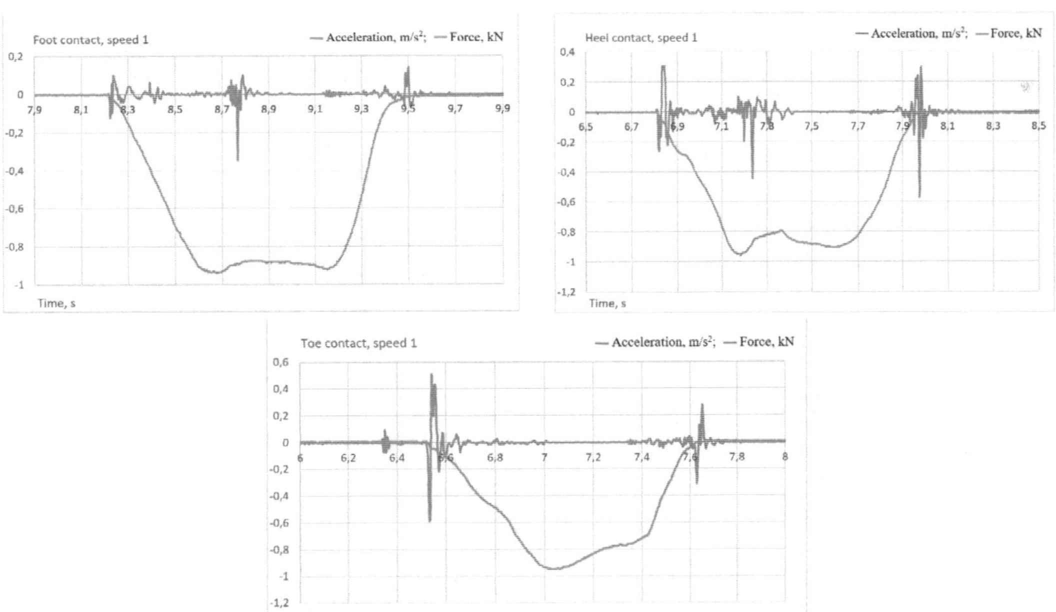

Figure 7. Force and acceleration diagrams for characteristic contacts at speed 1.

IOP Conf. Series: Materials Science and Engineering 1298 (2023) 012002 doi:10.1088/1757-899X/1298/1/012002

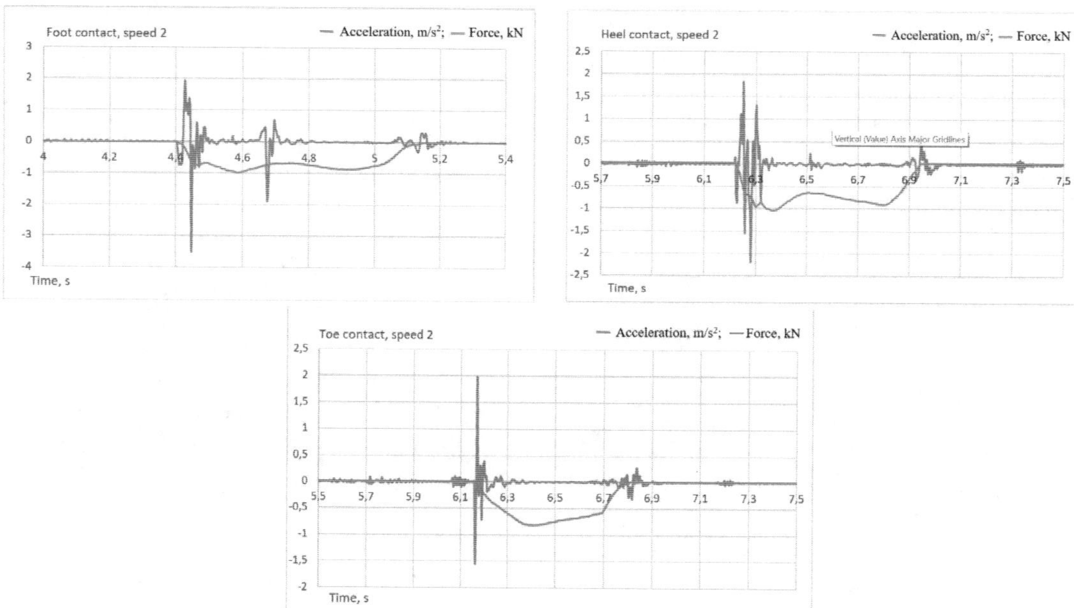

Figure 8. Force and acceleration diagrams for characteristic contacts at speed 2.

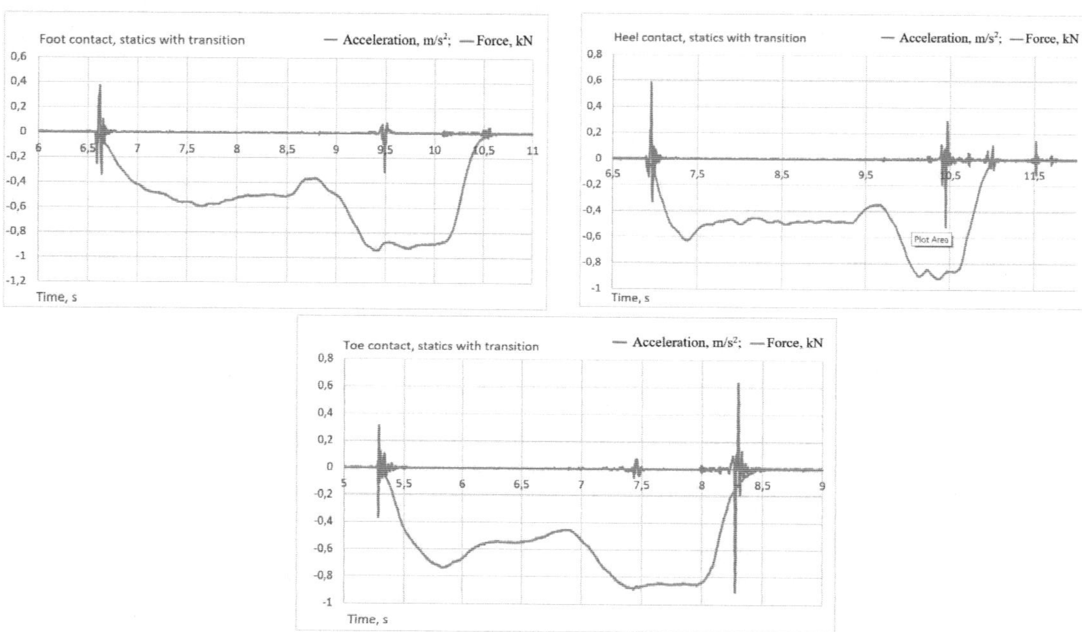

Figure 9. Force and acceleration diagrams for characteristic contacts.

Fig. 11 shows the force and acceleration diagrams for speed 1 and speed 2, but during this measurement the subject kept his leg stiff, trying to imitate as closely as possible a man with a prosthetic aid, in an effort to obtain as accurate and realistic results as possible. Comparing the diagrams from Fig. 7 and Fig. 8 and the diagrams from Fig. 11, an increase in force and an extremely large increase in the value of acceleration in the case of a 'stiff leg' is noticeable, which is logical because a person with a prosthesis hits the ground harder when leaning and thus causes greater force.

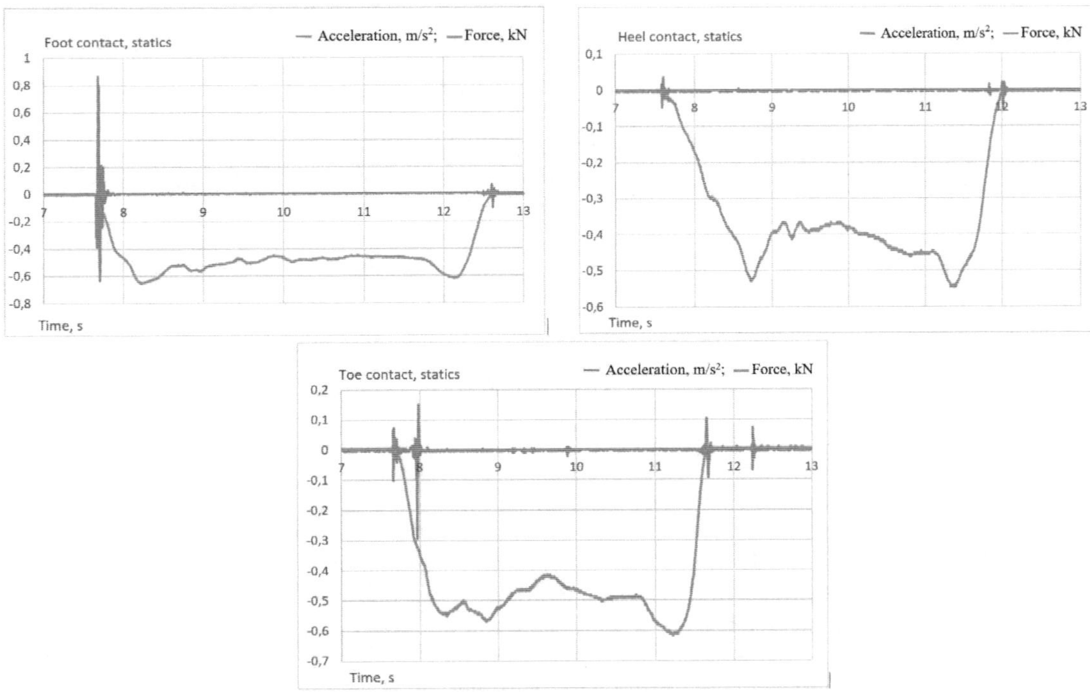

Figure 10. Force and acceleration diagrams for characteristic contacts, ‘statics’.

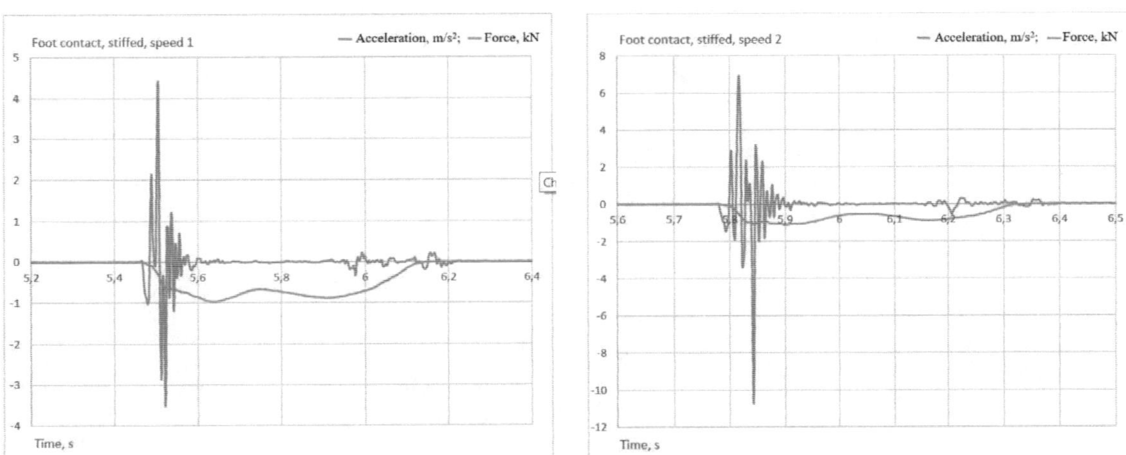

Figure 11. Force and acceleration diagrams for characteristic contacts with a ‘stiff leg’.

6.2. Defining characteristic positions and support surfaces

The simulation was performed for 3 characteristic positions, which are exactly defined and shown in Fig. 12.

In the first case, the force enters the construction on the rear part of the foot via surface A1, in the second case the force acts on surface A3, while in the third case the force acts on the front part of the foot on surface A2.

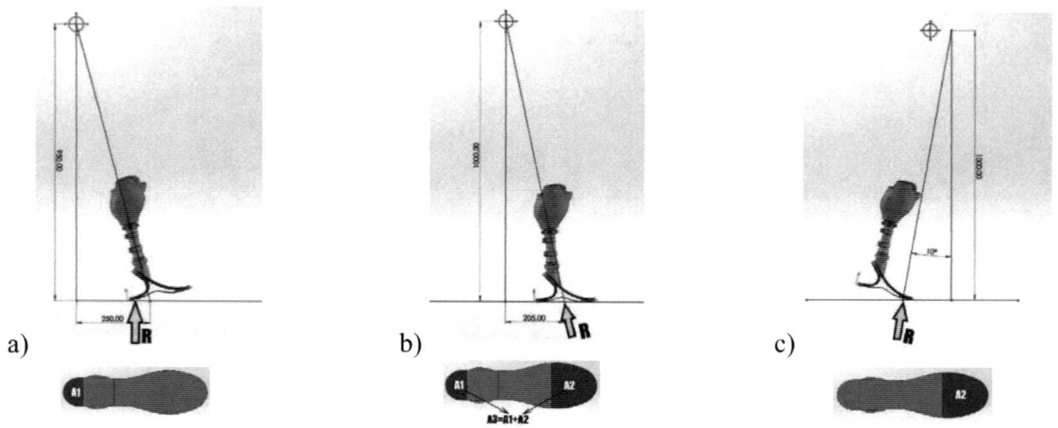

Figure 12. Characteristic positions: a) "Heel Strike"; b) „Mid Stance"; c) „Terminal Stance".

6.3. Defining the conditions for the simulation setting

For the purposes of analysis, different materials on different parts of the model are specially defined, such as: Carbon fiber, Polypropylene, Rubber. According to Fig. 13, that is, taking into account the zones that need to transfer the load from the 'socket' to the remaining part of the limb, and the critical zones that must not transfer the load, zones on the 'socket' were created that represent supports for the realized simulation. After that, the force values obtained in the diagrams presented in the previous chapter were applied. The obtained values of forces represent the vertical component of forces, and on the basis of Fig, 12, the values of forces required for numerical analysis are calculated.

Table 1. Basic material characteristics used in numerical analysis.

No.	Description	Elastic modulus: (N/mm^2)	Poisson's ratio (-)	Mass density (kg/m^3)	Tensile Strength (N/mm^2)	Yield Strength (N/mm^2)
1	Carbon fiber	16500	0.2	1800	255	200
2	Polypropylene	1500	0.3	1000	40	32
3	Rubber	1900	0.48	2100	165	145

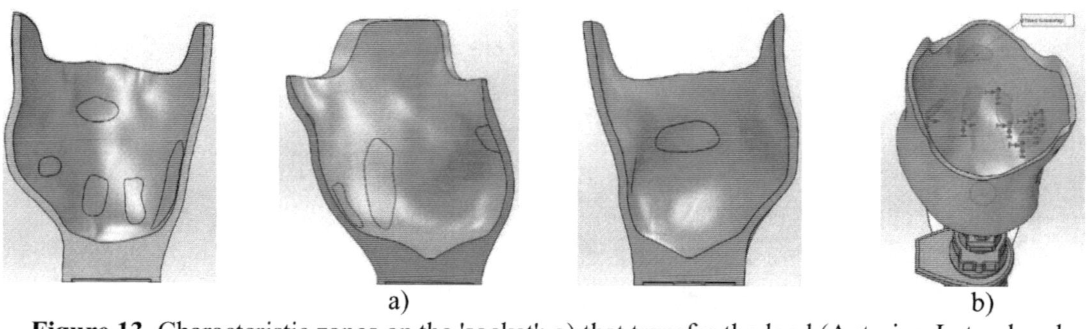

Figure 13. Characteristic zones on the 'socket': a) that transfer the load (Anterior, Lateral and Posterior); b) for defining supports.

Fig. 13 shows the zones on the 'socket' that transmit the load, as well as the zones where the support points are defined for the purpose of numerical analysis.

When defining the fixed supports, according to [9], that is considering the zones that should transfer the load from the 'socket' to the remaining part of the limb, and the critical zones that must not transfer the load, zones were created on the 'socket' that represent the fixed supports for this simulation.

IOP Conf. Series: Materials Science and Engineering 1298 (2023) 012002 doi:10.1088/1757-899X/1298/1/012002

Load zones include areas: Anterior (Patellar ligament, Medial tibia flare, Anterior compartment, Medial shaft of tibia), Lateral (Lateral shaft of fibula), Posterior (Posterior compartment). Static Linear simulation was used in this case. A curvature-based mesh with a maximum element size of 3 mm was created. The Mesh control option was applied at the joints in order to obtain the most accurate result and the size of the element was set to 1.5 mm.
When it comes to defining the contact between components, Global Contact – Bonded was defined. This allows two touching faces or faces with a specific gap between them to remain touching or to force the gap to remain the same size no matter what loads are placed upon the two faces during the analysis.

6.4. *Numerical analysis*
After the simulations for 3 characteristic positions, Fig. 14 shows the distribution of pressure, tension and deformations in the entire construction of the lower leg prosthetic device. Fig. 15 shows the stress state on the segment of the 'socket', which is very important when analyzing the behavior of this part of the structure.

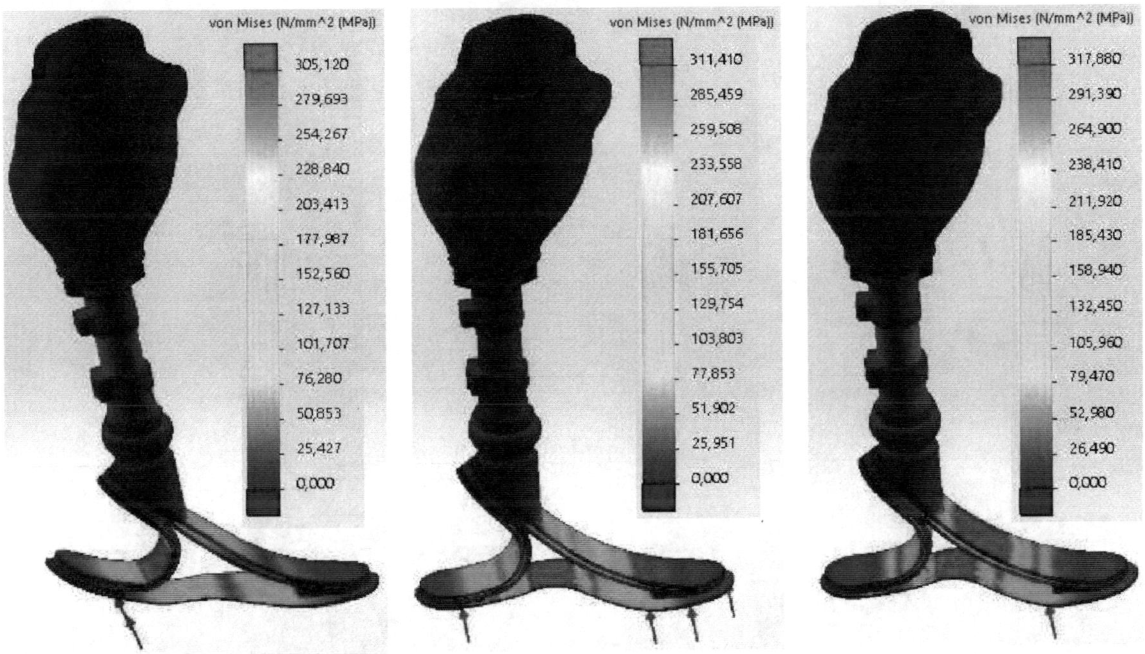

a) b) c)

Figure 14. Numerical analysis: a) position 1; b) position 2; c) position 3.

6.5. *Results analysis*
Figure 14.b) shows the stress state in the construction of the entire prosthesis for position 1, when the subject makes contact with the ground through the heel. In this case, the maximum stress that occurs is 305.12 MPa, and that is on the foot.
Figure 15.a) shows the stress state in the 'socket' and the stress values in the 'socket' do not exceed 10 MPa. Figure 14.c) shows the stress state in the structure of the prosthesis for position 2, i.e. the situation when the subject rests his whole foot on the ground, i.e. when the force enters the structure on both the front and back of the foot (toes and heel). The maximum stress on the entire structure of the prosthesis in this case is 311.41 MPa, and that is on the central part of the foot. Figure 15.b) shows the distribution of stress in the 'socket', when resting with the foot. For case 3, i.e. toe contact, Fig. 14.d) the maximum stress that occurs is 317.88 MPa, and the stress state on the 'socket' is shown in

IOP Conf. Series: Materials Science and Engineering 1298 (2023) 012002 doi:10.1088/1757-899X/1298/1/012002

Fig. 15.c). If all three cases are taken into account, it can be observed that the highest stresses and deformations occur in case 3 - contact with toes (Terminal Stance). Therefore, the most critical position for the structure is position 3, because then the highest stresses and deformations occur. However, these values do not differ to a large extent from those of the previous 2 positions. On the basis of the analysis presented above, it should be noted that slightly higher stress values were obtained, as well as deformation in the "foot" zone in all three cases, which is most influenced by the geometry of the foot itself, but also by the characteristics of the materials, which are in the tabular recommendations in a very wide range. In this sense, the values of the stress state on the foot element indicate that a 'non-linear analysis' should be included in some part analysis, in addition to the very complex methodology implemented, where further directions for research are opened.

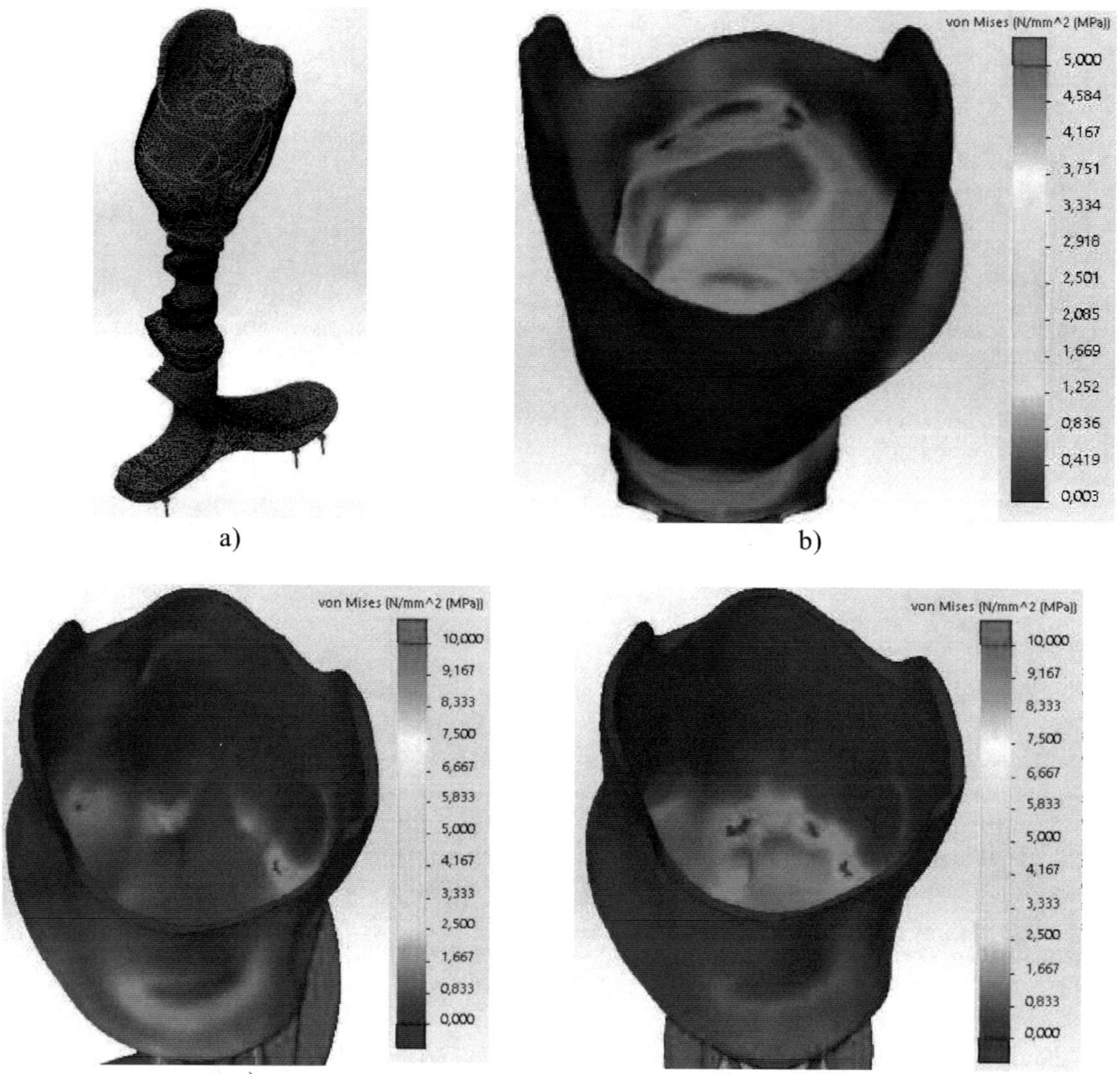

Figure 15. Distribution of stress in the 'socket': a) model discretization; b) when supporting with the heel for position 1; c) when supporting with the foot for position 2; d) when supporting with the foot for position 3.

In addition to other effects, it is necessary to determine the exact properties of the material, and to apply a combined numerical analysis of 'solid-plane' finite elements.

According to [1] (page 9, Fig. 13) it can be seen that the highest values of von Mises stress is concentrated in the lateral and basal planes of socket and amount to 6.08 MPa. In this research the socket is analysed at the most extreme load conditions, which exist at heel strike in Gait cycle. Different pressure values were applied on sockets plane for approximately 75.5 kg in body mass. Different types of materials were combined: Epoxy resin, Carbon Fiber stockinet, Hybrid (Fiber glass + Carbon) stockinet etc. Changing the type of reinforcement and matrix has a great influence on tensile properties. From that analysis it can be seen that reinforcing with Carbon gives the optimum mechanical properties.

According to [2] the interface pressure at the prosthetic socket/lower limb region was recorded experimentally using an F-socket device during the gait cycle. The maximum values of maximum principal stress are 26.485 MPa (page 11, Fig. 16).

According to results of numerical analyses of this paper, the maximum stresses in the socket do not exceed the values of 12-15 MPa. The main reason for the present deviation of the results is in the settings of the numerical simulation itself. If results of this analysis are compared with the results from the mentioned researches, the main reason for the mentioned difference is the boundary conditions, i.e. the zones that are taken as fixed supports, that is, the places that are in the tightest connection with the remaining part of the human body (residuum).

7. Conclusion

Based on the research conducted, it can be concluded that the creation of an adequate prosthetic device is a very complex and multidisciplinary process. It implies the use of modern and traditional knowledge and skills. For example when defining a reliable starting model of the socket, as one of the most important segments of the construction of the prosthetic aid, it is only possible to apply the measurement technology of 3D scanning, which is the most reliable for complex spatial curved surfaces, and for the needs of further creation of 3D models. When it comes to the numerical analysis of the integrity of the structure, an analysis of the structure of the prosthesis in three characteristic positions (heel contact, foot contact and toe contact) was performed, based on the diagram of forces and acceleration during the human walking cycle. On the 3D model of the 'socket', the zones that transmit the loads are marked, which meant defining the boundary conditions for the analysis. A numerical simulation was performed for all three characteristic positions. Analyzing the results, it was observed that the highest stresses and deformations occur in case 3 - toes contact. However, these values are not significantly higher compared to the previous 2 positions. The values of the stress states in the foot zone should be taken with caution because it is a combination of different materials for which the exact properties need to be known, and the principles of 'non-linear analysis' must also be applied. Also, in order to improve the numerical analysis, it is necessary to include the combination of "solid-plane" finite elements. When considering the stress state of the 'socket', it can be observed that the stresses do not exceed the values of 12-15 MPa. Changing the characteristics of the material (from the selected wide spectrum), such as the modulus of elasticity and Poison's coefficient, have no significant effect on the results. Changing the supports, i.e. reducing/increasing the surface of the supports also does not significantly affect the results. Therefore, the main cause of the stress state of the 'socket' lies in the thickness of its wall. The specified values leave space for reducing the wall thickness and reducing the mass. All of the above points to the great complexity of analyzing this type of construction, and the possibility of further research in the area of redesign and further optimization.

References

[1] Al-Khazraji K at al. 2012 Tensile and Fatigue Characteristics of Lower-Limb Prosthetic Socket Made from Composite Materials Proceedings of the 2012 International Conference on Industrial Engineering and Operations Management Istanbul, Turkey, pp 843 - 852

[2] Jweeg J M at al. 2018 Optimised Analysis, Design, and Fabrication of Trans-Tibial Prosthetic Sockets IOP Conf. Series: Materials Science and Engineering 433, 012058

[3] Colombo G at al. 2014 Virtual patient to assess prosthetic devices Conference: 3rd International Digital Human Modeling Symposium (DHM2014) Odaiba, Tokyo, Japan

[4] Vitali A at al. 2017 Design and Additive Manufacturing of Lower Limb Prosthetic Socket Proceedings of the ASME 2017 International Mechanical Engineering Congress & Exposition IMECE, Tampa, Florida, USA

[5] Colombo G at al. 2013 A digital patient for computer-aided prosthesis design Interface Focus 3: 20120082. Royal Society Publishing,

[6] Mohd S J at al. 2019 Analysis of Pressure Distribution in Transfemoral Prosthetic Socket for Prefabrication Evaluation via the Finite Element Method, Bioengineering 2019, 6, 98

[7] Henao C S at al. 2020 Influence of Gait Cycle Loads on Stress Distribution at The Residual Limb/Socket Interface of Transfemoral Amputees: A Finite Element Analysis Scientific Reports 10:4985

[8] Colombo G 2007 ICT Methodologies to Model and Simulate Parts of Human Body for Prosthesis Design Digital Human Modeling, HCII 2007, LNCS 4561, pp 559–568

[9] M. Buzzi at al. 2012 3D modelling and knowledge Tools to automate prosthesis development process International Journal on Interactive Design and Manufacturing (IJIDeM), Volume 6, Number 1, pp 41,

[10] Ali I at al. 2014 Finite Element Modelling and Analysis of Trans-Tibial Prosthetic Socket, Global Journal of Researches in Engineering: A Mechanical and Mechanics Engineering Volume 14 Issue 4 Version 1.0, pp 43-50

[11] Colombo G at al. 2010 A new design paradigm for the development of custom-fit soft sockets for lower limb prostheses, Computers in Industry Computers in Industry vol 61, pp 513–523

[12] Oldfrey B at al. 2020 Digital Fabrication of Lower Limb Prosthetic Sockets, AT2030 Innovation Insights

[13] Krajňáková V at al. 2019 Application of the ACTEC EVA scanner for orthotics in practice, Lekar a technika – Clinician and Technology, Vol 49, No. 3, pp 92-96

[14] Sanders, J. E., McLean, J. B., Cagle, J. C., Gardner, D. W., Allyn, K. J. (2016). Technical Note: Computer-Manufactured Inserts for Prosthetic Sockets, Med Eng Phys., 38 (8), pp 801- 806.

[15] Pitkin M R 2010 Biomechanics of Lower Limb Prosthetics, Springer-Verlag Berlin Heidelberg,

[16] Alessandro F at al. 2015 Experimental Stress Analysis for Materials and Structures, Springer International Publishing Switzerland

[17] Xiaohong J at al. 2003 Load Transfer Mechanics Between Transtibial Prosthetic Socket and Residual Limb – Dynamic Effects, Journal of Biomechanics, 37(9):1371-7

[18] Rajtukova V at al. 2014 Biomechanics of Lower Limb Prostheses, Procedia Engineering 96, pp 382 – 391,

[19] Toshiki K at al. 2015 Socket reaction moments in transtibial prostheses during walking at clinically perceived optimal alignment, International Society For Prosthetics And Orthotics, 40(4):503-8,

[20] Ramirez J F at al. 2012 Algorithm for the prediction of the reactive forces developed in the socket of transfemoral amputees, Dyna (Medellin, Colombia) 79(173): pp 89-95.

IOP Conf. Series: Materials Science and Engineering 1298 (2023) 012003 doi:10.1088/1757-899X/1298/1/012003

Testing the impact energy and determination of the brittleness temperature of the welded joint of steel A516 Gr.70

Ermin Bajramović and Fadil Islamović

Faculty of Technical Engineering, University of Bihać, Bihać, Bosnia and Herzegovina

E-mail: ermin.bajramovic@unbi.ba

Abstract. Most problems with steel structures occur in the welded joint or its immediate vicinity. For this reason, special attention is paid to testing the properties of the welded joint. Fractures of welded steel structures in operating conditions have led to the need to study the influence of temperature on the behavior of steel and welded joints. For the purpose of these tests, the critical test conditions (temperature) that cause brittle fracture are defined. The paper will present the results of experimental tests of the impact energy of the welded joint of steel A516 Gr.70. Impact energy testing at different temperatures was done with the aim of determining the brittleness transition temperature of the base metal and components of the welded joint. Also, these tests should give an assessment of the influence of temperature on the ratio of the crack formation energy W_{iu}, and the crack propagation energy W_a in the total fracture energy W_t. This test procedure determines the tendency to brittle fracture, that is, the tendency to increase brittleness during exploitation. The obtained results are most often used for ongoing quality control and evaluation of material homogeneity.

1. Introduction

Every welded structure must meet the basic condition, which is safety and reliability conditions of exploitation. The most important characteristics of the operational safety of welded structures and the evaluation of the integrity and remaining operational life are the characteristics that describe the appearance and crack propagation under the influence of variable loads. The formation and propagation of cracks most often occur in welded joints or their immediate vicinity. More information on the behavior of the welded joint are obtained by impact testing (fracture energy of the test tube with a notch). The fracture energy of the test tubes with a notch is used as a comparative value in the selection of materials, as well as for the evaluation of the brittleness of the base material and welded joint components at reduced and elevated temperatures.

The successful application of the steel from which the welded structure is made depends on the degree of deterioration of the properties of the base metal (BM) during welding. Furthermore, the heat

IOP Conf. Series: Materials Science and Engineering 1298 (2023) 012003 doi:10.1088/1757-899X/1298/1/012003

affected zone (HAZ) and weld metal (WM) are the most common places of reduced impact energy with transitional brittleness temperature. The problem is how to evaluate the impact energy in the critical areas of the welded joint. The dispersion of the results of the impact energy determined in this way is a consequence of the heterogeneity of the microstructure and mechanical properties of the area where the tip of the notch is located, whereby the width of the HAZ depends on the parameters of the welding regime.

How the material behaves during stress concentration, i.e., during hindered deformation, can be explained by bending tests with impact force on test tubes with a V or U notch. These tests are performed with the aim of determining the absorbed energy required to break the test tube under the established test conditions. For a more complete assessment of the material's behavior under impact loading, it is important to know which part of the energy is spent for crack formation and which part for crack propagation. If the value of the total impact energy (W_t) is known for a material, and the energy used to break it is the crack growth energy (W_a), then the crack initiation energy (W_{iu}) is:

$$W_{iu} = W_t - W_a \tag{1}$$

In addition to learning about the participation of the energies for the creation and growth of adhesion in the total fracture energy, it is very important to know the temperature of the transition from the ductile to the brittle state.

The importance of the determination of the transition temperature from the ductile to the brittle state is shown by the fact that a large number of authors were engaged in research into what is the temperature of the transition from the ductile to brittle state for different conditions and different materials. How important the ductile to brittle transition temperature (DBT) and brittle fracture stress are as toughness criteria for structural materials, the authors [1] say in their work. They test a low-carbon steel with an ultrafine elongated grain structure (grain size 1.2 μm) and two ferrite-pearlite structures with grain sizes of 10 μm and 18 μm. They show the relationship between stress and grain size, and how grain size and temperature cause DBT fracture in low carbon steel.

The authors [2] observe that the loss of all toughness causes the most catastrophic brittle fracture of ferritic steels when the temperature drops below their ductile to brittle transition temperature (DBT). They make an important attempt to model coupled ductile/brittle fracture using mixed micro-cavities and micro-cracks. Using the resulting models as a starting point, they look for the fracture response in a small impact fracture test at temperature of 22 °C and -196 °C, where the fracture is characterized as ductile and brittle, respectively.

The research aimed at determining the impact of the heterogeneity of the material structure on the behavior of the welded joint by means of impact tests on test tubes with a notch removed from the parts of the structure of the tank that is in service for several years, proved to be very useful from the aspect of evaluating the usability and the remaining life of the tank that serves for storage of petroleum products. [3]

The authors [4] in their work Behavior of welded joints of steam pipelines made of high-alloy steel under the influence of temperature and time of exploitation present experimental results of the assessment of the welded joint with the ultimate goal of looking at the possibilities of revitalization and extension of the working life of vital components of thermal power plants made of high-alloy steel for working at elevated temperatures.

The behavior of 12Cr–0.6Mo–2.2W–4Co–0.8Cu-VNb steel was investigated by tensile and Charpy impact tests in the temperature range of 133–473 K. At temperatures below 213 K, the steel shows a level of yielding. Tests show that the ductile-brittle transition occurs at 323 K. It becomes completely brittle at a temperature of 273 K. [5]

The authors [6] say how catastrophic and expensive brittle fractures of pressure vessels can be. They give recommendations to avoid such problems, identify some of the causes of the danger of cold embrittlement and brittle fracture in pressure vessels. They provide recommendations on how to evaluate existing pressure vessels subject to cold conditions and brittle fracture using API 579-1/ASME FFS-1, Part 3.

IOP Conf. Series: Materials Science and Engineering 1298 (2023) 012003 doi:10.1088/1757-899X/1298/1/012003

What is the importance of knowing the parameters based on the application of fracture mechanics, the authors [7] provide a general analysis and presentation of the impact of cracks type errors that occur in welded structures, and methods of assessing the integrity and life of welded steel structures.

This paper will present the results of experimental tests of the impact energy of the welded joint of steel A516 Gr.70 [8-9]. Testing the total impact energy of test tubes with a notch in the base metal (BM), weld metal (WM) and heat affected zone (HAZ) of the welded joint, at different temperatures, was done with the aim of determining the transition temperature from ductile to brittle state. Furthermore, these tests should give an assessment of the effect of temperature on the ratio of the crack formation energy (W_{iu}) and the crack propagation energy (W_a) in the total fracture energy (W_t).

2. Brittle fracture temperature

It is known that by lowering the temperature, the strength of the metal increases and the ability to deform, i.e., the toughness, decreases. In some metals, the toughness decreases gradually up to a certain temperature, and then decreases abruptly. The temperature at which there is a sudden decrease in the ability to deform, that is toughness, is called the transitional fracture temperature or the brittle fracture temperature.

The temperature changes from the ductile to the brittle state, Figure 1, is the temperature at which there is a pronounced decrease in the ability of the material to absorb force without breaking. Brittle fracture is only possible when cracks can propagate at high speeds.

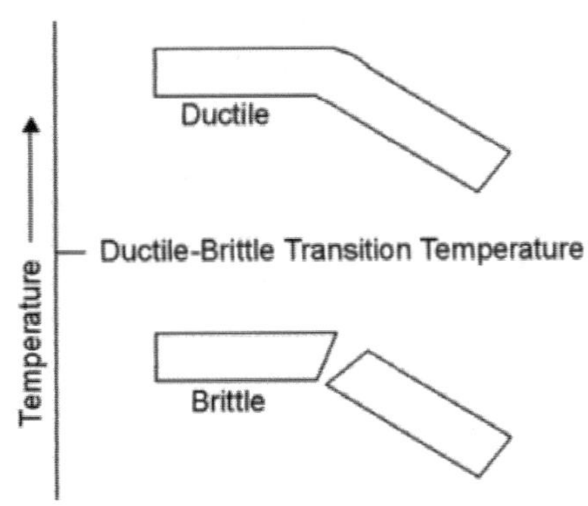

Figure 1. Ductile brittle transition temperature.

Whether the transition from tough to brittle fracture will occur depends, on the one hand, on the properties of the steel at the test temperature, and on the stress distribution conditioned by the method of stressing and the shape of the test tube, on the other hand. Of particular importance is the effect of the notch.

A sudden drop in the energy required for fracture, i.e., toughness, generally does not occur all at once, but in one, more or less, wider temperature range (transitional range). The position of this critical temperature range depends, as schematically shown in Figure 2, on the test conditions and on the properties of the steel. The height of the temperature range, made under certain conditions, is an important characteristic of the quality of steel and its treatment.

IOP Conf. Series: Materials Science and Engineering 1298 (2023) 012003 doi:10.1088/1757-899X/1298/1/012003

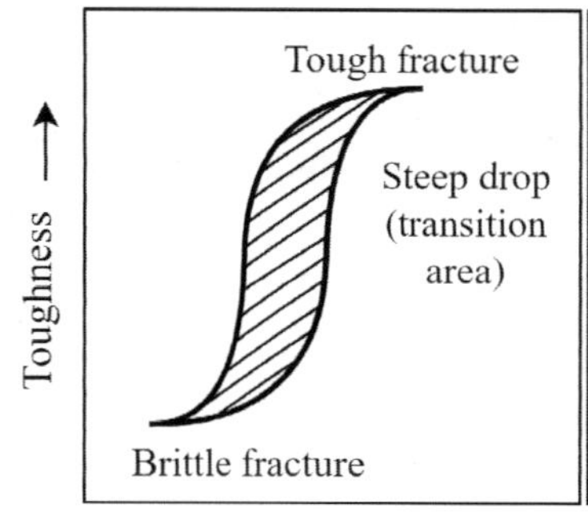

The direction of movement of the steep slope	With increasing
→	Notch sharpness
← →	Notch depth
→	Notch width
→	Notch size
→	Test speed
← →	Thermal treatment
← →	Metallurgic processing
→	Cold deformation

Figure 2. Scheme of the influential quantities during the testing of the impact energy.

In the case where there is no sudden drop in toughness, the critical temperature is not clearly defined. According to some authors, the critical temperature is the one at which the work required for fracture drops to half of its maximum value or at which a mixed fracture occurs with 50 % brittle fracture. According to other authors, the lowest test temperature at which there is no sign of brittle fracture is considered the critical temperature, which means the upper limit of the transition region. Today, the most accepted criterion for determining the transition temperature from the ductile to the brittle state is an impact energy of 27 J, i.e., the temperature at which the impact energy is at least 27 J.

Whether the steel is susceptible to brittle fracture is determined by the dependence of the impact energy required for fracture and the test temperature The ductile-brittle transition temperature, Figure 3, is determined by monitoring the change in impact energy by testing test tubes with V notch, performed at different temperatures.

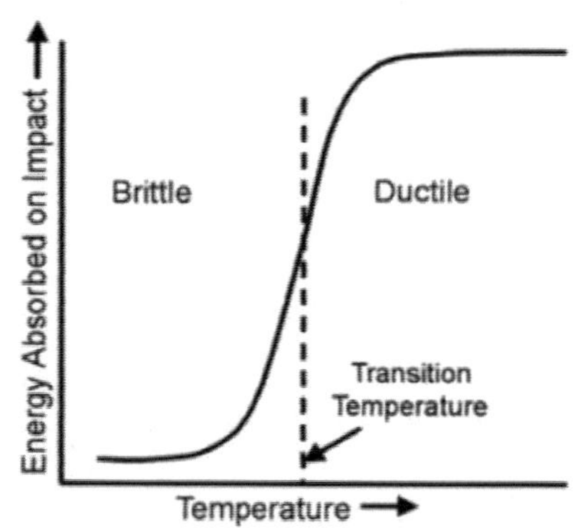

Figure 3. Ductile-brittle transition curve.

Impact testing is based on the theory that "a material breaks after absorbing a certain amount of energy". This temperature depends on the factors that determine the transition temperature from ductile-brittle fracture:

- A lower temperature that reduces the vibrations of atoms causes brittleness in the material.
- Higher dislocation density causes the material to be more brittle.
- Smaller grain size reduces the transition temperature.
- Higher carbon content increases brittleness.

For applications at low temperatures, the required level of impact energy is a very important piece of information and is therefore included in the recommendations for material selection. It should be noted that the required impact energy value is high even at very low temperatures, which indicates the importance of this parameter. Information about the ductile-to-brittle transition temperature for a particular material can prevent potential disasters.

3. Methods and research results

For the purpose of these studies, three groups of test tubes were examined in the temperature range from -60 °C to 540 °C depending on the location of the V-2 notch, namely, test tubes with notch in the base metal (BM), in the weld metal (WM), and in the heat-affected zone (HAZ). The experimental test was performed on an instrumented Charpy pendulum of the brand WOLPERT 150/300 J.

According to the EN ISO 148 Standard [10], the value of the absorbed impact energy KV was obtained by reading the deflection of the scale on the pendulum. As the testing of the base metal (BM) and the components of the welded joint (WM and HAZ) was done according to the EN ISO 14556:2016 Standard [11], the value of the total impact energy (W_t), but also its components, crack formation energy (W_{iu}) and propagation energy was obtained cracks (W_a) by calculating the area under the force-time curve.

Testing on an instrumented Charpy pendulum yielded two types of diagrams, force-time and energy-time. Representative diagrams obtained by impact testing at a temperature of -60 °C for test tubes with a notch in BM, WM, and HAZ are shown in Figure 4.

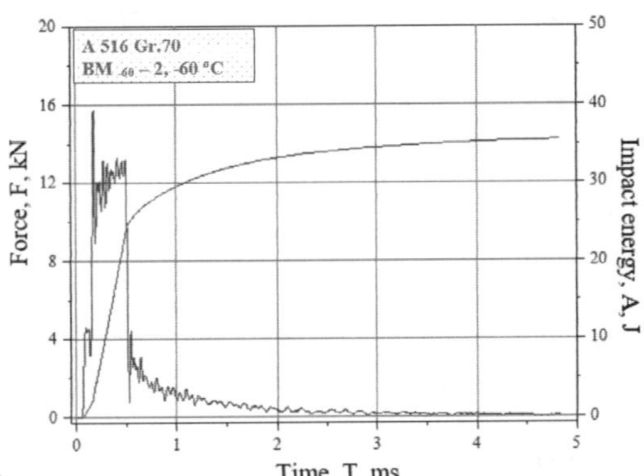

a)

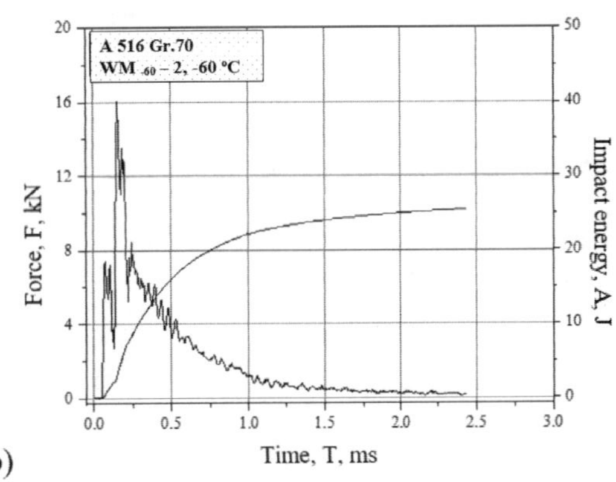

b)

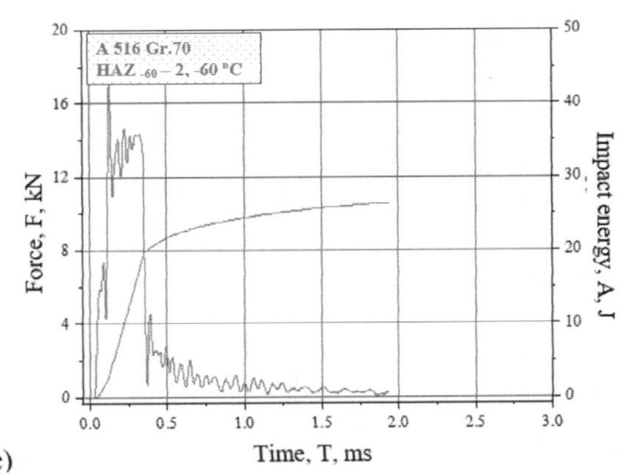

c)

Figure 4. Force-time and energy-time diagrams for temperature -60 °C for test tubes with notch of steel A 516 Gr.70: a) BM, b) WM, c) HAZ.

With this, it was possible to evaluate how the test temperature affects the values of the total impact energy (W_t), as well as its components W_{iu} and W_a, and the obtained values are shown in Tables 1, 2 and 3.

Table 1. The values of the energies for BM.

Sample mark	Test temperature T (°C)	Absorbed impact energy* KV (J)	Total impact energy W_t (J)	Crack formation energy W_{iu} (J)	Crack propagation energy W_a (J)
BM_{-60} - 1	-60	37	36	28	8
BM_{-60} - 2		37	35	26	9
BM_{-60} - 3		36	35	26	9
BM_{-40} - 1	-40	66	64	53	11
BM_{-40} - 2		62	60	53	7
BM_{-40} - 3		64	63	54	9
BM_{-20} - 1	-20	123	121	58	63
BM_{-20} - 2		127	125	59	66
BM_{-20} - 3		131	128	60	68
BM_{20} - 1	20	188	185	60	125
BM_{20} - 2		184	181	60	121
BM_{20} - 3		190	188	61	127
BM_{200} - 1	200	201	198	58	140
BM_{200} - 2		205	202	59	143
BM_{200} - 3		206	204	59	145
BM_{400} - 1	400	142	139	54	85
BM_{400} - 2		133	131	51	80
BM_{400} - 3		146	143	52	91
BM_{540} - 1	540	90	88	23	65
BM_{540} - 2		86	84	23	61
BM_{540} - 3		95	93	25	68

Table 2. The values of the energies for WM.

Sample mark	Test temperature T (°C)	Absorbed impact energy * KV (J)	Total impact energy W_t (J)	Crack formation energy W_{iu} (J)	Crack propagation energy W_a (J)
WM_{-60} - 1	-60	24	23	20	3
WM_{-60} - 2		26	25	21	4
WM_{-60} - 3		26	25	21	4
WM_{-40} - 1	-40	46	45	40	5
WM_{-40} - 2		43	41	34	7
WM_{-40} - 3		48	47	41	6
WM_{-20} - 1	-20	86	84	39	45
WM_{-20} - 2		87	85	38	47
WM_{-20} - 3		85	84	36	48
WM_{20} - 1	20	145	143	48	95
WM_{20} - 2		127	125	48	77
WM_{20} - 3		143	140	50	90
WM_{200} - 1	200	147	144	43	101
WM_{200} - 2		144	141	43	98
WM_{200} - 3		140	137	41	96
WM_{400} - 1	400	84	82	26	56
WM_{400} - 2		95	92	30	62
WM_{400} - 3		89	86	30	56
WM_{540} - 1	540	54	53	14	39
WM_{540} - 2		59	58	16	42
WM_{540} - 3		60	59	16	43

Table 3. The values of the energies for HAZ.

Sample mark	Test temperature T (°C)	Absorbed impact energy * KV (J)	Total impact energy W_t (J)	Crack formation energy W_{iu} (J)	Crack propagation energy W_a (J)
HAZ_{-60} - 1		32	31	24	7
HAZ_{-60} - 2	-60	28	27	20	7
HAZ_{-60} - 3		29	28	22	6
HAZ_{-40} - 1		59	57	49	8
HAZ_{-40} - 2	-40	50	48	41	7
HAZ_{-40} - 3		59	57	50	7
HAZ_{-20} - 1		105	102	57	45
HAZ_{-20} - 2	-20	107	105	57	48
HAZ_{-20} - 3		109	106	59	47
HAZ_{20} - 1		182	179	64	115
HAZ_{20} - 2	20	166	162	61	101
HAZ_{20} - 3		185	182	65	117
HAZ_{200} - 1		192	189	42	147
HAZ_{200} - 2	200	204	201	49	152
HAZ_{200} - 3		186	183	42	141
HAZ_{400} - 1		129	126	47	79
HAZ_{400} - 2	400	118	116	44	72
HAZ_{400} - 3		125	123	45	78
HAZ_{540} - 1		81	79	31	48
HAZ_{540} - 2	540	81	79	31	48
HAZ_{540} - 3		83	80	32	48

The change in the absorbed impact energy (KV), which can be viewed as the total impact energy (W_t), as well as the change in its components W_{iu} and W_a, depending on the test temperature is shown diagrammatically in Figure 5 for test tubes with a notch in BM, for test tubes with with a comma in WM in Figure 6 and for test tubes with a comma in HAZ in Figure 7.

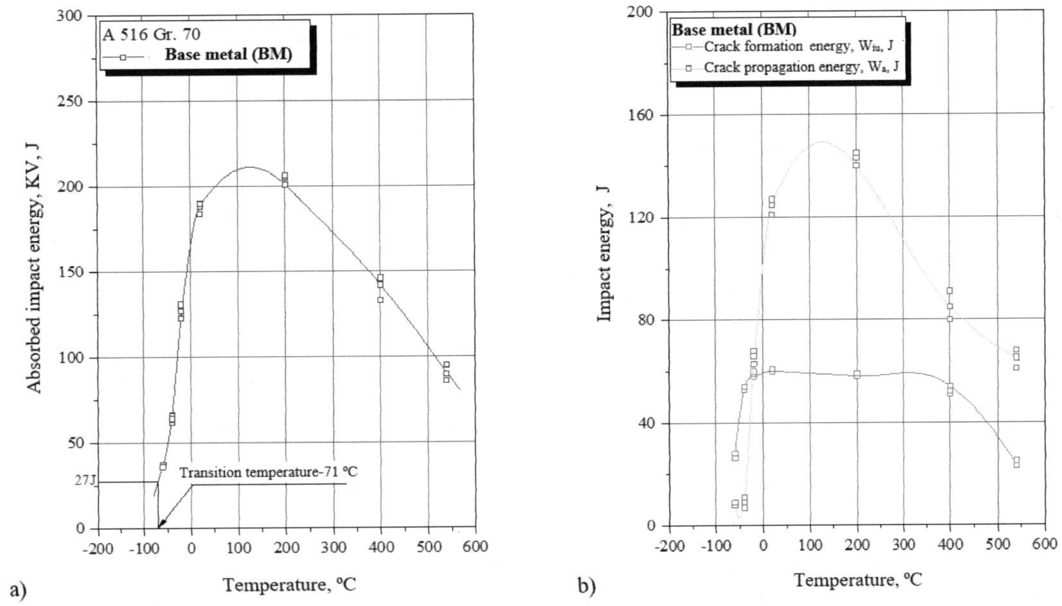

Figure 5. Change in energy depending on the test temperature for test tubes with notch in BM, a) KV (W_t), b) W_{iu} and W_a.

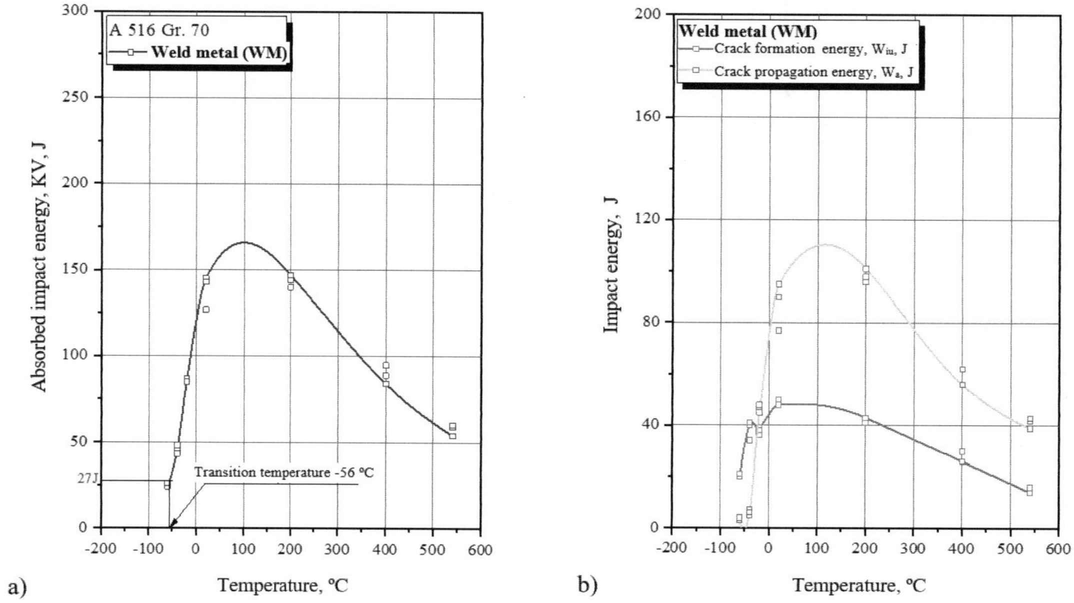

Figure 6. Change in energy depending on the test temperature for test tubes with notch in WM, a) KV (W_t), b)W_{iu} and W_a.

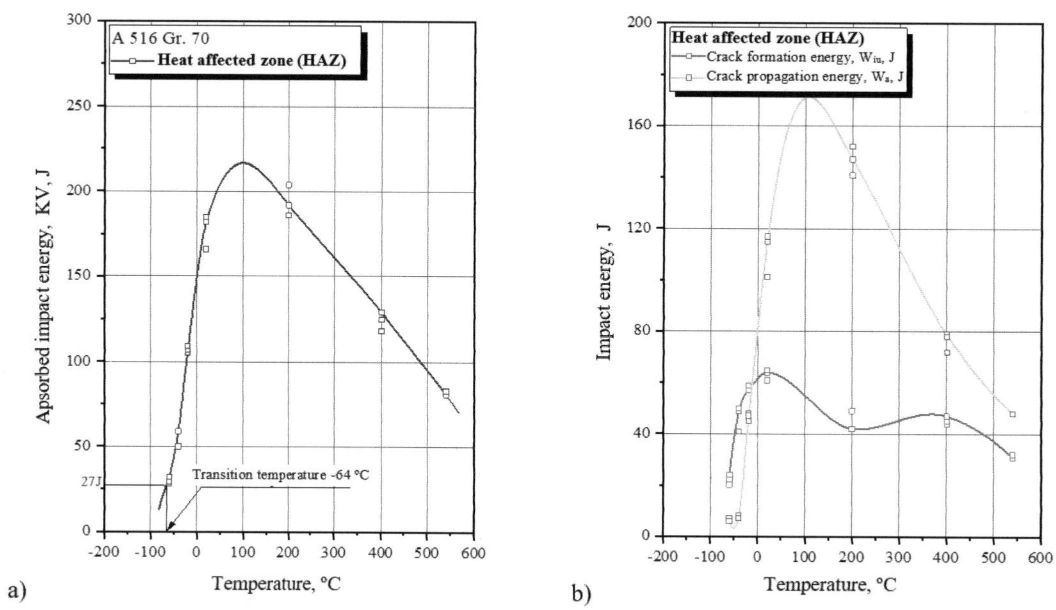

Figure 7. Change in energy depending on the test temperature for test tubes with notch in HAZ, a) KV (W_t), b) W_{iu} and W_a.

4. Discussion of results

Based on the obtained results of impact tests (Tables 1, 2 and 3), it is clearly observed that the total impact energy (W_t) depends on the location of the V-2 notch (BM, WM or HAZ), and the test temperature. Also, diagrammatically, it is clearly seen that the place of V-2 notch preparation and the test temperature have an influence on the values of energy for crack formation (W_{iu}) and energy of crack propagation (W_a).

Tests of test tubes with a V-2 notch in BM (Figure 5) showed that the ratio of W_{iu} and W_a energies changes significantly with the change in temperature. As the temperature increases, the percentage of the ductile component (crack growth energy, W_a) increases up to 200 °C, and then decreases up to 540 °C. The participation of the brittle component (crack formation energy, W_{iu}) has a growing tendency up to 20 °C, it is almost constant when tested at a temperature of 20 °C and up to 400 °C, and then it decreases up to 540 °C. The ratio of W_{iu} and W_a energies is 29:71%.

When testing test tubes with a V-2 notch in WM (Figure 6), also with increasing temperature, the percentage of the ductile component (crack growth energy, W_a) increases up to 200 °C, and after that it decreases until 540 °C. The share of crack formation energy (W_{iu}) increases with increasing temperature up to 20 °C, and then slowly decreases up to 540 °C. The ratio of W_{iu} and W_a energies is 33:67%.

The biggest influence on the change in the ratio (W_{iu}) and (W_a) is in the test tubes with a V-2 notch in the HAZ (Figure 7). The share of crack growth energy (W_a) increases with increasing temperature, and has the highest value during the test at 200 °C, and then decreases until 540 °C. The share of crack formation energy (W_{iu}) increases with increasing temperature up to 20 °C, and then slightly decreases up to 200 °C, is constant up to 400 °C and then slightly decreases up to 540 °C. The ratio of W_{iu} and W_a energies is 23:77%.

Furthermore, these tests for steel A516 Gr.70, i.e. for all components of the welded joint, also determined the transition temperature that defines the transition from tough to brittle fracture.

Table 4. Transition temperature BM, WM and HAZ of steel A516 Gr.70.

Notch placement	Transition temperature (°C)
BM	- 71
WM	- 56
HAZ	- 64

In order to determine the value of the transition temperature in this paper, the criterion 27 J was used, the results of which for the examined low-alloy carbon steel A516 Gr.70 are shown in Table 4.

5. Conclusion

The impact energy depends on the location of the notch and the test temperature. Test tubes with V-2 notch in BM have the highest value of total impact energy, test tubes with V-2 notch in HAZ have a slightly lower value, and test tubes with V-2 notch in WM have the lowest value of total impact energy.

The total impact energy (W_t), increases from -40 °C to 200 °C, where the highest values were obtained, and then decreases with increasing temperature up to 540 °C. It also has a significant influence on its components, the energy of crack formation (W_{iu}) and the energy of crack propagation (W_a). It can be seen that the ratios of participation of W_{iu} and W_a energies are different.

In general, all the diagrams show that the test tubes tested at room temperature 20 °C up to a temperature of 540 °C correspond to the character of ductile materials, with a relatively small share of crack formation energy (W_{iu}), and a large share of crack propagation energy (W_a). When tested at temperatures from -20 °C to -60 °C, it corresponds to the character of ductile-brittle materials, with an approximately equal share of crack formation energy (W_{iu}) and crack propagation energy (W_a) in test tubes tested at -20 °C, and brittle materials with the dominant participation of crack formation energy (W_{iu}) at a temperature of -60 °C.

The data obtained in this research are very important when choosing materials for specific operating conditions.

6. References

[1] Inoue T, Qiu H, Ueji R and Kimura Y 2021 Ductile-to-Brittle Transition and Brittle Fracture Stress of Ultrafine-Grained Low-Carbon Steel *Materials* **14** 1634 doi.org/10.3390/ma14071634

[2] Deliktaş B, Turtuk I.C and Voyiadjis G.Z 2019 Modeling Temperature-Driven Ductile-to-Brittle Transition Fracture in Ferritic Steels In: Voyiadjis, G. (eds) *Handbook of Nonlocal Continuum Mechanics for Materials and Structures* doi.org/10.1007/978-3-319-58729-5_6

[3] Islamović F, Bajramović E, Bajramović E and Islamović B 2023 *Impact Testing of Steel Welded Joint of Exploited Vertical Cylindrical Tank of Large* Lecture Notes in Networks and Systems **539** doi.org/10.1007/978-3-031-17697-5_20

[4] Bajramović E, Gačo Dž and Islamović F 2022 Behaviour of High-Alloy Steel Welded Joints of Steam Pipelines under the Influence of Temperature and Exploitation Time *Transactions of FAMENA* **46(3)** pp 103-113 doi.org/10.21278/TOF.463039822

[5] Dudko V, Borisova J and Kaibyshev R 2018 Ductile-Brittle Transition in Martensitic 12%Cr Steel **134** ACTA PHYSICA POLONICA A No. 3 Proceedings of the International Symposium on Physics of Materials (ISPMA 14) doi: 10.12693/APhysPolA.134.649 PACS/topics: 81.70.Bt, 81.40.Np

[6] Benac D.J, Cherolis N and Wood D 2016 Managing Cold Temperature and Brittle Fracture Hazards in Pressure Vessels *J Fail. Anal. and Preven.* **16** 55-66 doi.org/10.1007/s11668-015-0052-3

[7] Bajramović E and Islamović F 2021 Assessment of integrity and remaining working life of welded steel structures *IOP Conf. Ser.: Mater. Sci. Eng.* **1208** 012011 doi: 10.1088/1757-899X/1208/1/012011

[8] ASTM E23-05 Standard test methods for notched bar impact testing of metallic materials *Annual Book of ASTM Standards* 03.01 ASTM International Pennsylvania USA 2005

[9] ASTM A516/A516M-10 2015 *Standard Specification for Pressure Vessel Plates, Carbon Steel, for Moderate- and Lower-Temperature Service*

[10] EN ISO 148-1:2017 *Metallic materials - Charpy pendulum impact test - Part 1: Test method*

[11] EN ISO 14556:2016 *Metallic materials - Charpy V-notch pendulum impact test - Instrumented test method*

IOP Conf. Series: Materials Science and Engineering 1298 (2023) 012004 doi:10.1088/1757-899X/1298/1/012004

Modeling of Stainless Steel Bending Force on CNC Press Brake

Hasan Cinac[1], Sanel Gredelj[2]

[1]Airvent doo, Velika Kladuša, Bosnia and Herzegovina
[2]Faculty of Technical Engineering, University of Bihać,Bosnia and Herzegovina

E-mail addresses: hasancinac2@gmail.com , sanel.gredelj@unbi.ba

Abstract. Bending processing is one of the most common plastic deformation processing procedures. It is widely represented in the metal processing industry. In this paper, sheet metal bending on a CNC press with a tool that has two basic parts is analysed: the lower-die and the upper-punch. The sheet metal material is stainless steel, which is one of the most commonly used materials. During the bending process on presses, one of the main challenges is the experimental determination of the bending force. The methodology used in this work is based on the experimental stochastic method, that is, design of experiment. A rotatable plan for three input parameters: material thickness, bending angle and bending line length, are varied at five levels. The output parameter is precisely the bending force. No special devices are needed to measure the bending force as an output parameter, because the bending force is measured and registered on the CNC press via the control panel. After the experiment, using the experimental design matrix, the coefficients of an adequate nonlinear mathematical model were determined. Such a mathematical model is suitable for analysis and optimization of the bending force.

1. Introduction

Bending is one of the most common methods of processing by deformation, and the application of bending is very wide, from the automotive, food, aircraft industry, shipbuilding etc. With this processing procedure, it is possible to bend sheets, pipes and other profiles of different thicknesses and materials. Bending is performed on CNC press where, after entering data via the control panel, the appropriate tool is placed on the machine according to the dimensions and material of the workpiece, which makes the processing efficient and optimal. [1]

Due to its range of mechanical properties, resistance to corrosion and high temperature, easy forming and processing, as well as aesthetic appearance, stainless steel is used in almost all constructions and products. In bending processes, stainless steel is one of the most commonly used materials, so it is necessary to know as much as possible about its behavior during processing and, accordingly, to plan and ensure the optimal performance of the process. [2]

During the press bending process, one of the main challenges is the analytical or experimental modeling of the bending force. The mathematical model of bending forces is the basis for their optimization, which can be used to reduce the load on the press and the possibility of overloading, reduce maintenance costs, and thus increase the effectiveness of the entire processing system. [3]

IOP Conf. Series: Materials Science and Engineering 1298 (2023) 012004 doi:10.1088/1757-899X/1298/1/012004

2. Research Equipment

For the experiment, the CNC press „Amada HFP 10-30 NT" (figure 1.) was selected, the characteristics of which are given in table 1.

Figure 1. CNC press brake "Amada HFP 100-30 NT".

Table 1. Characteristics of the CNC press.

Characteristic	Unit	Value
Maximum pressure force	kN	1000
Maximum bending length	mm	3000
Distance between cylinders	mm	2705
Maximum stroke length of the punch	mm	200
Punch lowering speed	mm/s	10
Dimensions	mm	L 4498 × B 2450 × H 2700
Mass	kg	6750
Drive power	kW	12,5
Machine brand	-	Amada
Made in	-	Japan
The year of production	-	2007

The sheet metal bending tool consists of the upper and lower parts placed in their supports on the machine, figure 2. The tools are mechanically connected to the supports on the machine (the upper tool is fixed by turning the handle, while the lower tool is connected to its support with screws). The upper part of the tool – the punch performs the working stroke and transfers adequate force on the workpiece, while the lower part of the bending tool – die is fixed. Both tools are segmented, i.e. depending on the required lenght of the bending line and the geometry of the workpiece that needs to be achieved, it is possible to combine segmets of different lenghts. The characteristics of the tools are given in table 2.

a) b)

Figure 2. Bending tool parts: a) lower tool - die, b) upper tool – punch.

Table 2. Characteristics of the tool used for sheet metal bending.

Characteristic	Unit	Value
Manufacturer	-	Amada
Tool material	-	42CrMo4
Punch radius	mm	0,65
Punch angle	°	30
Maximum force of the upper tool	kN/m	1000
Maximum force of the lower tool	kN/m	1000
Die opening width	mm	12

The workpiece is a stainless steel sheet with a constant widht of 50 mm and a lenght that depends of the design experiment matrix. The samples used in the experiment were obtained by laser cutting from a sheet metal plate 3000 × 1500 mm to the required size and are shown in figure 3.

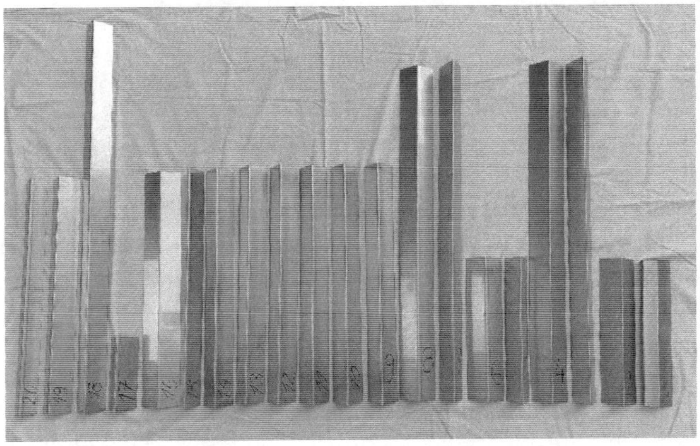

Figure 3. Samples used in the experiment.

Table 3. Characteristics of the workpiece.

Characteristic	Unit	Value
Material – stainless steel	-	1.4301 (DIN EN 10088-3)
Length	mm	179,6-1024,4
Width	mm	50
Thickness	mm	0,66-2,36
Density	kg/m^3	7900
Tensile strength	MPa	520÷720
Yield strength	MPa	230

Since CNC press brake was used, for each working stroke the machine automatically measures the bending force and shows it on the control panel display. The force required for bending the press achieves by two cylinders which, via the upper tool (punch), transfer the load to the workpiece supported on the lower tool (die). The force that the machine shows on the display is instantaneous and is updated in a very small time interval so that is easy to follow the change in its value, starting from the first contact of the punch with the workpiece until complete unloading, i.e. completion of the bending process. The machine display trough which the force is presented is shown on figure 4.

Figure 4. Display of the control panel.

The total value of the achieved force F is presented as sum of the values of the individual forces on first F_1 and the second F_2 cylinder, i.e. $F = F_1 + F_2$. Also, changes of these two individual forces on both cylinders are measured, updated and showed on the display of the control panel.

3. Experimental Plan and Results

The goal of this experiment is to output a non-linear mathematical model, so a rotatable plan is used as a special form of the central compositional plan. Rotatable plans have properties of applicability and optimality, so they are suitable for modeling manufacturing processes that need to be optimized. The rotatable plan contains the base part of the plan 2^k, where k is number of independently variable factors of symmetrically placed point n_α around the center of the plan and the repetition point n_0 in the center of the plan.

In this experiment, three independetly variable factors are varied, i.e. $k = 3$. The total number of measurements and repetitions at the central point depends of the number of variables k, value α and value $p = 0$. The listed and other relevant quantities for the experiment are shown in table 4.

Table 4. Parameters of the second-order rotatable plan for $k = 4$, $n_0 = 7$ and $N = 31$ [4].

Parameter	Value
α	2
λ_4	0,8611
$a_1 = a_{11}$	0,1428
$a_2 = a_7 = a_{17}$	– 0,0357
$a_3 = a_{12}$	0,0416
$a_4 = a_{13}$	0,0625
a_5	0,0312
$a_6 = a_{18}$	0,0037
a_{14}	0,0349

The required number of measurements is calculated according to the formula [4]:

$$N = 2^k + 2k + n_0 = 2^3 + 2\cdot 3 + 6 = 20. \tag{1}$$

The input parameters of the experiment are: the bending angle α, the length of the bending line L mm and the sheet thickness s mm. The output parameter is the bending force F kN·10^{-1}. The force unit kN·10^{-1} is used because it is shown on the display of the CNC machine in that form.

Coding is performed according to the formula [5]:

$$X_i = \frac{x_i - x_{0i}}{\Delta x_i}, \quad i = 1,2,3, \tag{2}$$

where there is the mean level is the value of the i-th factor:

$$x_{0i} = \frac{x_{i\max} + x_{i\min}}{2}, \tag{3}$$

and variation interval:

$$\Delta x_i = \frac{x_{i\max} - x_{i\min}}{2}. \tag{4}$$

After coding the independently variable values, table 5. (basic factors), the plan matrix of the experiment with the results has the form shown in table 6.

Table 5. Physical and coded values.

Influential sizes		Coded and physical values of input factors				
Physical values	$x_1 = \alpha°$	48	65	90	115	132
	$x_2 = L$ [mm]	179,6	350	600	850	1020,4
	$x_3 = s$ [mm]	0,66	1	1,5	2	2,34
Coded values	X_i	−1,6817	−1	0	+1	+1,6817

Table 6. Design matrix and results of the experiments.

Ordinal number of measurements N_j		Coded values X_1 ($\alpha°$)	X_2 (L mm)	X_3 (s mm)	F kN·10^{-1}
1	A set of first-order points $n_k = 2^k = 2^3 = 8$	−1	−1	−1	4
2		+1	−1	−1	3,7
3		−1	+1	−1	9,5
4		+1	+1	−1	9,3
5		−1	−1	+1	8,4
6		+1	−1	+1	8
7		−1	+1	+1	20
8		+1	+1	+1	19,4
9	A set of central plan points $n_0 = 6$	0	0	0	6,3
10		0	0	0	6,4
11		0	0	0	6,12
12		0	0	0	6,4
13		0	0	0	6,3
14		0	0	0	6,3
15	A set of plan points on the central axes $n_\alpha = 2k = 2 \cdot 3 = 6$	−1,6817	0	0	6,3
16		+1,6817	0	0	5,6
17		0	−1,6817	0	1,8
18		0	+1,6817	0	10,5
19		0	0	−1,6817	2,2
20		0	0	+1,6817	26

IOP Conf. Series: Materials Science and Engineering 1298 (2023) 012004 doi:10.1088/1757-899X/1298/1/012004

4. Homogeneity of the results of the experiment

After the experiment, the homogeneity of the dispersions, that is, the homogeneity of the experiments, is checked. By repeating the experiment for constant values of the input parameters, the difference in the obtained output numerical results can be determined. Checking the homogeneity of variances for a certain level of confidence ($P = 0{,}95$) can be performed by Cochran's criterion or Fisher's criterion for distributions that are approximately normal. Verification of the homogeneity of dispersions will be performed according to the Cochrane criterion for the level of confidence $P = 0{,}95$ [6]:

$$K_h = \frac{\max S_j^2}{\sum_{j=1}^{N} S_j^2} \le K_t\left(f_j, N_u\right), \tag{5}$$

where there are:
K_t – tabular value according to Cochran's criterion for degrees of freedom f_j and $N = n_0$,
f_j – degree of freedom ($f_j = n_0 - 1$),
n_0 – the number of repetitions in the sample,
S_j^2 – sample variance:

$$S_j^2 = \frac{1}{n_0 - 1}\sum_{i=1}^{n}\left(y_{ji} - \bar{y}_j\right)^2. \tag{6}$$

Sample mean:

$$\bar{y}_j = \frac{1}{n_0}\sum_{i=1}^{n} y_{ji}, \; j = 1, 2, ..., N_u; \; i = 1, 2, ..., n. \tag{7}$$

As can be seen from table 6., the repetition was carried out in measurements 9-14 for the mean values of the varied parameters, i.e. at the zero point of the plan.

The degree of freedom is calculated according to the equation:

$$f_j = n_0 - 1 = 6 - 1 = 5. \tag{8}$$

Cochrane criterion:

$$K_h = \frac{\max S_j^2}{\sum_{j=1}^{N} S_j^2} = \frac{0{,}0027}{0{,}00566} = 0{,}48. \tag{9}$$

In table I – Cochran's criterion [4] there is $K_t(f_j, n_0)$, where there are:

$f_j = 5$ – degree of freedom,
$N_u = n_0 = 7$ – number of samples,

so it is for the level of risk $\alpha = 0{,}05$:

$$K_t(f_j, n_0) = K_t(5{,}6) = 0{,}445. \tag{10}$$

Table 7. Results of the experiment at repeating in the central plan point.

Ordinal number of attempt	Input values			Measurement result	Mean value	Sum of squares	Variance	Degree of freedom
	α	L	s	F	$\bar{y}_j$	$(y_{ji} - \bar{y}_j)^2$	S_j^2	f_j
9	90	600	1,5	6,3	6,3167	0,000278	$5{,}56\cdot10^{-05}$	5
10	90	600	1,5	6,4	6,3167	0,006944	0,001389	5
11	90	600	1,5	6,2	6,3167	0,013611	0,002722	5
12	90	600	1,5	6,4	6,3167	0,006944	0,001389	5
13	90	600	1,5	6,3	6,3167	0,000278	$5{,}56\cdot10^{-05}$	5
14	90	600	1,5	6,3	6,3167	0,000278	$5{,}56\cdot10^{-05}$	5
						$\sum_{i=1}^{n} S_j^2 =$	0,00566	

Since it is valid:

$$K_h = \frac{\max S_j^2}{\sum_{j=1}^{N} S_j^2} = \frac{2{,}01\cdot10^{-5}}{7{,}35\cdot10^{-5}} = 0{,}48 > 0{,}445 = K_t(6{,}7) = K_t(f_j, N_u), \tag{11}$$

the experiment is at the limit of control, because the dispersion is not homogeneous for the level of risk $\alpha = 0{,}05$.

However in table I – Cochran's criterion [4] there is $K_t(f_j, n_0)$ for the level of risk $\alpha = 0{,}01$:

$$K_t(f_j, n_0) = K_t(5{,}6) = 0{,}520. \tag{12}$$

Since it is valid:

$$K_h = \frac{\max S_j^2}{\sum_{j=1}^{N} S_j^2} = \frac{2{,}01\cdot10^{-5}}{7{,}35\cdot10^{-5}} = 0{,}48 < 0{,}52 = K_t(6{,}7) = K_t(f_j, N_u), \tag{13}$$

the experiment is under control (it can be continued if the first measurements were made at the central poin of the plan), because the dispersion is homogeneous for the level of risk $\alpha = 0{,}01$.

IOP Conf. Series: Materials Science and Engineering 1298 (2023) 012004 doi:10.1088/1757-899X/1298/1/012004

5. Mathematical model and calculation of coefficients

The bending force is modeled by a non-linear model, i.e. second-order polynomal for $k = 3$:

$$F = b_0 + b_1 X_1 + b_2 X_2 + b_3 X_3 + b_{11} X_1^2 + b_{22} X_2^2 + b_{33} X_3^2 + \\ + b_{12} X_1 X_2 + b_{13} X_1 X_3 + b_{23} X_2 X_3 . \quad (14)$$

The coefficients for the mathematical model are calculated according to the following equations [4]:

$$\left. \begin{aligned} & b_0 = a_1 \sum_{j=1}^{N} Y_j + a_2 \sum_{i=1}^{k} \sum_{j=1}^{N} X_{ij}^2 Y_j , \\ & b_i = a_3 \sum_{j=1}^{N} X_{ij} Y_j , i = 1,\ 2,\ ..., k , \\ & b_{im} = a_4 \sum_{j=1}^{N} X_{ij} X_{mj} Y_j ,\ 1 \le i < m \le k , \\ & b_{ii} = a_5 \sum_{j=1}^{N} X_{ij}^2 Y_j + a_6 \sum_{i=1}^{k} \sum_{j=1}^{N} X_{ij}^2 Y_j + a_7 \sum_{j=1}^{N} Y_j ,\ i = 1,\ 2,\ ..., k , \end{aligned} \right\} \quad (15)$$

where are:

$b_0, b_1, b_2, b_{12}, b_{11}, b_{22}$ – coefficients of the mathematical model;

X_{ij}, X_{mj} – coded values;

a_i $(i = 1,\ 2,\ ..., 7)$ – parameters determined from table 4.

According to the equations (15), using table 6. and data from table 4. coefficients of the assumed mathematical model of the bending force $y_j = F$ are:

$b_0 = 6{,}2712$,	$b_1 = -\ 0{,}1960$,	$b_2 = 3{,}5683$,	$b_3 = 5{,}0763$,
$b_{12} = -\ 0{,}0125$,	$b_{13} = -\ 0{,}0625$,	$b_{23} = 1{,}4875$,	
$b_{11} = 0{,}1674$,	$b_{22} = 0{,}2381$,	$b_{33} = 3{,}0489$,	

The Fisher test (F-test) is used to evaluate the significance of the coefficients, the criterion of which is determined by the condition [6]:

$$F_{ri} = \frac{S_{bi}^2}{S_0^2} \ge F_t(1, f_0),\ F_{rij} = \frac{S_{bij}^2}{S_0^2} \ge F_{(1, f_0)}, \quad (16)$$

where there are:

S_{b0}^2, S_{bi}^2, S_{bij}^2, $i, j = 1,\ 2,\ 3$ - error estimates of model coefficients.

As the degrees of freedom of the model coeficients are $f_{b0}, f_{b1}, ..., f_{b33} = 1$, it is:

$$F_{r0} = \frac{N b_0^2}{S_0^2}, \; F_{ri} = \frac{(N - n_0) b_i^2}{S_0^2}, \; F_{rij} = \frac{(N - n_0) b_{ij}^2}{S_0^2}, \; i, j = 1, \; 2, \; 3. \quad (17)$$

Error estimate in the center point of the plan:

$$S_0^2 = \frac{\sum_{j=1}^{n_0} (y_{0j} - y_0)}{f_0}, \quad (18)$$

where there are:

$f_0 = n_0 - 1 = 6 - 1 = 5$ - degree of freedom in the central point of the plan,

$\bar{y}_0 = \frac{\sum_{j=1}^{n_0} y_{0j}}{n_0} = 6{,}3167$ - arithmetic mean of the experimental measurements results values in the central point of the plan.

In table III – Ficher's distribution [4] it is located $F_t(1, f_0) = F_t(1{,}5) = 6{,}61$. The results of the evaluation of the significance of the coeficients, after the necessary calculations according to the formulas (16), (17) and (18), are given in table 8.

Table 8. Evaluation of the significance of the coefficients.

Coefficient	Value	F_r-value	$F_{(1,f0)}$	Estimate
$\boldsymbol{b_0}$	**6,2712**	13,88	6,61	significant
b_1	– 0,1960	0,0967	6,61	not significant
$\boldsymbol{b_2}$	**3,5683**	32,03	6,61	significant
$\boldsymbol{b_3}$	**5,0763,**	64,83	6,61	significant
b_{12}	– 0,0125	0,0002	6,61	not significant
b_{13}	– 0,0625	0,0058	6,61	not significant
b_{23}	1,4875	3,26	6,61	not significant
b_1^2	0,1674	0,0744	6,61	not significant
b_2^2	0,2381	0,1505	6,61	not significant
b_3^2	**3,0489**	24,88	6,61	significant

By including the values of significant coefficients in the assumed coded mathematical model (14), the following form of the mathematical model is obtained:

$$F = 6{,}2712 + 3{,}5683 \cdot X_2 + 5{,}0763 \cdot X_3 + 3{,}0489 \cdot X_3^2. \quad (19)$$

By including the values X_1, X_2, and X_3 into the mathematical model (14) for each of the 20 measurements, the calculated values of y_R are obtained as shown in the table 9.

By including the expression:

$$\left.\begin{aligned} X_2 &= 2\frac{L-L_0}{L_{+1}-L_{-1}} = 2\frac{L-600}{850-350} = \frac{L-600}{250}, \\ X_3 &= 2\frac{s-s_0}{s_{+1}-s_{-1}} = 2\frac{s-1,5}{2-1} = \frac{s-1,5}{0,5}, \end{aligned}\right\} \tag{20}$$

into the coded mathematical model (19), a mathematical model in physical form is obtained:

$$F = 9,9185 + 0,01427 \cdot L - 26,4342 \cdot s + 12,1956 \cdot s^2 . \tag{21}$$

According to the above, the value of the bending force is most influenced by the thickness of the sheet and, to a lesser extent, the lenght of the sheet. The bending angle turned out to be insignificant, i.e. without influence on the value of the bending force.

6. Adequacy and analysis of the mathematical model

In order for the obtained coded model to be fully accepted, it is necessary to examine the adequacy of the mathematical model itself, that is, whether the output parameters y_R calculated from the model correspond to the output parameters y_E of the experiment. Adequacy of the coded mathematical model is checked by comparing the experimentally obtained values of y_E with the values calculated from the model y_R.

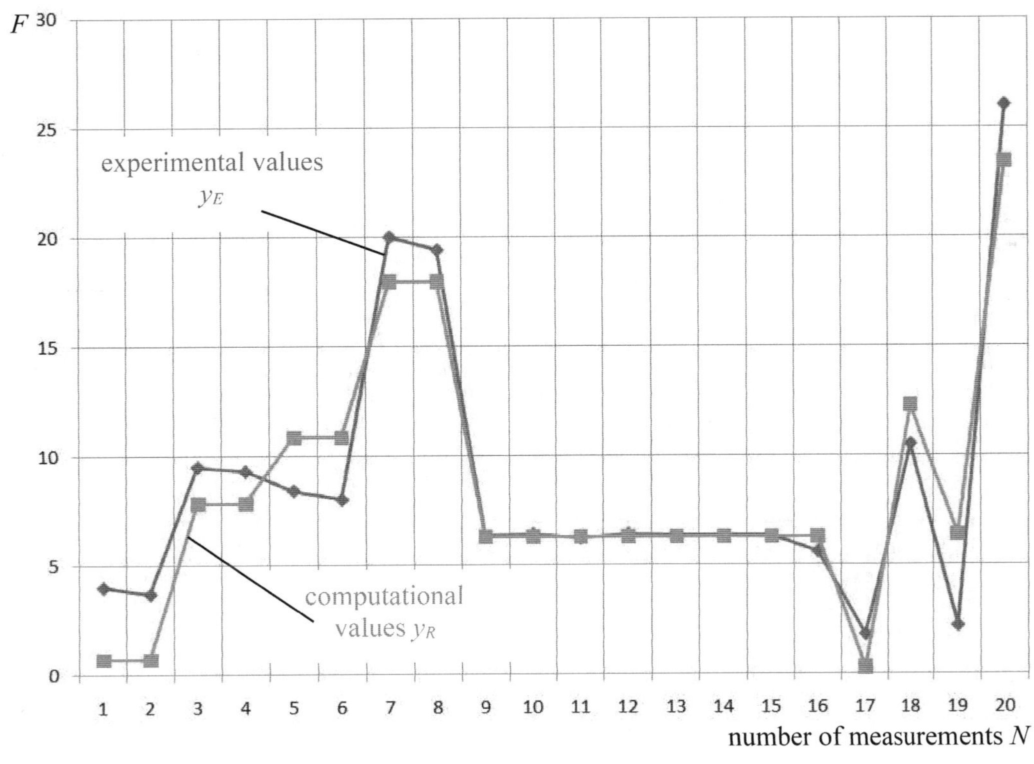

Figure 5. Graphical comparison of experimentally and computationally obtained values.

Table 9. Values of experimental and computational results.

N_j	Coded values			Results		$(y_R - y_E)^2$	$(y_E - \bar{y}_E)^2$
	X_1	X_2	X_3	y_E	y_R		
1	−1	−1	−1	4	0,675	11,054	21,437
2	+1	−1	−1	3,7	0,675	9,149	24,305
3	−1	+1	−1	9,5	7,812	2,850	0,757
4	+1	+1	−1	9,3	7,812	2,215	0,449
5	−1	−1	+1	8,4	10,828	5,895	0,053
6	+1	−1	+1	8	10,828	7,997	0,397
7	−1	+1	+1	20	17,965	4,143	129,277
8	+1	+1	+1	19,4	17,965	2,061	115,993
9	0	0	0	6,3	6,271	0,001	5,429
10	0	0	0	6,4	6,271	0,017	4,973
11	0	0	0	6,2	6,271	0,005	5,905
12	0	0	0	6,4	6,271	0,017	4,973
13	0	0	0	6,3	6,271	0,001	5,429
14	0	0	0	6,3	6,271	0,001	5,429
15	−1,68	0	0	6,3	6,271	0,001	5,429
16	+1,68	0	0	5,6	6,271	0,451	9,181
17	0	−1,68	0	1,8	0,271	2,337	46,649
18	0	+1,68	0	10,5	12,271	3,137	3,497
19	0	0	−1,68	2,2	6,355	17,265	41,345
20	0	0	+1,68	26	23,427	6,623	301,717

The most important criterion for choosing a mathematical model is the multiple regression coefficient R (coefficient of determination). When two or more models describe the process well, the decision about the best model is made based on the multiple regression coefficient. The regression coefficient is also used to examine the agreement (fitting) between y_R and y_E (table 9.) and is calculated according to the formula [6]:

$$R = \sqrt{1 - \frac{\sum_{j=1}^{N}(y_{Rj} - y_{Ej})^2}{\sum_{j=1}^{N}(y_{Ej} - \bar{y}_E)^2}}, \tag{22}$$

where there is:

$$\bar{y}_E = \frac{\sum_{j=1}^{N} y_{Ej}}{N}, \tag{23}$$

IOP Conf. Series: Materials Science and Engineering 1298 (2023) 012004 doi:10.1088/1757-899X/1298/1/012004

arithmetic mean of all experimental results.

According to the value of the experimental and computational results (table 9.) it is obtained:

$$R = \sqrt{1 - \frac{\sum_{j=1}^{N}(y_{Rj} - y_{Ej})^2}{\sum_{j=1}^{N}(y_{Ej} - \bar{y}_E)^2}} = \sqrt{1 - \frac{75{,}217}{732{,}622}} = 0{,}947\,. \quad (24)$$

The value $R = 0{,}947$ means that the mathematical models (19) and (21) with a reliability of 94,7 % describe the results of the experiment, which is an excellent result, that is, it can be concluded that the mathematical models are adequate and can be the basis for the optimization procedure.

7. Conclusion

Press bending is a widely used metalworking process. The bending force is one of the most important features of processing and it is very important to determine the mathematical model of the bending force.

One of the ways of modeling the bending force is the methodology shown in this paper, which is based on the experimetal stochastic method, that is, the design of the experiment. A rotatable plan with three input and one output parametes was used as the basic model. Input parameters: bending angle, bending line lenght and sheet thickness, are varied at five levels. The output parameter is the bending force *F*, which is sensitive to varying input parameters. After the experiment, using the plan-matrix, the coefficients of the nonlinear mathematical model were determined. The adequacy of this model is excellent and the model can be used for optimization.

References

[1] Hosford W F and Caddel R M (2011) Metal Forming: Mechanics and Metallurgy, *Cambrige University Press GB*

[2] Di Schino A (2020) Manufacturing and Applications of Stainless Steels *MDPI – Multidisciplinary Digital Publishing Institute*

[3] Karabegović E (2020) Eksperimentalne metode obrada deformisanjem *Univerzitet u Bihaću Tehnički fakultet BiH*

[4] Jurković M (1999) Matematičko modeliranje inženjerskih procesa i sistema *Mašinski fakultet Bihać BiH*

[5] Montgomery C D (2015) Design and Analysis of Experiment *Arizona State University John Wiley and Sons USA*

[6] Hodžić D and Džanić A (2020) Planiranje eksperimenta *Univerzitet u Bihaću Tehnički fakultet BiH*

IOP Conf. Series: Materials Science and Engineering 1298 (2023) 012005 doi:10.1088/1757-899X/1298/1/012005

Testing of Mechanical Properties of Polymer PA6 – 30% Glass Fibres

Amir Ćoralić[1], Anel Maslo[1], Mirza Manjgo[2], Vesna Raspudić[1], Mersida Manjgo[3] and Tomaž Vuherer[2]

[1] University of Mostar, Faculty of Mechanical Engineering, Computing and Electrical Engineering, B&H
[2] University of Maribor, Faculty of Mechanical Engineering, Slovenia
[3] University Džemal Bijedić, Faculty of Mechanical Engineering, B&H

E-mail: amir@ebi.ba

Abstract. Polymers and polymer composite materials, due to their properties, are becoming increasingly important materials in industry. They are used in a wide range of applications in space and aerospace, aviation, marine and automotive industry, as well as in production of sports equipment, household appliances, medical equipment, etc. They are expected to withstand loads during their lifetime, which requires predicting the reliability of structural elements, i.e. knowledge of material properties. To achieve better properties, polymer materials are reinforced with different percentages of fibre-reinforced materials, including glass fibres. The paper presents the results of experimental determination of the mechanical properties of injection moulded PA6 polymer reinforced with 30% glass fibres, determined by tensile and impact tests.

1. Introduction

In recent years, the consumption of iron, steel, lead and copper has reached a significant level of saturation in industrially developed countries, while the consumption of relatively new materials such as aluminium, plastics and composites has been growing rapidly [1]. Due to their properties, polymers and polymer composites are becoming increasingly important materials in industry. They are used in a wide range of applications in space, aviation, marine and automotive industry, as well as in production of sports equipment, household appliances and medical equipment. There are many advantages that these materials offer compared to traditional metal materials such as easy manufacturing, corrosion resistance, favourable strength-to-weight ratio, multipurpose use, etc. They are also expected to have appropriate load carrying capacity during their lifetime, which requires estimating of material properties to predict the reliability of structural elements made of them [2].

The properties of the final product are also largely influenced by the manufacturing process itself. The basic difference between the manufacturing of parts made of metal and polymer materials is that metals are usually shaped in a series of successive operations until the final shape and condition of the product is reached, while in the case of polymer materials it is usually only one operation that is required, which makes them easy to process into components and structures of complex shapes in an automated fashion, using traditional moulding techniques or extrusion processes.

Fibre-reinforced polymer composites are a growing class of composite materials, defined by the combination of at least two constituents: the matrix material and the reinforcing fibres, which can be continuous and discontinuous. Engineering thermoplastics, such as polyamide (PA) or polypropylene

IOP Conf. Series: Materials Science and Engineering 1298 (2023) 012005 doi:10.1088/1757-899X/1298/1/012005

(PP), represent many matrix materials used for fibre-reinforced composites due to their superior properties [3]. Glass fibres are frequently used as reinforcements due to their availability, low cost, and high strength, but many other materials are also possible, like aramid and carbon fibres. In the case of short fibre-reinforced polymers, it is considered that the glass fibres are shorter than a few millimetres and discontinuously distributed in the polymer matrix. They do not have such a high level of stiffness and strength as composites reinforced with continuous fibres, but the key advantages of discontinuous fibres are better surface quality, lower manufacturing costs and potential of high-volume production.

A possible disadvantage is thereby that the manufacturing process may strongly affect the final state of the fibres in the manufactured part. The fibre orientation, fibre length and fibre concentration can show a large degree of heterogeneity throughout the moulded part, especially for complex shaped structures [4,5]. These variations of the fibre microstructure have a profound impact on the performance of the finished product, evident in the heterogeneous and anisotropic structural properties, which affect the reliable and safe operation of the product in exploitation and the expected lifetime [6].

2. Injection moulding of short fibre-reinforced polymers

Injection moulding is the most important cyclic process of polymer processing. The material is injected and forced through a gate changing the material's configuration and causing the fibre alignment in injected components based on the flow conditions during mould filling. This process can cause a local variation of the fibre concentration, which may lead to a change along the flow path of the material as well as through the thickness of the moulded part. A high shear flow gradient in the edge regions leads to the formation of a boundary layer directed parallel to the flow direction, while the fibre orientation in the core region is perpendicular to the global flow direction [7]. Thus, in simplified terms, it can be assumed that injection-moulded thermoplastics reinforced with short fibres can be described by the three-layer model shown in Figure 1.

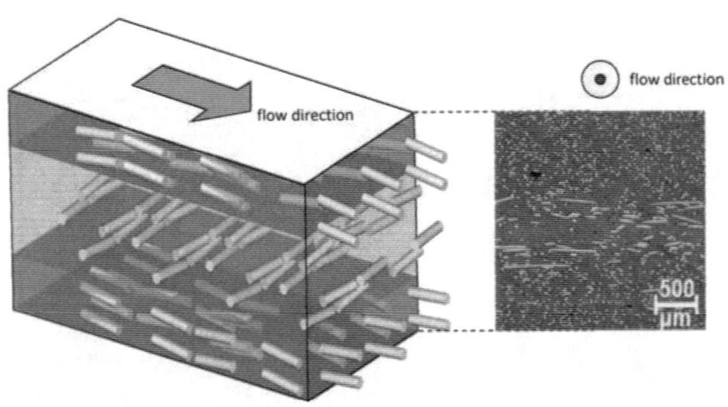

Figure 1. The three-layer model of fibre orientation [8]

However, in the vicinity of injection gates, as well as in the places where materials are joined from different directions forming weld lines, there are changes in the fibre orientation resulting in weak places with poorer mechanical properties, which cannot be described by the previously mentioned model (Figure 2).

The mechanical properties are higher in the direction in which more fibres are aligned, while they are reduced in the direction where fibres are not well aligned. Materials with fibres oriented parallel to the loading direction have much higher stiffness and strength compared to vertical fibres at weld lines, where external stresses are transmitted mainly by the matrix material. In addition, the vertical fibres in these places increase the negative effects further, because they prevent the mutual diffusion of the macromolecules of the matrix, which leads to an additional decrease in the strength of the weld line. In addition, air traps often occur, because the air cannot escape between the melt streams. An additional crack initiation can occur in the weld line area due to a notch which can be formed if the two melt streams freeze too fast at the mould wall [10]. All above mentioned indicates the need to study and characterize the mechanical and fracture-mechanical properties of the weakened part of the weld line, with the aim of reliably predicting the performance of structural elements.

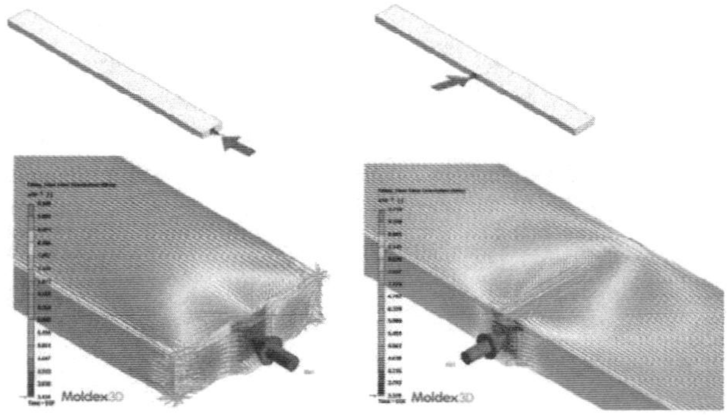

Figure 2. Fibre alignment for different injection gates [9]

3. Experimental tests

This paper presents the results of the tensile and Charpy impact tests of PA6 polymer reinforced with 30% glass fibres. The test samples were produced by injection moulding with the gate location placed in the middle of their geometry. For the analysed gate location, a computer simulation has also been conducted to get numerical representation of fibre alignment, whereby the commercial software Moldex3D has been used. The results for fibre distribution in the longitudinal direction are presented in Figure 3a, and there is the corresponding distribution in the cross section in the middle of the test samples in Figure 3b.

a)

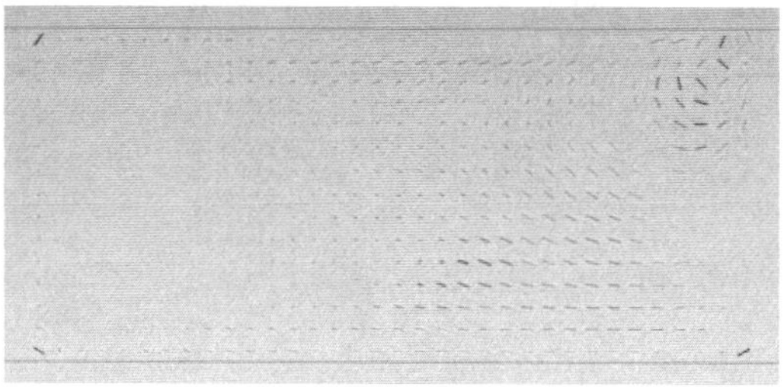

b)

Figure 3. Fiber alignment obtained by computer simulation Moldex3D

3.1. Tensile test

Mechanical strength data measured under a quasi-static loading condition describe the deformation behaviour of the tested sample under a short-term loading condition. Too high a test speed leads to embrittlement i.e., the strength values are increased, and ductility is reduced. The tests are usually performed at a temperature of 23°C and 50% relative humidity since the temperature and the moisture content also influence the mechanical properties. Polyamides are hygroscopic and absorb water from the environment. As the temperature and humidity increase, the ductility increases while the strength properties decrease.

The tensile test has been performed on a Zwick/Roell, Vibrophore 100 electro-mechanical testing machine (Figure 4a). The test procedure is defined by the standard ASTM 638-10 [11]. The test sample is stretched until fracture and the associated force–displacement curve is recorded, from which the stress–strain curve is calculated. The fractured samples are presented in Figure 4b. The results for the mechanical characteristics determined by the tensile tests are given in table 1.

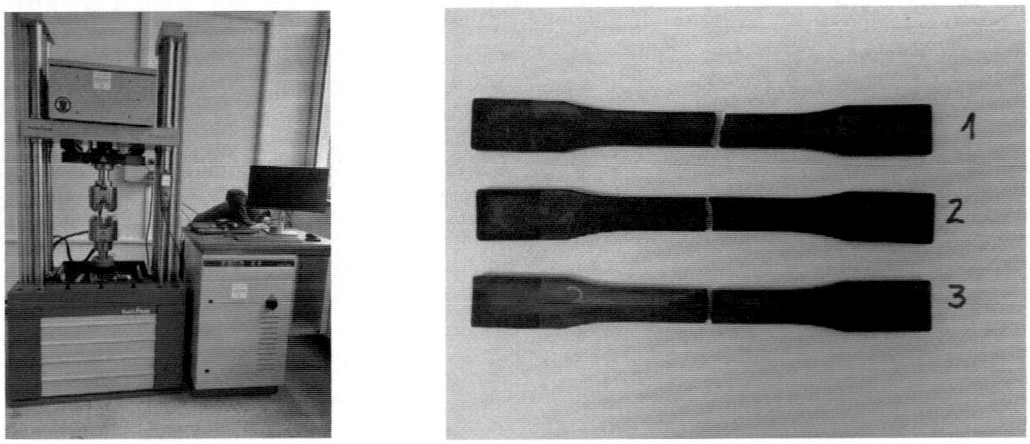

a) b)

Figure 4. a) Testing machine Zwick/Roell, Vibrophore 100
b) Injection moulded samples after rupture

Table 1. Mechanical characteristics determined by the tensile test.

Specimen	R_m MPa	$R_{p0.2}$ MPa
PA6-30-1	101	82
PA6-30-2	99	81
PA6-30-2	97	80

Figure 5 shows the force-strain and stress–strain behaviour of a tested sample under a quasi-static test.

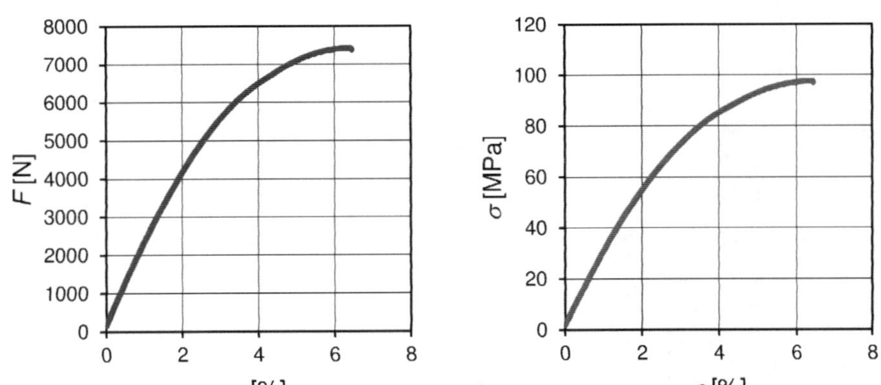

a) b)

Figure 5. The tensile test results for specimen PA6-30-3
a) Force – strain curve, b) Stress – strain curve

IOP Conf. Series: Materials Science and Engineering 1298 (2023) 012005 doi:10.1088/1757-899X/1298/1/012005

After the tensile test, the fracture surface of the test samples was recorded with a microscope LEICA WILD M10 (Figure 6a), and the fracture surface is shown in Figure 6b. The fibres are arranged in the manner predicted by the numerical simulation performed using Moldex3D software.

a)

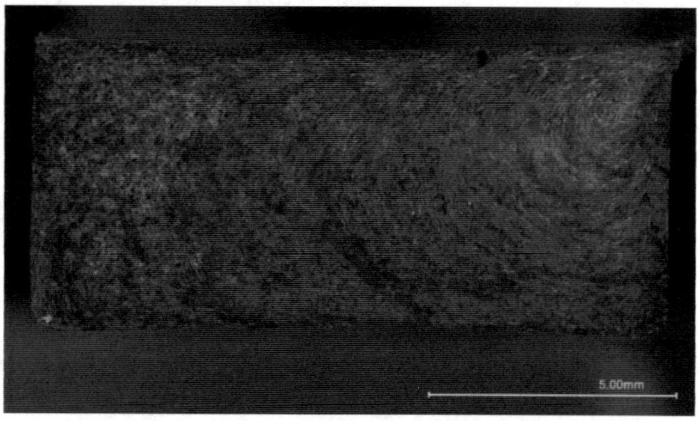

b)

Figure 6. a) Microscope LEICA WILD M10 when examining the fracture surface of the samples b) Fracture surface of the sample

3.2. Charpy impact test

Impact strength or toughness expresses the material resistance under impact loading and serves to evaluate the tendency of the material to brittle fracture under dynamic load. The results of these tests depend on the sample geometry, the loading speed and the temperature, which is why these tests are carried out on samples of the prescribed shape and dimensions and under the prescribed conditions. The impact test procedure was performed according to the ASTM D6110 standard [12]. In the case of polymers, the test specimen can be with or without a notch. In this case, the test was performed on specimens without a notch.

The Charpy test was performed on the instrumented Charpy pendulum AMSLER RPK300 (Figure 7a). The results of these tests show the change in force with time i.e., the force-time and energy-time curves (Figure 7b).

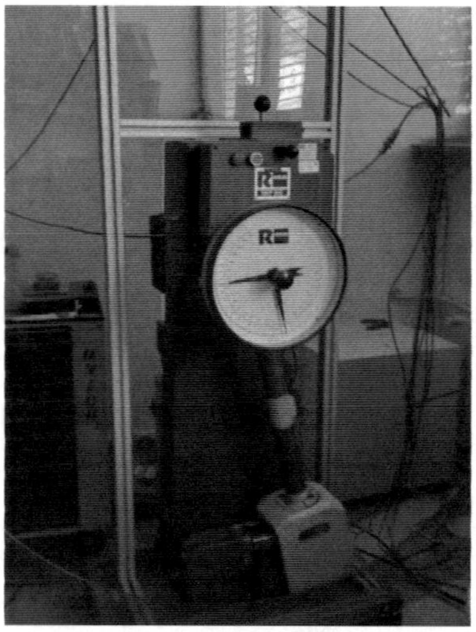

a)

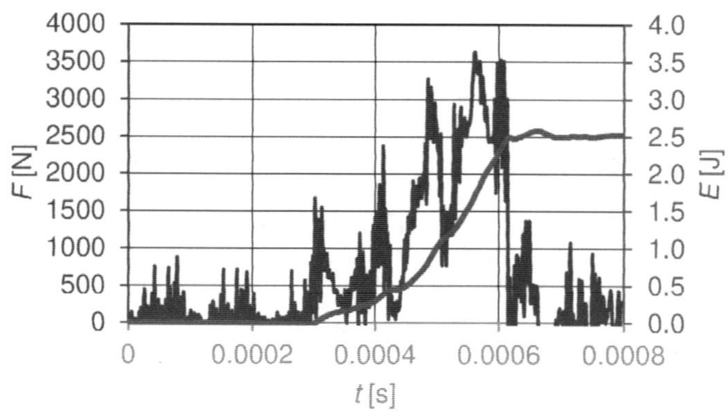

b)

Figure 7. a) Charpy impact testing machine AMSLER RPK300
b) Force – time and energy – time curves

Based on the results of the impact tests, we can conclude that the PA6 material with a fibre content of 30% is relatively brittle. The total energy (2.5 J) for the fracture is mainly the energy for the creation/initiation of the crack (2.5 J) and the energy for propagation is almost (0 J). This indicates the brittleness of the material because the material, when the crack is created, is not able to resist the crack expansion, so that its rapid expansion occurs.

4. Conclusion

Mechanical properties of the PA6 reinforced with 30% glass fibres have been determined. The tested samples were made by the injection moulding process with the gate location in the middle of the samples. The tensile strength values range up to 99 MPa, and the yield stress is 81 MPa. The fracture surface of the test samples has been recorded with a microscope and it can be concluded that there is a good agreement in the fibre distribution with the model established using Moldex3D simulation.

Based on the results of the impact tests, it can be concluded that the PA6 material reinforced with 30% of glass fibres is relatively brittle. The total energy for fracture is mainly the energy for creation (crack initiation) and the energy for propagation is almost 0 J, which indicates the brittleness of the material.

References

[1] Pagano C, Surace R, Bellantone V, Baldi F and Fassi I 2018 *Mechanical characterisation and replication quality analysis of micro-injected parts made of carbon nanotube/polyoxymethylene nanocomposites*, Journal of Composite Materials, pp 645-657. doi:10.1177/0021998317713258

[2] Pintarič T 2017 *Predelava plastičnih mas*, Elektronska izdaja, Novo mesto, http://www.razvoj-upd.si/wp-content/uploads/2017/07/17.-PREDELAVA-PLASTICNIH-MAS-Tomaz-Pintaric.pdf

[3] AVK Industrievereinigung Verstärkte Kunststoffe 2010 *Handbuch Faserverbundkunststoffe: Grundlagen, Verarbeitung, Anwendungen*, Vieweg+Teubner, Wiesbaden

[4] Sasayama T, Okabe T, Aoyagi Y and Nishikawa M 2013 *Prediction of failure properties of injection-molded short glass fiber-reinforced polyamide 6,6*, Composites Part A: Applied Science and Manufacturing, Vol 52, pp 45-54

[5] Holmstroem P H, Hopperstad O S and Clausen A H 2020 *Anisotropic tensile behaviour of short glass-fibre reinforced polyamide-6*, Composites Part C: Open Access, Vol 2, 100019, ISSN 2666-6820, https://doi.org/10.1016/j.jcomc.2020.100019.

[6] Lee C S, Kim H J, Amanov A, Choo J H, Kim Y K and Cho I S 2019 *Investigation on very high cycle fatigue of PA66-GF30 GFRP based on fiber orientation*, Composites Science and Technology, Vol 180, pp 94-100, ISSN 0266-3538, https://doi.org/10.1016/j.compscitech.2019.05.021.

[7] Willems F, Reitinger P and Bonten C 2020 *Calibration of Fiber Orientation Simulations for LFT—A New Approach*, Journal of Composites Science. 4(4):163. https://doi.org/10.3390/jcs4040163

[8] Korte W 2022 *Back-to-Basics: Fiber Orientation and Component Behavior*, https://www.partengineering.com/en/blog/article/back-to-basics-fiber-orientation-and-component-behavior

[9] Wang M L, Chang R Y and Hsu C H 2018 *Molding Simulation: Theory and Practice*, Carl Hanser Verlag, Munich

[10] Geyer A and Bonten C 2019 *Enhancing the weld line strength of injection molded components*, AIP Conference Proc. 22 2055 (1): 070023. https://doi.org/10.1063/1.5084867

[11] ASTM 638-10

[12] ASTM 6110

IOP Conf. Series: Materials Science and Engineering 1298 (2023) 012006 doi:10.1088/1757-899X/1298/1/012006

The use of modern 3D technologies for the development of a prototype mountain bike brake lever within the concept of personalized production

Sebastijan Žibert, Maša Mamić, Elvis Hozdić

University of Novo mesto, Faculty of Mechanical Engineering, Na Loko 2, SI-8000, Novo mesto, Slovenia

E-mail: elvis.hozdic@fs-unm.si

Abstract. Mountain biking (MTB) is a popular sport for more and more people. Control over the bike is very important for a successful ride, especially when the rider gets off the mountain bike in rough terrain. Control over the bike is ensured by appropriate braking systems, which differ depending on the type of mountain biking. During a challenging descent, the cyclist consumes energy to overcome obstacles on the trail, so using the right brakes is crucial. The rider performs the braking process with the brake lever of the bicycle. On difficult descents, for example in bad weather (rain), it is important that the finger does not "*slip*" on the brake lever. Also important is the feeling of contact with the brake lever that the rider has when braking. For this reason, from an ergonomic point of view, the shape of the brake lever is crucial for a safe and controlled mountain bike ride. This paper presents the results of the use of advanced 3D technologies for the development of a prototype mountain bike brake lever within the concept of personalised production. The common thread of the presented research is the process of developing a prototype mountain bike brake lever from an ergonomic point of view, supported by modern 3D technologies such as 3D scanning, 3D modelling and 3D printing.

1. Introduction

Mountain biking (MTB) is experiencing a growing popularity, attracting a diverse range of enthusiasts. A key factor for a successful and enjoyable ride is the rider's ability to maintain control over the bike, particularly when navigating through challenging and rough terrains. To ensure optimal control, the choice of appropriate braking systems becomes paramount, as they can vary depending on the specific type of mountain biking involved.

The brake lever is the means by which the rider performs the braking action. During difficult descents, such as in bad weather, it is important to prevent fingers from slipping on the brake lever. It is also important that the rider has the feeling of touching the brake lever when braking. From an ergonomic standpoint, the shape of the brake lever is crucial for safe and controlled mountain biking.

The shape of the brake lever is a crucial characteristic, as it represents the contact surface between the bike and the cyclist. By shape, we primarily mean the ergonomic design, which allows the cyclist to have an "ideal" finger position and provides a good sense of control when braking. Ergonomics is a subjective matter, and there is no universally ideal shape of brake lever that would be suitable for every cyclist from an ergonomic perspective [1].

An example of brake lever ergonomics comes from the Company "NN" (a globally-renowned manufacturer of hydraulic bicycle braking systems). This company, in collaboration with a multiple world champion in downhill mountain biking, has developed a brake lever specifically tailored to his preferences. By altering the shape of the brake lever, the modulation of the braking system was influenced. The modulation provided by the new lever design offered an "on/off" feel on the brake, meaning that only a small lever stroke was required to achieve maximum braking power. The maximum braking power is achieved at a relatively large distance from the handlebars. Figure 1 illustrates [1]: a) a prototype brake lever, manufactured using 3D-printing technology with titanium, b) the current (existing) brake lever.

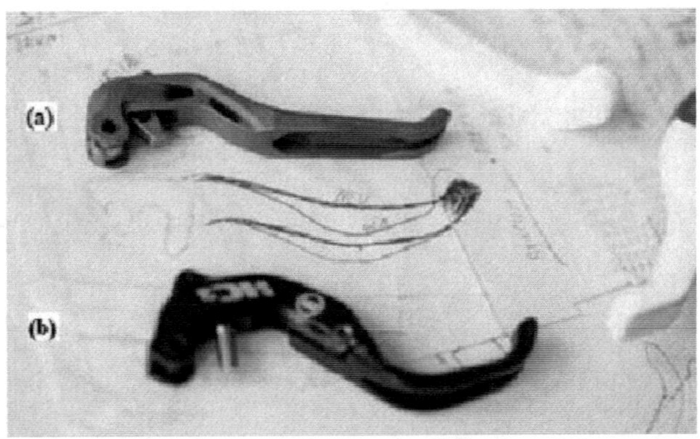

Figure 1. A prototype brake lever, manufactured using 3D printing technology [1]

A similar example also comes from the aforementioned company, where they collaborated with a top cyclist in the "trail bike" category to develop a brake lever that was tailored to his riding style [2]. He wanted the lever to be as short as possible and close to the handlebar grip. This design allows a firm grip on the handlebar, providing good control of the bike, as the finger on the brake lever remains bent rather than stretched. The aforementioned lever allows for modulation adjustment of the braking system (changing the mechanical ratio) and reach adjustment (distance from the handlebar). This allows the cyclist to fine tune the adjustment according to their riding preferences, see Figure 2.

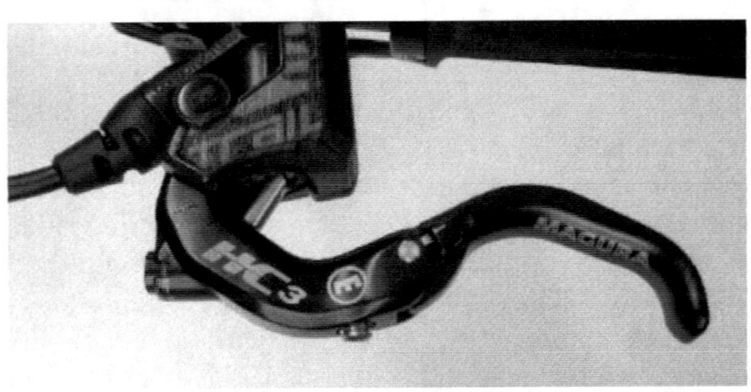

Figure 2. *Magura* – HC3 brake lever [2]

An interesting example is the development of a prototype brake lever created for a video project. The ride took place on a steep rocky slope in Iceland. The goal was to create a brake lever that would provide high braking performance with minimal finger force applied to the lever. This was crucial as the descent was challenging and allowed the rider to save a significant amount of energy normally expended in braking. They achieved this by lengthening the existing lever by 9 mm, achieving a new mechanical ratio and consequently increasing the braking power [3], see Figure 3.

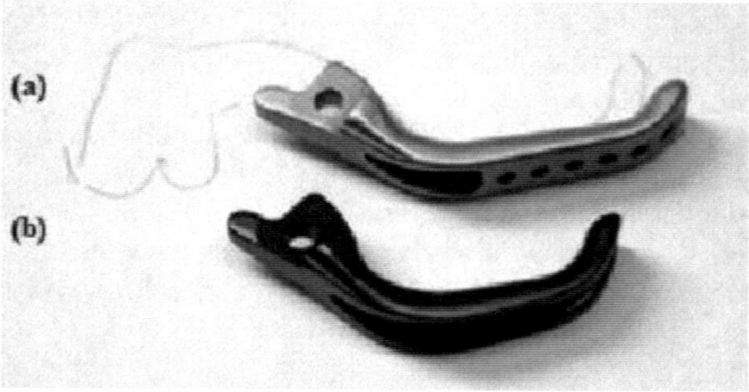

Figure 3. *Magura* – HC3 brake lever, the existing lever by 9 mm [3]

The brake lever of the company "MM" is also an interesting example, see Figure 4. The lever is made of aluminium alloy and is processed using milling technology. This particular lever allows for several adjustments. Compared to the previously mentioned examples, it has a relatively short contact surface for the finger. Furthermore, the contact surface is additionally treated with perforations for improved grip [4].

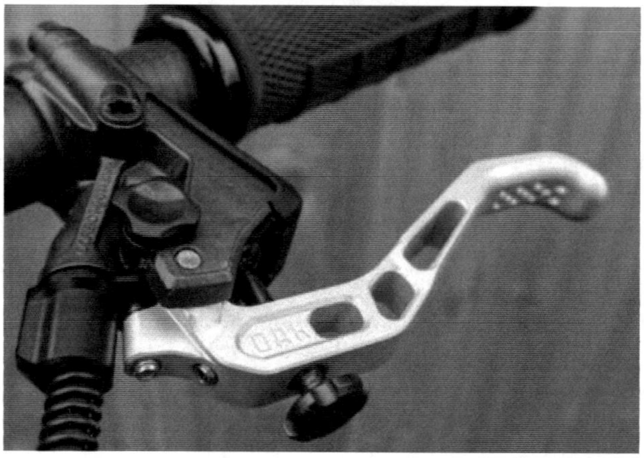

Figure 4. The brake lever of the company "MM" [4]

Through a review of relevant literature focusing on the ergonomics research of bicycle brake levers, two design directions for mountain bike brake levers have been identified [5-11]: 1) ensuring a comfortable grip of the brake lever during rides, reducing braking issues, and enhancing braking comfort, 2) improving braking efficiency while prioritizing safety during rides.

In the quest for optimal brake lever design, manufacturers are continuously exploring different shapes, materials, and features to enhance rider comfort, safety, and control during mountain biking. However, it remains essential to recognize the subjective nature of ergonomics and tailor brake lever designs to suit individual riders' preferences and needs.

This paper presents the development of a brake lever model explicitly designed according to the ergonomic considerations and preferences of downhill mountain bikers. This study addresses the inadequate ergonomic features of the existing brake lever (specifically the design of the brake lever) highlighted by a mountain bike rider during a discussion. The cyclist pointed out the following issues observed while riding: 1) finger slipping on the brake lever surface during rides, particularly where the finger makes contact during braking, and the angle of the brake lever tip (α_r), as shown in Figure 5, 2) the surface of the lever in contact with the finger lacks additional elements to improve friction between the finger and the lever during braking, 3) the ergonomic design of the lever is not optimal, and the rider suggested possible improvements in the shape of the lever, specifically intended for one-finger braking.

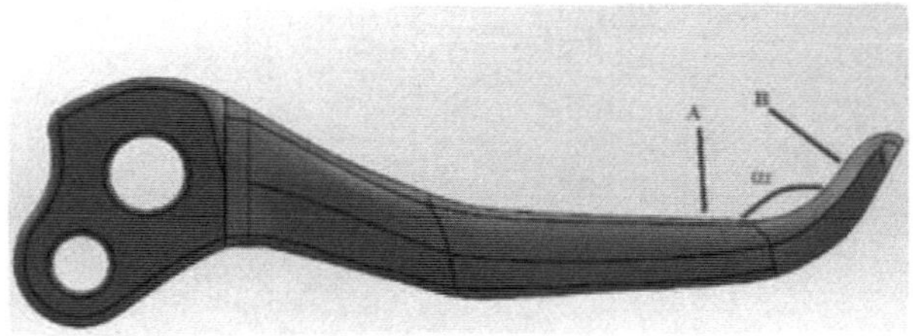

LEGEND:
A – Contact surface of the finger
B – The tip of the brake lever
α_r – Brake lever tip angles

Figure 5. The current brake lever

The paper presents the application of modern reverse engineering and rapid prototyping technologies in the development of a prototype model for a mountain bike brake lever.

2. Theoretical background

2.1. Mountain bike brake lever

Mountain biking is becoming an increasingly popular sport, engaging a growing number of people. It encompasses various disciplines, ranging from casual off-road cycling to challenging downhill rides on challenging terrain. This sport has developed greatly in the last decade. The cycling industry is rapidly advancing, launching new and improved bike models every year. The constant development of manufacturing technologies allows the production of high-quality mountain bikes and their components. This also applies to the field of mountain bike braking systems.

For successful riding, especially in downhill mountain biking in rough terrain, control over the bike is crucial and it is provided by appropriate braking systems, which vary depending on the type of mountain biking. Hydraulic braking systems are most commonly used on mountain bikes, providing maximum braking power without the need for excessive force at the lever. The use of these brakes is essential for challenging descends that require energy to overcome obstacles on the trail. Brakes are crucial components of mountain bikes as well as bicycles in general [12, 13]. They allow to control the speed and bring the bike to a stop. They operate on the principle of friction between two braking surfaces (brake disc - brake pads). The braking systems are generally divided into two main groups: mechanical and hydraulic.

In mechanical braking systems, pressure on the brake lever creates tension in the cable surrounded by the brake line. The forces are transmitted through the cable to the mechanical brake module. The tension in the cable causes two brake pads, which are attached to the brake module, to move toward the braking surface of the bicycle rim. Friction is generated between the surfaces, which allows the speed to be reduced or the bicycle to stop [14].

The second group of braking systems are hydraulic disc brakes. When we operate the brake lever of a hydraulic braking system, pressure is generated in the brake fluid, which is located in the airtight brake hose. The pressure in the brake fluid pushes on the pistons in the brake calliper, which are in direct contact with the brake pads. When the surfaces of the brake pads and the brake disc come into contact, friction is generated, allowing for the speed to be reduced or the bicycle to stop [14].

The hydraulic braking system consists of the following components: 1) hydraulic brake lever mechanism, 2) hydraulic brake hose, 3) hydraulic brake calliper, 4) brake pads, and 5) brake disc.

The brake lever is a crucial component of the hydraulic braking system. Generally, they are designed based on the movement mode of the hydraulic piston in the brake cylinder. The brake lever has a significant effect on the braking power or torque during braking [15].

The brake lever also has a decisive influence on the physical background of the braking system's operation. It determines the mechanical ratio (X_{meh}). In this particular ratio, three points (T_1, T_2, T_3) are significant, as shown in Figure 6.

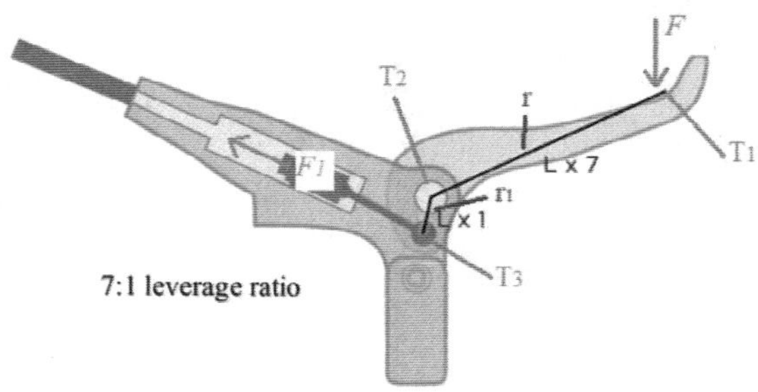

Figure 6. The mechanical leverage of the brake lever [15]

At point (T_1), the cyclist exerts a force ($\boldsymbol{F}$) on the brake lever during braking. The point (T_2) is the pivot point. Point (T_3) represents the mechanism that pushes the hydraulic piston (push rod axis). The mechanical ratio is the ratio between the lengths of the levers (r and r_1), which can be expressed by equation 2.1 [15].

$$X_{meh} = \frac{r}{r_1} \tag{2.1}$$

The ratio mentioned provides us with the magnitude of the force on the pushrod of the hydraulic piston ($\boldsymbol{F}_1$) in relation to the force ($\boldsymbol{F}$) that the cyclist exerts on the brake lever when braking with their finger. This ratio is represented by equation 2.2 [15].

$$\boldsymbol{F} * r = \boldsymbol{F_1} * r_1 \tag{2.2}$$

In general, brake levers are designed to minimise the distance between the pivot point (T_2) and the pushrod axis of the hydraulic piston (T_3). This results in greater mechanical leverage because the movement of the brake lever on the force-applied side ($\boldsymbol{F}$) causes less displacement of the pushrod axis of the hydraulic piston (point T_3) [15].

Since the total braking ratio depends on both the mechanical and hydraulic ratio (X_{hyd}), it is necessary to consider the hydraulic ratio as well. The hydraulic ratio refers to the ratio of the surface areas between the hydraulic piston and the pistons in the brake calliper (piston ratio). This ratio determines the displacement of the hydraulic piston, which affects the movement of the pistons in the brake calliper. The surface area of the hydraulic piston (A) is smaller than the surface area of the piston in the brake calliper (A_1) [15].

Equation 2.3 represents the hydraulic braking ratio. The hydraulic braking ratio (X_{hyd}) has a decisive influence on the braking performance. Figure 7 illustrates the hydraulic braking ratio (X_{hyd}) [15].

$$X_{hid} = \frac{A_1}{A} \tag{2.3}$$

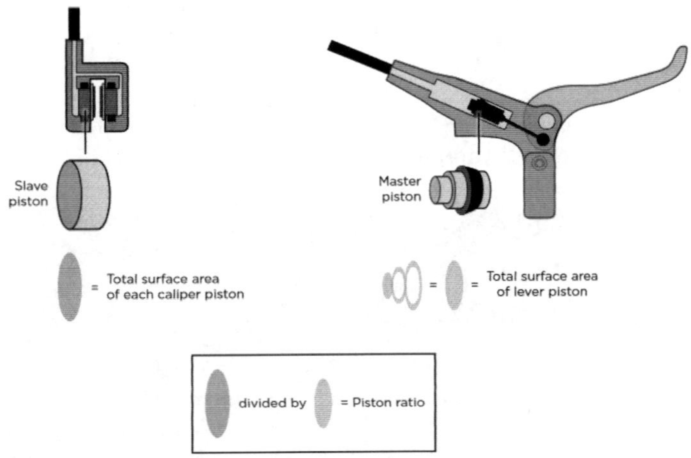

Figure 7. Hydraulic braking ratio [15]

Based on the mechanical ratio (equation 2.1) and the hydraulic ratio (equation 2.3), the braking system ratio (X_{BS}) is described by equation 2.4 [15]. Figure 8 illustrates the braking system ratio.

$$X_{BS} = X_{meh} * X_{hid} = \frac{r}{r_1} * \frac{A_1}{A} \tag{2.4}$$

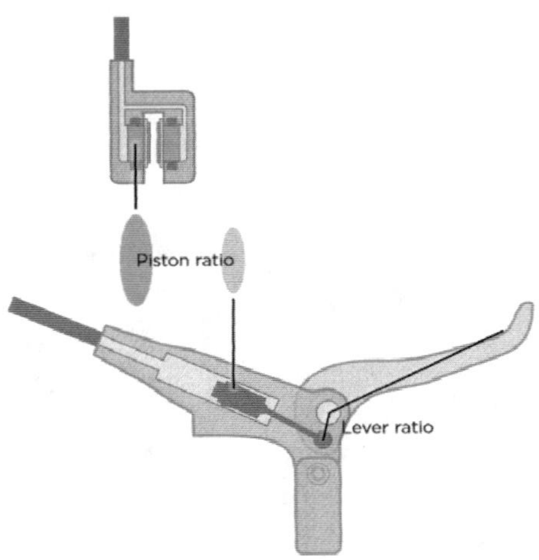

Figure 8. The braking system ratio [15]

Understanding the braking system ratio (mechanical and hydraulic ratio) allows us to calculate the clamping force of the brake calliper pistons ($\boldsymbol{F_2}$) by knowing the force exerted by the hydraulic piston ($\boldsymbol{F_1}$), which can be calculated from the ratio of the lever lengths (r and r_1) and the force exerted by the cyclist on the brake lever ($\boldsymbol{F}$). The calculation of the brake calliper pistons ($\boldsymbol{F_2}$) is represented by equation 2.5 [16].

$$\boldsymbol{F_2} = X_{BS} * \boldsymbol{F_1} = \left(\frac{r}{r_1} * \frac{A_1}{A}\right) * \boldsymbol{F_1} \tag{2.5}$$

Equations 2.1 through 2.5 represent an example of a linear braking system in which the braking force (clamping force of the brake calliper pistons – $\boldsymbol{F_2}$) varies linearly with lever stroke ($X_{meh} = const.$).

In the following sections of this paper, we will discuss a braking system where the mechanical ratio (X_{meh}) varies with the lever stroke (consequently, the clamping force of the brake calliper pistons – $\boldsymbol{F_2}$) does not vary linearly. In the system shown in Figure 9, the mechanical transmission (X_{meh}) depends on the curved shape of the connecting link (B), which is in contact with the rotating module (highlighted in red) to which the pushrod of the hydraulic piston (C) is attached [17].

Figure 9. Presentation of individual modules of a progressive brake system [17]

The mechanical ratio (X_{meh}) increases progressively with the movement of the lever in the mentioned mechanism, which means that the braking force (clamping force of the brake calliper pistons – $\boldsymbol{F_2}$) increases progressively with the lever movement. The mechanical ratio of the mentioned braking system (X_{meh}) is represented by the red curve in Figure 10.

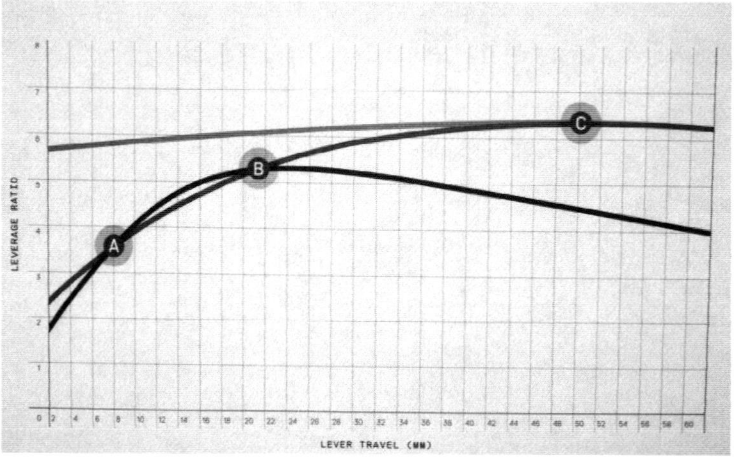

Figure 10. Mechanical ratio in relation to the lever travel [18]

The advantage of such a mechanism is that the brake is activated more quickly (the initial lever movement causes a greater displacement of the brake calliper pistons with the brake pads towards the brake disc due to a small initial mechanical ratio (X_{meh}). This reduces the so-called "*dead*" travel of the brake lever. When the brake pads make contact with the brake disc, the mechanical ratio (X_{meh}) increases and modulates the force at the same time. Higher mechanical ratio means less displacement of the pushrod of the hydraulic piston, higher force on the pushrod, and better control over braking – modulation) [17, 18].

2.2. Modern 3D technologies

Industrial production in the 21st century, under the constant pressure of globalization, is characterized by regionalization, ubiquitous presence and increasing personalization of products, with each customer seeking a product according to his own preferences. As a result, the number of variations of individual products and their diversity in terms of required production volumes increases significantly.

In the field of modern product development, companies are using advanced technologies such as computer-aided design (CAD), simulations, 3D printing and digital modelling in the context of personalised production to enable customers to design and customize products according to their needs. This can include choosing different features, designs, colours, materials or even resizing.

Modern technologies such as 3D scanning and 3D printing allow us to create conceptual and functional prototypes using appropriate materials. These technologies allow for rapid prototyping (RP) and rapid manufacturing (RM) of new end products [19, 20]. The time from the creation of a 3D virtual model to the printed prototype is significantly shorter compared to conventional methods. This enables rapid testing of printed prototypes and the ability to iterate and produce a larger number of iterations within a relatively short period of time.

The use of modern 3D technologies [21-25] enables us to develop and create both virtual and physical models of future products. In the following sections, the modern reverse engineering technologies used in the development and production of a mountain bike brake lever model are presented in detail.

2.2.1. 3D scanning and shape capturing technologies. 3D scanning and shape capturing technologies are used to convert physical objects into 3D digital models. There are various methods for shape capturing, such as optical scanning, laser scanning, and the use of sensors and cameras. Scanning is used to gather data about the shape and texture of objects, which can then be further used to design, analyse, or reproduce the objects in digital form.

3D scanning is the first stage of the so-called reverse engineering process. It involves capturing data from a physical model (using 3D scanners) and representing it in digital form, usually as a point cloud. In 3D scanning, it is crucial to obtain high-quality data, including the number, distribution, and density of points. The quality and quantity of the input data has a significant impact on subsequent processing, such as the creation of a CAD model [26].

Data digitisation technology (3D scanning) can be categorised into contact and non-contact methods. Contact technologies involve touching the object with a probe or sensor to gather data. Non-contact methods use optical methods (coherent and non-coherent radiation) and triangulation (analysis of the reflected beam from the object) is used to determine the measured points in space. The technology of data digitisation is used in dimensional inspection (comparison of real products with CAD models), reverse engineering, 3D printing, medical purposes (prosthetics), archiving, and many other fields [26].

Digitisation (3D scanning) represents the geometry of the measured object in a point cloud. To obtain the complete geometry of the measured object, several measurement perspectives are usually required. Therefore, it is necessary to rotate the measured object. The collection of all 3D scans (point clouds) represents the complete geometry of the measured object (a merged point cloud).

The digitised data often contains "noise" in the form of poorly defined geometric features of the measured object. Editing of these geometric features is possible with special software and tools [26], [27].

The processing of digitised data in reverse engineering with specialised software is usually done in the following order [26, 27]: 1) removal of point clouds describing the positioning surface of the measured object, 2) alignment of individual point clouds into a merged point cloud (representing the measured object), 3) registration to ensure precise alignment of individual point clouds into a merged point cloud and removing duplicate points in overlapping areas, 4) removing "noise" in the data, 5) generating a mesh file from the point cloud (with important characteristics such as resolution and accuracy), and 6) exporting the mesh file in STL (Standard Tessellation Language) format.

IOP Conf. Series: Materials Science and Engineering 1298 (2023) 012006 doi:10.1088/1757-899X/1298/1/012006

2.2.2. 3D modelling. 3D modelling is the process of creating a three-dimensional digital representation of an object or a scene using specialised software. It involves the creation, manipulation, and visualisation of geometric shapes, textures, and materials to generate a virtual model that closely resembles the real object or environment. In 3D modelling, designers use various techniques to construct and refine the digital model. These techniques may include: 1) creating basic shapes such as cubes, spheres, or cylinders and modifying them into more complex objects (primitive modelling), 2) constructing objects by defining their surfaces with interconnected polygons (polygonal modelling). This technique allows for detailed control over the shape and structure of the model, 3) the use of Non-Uniform Rational B-Splines (NURBS modelling) to create smooth and curved surfaces. NURBS provide precise control over shape and allow smooth transitions between different parts of the model, and 4) the use of sculpting tools to shape and mould the digital model as if working with virtual clay. This technique is particularly useful for creating organic and detailed shapes.

After the 3D model is created, additional steps may be taken to enhance its realism, such as adding textures, materials, lighting, and animation. The finished 3D model can be used for various purposes, including visualisations, virtual simulations, video games, movies, product design, architectural visualisation, and more.

Overall, 3D modelling enables designers and engineers to bring their ideas to life in a virtual environment, allowing for accurate visualisation, analysis, and communication of complex objects or concepts.

2.2.3. 3D printing. 3D printing, also known as additive manufacturing [28-32], is a revolutionary technology that allows the production of physical objects from a 3D digital model. It is a process in which successive layers of material are deposited or solidified to build the final object. 3D printing offers several advantages over traditional manufacturing methods, such as the ability to create complex shapes, customisation, rapid prototyping, and on-demand production.

The process of 3D printing typically involves the following steps [30]: 1) creation of a 3D model: the first step is to create a 3D digital model of the desired object using computer-aided design software (CAD) or by scanning an existing object using 3D scanning techniques, 2) slicing: the 3D model is then sliced into thin cross-sectional layers using slicing software. Each layer is a 2D representation of the object's geometry, 3) material selection: the appropriate material for 3D printing is selected based on the object's requirements, such as strength, flexibility, or heat resistance. Various materials can be used for 3D printing, including plastics, metals, ceramics, resins, and even food grade materials, 4) printing process: the 3D printer follows the instructions of the cut file and begins the printing process. There are different types of 3D printing technologies, including fused deposition modelling (FDM), stereolithography (SLA), selective laser sintering (SLS), and powder bed fusion (PBF). Each technology has its own unique way of depositing or solidifying the material layer by layer to build up the object, and 5) post-processing: after the printing process is complete, the object may require post-processing to improve its surface or add additional features. This can include removing support structures, sanding, polishing, painting, or applying other finishing techniques.

3D printing has revolutionised the way we design, create and manufacture objects, enabling greater creativity, efficiency and flexibility in the production process. The process is constantly evolving and finding new applications that push the boundaries of what is possible in the world of manufacturing.

3. Methodology

The following paper presents the development of a brake lever model explicitly designed according to the ergonomic preferences of the rider (downhill mountain bike competitor).

In the first phase of model development, the existing brake lever was scanned using 3D scanning technology. 3D models of the brake lever were then created using the point cloud data and further processing in a modeling environment (SolidWorks). The individual models were 3D printed and physically tested on the bike. Together with the rider, the confirmed model was finally selected as the most suitable from an ergonomic perspective for safe and controlled downhill mountain biking.

Modern 3D technologies, additive manufacturing technologies and advanced materials enable the production of conceptual and functional prototypes in a relatively short time frame, from the creation of a 3D model to obtaining the final printed prototype. The process of creating a prototype model of a mountain bike brake lever is illustrated in the IDEF diagram (see Figure 11).

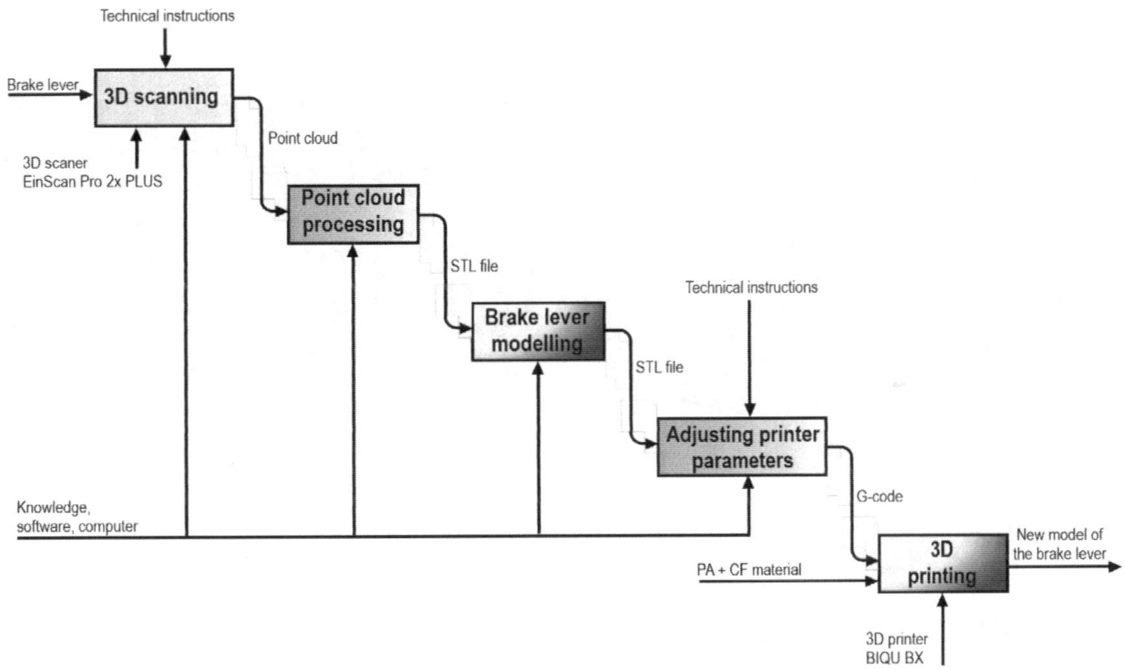

Figure 11. IDEF diagram of the mountain bike brake lever prototype model production

The process diagram consists of a 3D scan of the existing brake lever, point cloud processing, creation of a model of the new brake lever design, generation of G-code, and production of a physical model using FDM 3D printing technology.

During the experimental phase, the SHINING 3D scanner (EinScan Pro 2x PLUS) was used (see Figure 12a). The process of 3D scanning the mountain bike brake lever is illustrated in Figure 12b.

a) b)

Figure 12. IDEF diagram of the mountain bike brake lever prototype model production

IOP Conf. Series: Materials Science and Engineering 1298 (2023) 012006 doi:10.1088/1757-899X/1298/1/012006

After the completion of the 3D scanning process of the brake lever (top and bottom of the brake lever), both point clouds had to be merged into a single point cloud describing the entire geometry of the brake lever.

The point processing procedure consisted of two steps: 1) in the first step, point clouds representing markers were removed from both 3D scans, and 2) in the second step, the merging of the point clouds (top and bottom of the brake lever) was performed. The program connected the two point clouds into a unified cloud based on the defined points. This is illustrated in Figure 13.

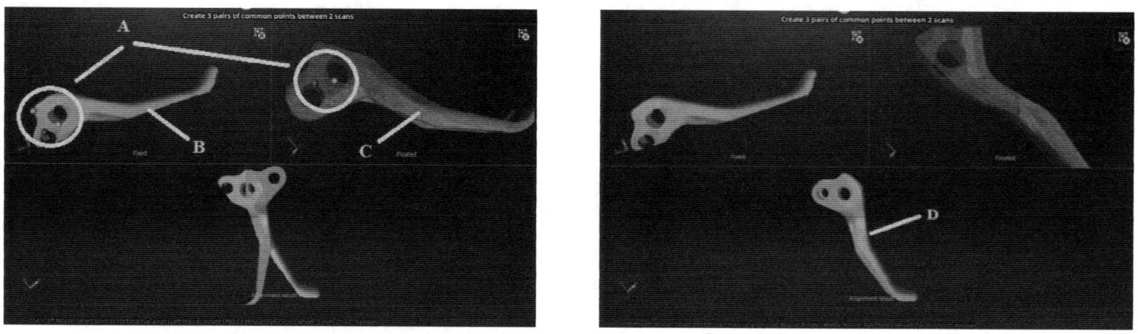

Figure 13. Establishing common points during the merging of point clouds

The generated point cloud (in STL format) is further used in the modelling process of the mountain bike brake lever in the SolidWorks software environment as a reference for creating a new 3D model of the brake lever (see Figure 14).

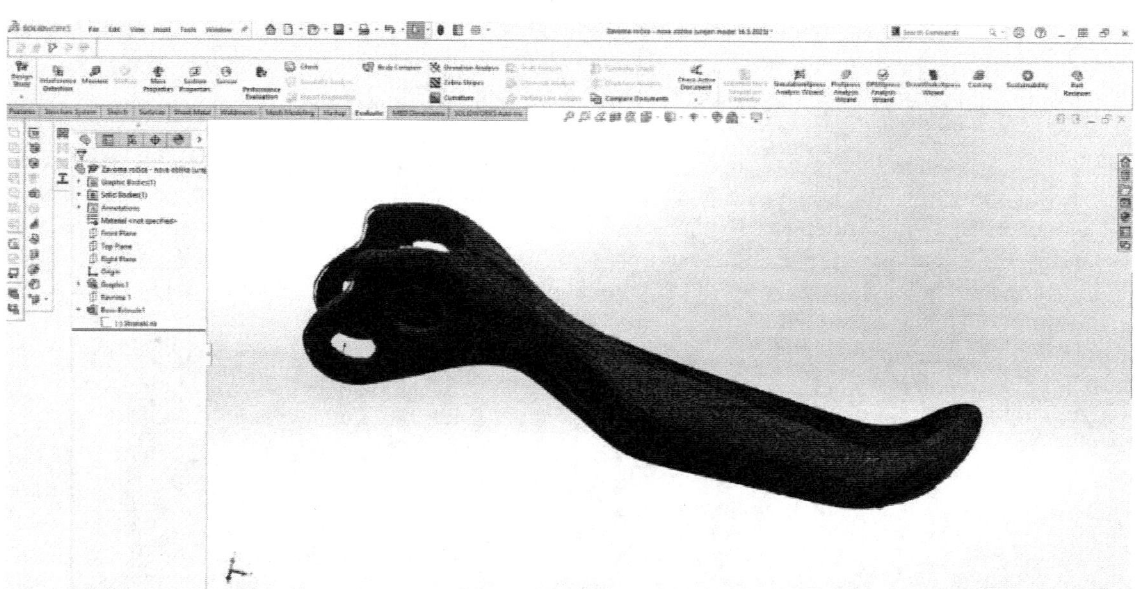

Figure 14. The STL file of the brake lever is imported into the Solid Works software as part of the modelling process

In the next modelling step, the side view and plan view of the new shape of the brake lever are defined on the basis of the point cloud from the STL file (see Figure 15).

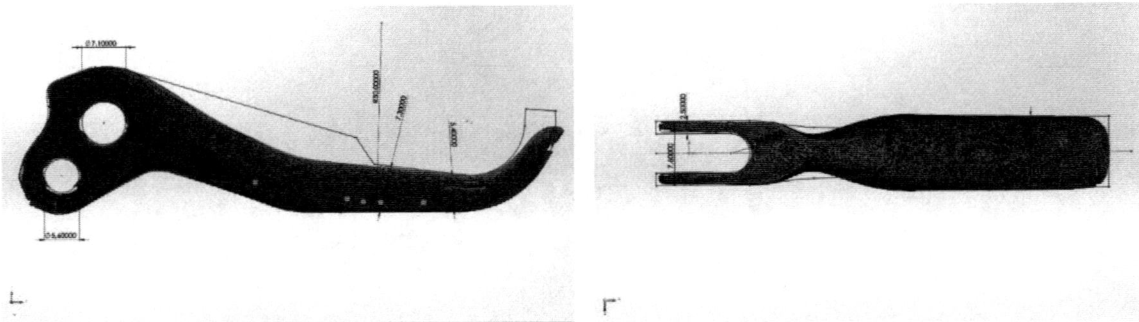

Figure 15. The side view and plan view of the new shape of the brake lever

Using the *Extruded Boss* feature, the side view and plan view were converted into solid bodies. As a result, two bodies (Body 1 and Body 2) were obtained, overlapping each other. Using the *Combine* feature, both bodies were merged into a single body (the combined volume of both bodies). After performing this operation, a rough shape of the new brake lever was created, as shown in Figure 16a.

Further processing of the rough model followed, in particular the creation of additional recesses at the locations of the holes where the brake modules and screws are attached to the lever. To do this, Plane 3 was defined and a 2D cutout was drawn based on the geometry of the existing brake lever (STL file). Then, the *Extruded Cut* feature was used to create *Recess 1*, as shown in Figure 16b.

In the next step, *Recess 2* was defined on *Plane 1*, and the *Extruded Cut* feature was used to remove the specified volume of the brake lever, as shown in Figure 16c. In the next step, the *Fillet* feature was used to define edge fillets on the model of the new mountain bike brake lever, as shown in Figure 16d.

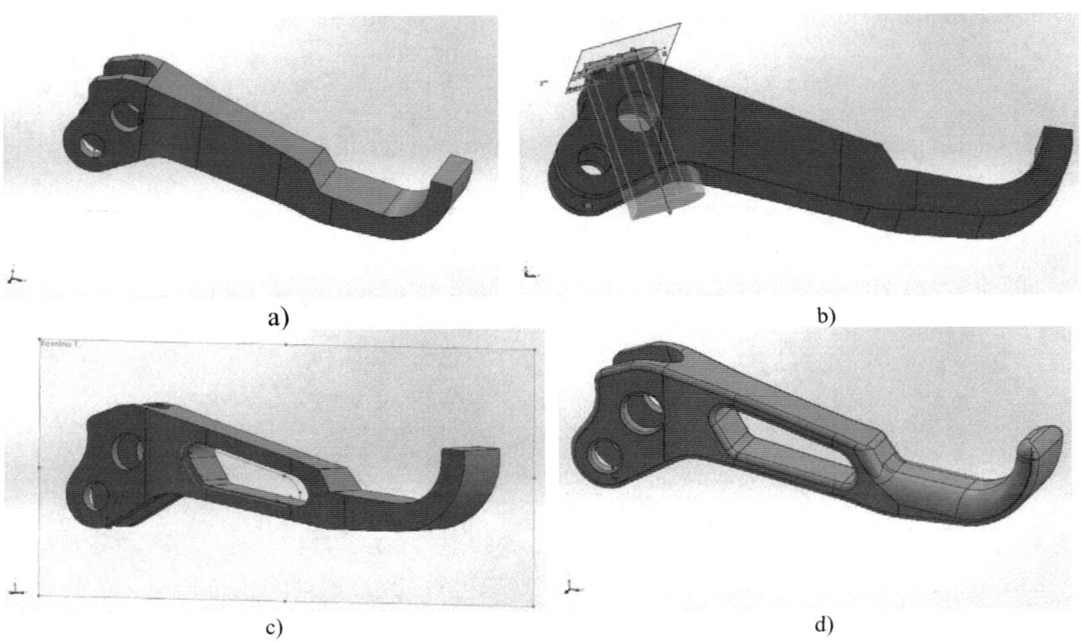

Figure 16. Modelling of the mountain bike lever

The final step in creating the new model of the mountain bike brake lever was to define additional elements to enhance the grip between the rider's fingers and the brake lever during braking.

The final model of the mountain bike brake lever is shown in Figure 17.

IOP Conf. Series: Materials Science and Engineering 1298 (2023) 012006 doi:10.1088/1757-899X/1298/1/012006

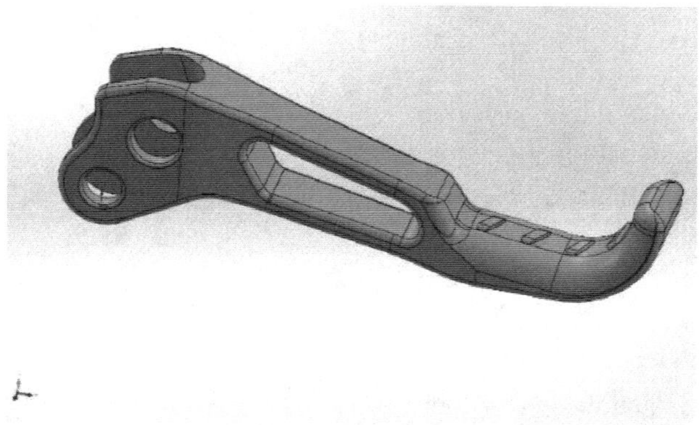

Figure 17. Defining additional elements for improved grip

The final 3D model of the new shape of the mountain bike brake lever was exported in STL format to the *Ultimaker Cura* software environment. In this software environment, the printing parameters were defined taking into account the type of material used for the production of the new mountain bike brake lever (Nylon reinforced with carbon fibers).

Nylon reinforced with carbon fibers (PA6-CF) is a composite material that combines the properties of nylon, a synthetic polymer known for its toughness and flexibility, with the strength and stiffness of carbon fibers. In this composite, the carbon fibers are embedded or woven into the nylon matrix, resulting in a material with enhanced mechanical properties.

The carbon fibers provide high tensile strength, excellent stiffness, and low weight, making the material ideal for applications where lightweight and high strength are crucial. Additionally, the nylon matrix provides impact resistance and toughness, improving the overall durability of the composite.

The combination of these two materials results in a composite with a high strength-to-weight ratio, making it suitable for use in various industries, including automotive, aerospace, sporting goods, and consumer electronics. PA6-CF is commonly used to manufacture components such as bicycle frames, automotive parts, aircraft components, and sports equipment, where the focus is on achieving a balance between strength, weight, and performance.

Material specification of PA6-CF material defined by *Flashforge* manufacturer is presented in Table 1.

Table 1. Material specifications of PA6-CF material defined by *Flashforge* manufacturer

Material properties	PA6-CF
Colour	Black
Diameter [mm]	1.75
Nozzle temperature [°C]	270-300
Bed temperature [°C]	90-120
Water absorption (equilibrium in water, 23°C)	<1
Tensile Strength [MPa]	83,5-88,5
Flexural Strength [MPa]	140-145
Cost [euro/kg]	109.00

In 3D printing, a desktop printer BIQU BX was used. The main parameters of the 3D printing process for the brake lever are presented in Table 2.

Table 2. Main 3D printing parameters for the brake lever

3D printing parameter	Value
Nozzle size	0.4
Nozzle (extruder) temperature [°C]	280
Bed (platform) temperature [°C]	95
Layer height [mm]	0.20
First layer height [mm]	0.40
Base print speed [mm/s]	35
Travel speed [mm/s]	100
Fill density [%]	100
Fill angle [°]	45
Enable raft	+

4. Results

The printed model of the new mountain bike brake lever is shown in Figure 18.

Figure 18. The printed model of the new mountain bike brake lever

The printed model of the mountain bike brake lever was mounted on the bike and tested by the rider. During the process of modelling and 3D printing the new shape of the brake lever, several iterations were made and tested by the mountain bike rider, as shown in Figure 19.

In the first iteration of the brake lever model (Figure 19a), the same section with holes was retained (this applies to all iterations) as for the existing brake lever (Figure 16a). Based on the established criteria for the new lever design (intended for one-finger braking), a new section was constructed as shown in Figure 19a. The lever has a specially designed surface where the finger comes into contact. However, on the physical model, it was found that the lever is not suitable because the radius at the finger contact point is too small. This sharp edge causes discomfort when the finger rests, which is not ergonomically appropriate.

The second iteration of the brake lever model (Figure 19b) differs from the previous model (Figure 19a) primarily in the surface with which the finger comes into contact during braking. In this version, the surface is rounded and no longer flat as in the previous version. From an ergonomic standpoint, this has proven to be more suitable after tests on the bike (the finger rests more comfortably on the surface).

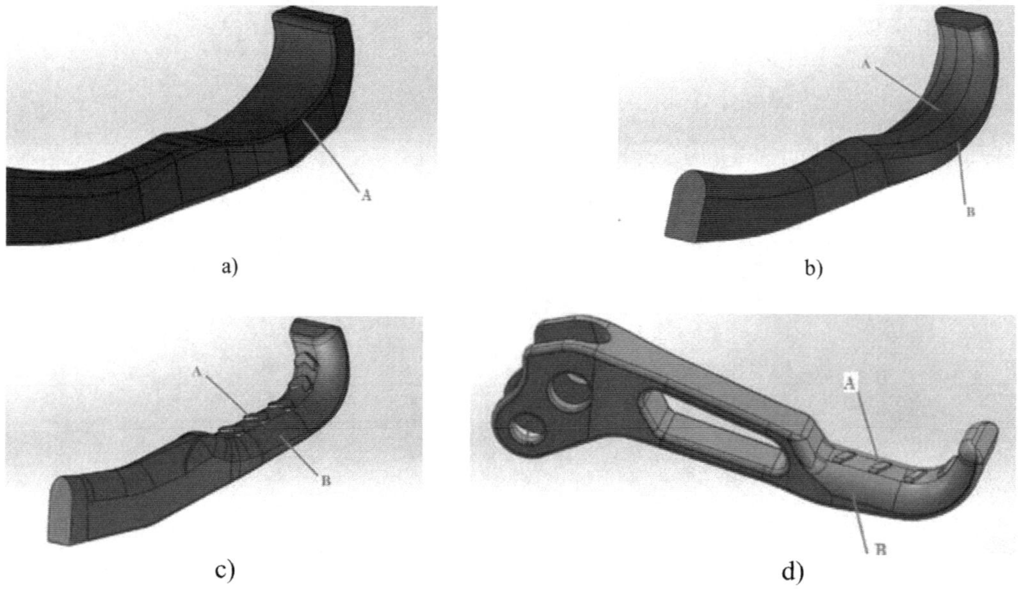

Figure 19. The iteration of new brake lever model

The third iteration of the brake lever model (Figure 19c) is completely redesigned and aesthetically refined compared to the previous versions. In addition, elements for better grip in the finger contact area have been defined. The cyclist confirms that this version is ergonomically more suitable compared to the previous versions (good feeling when braking). In terms of ergonomics, the most influential factors are the rounding of the edge, which is appropriate with a sufficient radius, and the shape of the brake lever tip (angle and height).

The final design of the brake lever (Figure 19d) is completely unique and different from all previous iterations. In this version, the positions of the hole of the existing brake lever and the shape of the contact surface for the finger (Figure 19c) were adopted. After conducting a test on the bike, the lever proved to be the most ergonomically suitable, which was also confirmed by the rider. The elements for better grip are smaller and slanted in this version. The newly installed model of the mountain bike brake lever is presented in Figure 20.

Figure 20. The installed model of the new mountain bike brake lever

IOP Conf. Series: Materials Science and Engineering 1298 (2023) 012006 doi:10.1088/1757-899X/1298/1/012006

After completing the test ride, the rider found that the new design of the brake lever was superior to the previous one in terms of ergonomics. It provides greater stability for the finger and a comfortable feel during the braking process. The rider also highlighted that new brake lever design allows for a more consistent finger position on the surface of the lever (no finger slippage) during high-speed rides on rough terrain, where precise control of the bike in crucial. This enables high-quality braking performance.

5. Conclusion

The paper describes the comprehensive process of additive manufacturing based on reverse engineering from the idea to the final product. It takes a closer look at the integration of modern 3D technologies, from data generation to conversion into digital form to the production of the physical product. The digitization of data enables the process of reverse engineering, which paves the way for modern model creation methods and ultimately simplifies the path to the desired product, which is often of an individual nature.

Modern 3D scanning, modelling and printing technologies have revolutionized the way we design, manufacture and customize products. They bring personalization, speed, precision, material diversity, flexibility, accessibility, and innovation to production processes, enabling the creation of better products tailored to individual user needs. These observations were confirmed in the paper presented. Further development of modern 3D technologies, as well as the advancement of modern materials in the future, will significantly expand the application field of 3D technologies in the production of custom-made products.

This paper presents the use of modern 3D scanning, modelling and printing technologies for the development and production of a prototype mountain bike brake lever.

Throughout the theoretical and experimental work, certain shortcomings of the process itself were noted and specific improvements were identified that could be made with the aim of enhancing the overall process of individual production of mountain bike brake levers.

Future research in the field of this article's topic will focus on specific improvements in the domain of design and ergonomics of mountain bike brake levers, as well as exploring potential opportunities for using different materials and various technologies for the personalized production of physical models of mountain bike brake levers. Additionally, it will focus on the use of modern 3D technologies for prototype testing and validation.

References

[1] Bruni L, MacAskill D 2018 Montain Bike News – Loic Bruni and Danny MacAskill's Custom MAGURA Levers, International Mountainbike Magazine. https://www.imbikemag.com/mountain-bike-news/2018/06/loic-bruni-danny-macaskills-custom-magura-levers/ Visited 06.08.2023

[2] Cunningham R 2017 Magura HC3 Adjustable Brake Lever – Review. Pinkbike. https://www.pinkbike.com/news/magura-hc3-adjustable-brake-lever-review.html, Visited 06.08.2023

[3] MAGURA 2022 Bike & Motorcycle Brakes. https://www.magura.com/en/components/news/2021/prototype-levers-for-danny-macaskill-s-the-slabs/ Visited 14.11.2022

[4] Beer M 2022 Review: Oak Components Root Lever - An Aftermarket Upgrade for Magura Brakes. https://www.pinkbike.com/news/review-oak-components-root-lever-magura-brakes.html Visited 16.11.2022

[5] Ling D 2015 Research on risk perception of unsafe behavior of non motor vehicle riders. Beijing Jiaotong University, 2-5.

[6] Yuanjie G, Xiao J, Yan Y 2020 Design of bicycle turn signal based on safe travel. *Design*, **33**(03), 120-121.

[7] Xiyuan W, Qingyang Z, Difei Z 2018 One handed barrier free product design method based on

humanized design. *Packaging engineering*, **39**(22), 235-240.
[8] Yulan D 1991 Ergonomics. Beijing University of Technology Press, 111-113.
[9] Zeng X, Zheheng E 2021 Research on the Design of Brake Auxiliary Device based on the improvement of riding safety. *Journal of Physics: Conference Series*, **1948**(1). IOP Publishing.
[10] Covill D, Elton E, Mile M, Morris R 2014 Analysis of human factors issues relating to the design and function of bicycles. In Contemporary Ergonomics and Human Factors, Proceedings of the international conference on Ergonomics & Human Factors 2014, Southampton, UK, 7-10. CRC Press.
[11] Siebert, F W, Ringhand M, Englert F, Hoffknecht M, Edwards T, Rötting M 2021 Braking bad–Ergonomic design and implications for the safe use of shared E-scooters. *Safety science*, **140**, 105294.
[12] Parra M 2018 What is Mountain Biking? Equipment, Types, Top Spots - Explore-Share.com https://www.explore-share.com/blog/mountain-biking-equipment-best-season-top-spots/ Visited 30.10.2022
[13] Simonovich R 2021 Types of mountain bikes: all the different mountain bike categories explained https://www.bikeperfect.com/features/types-of-mountain-bikeA Visited 30.10.2022
[14] Bikelockwiki 2022 Types of Bike Brakes Explained | No.1 Bicycle Brake Guide. https://www.bikelockwiki.com/types-of-bike-brakes/ Visited 01.11.2022
[15] Sram 2012 Hydraulic Disc Brakes Overview. https://cdn.sram.com/sites/default/files/techdocs/gen.0000000004234_rev_a_hydraulic_disc_brakes_overview.pdf Visited 04.11.2022
[16] Morgan J M 2020 AASQ #92: Braking Force vs Rotor Size, Lever Leverage, and fixing Sticky Pistons. https://bikerumor.com/aasq-92-braking-force-vs-rotor-size-lever-leverage-more-hydraulic-brake-answers/ Visited 04.11.2022
[17] Sram SwingLink 2018 The Secret to Better Brake Modulation - News & Press « Live to Play Sports https://livetoplaysports.com/news-press/2018/07/secret-better-brake-modulation-sram-swinglink/ Visited 20.11.2022
[18] Richardson T 2014 First Impression: Shredding SRAM's New Guide RSC Disc Brakes – Bikerumor. https://bikerumor.com/first-impression-shredding-srams-new-guide-rsc-disc-brakes/ Visited 22.01.2023
[19] Pham D, Dimov S 2001 *Rapid Manufacturing: The Technologies and Applications of Rapid Prototyping and Rapid Tooling*, Springer Verlag, London.
[20] Kruth J-P, Leu M C, Nakagawa T 1998 Progress in Aditive Manufacturing and Rapid Prototyping. *Annals of the CIRP*, **4** (7/2).
[21] Gibson I, Rosen D, Stucker B 2014 *Additive Manufacturing Technologies: 3D Printing, Rapid Prototyping, and Direct Digital Manufacturing*. Springer.
[22] Campbell R I 2017 *Additive Manufacturing: Materials, Processes, Quantifications and Applications*. Butterworth-Heinemann.
[23] Kruth J P, Leu M C, Nakagawa T 2010 *Additive Manufacturing of Metals: The Technology, Materials, Design and Production*. Springer.
[24] Campbell I 2019 *3D Printing and Additive Manufacturing: Principles and Applicatio*ns (5th ed.). CRC Press.
[25] Singamneni S, Singamneni S B 2021 A systematic review of 3D printed drugs. *Current Drug Delivery*, **18**(2), 198-209.
[26] Godec D, Šercer M 2015 *Aditivna proizvodnja*, Faculty of Mechanical Engineering and Naval Architecture, University of Zagreb.
[27] Green C 2022 Post-Processing a 3D Scan with Artec Studio and Geomagic Control X – TriMech. https://trimech.com/blog/post-processing-a-3d-scan-with-artec-studio-and-geomagic-control-x Visited 02.12.2022
[28] Hsueh M H, Lai C J, Wang S H, Zeng Y S, Hsejh C H, Pan C Y, Huang W C 2021 Effeect of

printing parameters on the thermal and mechanical properties of 3D-Printed PLA and PETG, using fused deposition modeling. *Polymers* **13** 1758

[29] Valaean C, Marsavina L, Marghitas M, Linul E, Razavi J, Berto F 2020 Effect of manufacturing parameters on tensile properties of FDM printed specimens. *Procedia Structural Integrity* **26** 313-320

[30] Hodzic D, Pandzic A, Hajro I, Tasic P 2020 Strength comparison of FDM 3D printed PLA made by different manufacturers. *TEM Journal* **9** 966-970

[31] Gibson I, Rosen D, Stucker B 2015 Introduction and basic principles. *Additive manufacturing technologies*. Springer, New York, USA.

[32] Zhakeyev A. Wang P, Zhang L, Shu W, Wang H, Xuan J 2017 Additive manufacturing: unlocking the evolution of energy materials. *Advanced Science* **4**(10), 1-44

IOP Conf. Series: Materials Science and Engineering 1298 (2023) 012007 doi:10.1088/1757-899X/1298/1/012007

Calculation of convection coefficient during steel hardening of axisymmetric probes in thermal oils

Šefik Behrem[1] and Bahrudin Hrnjica[1]
[1]University of Bihac, Tehnical faculty Bihac Dr. Irfana Ljubijankića bb, 77000 Bihac, Bosnia and Herzegovina

E-mail: sefik.behrem@gmail.com.

Abstract. The paper presents the results of the convection coefficient during the process of two-dimensional axisymmetric steel hardening of in two different thermal oils used in the area of thermal treatment of steel. One thermal oil was Isorapid 277 HM preheated to the set temperature, and the other one was Marquench 722 also preheated to the set temperature. The experimental work setup consists of six quenches, three dimensionally different axisymmetric probes in two different thermal oils. Steel hardening was done under strictly controlled conditions: heating the sample to the tempering temperature of most steels in a nitrogen atmosphere, vertical slow immersion of the test sample in the opposite flow of thermal oil. The temperature of the thermal oil was kept constant, stopping the test sample in the sample steel hardening area and non-stationary data collection until thermal oil equilibrium temperature with interruption of data collection. The data collection rate was two probes/second in parallel from four points located within the test cylinder. Three measuring points were located directly below the surface of the cylinder distributed along the height of the cylinder, while the fourth one was in the centre of gravity of the cylinder. Time-measured temperature values in the indicated four points of each probe were written in a text file. The set mathematical model describes the non-linear two-dimensional axisymmetric non-stationary heat transfer through the cylinder. The set mathematical model requires the inverse solution of the heat conduction problem, that is, ill-conditioned problems, so the solution of the problem came down to a sufficiently accurate estimation of the unknown convection coefficient imposed on the outer surface of the probe. The chosen solution algorithm was the one from previously published works. The solution algorithm was based on a iterative combination of the finite element method (FEM), while the second part of the solution algorithm is the Levenberg Marquardt method (LMM) from the line of optimization deterministic algorithms, which takes the calculation results of the temperatures at the points of the computational domain from the FEM, and for comparison, that is, error minimization, using the experimental results of temperature measurements at the same place.

1. Theoretical description

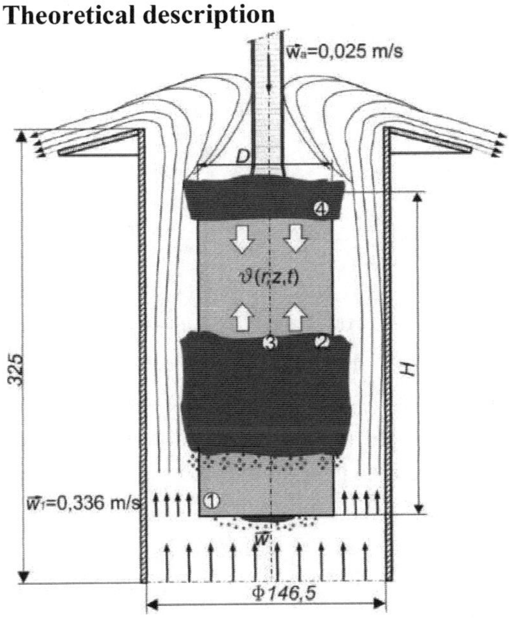

Figure 1. Schematic view of the described problem.

The problem consists of a cylinder of diameter $D \in \{25, 50, 75\}$ mm, height $H \in \{100. 150, 225\}$ mm, initial temperature ≈ 850 ^{0} C which descends vertically, at a speed of 0.025 m/s, and thermal oil, which flows in the opposite direction, at a speed of 0.336 m/s. During the lowering and stopping of the cylinder in the quenching area, four temperatures inside the cylinder were continuously measured in time until they reached their temperature equilibrium with the temperature of the thermal oil, using four built-in thermocouples (1,2,3,4). Figure 1 schematically shows the described problem along the direction of movement of two steam fronts that were formed on the surfaces of the base of the test cylinder. Based on the obtained values of the temperature in the probe, the convection coefficient from the surface of the immersed probe to the thermal oil was calculated.

2. Experimental description

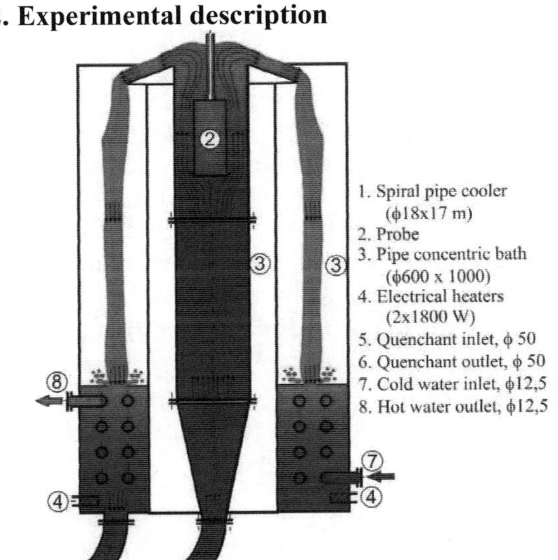

Figure 2. Schematic view of the described problem.

The experimental setup of the problem consists of steel hardening three axisymmetric probes of different dimensions: ϕ25x50, ϕ50x150 i ϕ75x225 mm. The steel quenchant were two thermal oils. One thermal oil was Isorapid 277 HM, physical properties: density at 20°C was 866 kg/m^3, ignition temperature of heated oil was 235°C. The second thermal oil was Marquench 722, physical properties: density at 20 °C was 880 kg/m^3, ignition temperature of heated oil was 276°C. Viscosity and specific heat capacity of thermal oils as a function of temperature were given in detail in [1]. During the experiments, Isorapid 277 HM was heated and maintained at a temperature of 40 ^{0}C, while Marquench 722 was heated and maintained at a temperature of 90 ^{0}C. During the experiments, Isorapid 277 HM was heated and maintained at a temperature of 40 °C, while Marquench 722 was heated and maintained at a temperature of 90 °C. Figure 2 schematically shows the described problem of the measuring line is given.

3. Mathematical formulation of the problem

3.1. Thermodynamic equations of the given problem

According to the above assumptions, discussed in detail in papers [1], [3] and [6], the general form of the differential equation of heat conduction for this problem is

$$\frac{\partial \vartheta}{\partial t}=\frac{\lambda\ (\vartheta)}{\rho(\vartheta)\cdot c(\vartheta)}\left[\frac{1}{r}\cdot\frac{\partial}{\partial r}(r\cdot\frac{\partial \vartheta}{\partial r})+\frac{\partial}{\partial z}(\frac{\partial \vartheta}{\partial z})\right]. \tag{1}$$

In this equation: $\lambda(\vartheta), \rho(\vartheta)$ and $c(\vartheta)$ are the physical properties of the probe as a function of temperature, which make this differential equation nonlinear. To solve the set equation, it is necessary to know the initial condition at each point of the computational domain and for each test probe is $\vartheta(r,z,t=0) \approx 850\ ^0\mathrm{C}$. According to the assumptions from [1] the computational domain is axisymmetric, so the calculation is reduced to a sufficiently simplified rectangular computational domain in Figure 3 taken from [1] with limitation conditions according to equations (2), (3), (4) and (5)

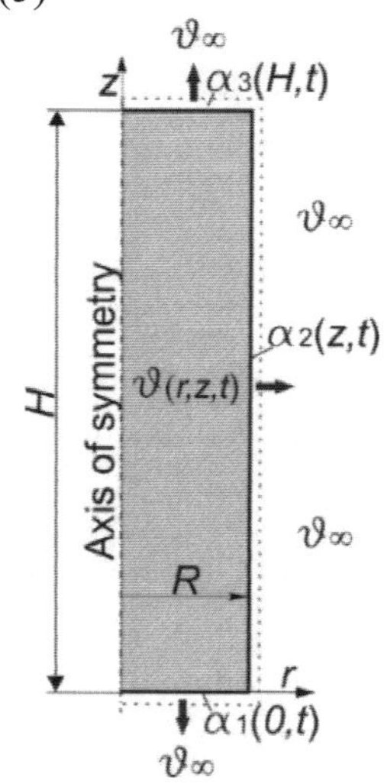

Figure 3. Computing domain.

$$\left(\frac{\partial\vartheta(r,z,t)}{\partial r}\right)_{r=0}=0, \tag{2}$$

$$-\lambda(\vartheta)(\frac{\partial\vartheta}{\partial z}) = \alpha_1(0,t)\cdot(\vartheta(r,0,t)-\vartheta_\infty), \qquad z=0,\ \mathrm{i}\ 0 \le r < R, \tag{3}$$

$$-\lambda(\vartheta)(\frac{\partial\vartheta}{\partial r}) = \alpha_2(z,t)\cdot(\vartheta(R,z,t)-\vartheta_\infty), \qquad r=R,\ \mathrm{i}\ 0 < z < H, \tag{4}$$

$$-\lambda(\vartheta)(\frac{\partial\vartheta}{\partial z}) = \alpha_3(H,t)\cdot(\vartheta(r,H,t)-\vartheta_\infty), \qquad z=H,\ \mathrm{i}\ 0 \le r < R. \tag{5}$$

In the above equations, the unknown convection coefficients ($\alpha_1(0,t)$, ($\alpha_2(z,t)$ and $\alpha_3(H,t)$ on the corresponding axisymmetric surfaces, shown in Figure 1, are described by a single function of the form, $\alpha(z,t)$ ($0 \le z \le H$), which defines the boundary condition of the third kind. Equation (2) defines a boundary condition of second kind or an adiabatic boundary condition. The inverse method of solving is an approximate estimation of the unknown convection coefficient. Regression forms of the functions of physical properties of the probe from temperature: $\lambda(\vartheta), \rho(\vartheta)$ and $c(\vartheta)$, are from [1].

3.2. Description of the solving algorithm

The second author of the paper made his own computer code for solving the set algorithm in C# and his pseudocode is visible [8]. The algorithm is a modification of the algorithm used in works [1], [3] and [6]. The modified algorithm is based on an iterative combination of the optimization Levenberg Marquardt method (LMM) and the numerical finite element method (FEM). The basic requirement for the LMM method is the minimization of the error between the calculated and measured temperature at the same point. The instantaneous convection coefficient $\alpha(t,z)$ is determined by marked three points on the outer surface of the probe horizontally near the measuring point. Calculation parameters are probe height $z = \{4.5, H/2, H\text{-}4.5\}$. At the mentioned heights, the current convection coefficients are then only time functions: $f_1(t), f_2(t)$ and $f_3(t)$. In the first step, through the selected three points, a three-fold 1D horizontal axisymmetric numerical grid is formed with boundary conditions: $f_1(t), f_2(t)$ and $f_3(t)$, which should be evaluated iteratively using a combination of FEM and LMM methods. The LMM uses convection coefficient optimization only as a function of time t (single-objective optimization). In the second step, based on the estimated 1D boundary condition (convection coefficient) at three points, a boundary condition for the 2D axisymmetric domain is formed. Function of two variables, $f(z,t)$ formed by bilinear interpolation (extrapolation) of functions $f_1(t), f_2(t)$ and $f_3(t)$, on the outer surface of the cylinder. The function formed in this mode is now the boundary condition of the third kind for the 2D axisymmetric domain, which should be estimated again accurately enough by the iterative combination of FEM and LMM with the condition of minimizing the error of measured and calculated results in the same point of the domain. During this iterative evaluation, LMM uses convection coefficient optimization as a function of two parameters: probe height z and time t (multi-objective optimization). If the comparison is not satisfactory, the calculation returns to the level of one-dimensional function estimation. A more detailed description of the algorithm is given in [8]. An example of the quantitative results of the iterative evaluation is the text file in Figure 4. The first row is manually carefully selected time moments from the experimental cooling curves, in seconds, while the second row represents three probe heights z in meters. The last three rows represent the time functions $f_1(t), f_2(t)$ and $f_3(t)$ in W/(m2K).

```
% Grid time htc                    x
0 12 14 16 18 20 22 23 24 25 27.5 30 40 50 70 100 150 200 300 1000
0.0045 .1125 0.2205
% data
70 90 400 1750 1300 1320 1200 1175 1150 1000 1050 1100 900 925 825 775 740 625 625 625
10 10 90 110 250 2450 2600 2425 2250 1950 1700 1250 1250 1000 900 800 750 500 375 300
30 30 30 50 90 100 120 2050 2200 2050 1700 900 825 750 485 485 425 300 350 300
```

Figure 4. Convection coefficient search results when switching off probe ϕ75x225 in Marquench 722 temperature 90 ^{0}C.

4. Comparison and discussion of paper results

Figures 5 and 6 show comparisons of the experimental results of temperatures and cooling rates in five different steel steel hardening agents. The results are shown for point 2, visible in Figure 5, of the 25x100 cylinder. In previously published works, [1], [3] and [6], the results of steel hardening in water and water solutions were given.

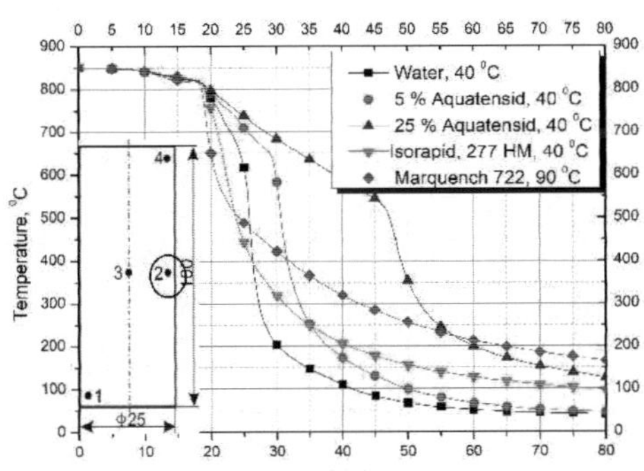

Figure 5. Comparing temperatures at marked point of probe

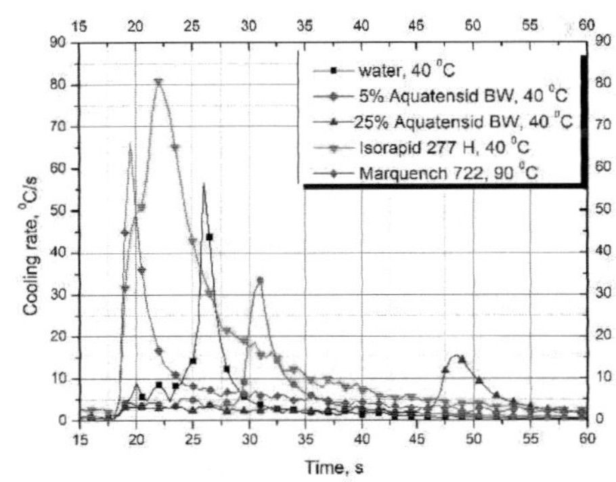

Figure 6. Comparing cooling rates at marked point of probe

The analysis of the cooling curves from Figure 6 in point 2 is given in the conclusion of the paper.

4.1. Convection coefficient for the probe ϕ25x100

Figures 7 and 8 show the results of the convection coefficient for three values of the z axis when steel hardening the probe in thermal oils.

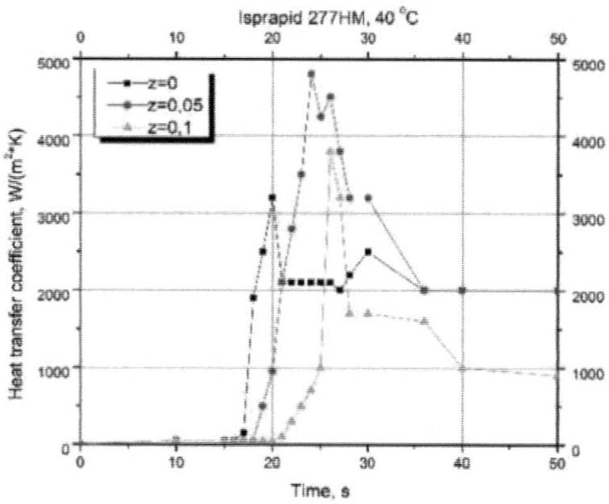

Figure 7. Convection coefficient in point 2 of probe in Isorapid 277 HM, 40 °C.

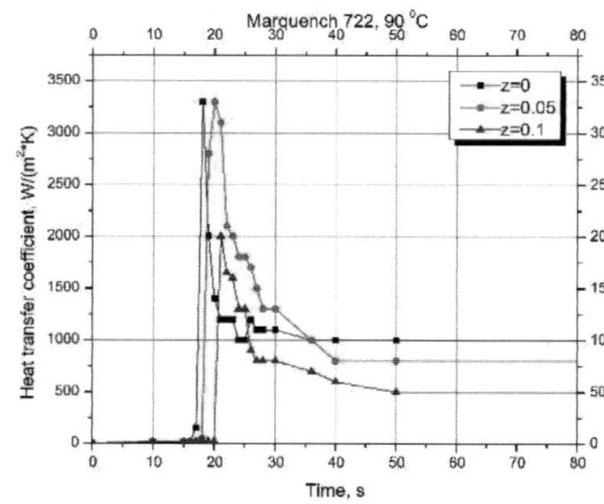

Figure 8. Convection coefficient in point 2 of probe in Marquench 722, 90 °C.

Figures 9 and 10 show the satisfactory accuracy of the convection coefficient $\alpha(t, z)$ by comparing the measured and calculated values of temperatures in time at the four marked points of the probe.

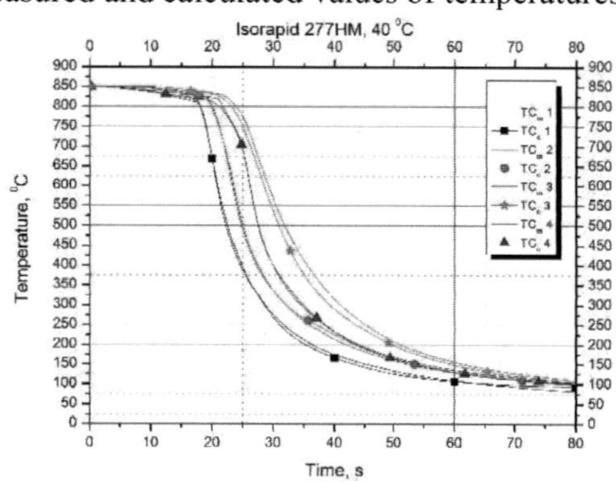

Figure 9. Comparison results in indicated points of.

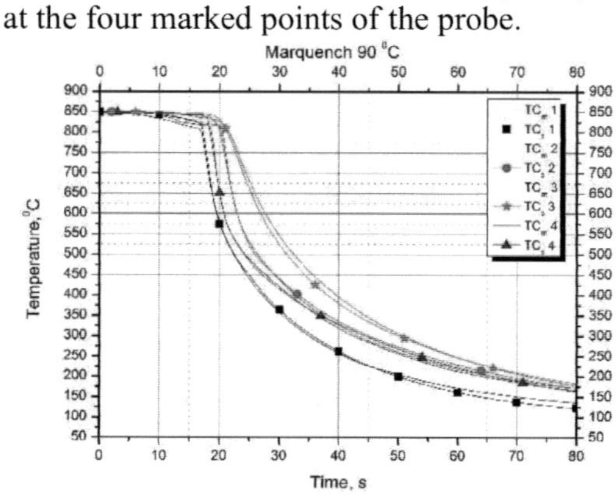

Figure 10. Comparison results in marked points.

In Figures 9 and 10, TC is the symbol of the thermocouple, and the indices m and c represent the measured and calculated temperature in the marked points of individual probes.

IOP Conf. Series: Materials Science and Engineering 1298 (2023) 012007 doi:10.1088/1757-899X/1298/1/012007

4.2. Convection coefficient for the probe ϕ50x150

Figures 11 and 12 show the results of the convection coefficient for the probe in thermal oils.

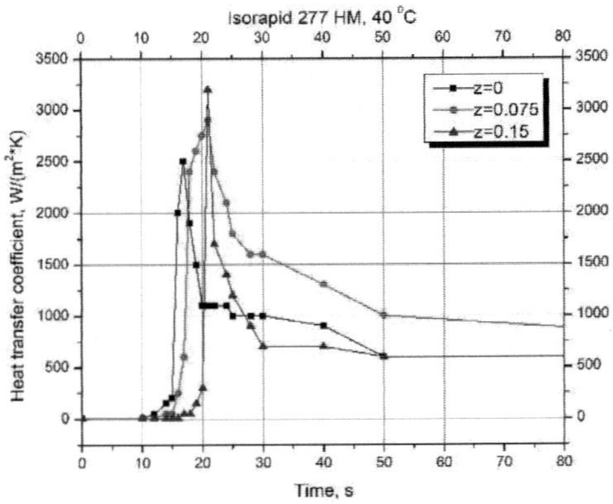

Figure 11. Convection coefficient in point 6 of probe in Isorapid 277 HM, 40 ^{0}C.

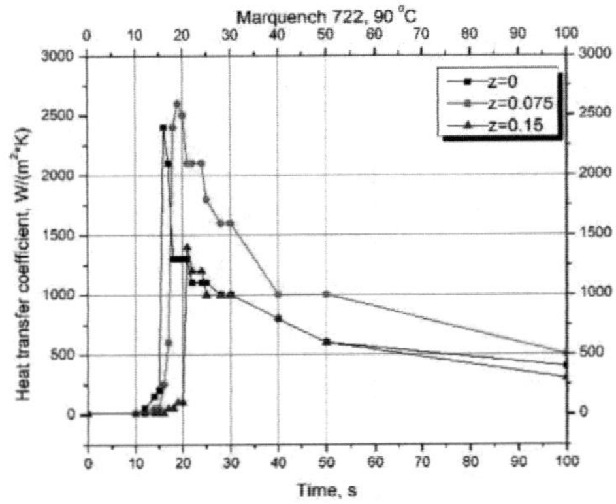

Figure 12. Convection coefficient in point 6 of probe in Marquench 722, 90 ^{0}C.

Figures 13 and 14 show the satisfactory accuracy of the convection coefficient $\alpha(t, z)$ by comparing the measured and calculated values of temperatures in time at the four marked points of the probe

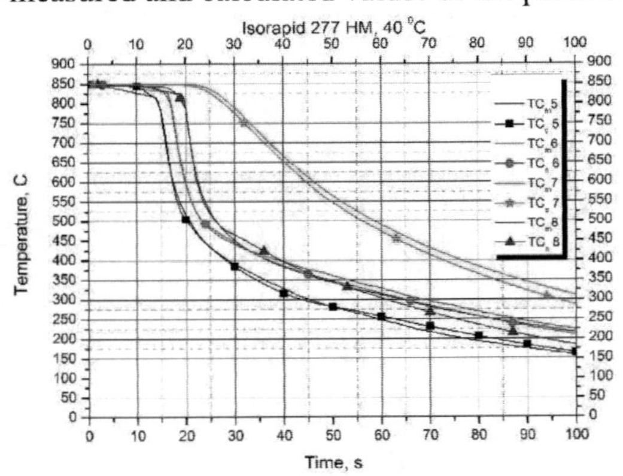

Figure 13. Comparison results in indicated points of.

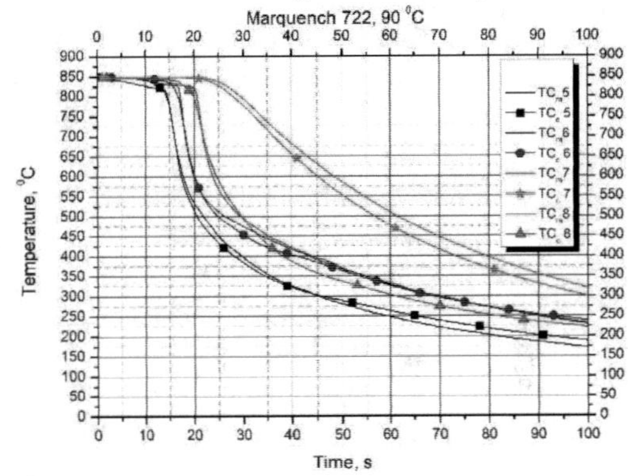

Figure 14. Comparison results in marked points.

4.3. Convection coefficient for the probe $\phi 75x225$

Figures 15 and 16 show the results of the convection coefficient for the probe in thermal oils.

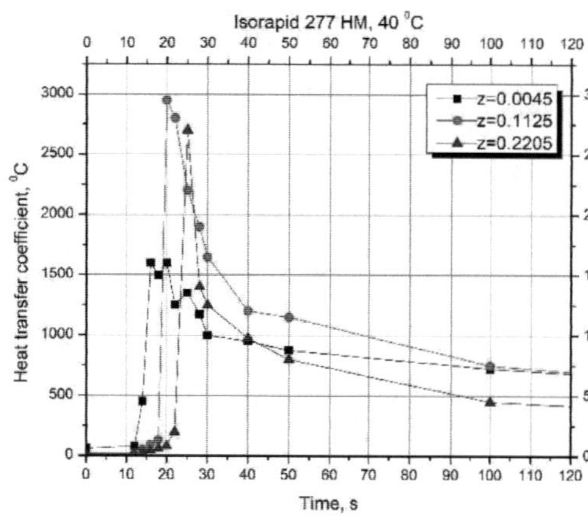

Figure 15. Convection coefficient in point 6 of probe in Isorapid 277 HM, 40 ^{0}C.

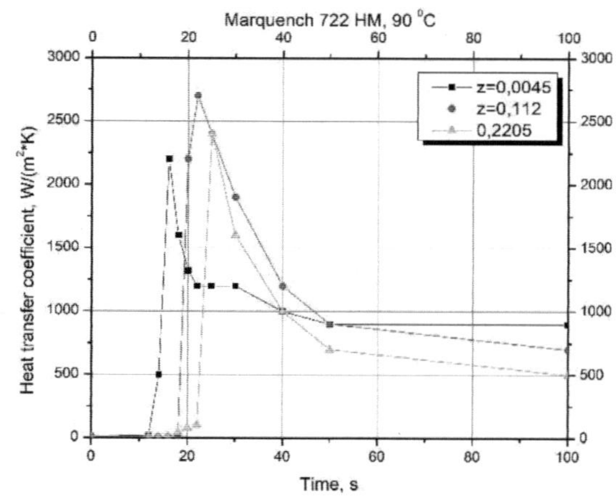

Figure 16. Convection coefficient in point 6 of probe in Marquench 722, 90 ^{0}C.

Figures 17 and 18 show the satisfactory accuracy of the convection coefficient $\alpha(t, z)$ by comparing the measured and calculated values of temperatures in time at the four marked points of the probe.

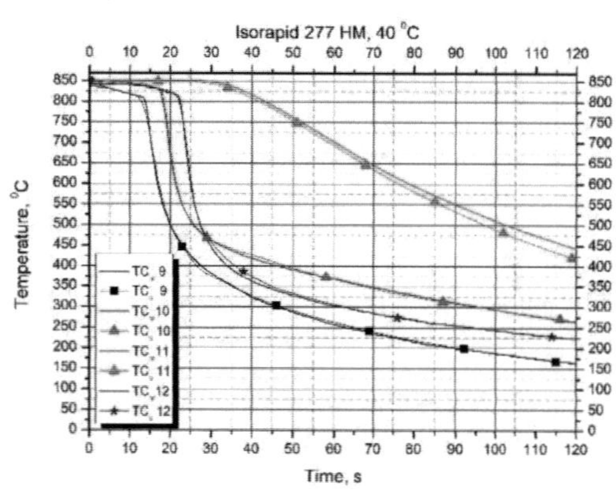

Figure 17. Comparison results in indicated points of.

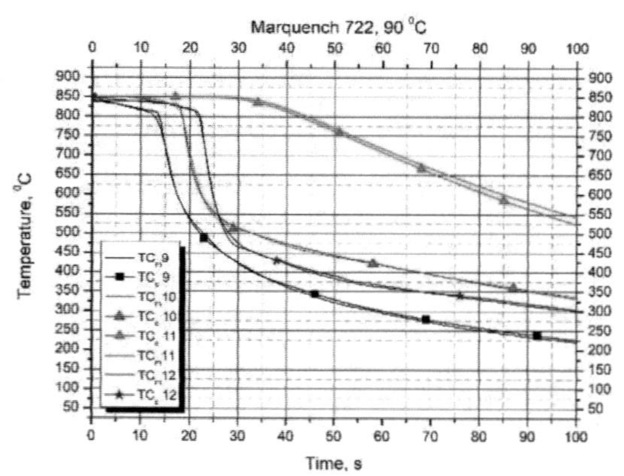

Figure 18. Comparison results in marked points.

5. Conclusion

Strictly controlled conditions of conducting each experiment, sampling rate and algorithm for solving inversely posed heat conduction problems are a prerequisite for obtaining an acceptable estimate of the convection coefficient. The validity of the obtained results of the convection coefficient is evaluated by the deviation between the experimental and calculated results of the temperature values in the marked points of the calculation domain. More significant deviations were shown in the central measurement point, for each simulation. In the figures showing the comparison of temperature results in the marked points, a maximum deviation of 3.5% can be seen in the center of gravity of each sample, while the deviations in the points relevant for the calculation of the convection coefficient are deviations below 3%. Larger deviations in the center of gravity are expected, because it is the farthest from the very rapid temperature changes that come from the outside of the sample surface. The reason for this is the assumptions introduced in the calculation model. The roughest assumption is the axisymmetric movement of the vapor film on the surface of the probe. In Figures 5 and 6 at point 2, a comparison of local maximum cooling rates shown that thermal oils cooled the test probe faster than water and aqueous solutions. Isorapid 277 HM cooled 1.4 times faster than water, Marquench cooled 1.2 times faster than water under exactly the same steel hardening conditions, which was contrary to previously known results. The reason for this was that pure water has very pronounced dynamics and instability of the vapour film at the moment of its breaking, [1] and [2], while thermal oils do not have the aforementioned instability of the vapour film, [7]. If they wanted to show the maximum temperature gradient or cooling rate in water, then they would have to increase the number of time points of the calculation around the location of the largest temperature gradients.. The future of this work is visible through the change in the conditions of conducting the experiment from the slow immersions covered in this work, to the fast ones that are more common in practice.

References

[1] Behrem Š 2015 *Impact kinematics wetting and diameter axisymmetric probe on Heat transfer coefficient during steel hardening of steel.* Phd Thesis, University of Bihac, Faculty of Technical Engineringg

[2] Behrem Š and Hrnjica B 2017 *Kinematics of steam film wetting while steel hardening axisymmetric probes in water.* 11th International Scientific Conference on Production Engineering 1 vol 1 (Bihać, Bosnia and Herzegovina: University of Bihać) pp 30–5

[3] Behrem Š and Hrnjica B 2018 *Estimate of convection coefficient during steel hardening steel in water.* Transactions of FAMENA XLII, pp 61–73, doi:10.21278/TOF.42Si106

[4] Bahrudin Hrnjica, Ali Danandeh Mer, Šefik Behrem and Necati Ağıralioğlu 2018 *Genetic programming for turbidity prediction: daily and monthly scenarios,* International Symposium on Urban Water and Wastewater Management, October 25-27, Denizli

[5] Behrem Š and Hrnjica B 2019 *Kinematics of steam film wetting while steel hardening axisymmetric probes in aqueous solutions.* 12th International Scientific Conference on Production Engineering 1 vol 1 (Bihać, Bosnia and Herzegovina: University of Bihać) pp 50–5

[6] Behrem Š and Hrnjica B 2021 *Results of convection coefficient estimation during the steel hardening of axisymmetric probes in aqueous solutions.* Transactions of FAMENA XLV, pp 49–58, doi:10.1520/JAI101981

[7] Behrem Š and Hrnjica B 2021 *Kinematics of steam film wetting while steel hardening axisymmetric probes in thermal oils.* 2021 *IOP Conf. Ser.: Mater. Sci. Eng.* 1208 011001

[8] Hrnjica B and Behrem Š 2022 A new multi-objective optimization approach for process parameters optimization during numerical simulation of steel hardening steel parts, Advances in Production Engineering & Management, Volume 17 | Number 1 | March 2022 | pp 16–32, ISSN 1854-6250

Instrumented microindentation – a versatile technique in the field of wood science and technology

Marko Petrič[1]

[1]University of Ljubljana, Biotechnical Faculty, Ljubljana, Slovenia

E-mail: marko.petric@bf.uni-lj.si

Abstract. Hardness is an important mechanical property of wood or other wood-based materials and it is also very important in the case of wood coatings. Different indentation methods for measurements of hardness and other mechanical properties have been utilised already for a long time. In the last 50 years the instrumented indentation was developed. Here, an indenter tests the surface of a specimen with a controlled load and movement, in order to assess its mechanical properties. Three ranges of instrumented indentation tests are distinguished. In the macro range, the force is higher than 2 N, in the micro range it is below 2 N and the depth of indentation must exceed 0.2 µm. When the indentation depth is lower than 0.2 µm, the indentation technique is called nanoindentation. There can be found thousands of scientific papers about applications of instrumented nanoindentation in research of a large variety of materials. Instrumented microindentation remained somehow in the shadow of nanoindentation, but yet, it is a highly versatile technique with potentials to study various (micro)mechanical properties. In this review, examples of possible applications of instrumented microindentation are presented, with a focus on investigations of wood, wood-based composites, surface coated wood and polymeric films. In the second part of this paper, as an illustration, investigations of a self-healing wood paint are shown. The cyclic method and a single-load indentation confirmed the different mechanical properties and good scratch resistance of the self-healing paint.

1. Introduction

Various definitions of hardness can be found in literature, but generally, it is agreed, that this characteristic is a measure of the material's resistance to a permanent or plastic deformation when another, harder material is pressed into a substance under investigation [1]. The hardness is closely related to other mechanical characteristics, for instance fatigue resistance, ductility, or strength [1]. Two different hardness determination methods can be distinguished: the non-cutting methods of deformation, and the scratching - cutting ones [2]. When we measure the so-called Brinell hardness, a small ball made of steel or WC is utilized to cause a plastic deformation of a material [3]. The other hardness tests that are nowadays also used, are for instance Vickers, Berkovich, Knoop, or Rockwell tests. These methods are differentiated by the material, size and the shape of the indenter's tip [4]. When hardness measurement is performed by one the mentioned methods, the result is a single hardness value, based on the depth of penetration of the probe into the specimen [5]. For hard metals and other hard and fragile substances it is more appropriate to apply higher loads by sharp (e.g. pyramidal) tips, and for soft materials, spherical indenters and lower forces should be selected [5]. The hardness values in the case of conventional methods (for instance Brinell, Vickers, or Rockwell) are calculated after removal of the tip. This means, that at the classical test methods elastic deformation is not considered [6]. An important finding is also the load-dependency of hardness [3]. The indentation size effect means that the measured / calculated hardness is influenced so by the indentation depth, as well as by the applied force the tip is pressed into the material with. The indentation size effect therefore opens the question whether hardness can be at all regarded as a basic material's mechanical property s [2]. In the case of complex materials (porous structures, bio-materials, small volumes, thin layers, etc.), interpretation of the results can be made only when the physics of the object under investigation is well known and with an expert knowledge of the indentation mechanics [7]. Also, when an indentation test is performed on a small scale, the limiting factor might be the requirement of accurate measurements of the defects caused by the indenter's tip [7]. All mentioned challenges led to development of various numerical models that were applied in hardness tests [7].

2. Instrumented indentation and some selected examples of usage of nanoindentation in wood science & technology

In order to overcome some of problems, related to classical indentation test methods, in the last decades, the depth-sensing indentation method, what is the other name for instrumented indentation, was developed [5]. At the depth-sensing method, the applied force and movement of the indenter are controlled and measured at the same time in the complete process of hardness or scratch measurement. During the experiment, so the indenter's movement, as well as the applied force are accurately controlled. Even the forces of few micro Newtons and the depth in the nano-meter range can be accurately controlled [4]. The fact that a single instrument can be applied in characterisation of very different material systems is based on high force and displacement resolutions (even in the case of very small loads and shifts, like ≈1 μN and ≈ 0.2 nm), and on the other hand, broad ranges of loads and movements are possible (from app. 10 μN to several hundreds of mN, and from μm range to 10 mm or more, respectively) [5]. To obtain information on elastic and plastic properties of a ductile substance, it is better to use sharp tips. On the other hand, when a brittle and more complex materials are studied, application of the spherical tip is recommended. We can obtain traditional hardness values from the depth-sensing indentation test, when the complete increasing-decreasing cycle of the test load is considered [6]. Instrumented indentation method allows also determination of some other characteristics of the material, such as, for instance, indentation modulus. An important advantage of the depth-sensing methods is that optical measurements of the indents are not a prerequisite [6]. Oliver and Pharr [8] developed and applied the instrumented indentation technique to assess elastic modulus and hardness in mechanical characterisation of substances at small scales. This is why their method became established in investigations of mechanical characteristics of films of the order 1 μm or even of few nanometers, and of small structures [8].

A typical graph of a depth-sensing indentation experiment, performed with a sharp indenter, is shown in Figure 1. Displacement, relative to the initial surface, is on the x-axis of the diagram labelled as "h", and the parameter P means the force (load) [7]. The represented curve shows the response that is characteristic for a homogeneous elastic–plastic substance. h_m and P_m mean maximum depth and load, respectively, and p_{ave} = is the average contact pressure, where A_m means the true projected contact area measured at h_m. P_m/A_m can be considered as the hardness (H). The P–h slope is during unloading initially linear. The elastic unloading stiffness, S = dP/dh, is defined as the slope of the upper portion of the unloading curve during the initial stages of unloading. The unloading force is labelled with P_u. Af complete unloading (zero load), The residual depth after complete unloading is h_r. The areas below the loading curve and unloading curves are W_t - the total work of loading, and W_e, the work, released by elastic unloading. Therefore, we can consider the area bounded by the unloading and loading curves as $W_p = W_t - W_e$, which is the plastic work performed by the indentation process [7].

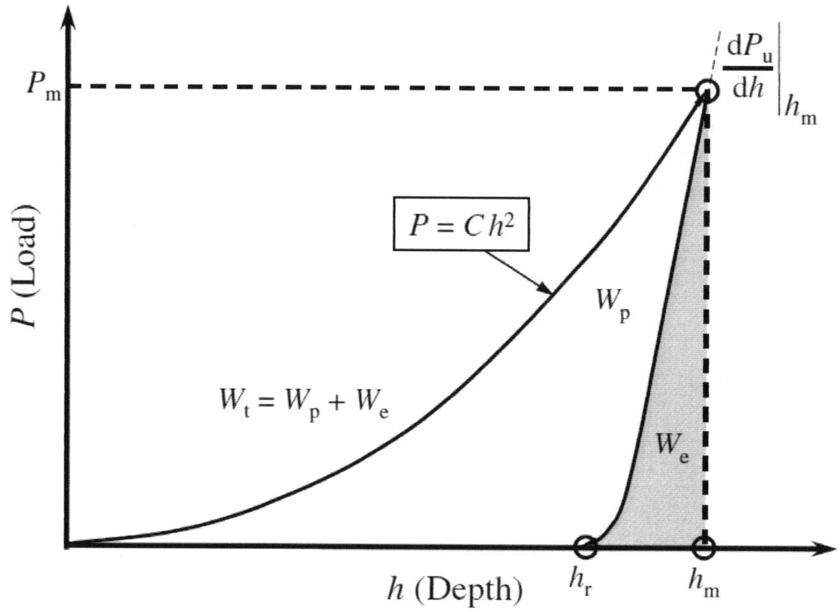

Figure 1: Graph, showing the indentation load–displacement test [7].

ISO issued a set of ISO 14577 standards [6] (4 parts) on depth-sensing (instrumented) indentation experiments, considering both plastic and elastic deformations. The standard determines three ranges of instrumented indentation test (Table 1).

In addition, in the Part 1 of the ISO 14577, special requirements are described for the nano range [6], however, we do not repeat them here, as this paper is a review of applications of microindentation tests. The ISO 14577 standard was actually prepared with regards to testing of metals, but it is stated, that it can be used also in the case of investigations of non-metallic materials and coatings.

Table 1: Ranges of instrumented indentation methods, as defined in [6].

The range	Load and displacement
Macro range	$2\ N \leq F \leq 30\ kN$
Micro range	$2\ N > F$; $h > 0{,}2\ \mu m$
Nano range	$h \leq 0{,}2\ \mu m$

With the development of atomic force microscopy, tests in the nano range – nanoindentation has attracted in recent years a lot of attention. If we type in the search string "nanoindentation" in a literature database search tool, for instance Clarivate Web of Science™ search tool, we get almost 29,000 hits! As written by VanLandingham [5] depth-sensing or instrumented indentation tests were performed in investigations of plentiful substances, so also in the nano range. To research tribological characteristics of surfaces, lateral probe motion was also applied. A typical example for this is a study of scratch resistance of coatings. Finite element modelling and dimensional analyses have also been frequently reported in studies of relations between load - displacement experimental results and the object's constitutive performance to indentation, and in analyses of experimental indentation results in prediction of the stress-strain performance.

If we limit the literature nanoindentation data search to wood and wood based material, by typing in the search string "nanoindentation AND wood" in the mentioned database, we still get a relatively high number of results, over 400. Although, the purpose of this paper is to review applications of instrumented microindentation in investigation of wood and wood based products, let us list some selected applications of instrumented nanoindentation in wood research. Outdoor wood coatings with nano TiO_2 and ZnO as UV absorbers were studied by Vlad Cristea and co-workers [9]. With the nanoindentation technique, they investigated the influence of the nanoparticles on mechanical characteristics of a selected outdoor wood finish prior and after artificial accelerated weathering. Similarly, hardness on a cell wall scale was assessed in the case of hydrophobisation with dimethyldichlorosilane, with an instrumented nanoindetation test [10]. Mechanical performance of multilayer wood waterborne coatings was investigated by nanoindentation by Wu *et al.* [11]. Obersriebnig *et al.* [12] applied a micromechanical assay based on nanoindentation, to study the adhesion of urea-formaldehyde and polyurethane adhesives to wood cell walls. The review of wood bond analysis methods, prepared by the group of Hunt *et al.* [13] reviewed basics of wood bonding and highlighted developments of the study and knowing of wood bonds. As shown, nanoindentations test are of a high importance in such studies. Analyses of of multiload nanoindentations in the S2 and corner compound middle lamellae of loblolly pine were performed by Jakes and Stone [14]. To develop nanoindentation protocols they analysed algorithm to improve the of accurateness of Young modulus measurements and quasistatic Berkovich hardness in wood cell walls. Nanoindentation 3D mapping in research of the influence of extractives on the mechanical characteristics of the water-based wood finishes was reported by Wu and co-workers [15]. Thermally modified wood can also be studied by instrumented nanoindentation. So, Wu *et al.* [16] compared nanomechanical performance of the wood cell wall, coated with a water-based polyacrylate lacquer. An interesting application of nanoindentation in the field of wood composites was shown by Zhou *et al.* [17]. Nanoindentation measurements showed that the adhesion energy of plasma-modified plastic bags, that were used as adhesive in production of plywood, and poplar cell walls was incrreased by 19.13% in comparison to the plasma untreated specimen. Nanoindentation can be successfully used also in the field of wood biodeterioration and biology studies of wood boring insects, in the case of the research by Kundanati and co-workers [18]. They found out that the mandibles of the larvae of the coffee wood stem borer beetle (*Xylotrechus quadripes*: Cerambycidae) have a sharp edge for initiation of fractures in Arabica wood and an appropriate form, the small wood chips can be made, that are suitable for digestion. An important part of their research represented nanoindentation investigations of the larval mandibles. In addition, hardness of the attacked wood was quantified with microindentation experiments. In some other work, a thin film of copper was prepared on the surface of pine wood veneer [19], and again, a nano-indenter was utilised for assessment of mechanical properties of the metallized wood surface [20]. Several other cases of application of nanoindentation tests of a great variety and diversity are reported as well: mechanical investigations of wet fibre surfaces of hardwood chemo-thermo-mechanical pulp and softwood kraft pulp [21], assessment of nanomechanical changes in elm wood from ancient timber constructions in connection to natural weathering [22], nanoindentation studies of mechanical properties creep performance of nanoclay-based wood–plastic composites [23], the mechanical characteristics of the transparent coated wood finished were observed [24], etc.

3. Microindentation

In spite of the fact that nanoindentation is predominant instrumented indentation technique when mechanical behaviour of wood and other lignocellulosics are investigated, the impression that microindentation in the area of wood science and technology is not of an interest, is misleading. This will be shown by selected cases in the following subsections. The focus is on microindentation of polymers and coatings, specifically wood coatings, and finally, this review is finished by some selected examples of microindentation studies of wood and wood-based composites.

3.1. Examples of microindentation studies of polymers, polymer films and surface coated wood

The work of Lu and Shinozaki [25] focused on measurements of the interfacial shear strength between a coating and a substrate. Here, they used the fact that the dimension of the tip diameter is of approximately the same size than the film thickness. A flat cylindrical microindenter with a 76.2 μm diameter generated interfacial failure in a soft polymer film on a solid substrate. The indenter was pushed into the polystyrene film on a solid substrate until interfacial fracture was induced. It was shown that the interface failure most likely initiated because of the mutual influence the opening orthogonal stress and of the shear stress.

Biobased coatings are nowadays intensively investigated, and so, Ouhichi and co-workers [26] used (1R,3S)-(+)-camphoric acid for UV-curing polyester resins, prepared from itaconic acid and 1,6-hexanediol. It was demonstrated, that the mechanical characteristics of the cured materials were better than those of a resin without cyclic building blocks. This was proven by depth-sensing microindentation tests for examination of the influence of the integration of camphoric acid and isophthalic acid on the mechanical performance of the UV-cured specimens.

The model for evaluation of the yield stress and the modulus of elasticity of paint layers in art paintings was developed by Leang and co-authors [27]. Discrepancy between the obtained values and the typical ones the time when the art paintings were made, could indicate the likelihood of an unoriginal painting. The model was developed on the basis of instrumented microindentation measurements.

Also, creep and elastic recovery of the polyurethane films were researched by microindentation [28]. A special viscoelastic model was applied to fit the evolved indentation depth through the microindentation creep trials. It was stated, that the network elasticity (and thus the stoichiometric ratio between -NCO and -OH) controls the creep instantaneous elasticity. Also, the authors found out that the capability of the network to disperse energy remained high in the case of non-continuous networks, with the ratio between -NCO and -OH lower than 1.

Adhesion of a coating is an extremely important characteristic of each surface finish, and Tendela and Kaufmann [29] combined a microindentation test with a radial speckle interferometer to investigate this issue. Similarly, microindentation has been performed on thin polymer coatings adhering to rigid substrates [30]. As the flat-ended cylindrical tip penetrated into the thin polymer film, a crack formation happened, initiated at the boundary (interface) at a critical applied stress. For hard coatings, this was directly detectable as a load drop in the force-displacement curve. For soft coatings, the point on the load-displacement curve corresponding to interface failure could be determined by direct observation with an optical microscope during testing.

The coating resistance against scratching can also be evaluated by scratch tests on a scratch tester, like it was done by Merino *et al.* [31] in the case of vegetable waste transformation into bio-based composites.

Wear performance, indentation, and scratch behaviour of various UV-curing wood coatings were effectively studied with the microindenter by Hermann *et al.* [32]. It was shown in this work, focused on paint composition, that the structures of monomers and oligomers are very important, and the most important physical properties and functionality were also confirmed. The hard and brittle coatings as a rule exhibited formation of defects at lower loads than the ductile and soft ones, when the scratch tests were carried out. Also, tearing and fissuring in the soft finishes under the surface was observed during the friction trials. More fractures were forme than in the case of the harder films.

Self healing wood coatings are of an increasing interest of researchers and their application is increasing. Such UV curable acrylate coatings were the research object of Paquet *et al.* [33,34], who among other instrumental techniques used also the Micro Combi Tester producing scratches with controlled depth. Self healing finishes, with the healing substances in microcapsules, was also the research topic of Schreiner *et al.* [35]. The forces for the destruction of the capsules were obtained from the force-indentation depth graphs.

Instrumented microindentation found its application also in the field of conservation of old artefacts. Kaszowska *et al.* [36] performed microindentation investigation for evaluation of the consolidation method of historic wall paintings. In their paper, the results of micromechanical investigations were discussed. Again, it was proven that the indentation experiments can be very suitable in evaluation of changes of the paint layer properties in historic wall paintings that are in the conservation process. In the same area of research, Krzemień *et al.* [37] focused on microscratching technique for the detection of stresses in the basic coat of panel paintings. The damage causing critical load was tightly correlated with the stress in the material. Such examinations are beneficial for the understanding of the influence of environmental changes on heritage objects.

3.2. Other microindentation application in research of wood and wood-based products

Lykidis *et al.* [38] modified the Brinell hardness determination method. In accordance with their results, it was found out that instead of visual measurements of indent diameter, Brinell hardness values of wood can be provided by instrumented indentation measurements and that in such a case, these values of hardness are correlated to density better. Also, this methodology led to alteration of values of Brinell hardness.

In order to investigate a suitable portable in-situ technique for the determination of mechanical characteristics of various wood species, Barletta [39] performed flat punch depth-sensing microindentation tests for evaluation of wood's strength properties. In this methodological work, the author discussed assessment of the obtained correlations with a statistic method. Valuable information from the performed instrumented microindentation experiments was obtained. Also, it was stated that the film on wood can interfere with the consistency of values of hardness in the case of the Brinell method. However, lower coating influence was observed upon the Janka hardness values.

Scharf and others [40] addressed the question on hardness of surface densified wood – is it a material or product property? Their starting point was a belief, that the density profile under the densified surface have to be considered, and therefore the recognised methods for determination of wood hardness can be questionable. They concluded that the interaction of the density profile and the test parameters caused high variability of the obtained values of hardness, irrespective of the method applied. Nevertheless, they reported that the performance of surface-densified wood can be assessed on the basis of the density profile itself. Its important influence on the material performance during determination of hardness is embedded in the continuous measurement data during hardness tests. So, the prediction of the density profile may be performed from indentation measurements.

The swelling effect of water in wood samples can be imitated by impregnation with the polyethylene glycol (PEG). PEG is used to keep wood in a swollen state and it is a favoured conservation mean when waterlogged archaeological wood is preserved. Otherwise, wood would shrink and crack when taken out of water and dried in air. The work of Fortino *et al.* [41] focused on scratch tests with a microindenter of small-scale samples of PEG impregnated wood. They studied the plastic and elastic deformation of wood with the scratch test, and could determine the influence of wood structure (for instance late- or early-wood), on the performance of wood.

Mechanical/abrasive wear performance and structure of the so-called "black-fiber" (which are vascular bundles consisting of longitudinal cells and vessels) palm wood natural composite was studied by Friedrich *et al.* [42]. In this case, again also by microhardness tests. Interestingly, microhardness tests with a Vickers indenter on individual cells of the vascular bundles gave more than 6x higher values than the literature-reported macro-hardness values.

One German company manufactures a biodegradable product that has a wood-like appearance and a woody structure, and can be cast as plastics. It is named a ''liquid wood'' [43]. This substance contains cellulose, hemp, flax and lignin. The research described in the paper of Nedelcu *et al.* [43] focused on the study of microindentation and differential calorimetry.

Mechanical properties of wood polymer composite (beech wood and low-density polyethylene) were investigated by the microindentation method by Dockal *et al.* [44]. The implementation of natural fillers - finely ground wood, coarsely ground wood and slate, increased the mechanical performance up to 100 %, e.g., hardness and modulus, when compared to the properties of the polyethylene without lignocellulosic fillers. The instrumented microindentation method was found useful also in the case of determination of hardness of the densified Yellow birch wood, as reported by Triquet and co-workers [45].

An interesting way of application of microindentation was shown by Adusumalli *et al.* [46]. In their study, an in-situ scanning electron microscope (SEM) microindentation technique was applied to study the deformation of single industrial pulp.

4. Instrumented microindentation investigations of a self-healing wood coating – a case study

At University of Ljubljana, Biotechnical Faculty, Dept. of Wood Science we have recently introduced the Anton Paar MST3 Micro Combi Tester (Figure 2). In the process of introduction of this instrument into our laboratory for surface finishing, the expert group of the producer determined scratch resistance and assessed the self-healing effect of a selected self-healing wood coating, and compared the results with mechanical properties of two types of conventional paint on parquet and walnut veneered wood. Herein, a short report of the producer's expert group, that has already been published [47] is provided.

4.1. Materials and methods

One self-healing paint on medium-density fiberboard (MDF) (the sample was provided by the furniture, using this commercial self-healing coating on their products) and two types of conventional paint on parquet and walnut veneered substrate. Both conventional paints consisted of two layers of transparent coating with total thickness of 75 μm. The self-healing coating on the MDF consisted of 60 μm pigmented coating and 60 μm of transparent coating. The wood blocks (dimensions 100 mm x 100 mm x 8 mm) were glued on an aluminum cylinder and placed in the vice on the Anton Paar MST3 Micro Combi Tester.

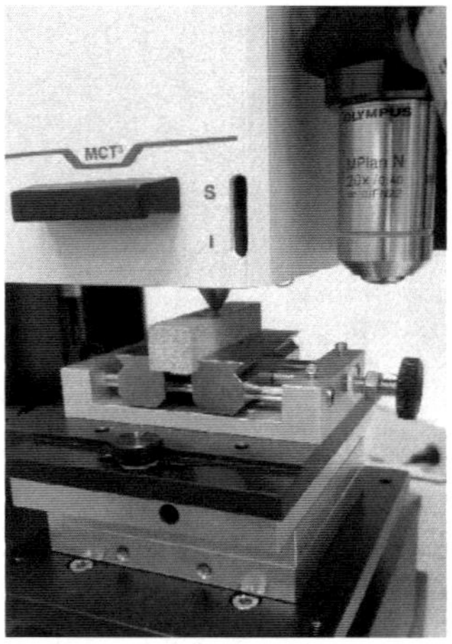

Figure 2: Micro Combi Tester MCT3 in the surface finishing laboratory of University of Ljubljana, Biotechnical Faculty, Dept. of Wood Science & Technology.

The scratch tests were done with 100 μm radius spheroconical diamond indenter on a 5 mm length with load linearly increasing from 0.05 N to 30 N. The true penetration depth and the residual depth were obtained using the prescan and postscan feature.

The microindentation experiments were done with Vickers indenter and cyclic load from load of 0.2 N and 10 N, which resulted in maximum indentation depth of ~8 μm and ~70 μm. Single-load indentation were done at 0.2 N and 10 N. At least three scratches and five indentations were done on each sample.

4.2. Results and discussion

Panorama images of the progressive load scratch on all three samples (not shown in this paper) did not exhibit any cracks. This means that no critical load could be defined, and that all three paints were able to withstand scratch damage simulated by a diamond indenter without tearing of the coating.

IOP Conf. Series: Materials Science and Engineering 1298 (2023) 012008 doi:10.1088/1757-899X/1298/1/012008

The differences in elastic recovery could easily be observed by comparing the penetration depth (Pd, sometimes called scratch depth) with the residual depth (Rd) (Figure 3). The smaller the residual depth, the better recovery. The walnut paint showed considerably larger residual depth than the self-healing paint. While the absolute value of residual depth indicates reasonably well the ability of the paint to recover, to properly characterize the elastic recovery, the residual depth should be normalised to the penetration depth. This is done by using the following formula to calculate the elastic recovery ER:

$$ER = \frac{|P_d - R_d|}{P_d}$$

Where P_d is penetration (scratch) depth and R_d is residual depth (depth after load removal). The value of ER shows which portion of the scratch has recovered. So, an ER value of 100 % means that the scratch has completely recovered and there is no defect left. Most paints have the ER value between 50% and 80% since the paint almost never recovers completely.

It was stated that the self-healing paint recovered up to ~80% while the two other paints had elastic recovery of only ~40 % to ~60 %. The self-healing paint has therefore much better scratch resistance.

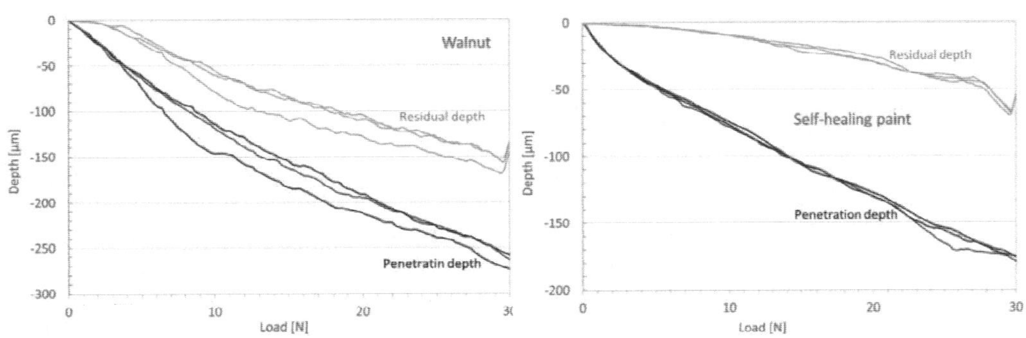

Figure 3: Penetration and residual depth on coating on walnut veneer (left) and on the self-healing coating [47].

It was stated that the self-healing paint recovered up to ~80% while the two other paints had elastic recovery of only ~40 % to ~60 %. The self-healing paint has therefore much better scratch resistance.

The results of the cyclic instrumented indentation test (modulus of elasticity vs. penetration depth) are shown in Figure 4. While both parquet and walnut paint showed a decrease in elastic modulus with increasing depth, followed by a plateau with E = 2 GPa, self-healing paint's elastic modulus first increased and then reached a plateau at ~2.8 GPa. The depth profile therefore confirmed different mechanical behavior of the self-healing paint as observed during the scratch test. On the other hand, single-load indentations can provide information about creep and elastic work of indentation since the single-load procedure contains the entire unloading part (in cyclic indentation the unloading is only partial to ~20% of the maximum load of each cycle). Single-load indentations at 0.2 N and 10 N performed on all three samples showed that while the creep is similar at both loads for all samples (~10%), the elastic to total work of indentation ratio is smaller at 10 N (~20%) for the coatings on parquet and walnut veneered samples than at 0.2 N (~40%). For the self-healing paint this ratio remains the same (~34%). This means that the self-healing paint maintains its ability to recover elastically irrespectively of the level of deformation which confirms its superior scratch resistance.

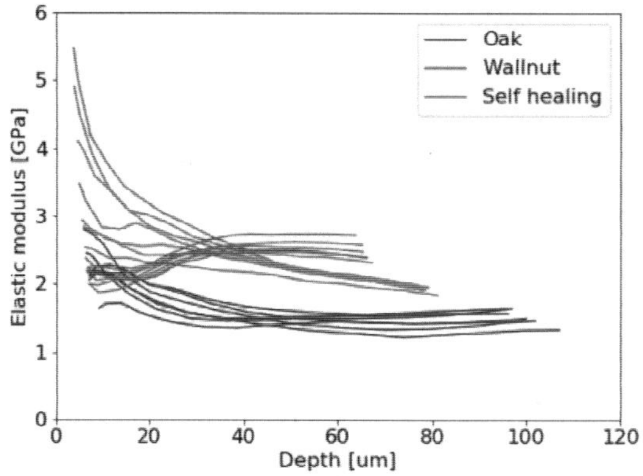

Figure 4: Depth profile of elastic modulus

To conclude this illustrative experimental part of the paper, it was shown that the among all three types of specimens (commercial coatings on parquet, walnut veneered MDF, and the sample with a self-healing paint) the self-healing paint performed the best, having the highest elastic recovery. This result was obtained using the comparison of residual and penetration depth. Instrumented indentation was a complementary method, done on the same instrument, providing elastic properties and hardness of the paints. Cyclic indentation is also able to provide depth profiles of mechanical properties. In this illustrative case the cyclic method and single-load indentation confirmed the different mechanical properties and good scratch resistance of the self-healing paint.

5. Conclusions

At University of Ljubljana, Biotechnical Faculty, Dept. of Wood Science we have recently introduced the Anton Paar MST3 Micro Combi Tester and a part of implementation of this instrument into our laboratory's investigation tasks, a literature review of possibilities in the fields of polymer films and wood science and technology has been prepared. While instrumented nanoindentation test results are far more frequently reported than the microindentation ones, the instrumented microindentation has also been shown to have interesting and useful applications: studies of mechanical properties of UV cured and bio-based coatings, evaluation of stress and elasticity modulus of art paintings, creep behaviour of polyurethane films, investigations of coating adhesion, mechanical properties of surface densified wood, thermally modified wood, and of wood-polymer composites, etc. In the second part, an illustrative research of the self-healing paint with the Micro Combi Tester, performed by the producer's expert team [47], on the coated samples, provided from Biotechnical Faculty Ljubljana, is shown. Instrumented microschratching and microindentation experiments clearly showed good scratch resistance and high elastic recovery of the self-healing paint.

References

[1] Broitmann E 2017 Indentation hardness measurements at macro-, micro-, and nanoscale: a Critical Overview *Tribol. Lett.* **65** 23

[2] Mukhopadhyay N K and Paufler P 2006 Micro- and nanoindentation techniques for mechanical characterisation of materials *Int. Mater. Rev.* **51** 209-45

[3] Liu M, Lin J, Lu C, Anh K, Zhou K and Koseki T 2017 Progress in indentation study of materials via both experimental and numerical methods *Crystals* **7** 258

[4] Cheng Y and Cheng C 2004 Scaling, dimensional analysis, and indentation measurements *Mater. Sci. Eng. R Rep.* **44** 91-149

[5] VanLandingham M R 2003 Review of Instrumented Indentation *J. Res. Natl. Inst. Stan. Technol.* **108** 249-65

[6] International standard ISO 14577-1:2015(E) (Second edition 2015-07-15) Metallic materials —

Instrumented indentation test for hardness and materials parameters — Part 1: Test method
[7] Gouldstone A, Chollacoop N, Dao M, Li J, Minor A M and Shen Y 2007 Indentation across size scales and disciplines: recent developments in experimentation and modeling *Acta Mater.* **55** 4015-39
[8] Oliver W M and Pharr G M 2004 Measurement of hardness and elastic modulus by instrumented indentation: advances in understanding and refinements to methodology *J. Mater. Res.* **19** 3-20
[9] Vlad Cristea M, Riedl B and Blanchet P 2011 Effect of addition of nanosized UV absorbers on the physico-mechanical and thermal properties of an exterior waterborne stain for wood *Prog. Org. Coat.* **72** 755-62
[10] Yang R, Wang S, Zhou D, Zhang J, Lan P and Jia C 2017 Construction of hydrophobic wood surface and mechanical property of wood cell wall on nanoscale modified by dimethyldichlorosilane *IOP Conf. Ser.: Mat. Sci. Eng.* **301** 012051
[11] Wu Y, Wu J, Wang S, Feng X, Chen H, Tang Q and Zhang H 2019 Measurement of mechanical properties of multilayer waterborne coatings on wood by nanoindentation *Holzforschung* **73** 871-7
[12] Obersriebnig M, Konnerth J and Gindl-Altmutter W 2013 Evaluating fundamental position-dependent differences in wood cell wall adhesion using nanoindentation *Int. J. Adhes. Adhes.* **40** 129-34
[13] Hunt C G, Frihart C R, Dunky M and Rohumaa A 2018 Understanding wood bonds–going beyond what meets the eye: a critical review *Rev. Adhesion Adhesives* **4** 369-440
[14] Jakes J E and Stone D S 2018 Best practices for quasistatic berkovich nanoindentation of wood cell walls *Forests* **12** 1696
[15] Wu Y, Zhang H, Yang L, Wang S and Meng Y 2021 Understanding the effect of extractives on the mechanical properties of the waterborne coating on wood surface by nanoindentation 3D mapping *J. Mater. Sci.* **56** 1401-12
[16] Wu Y, Wu X, Yang F, Zhang H, Feng X and Zhang J 2020 Effect of thermal modification on the nano-mechanical properties of the wood cell wall and waterborne polyacrylic coating *Forests* **11** 1247
[17] Zhou X, Cao Y, Yang K, Yu P, Chen W, Wang S and Chen M 2020 Clean plasma modification for recycling waste plastic bags: from improving interfacial adhesion with wood towards fabricating formaldehyde-free plywood *J. Clean. Prod.* **269** 122196
[18] Kundanati L, Chahare N R, Jaddivada S, Karkisaval A G, Sridhar R, Pugno N M and Gundiah N 2020 Cutting mechanics of wood by beetle larval mandibles *J. Mech. Behav. Biomed.* **112** 104027
[19] Li J, Wang R, Tian H, Wang Y and Qi D 2018 Research on the gradual process of the metallization structures and mechanical properties of wood veneer *Symmetry-Basel* **10** 550
[20] Li J, Wang Y, Zhang J, Wang Y and Qi D 2020 Physical characteristics and fluorescence effect of "environmentally friendly" metallized wood *AIP Adv.* **10** 045133
[21] Yan D and Li K 2013 Conformability of wood fiber surface determined by AFM indentation *J. Mater. Sci.* **48** 322-31
[22] Han L, Wang K, Wang W, Guo J and Zhou H 2019 Nanomechanical and topochemical changes in elm wood from ancient timber constructions in relation to natural aging *Materials* **12** 786
[23] Yadav S M and Yusoh K B 2019 Subsurface mechanical properties and subsurface creep behaviour of modified nanoclay-based wood–plastic composites studied by nanoindentation *Polym. Bull.* **76** 2179-96
[24] Wu Y, Wu J, Yang F, Tang C and Huang Q 2019 Effect of H_2O_2 bleaching treatment on the properties of finished transparent Wood *Polymers-Basel* **11** 776
[25] Lu Y and Shinozaki D M 2005 Effects of substrate constraint on micro-indentation testing of polymer coatings Wood *Mat. Sci. Eng. A-Struct.* **396** 77-86
[26] Ouhichi R, Bougarech A, Kluge M, Pérocheau Arnaud S, Abid S, Abid M and Robert T 2021

Camphoric acid as renewable cyclic building block for bio-based UV-curing polyhexylene itaconate Wood *Eur. Polym. J.* **152** 110423

[27] Léang M, Giorgiutti-Dauphiné F, Lee L-T and Pauchard L 2017 Crack opening: from colloidal systems to paintings *Soft Matter* **13** 5802

[28] Chagnon L, Arnold G, Giljean S and Brogly M 2013 Elastic recovery and creep properties of waterborne two-component polyurethanes investigated by micro-indentation *Prog. Org. Coat.* **76** 1337-45

[29] Tendela L P and Kaufmann G H 2012 Evaluation of coating adhesion using a radial speckle interferometer combined with a micro-indentation test *Opt. Laser Eng.* **50** 817-22

[30] Lu Y and Shinozaki D M 2002 Microindentation induced debonding of polymer thin films from rigid substrates J. *Mat. Sci.* **37** 1283-93

[31] Merino D, Simonutti R, Perotto G and Athanassiou A 2021 Direct transformation of industrial vegetable waste into bioplastic composites intended for agricultural mulch films *Green Chem.* **23** 5956

[32] Hermann A, Giljean S, Pac M-J, Marsiquet C, Beaufils-Marquet M, Burr D and Landry V 2021 Understanding indentation, scratch and wear behavior of UV-cured wood finishing products *Prog. Org. Coat.* **161** 106504

[33] Paquet C, Schmitt T, Klemberg-Sapieha J E, Morin J-F and Landry V 2020 Self-healing UV curable acrylate coatings for wood finishing system, part 1: impact of the formulation on self-healing efficiency *Coatings* **10** 770

[34] Paquet C, Brown S, Klemberg-Sapieha J E, Morin J-F and Landry V 2021 Self-healing UV curable acrylate coatings for wood finishing system, part 2: impact of monomer structure and self-healing parameters on self-healing efficiency *Coatings* **11** 1328

[35] Schreiner C, Scharf S, Stenzel V and Rössler A 2017 Self-healing through microencapsulated agents for protective coatings *J. Coat. Technol. Res.* **14** 809-16

[36] Kaszowska Z, Kot M, Białek-Kostecka D and Forczek-Sajdak A 2019 Application of micro-indentation tests to assess the consolidation procedure of historic wall paintings *J. Cult. Herit.* **36** 286-96

[37] Krzemień L, Kot M and Łukomski M 2018 Stress assessment in artistic materials using a micro-scratching technique *Exp. Techniques* **42** 473-9

[38] Lykidis C, Nikolakakos M, Sakellariou E and Birbilis D 2016 Assessment of a modification to the Brinell method for determining solid wood hardness *Mater. Struct.* **49** 961-7

[39] Barletta M 2006 Application of instrumented micro-indentations to 'in situ' mechanical characterization of wooden structures: part II—analysis of different species *Measurement* **39** 285-95

[40] Scharf A, Neyses B and Sandberg D 2022 Hardness of surface-densified wood. Part 1: material or product property? *Holzforscung* **76** 503-14

[41] Fortino F, Metsäjoki J, Ronkainen H, Bjurhager I, Heinemann S and Salminen L I 2020 Scratch resistance of PEG-impregnated green wood: a method for evaluation of swollen wood properties *Wood Sci. Technol.* **54** 715-35

[42] Friedrich K, Akpan E I Wetzel B 2017 Structure and mechanical/abrasive wear behavior of a purely natural composite: black-fiber palm wood *J. Mater. Sci.* **52** 10217–29

[43] Nedelcu D, Ciofu C and Lohan N M 2013 Microindentation and differential scanning calorimetry of ''liquid wood'' *Compos. Part B-Eng* **55** 11-5

[44] Dockal A, Ovsik M, Fluxa P, Stanek M and Senkerik V 2020 Implementation of natural fillers in polyethylene and the resulting mechanical properties *Mater. Tehnol.* **54** 341-3

[45] Triquet J, Blanchet P and Landry V 2021 Hardness of chemically densified Yellow birch in relation to wood density, polymer content and polymer properties *Holzforschung* **75** 114-25

[46] Adusumalli R-B, Raghavan R, Schwaller P, Zimmermann T and Michler J 2010 *In situ* SEM micro-indentation of single wood pulp fibres in transverse direction *J. Electron Microsc.* **59** 345–9

[47] Nohava J Scratch resistance and self-healing of wood paints Document No. XCSIA088EN-A https://www.anton-paar.com/corp-en/services-support/document-finder/application-reports/scratch-resistance-of-self-healing-paints-on-wood/ (6.7.2023)

Acknowledgments
Financial support of the Slovenian research and Innovation Agency (ARIS) through financing of the Research Programme "Wood and lignocellulosic composites" is greatly acknowledged.

IOP Conf. Series: Materials Science and Engineering 1298 (2023) 012009 doi:10.1088/1757-899X/1298/1/012009

Antifungal protection of beech wood using plant extracts

Mimica Stefanović[1], Nebojša Todorović[1], Goran Milić[1] and Aleksandar Lovrić[1]

[1]Department of Wood Processing, Faculty of Forestry, University of Belgrade, Kneza Višeslava street, Belgrade, Serbia

e-mail: nebojsa.todorovic@sfb.bg.ac.rs

Abstract. In this research, the antifungal efficacy of dry plant extracts of *Chinese cinnamon* bark (*Cinnamomum cassia* L.), Rtanjski tea herb (*Satureja montana* L.) obtained by water extraction and extract of Thyme herb (*Thymus serpyllum* L.) obtained by ethanol-water extraction was investigated against the fungi *Coniophora puteana* and *Serpula lacrymans*, which cause brown rot, and the fungus *Trametes versicolor*, which causes white rot of hardwood and conifers. Five different concentrations of plant extracts were tested, and water was used as a solvent. Samples of beech wood (*Fagus sylvatica* ssp. *moesiaca* (Maly) Czeczott.) were treated by the immersion method with solutions of plant extracts and exposed for 60 days to the action of the test fungi. The smallest mass loss (%) of beech wood when affected by each of the test fungi was in samples treated with *Thymus serpyllum* extract at a concentration of 25%, followed by *Satureja montana* at the same concentration. *Cinnamomum cassia* bark extract did not show significant antifungal activity against the test fungi at any concentration. The obtained results can affirm the use of dry plant extracts as ecologically acceptable agents in the preventive protection of non-resistant wood species against rotting fungi.

1. Introduction

The protection of wood from its destructors during the last century was based on the use of various chemical agents which have a negative impact on the environment and human health. Many agents are banned for use by legal regulations, and the use of some is restricted in most economically developed countries. Boron-based substances, copper, carbamates, triazoles, sulfamides, creosotes, and similar compounds are still in use [1]. A recent trend is the use of environmentally friendly methods in the preventive protection of less durable and non durable wood species, aiming to protect both the environment and human health. For this reason, numerous researchers have been dealing with this topic, and many proposing the application of natural substances for preventive protection of wood from biological destructors. The most successful agents were extracts of plants were the dominant active compounds are thymol and carvacrol [2, 3, 4, 5, 6, 7], as well as eugenol and cinnamon aldehyde [8, 9, 10, 11]. Plant species belonging to the genera *Cinnamomum*, *Thymus* and *Satureja* [10, 12] are known to be rich in these bioactive compounds. The extract of Cinnamomum cassia bark, as the main chemical compound, contains cinnamaldehyde (and its derivatives) and eugenol with proven antifungal activity against pathogenic fungi, even against xylophagous insects [7, 8, 9, 10]. Species from the *Thymus genus* are rich in active substances such as thymol and carvacrol [2, 3, 4, 5, 6, 7, 8]. The extract of mountain savory herb *(Satureja montana)* also contains carvacrol as the most dominant component, but it also includes p-cymene (a monoterpene hydrocarbon), borneol, trans-caryophyllene, caryophyllene oxide and γ-terpinene [3, 8, 9, 10, 13, 14, 15, 16, 17]. In this research, we investigated the extracts (with the use of water as a solvent) from *Cinnamomum cassia* (CC) with chemical composition [8], *Satureja montana* (SM) [2,3] and *Thymus serpyllum* (TS) [3] in the protecting beech wood against the rot fungi *S. lacrymans, C. puteana* and *T. versicolor*. Moesian beech (*Fagus sylvatica* ssp. *moesiaca*) is the most common species in Serbia and very important for domestic industry, but its usage is limited due to its natural lack of resistance to wood-decaying fungi. Treating wood with solutions of plant extracts can be an ecological alternative to conventional chemical treatments if their effectiveness is proven. Of course, there is also the question of whether, in addition to being environmentally friendly and effective, the treatment will meet the principle of cost-effectiveness, or if the production costs and the costs of transitioning from previous wood protection methods to this type will exceed an acceptable price. These are questions that require further research and collaboration with experts in the field of economics, as well as other areas.

2. Methods and material

Dry extracts (Extracta sicca) of *Cinnamomum cassia* L. bark J.Presl (MG 013), *Thymus serpyllum* L. herb (MG 040) and *Satureja montana* L. (MG 056/1) were obtained from the private company which produces and exports plant extracts to European markets. Water was used as the solvent for extracting CC and SM, while the TS extract was obtained through ethanol-aqueous extraction (60% v/v). The extraction of all three plant species was carried out by digestion – extraction at a temperature of 30 to 60° C at a pressure of 0.6 bar. The CC extraction temperature was approx. 40° C, with the duration of 24 h and the 8-10:1 drug-extract ratio; the SM and TS extractions were performed at a temperature of about 30° C for 24 h, while the drug-extract ratio was 4-6:1. Dry extracts contain 90% dry residue and are packed in hermetically sealed PP bags. Dry plant extracts were dissolved by mixing on a magnetic stirrer at a temperature of 35°C to 40°C in distilled, sterilized water at solution concentrations of 0.05 g/ml, 0.10 g/ml, 0.15 g/ml, 0.20 g/ml, and 0.25 g/ml, corresponding to 5.0%, 10.0%, 15.0%, 20.0%, and 25.0% by volume of 100 ml solution for each extract type.

The beech wood samples (*Fagus sylvatica* ssp. *moesiaca* (Maly) Czeczott.) were cut from air-dried timber to dimensions of 5 (tangential) x 20 (radial) x 20 (axial) mm. A total of 8 samples were prepared for each concentration, each type of extract and each test fungus, plus the control sample for each fungus. Samples of beech wood were prepared according to the procedure outlined in the European standard EN 113, except for the dimensions of the samples. The dimensions were adjusted to perform a modified disk-diffusion method, in which wood samples were used instead of filter paper soaked with the substance being tested for its effectiveness against specific pathogens. The oven-dried samples of beech wood were immersed in solutions of extracts of different concentrations for 24 hours until a constant mass was achieved. Each batch of 8 wood samples was immersed in 100 ml of a specific concentration solution of each extract. After that, they were placed on lattice girders and dried for 7 days under standard temperature 20±2°C and humidity conditions 65±5% until the mass of the samples remained unchanged for three consecutive measurements. Treated and control samples of wood were exposed to the test fungi for 60 days in a climatic chamber at a temperature of 22 ± 1° C and air humidity of 70 ± 5%. The exposure time was proportionally shorter compared to the 16 weeks (EN 113) due to the smaller dimensions of the wood samples. Following the exposure period, the surface of the samples was lightly brushed to remove the mycelium and the samples were dried in a kiln, until a constant mass (m_2) was achieved. Mass loss of the wooden samples (ML) was calculated using the difference between the mass of oven-dry samples of beech wood at the end of the test (after exposure to the test fungi) and before the test:

$$ML = \frac{\mathrm{m}_2 - \mathrm{m}_0}{\mathrm{m}_0} x\ 100\ (\%)$$

m_2 - mass of wood samples after the effect of the test fungi and after drying to an oven-dry state (g);
m_0 - mass of oven-dry samples of beech wood before the effect of test fungi and before immersion in solutions (g).

The retention level and percentage of weight gain of beech wood were determined after immersion in solutions of plant extracts at specific concentrations and calculated by [8, 9, 10]. Two-factorial analysis of variance (ANOVA) was used to analyze the obtained results. Significant differences between variables were determined by the Tukey post-hoc test ($p \leq 0.05$).

3. Research results and discussion

Table 1 gives the average ML of beech samples treated with extracts of the given concentrations after exposure to test fungi for 60 days.

Table 1. Average mass loss (%) of beech wood samples treated with different concentrations of plant extracts of *C.cassia, S.montana* and *T.serphyllum* bark after 60 days of exposure to the test fungi *C.puteana, S.lacrymans and T.versicolor* (standard deviation in brackets)

Plant extract	Concentration (%)	Mass loss (%)		
		C.puteana	*S.lacrymans*	*T.versicolor*
C.cassia	5	24.90 (1.13)	43.89 (1.28)	20.20 (0.54)
	10	22.82 (0.72)	41.08 (0.72)	21.35 (0.86)
	15	23.41 (0.63)	45.69 (1.37)	24.78 (0.56)
	20	21.28 (0.44)	45.11 (1.24)	22.22 (0.98)
	25	22.03 (0.52)	45.04 (2.29)	21.14 (1.50)
S.montana	5	17.73 (0.49)	41.81 (0.85)	21.75 (0.70)
	10	5.67 (0.38)	40.06 (0.65)	16.83 (0.47)
	15	4.76 (0.40)	26.71 (0.35)	18.91 (0.67)
	20	3.52 (0.08)	19.39 (0.39)	17.65 (1.50)
	25	2.65 (0.09)	6.30 (0.32)	12.77 (0.50)
T.serpyllum	5	13.69 (0.59)	44.11 (1.46)	18.71 (0.99)
	10	8.99 (0.74)	36.03 (0.50)	18.62 (0.39)
	15	2.59 (0.09)	18.03 (0.26)	17.23 (0.47)
	20	2.88 (0.29)	6.90 (0.32)	16.02 (0.56)
	25	2.11 (0.43)	6.10 (0.37)	11.91 (0.29)
Control	0	30.77 (0.76)	51.64 (1.69)	28.24 (0.65)

At all concentrations, the samples treated with CC did not exhibit a significant reduction in ML. It should be noted that the use of CC dry extract diluted with water as a preventive wood protection against decay has not been previously studied, making it challenging to directly compare these findings with those of other researchers. The results obtained here differ from previous studies that explored the antimicrobial and antifungal efficacy of the extracts of cinnamon oils or their active compounds (cinnamon aldehyde, coumarin, cinnamyl alcohol, cinnamic acid) against some pathogenic fungi, as well as rot fungi [18, 9, 10, 19, 11, 20]. The disparities in results could potentially be attributed to the choice of solvent. In earlier studies, pure or 10% DMSO (dimethylsulphoxide) or ethanol was used. Ethanol tends to dehydrate hyphae, while DMSO itself inhibits microorganisms and is used as an intermediate of active components in the production of antibiotics and antifungals, thereby reducing their minimum inhibitory concentration [21]. Only at a concentration of 1% and below, DMSO has no significant impact on fungi inhibition [22]. By using water, we avoided the inhibition of fungi with a solvent. Another reason for the divergent results could be the higher efficiency of es. oils compared to extracts, because they penetrate through the phospholipid layer of the microorganism cell membranes [23]. The third reason may be that the antifungal efficacy of the extracts was tested on wood and not on culture media. The dry extract of SM dissolved in water demonstrated good antifungal activity against *C.puteana* (at concentrations above 5%) and *S.lacrymans* (above 10%). The most significant activity was observed at a concentration of 25%, resulting in ML values of 2.65% and 6.30% for *C. puteana* and *S. lacrymans*, respectively. The best inhibitory activity of SM against *T.versicolor* was obtained at 25%, but it was less significant compared to the inhibitory activity against the other two test fungi. The reason may be the resistance of these fungi to phenolic compounds [24]. Based on the literature search, the effectiveness of dry extract of SM applied to wood against decaying fungi has not been investigated. However, studies have explored the antifungal effects of SM essential oil against various strains of pathogenic fungi, as well as the efficacy of compounds found in certain es. oils against *C.puteana* and *T.versicolor* [6].

The average ML under the influence of *C.puteana* fungus was significantly reduced with the TS treatment at concentrations higher than 10%, while the smallest ML of 2.11% was at the highest concentration of TS. As for *S.lacrymans,* the most substantial reduction in ML (6.10% and 6.90%) occured at concentrations of 25% and 20%. In the case of *T.versicolor,* the ML was lower than that of the control samples at all TS concentrations, while the lowest (11.91%) was found at a concentration of 25%. As for previous researches on the effectiveness of dry extract of TS against wood rotting fungi, we can rely on the results obtained by testing the antifungal efficacy of active compounds isolated from extracts and es. oils of species from the genus *Thymus*. However, the inhibitory effect of the TS extract solution applied to wood on the growth of wood rotting fungi has not been tested. The reason for the better antifungal effectiveness of the TS extract compared to the CC and SM extracts may be the extraction method used. Specifically, the TS extract was obtained through a 60% (v/v) ethanol-water extraction, while water was used for the CC and SM extracts. The effectiveness of plant extracts in providing antifungal protection to wood depends on the nature and quantity of the extract in the plant, extraction method, chemical composition, chemotype, etc. On the other hand, it also depends on the wood species, its physical and mechanical properties, and, of course, the type of fungus present. Factors influencing the choice of wood protection agents include the optimal extract concentration, type of solvent, method of application into the wood, ability to penetrate and fixate within the wood, the impact of the extract solution or chosen solvent on the treated wood's properties, durability of its protection, and similar considerations. A two-way ANOVA showed that both the fungi and the extract had a statistical significance ($p \leq 0.5$) on beech wood ML under the influence of the test fungi. Statistically significant differences between the mean values of ML under the influence of each factor were tested by the Tukey HSD test.

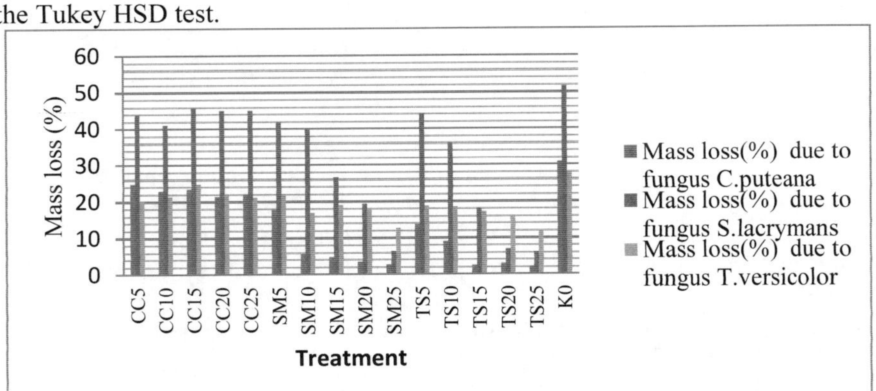

Figure 1. Effect of the type of treatment and fungus on mass loss.

Wood treated with extracts exhibited the highest effect against the *C.puteana* fungus, and the lowest one against the *S.lacrymans* fungus. As for the concentration of extracts, the most significant reduction in ML was achieved when wood was treated with 25% TS extract, followed by 25% SM, and the least significant reduction occured with CC at concentrations of 5% to 15%. The ML was highest after the effect of *S. lacrymans* on the control and CC-treated wood samples. The lowest ML was recorded after the effect of the *C. puteana* fungus on the wood treated with the TS extract, followed by the SM extract treatment. The TS and SM extracts exhibited a slightly lower impact on ML reduction after the exposure to the fungus *Trametes versicolor.* These findings are visually represented in Figure 1, which ilustrates the impact of beech wood treatment with CC, SM and TS of different concentrations on ML after 60 days of fungal exposure. The lowest amount of solution absorbed by the beech wood after 24 hours of immersion was observed for all concentrations of cinnamon bark extract compared to the solutions and concentrations of the other two extracts. Cinnamon extract had the least impact on mass loss of the wood compared to the control untreated samples when exposed to all three tested fungi. Higher amounts of absorbed substance were observed in wood treated with a solution of mountain savory herb extract, which resulted in a significant reduction in mass loss of the treated wood compared to the untreated wood, proportionally to the retention level. The same case was observed for beech wood samples treated with a solution of thyme extract. These differences in retention levels of the extract solutions, indicating their penetration efficacy into the wood, likely stem from the microscopic and submicroscopic structure of the wood itself, as well as the particle size of the dry extracts in the solution. Based on the data on the retention level or the amount of retained extract solution at specific concentrations, we can assume that the antifungal effect of certain extracts increases with the higher amount of extract solution retained in the wood samples.

4. Conclusion

Treating beech wood with a solution of *Thymus serpyllum* extract at concentrations of 25%, 20%, and 15% provides effective protection against the fungus C. puteana, as the average mass loss is less than 3% compared to the initial wood sample mass, in accordance with EN113:1996. This is followed by treating with a solution of *Satureja montana* extract at a concentration of 25% against the same fungus. However, against the fungi S. lacrymans and T. versicolor, the average mass loss of beech wood treated with the same extract solutions is greater than 3%, indicating that they cannot be considered effective for the preventive protection of beech wood against all decay-causing fungi. Treatment with *Cinnamomum cassia* extract at all concentrations did not significantly reduce the mass loss of beech wood compared to untreated wood when exposed to decay fungi. Considering the promising results of the research on extracts from the Cinnamomum genus against fungi and bacteria, it is desirable to continue investigating the efficacy of this extract at different concentrations using other suitable solvents. We can conclude that the tested dry plant extracts have the potential to be candidates for the preventive protection of beech wood against decay fungi, with further research taking into account different extract concentrations in other appropriate solvents that are environmentally friendly. Standardized extracts can also be included in the testing. To ensure the applicability of the research results in practice, it is necessary to examine the impact of these extracts on the changes in the physical and mechanical properties of beech wood.

5. REFERENCES

[1] Reinprecht, L., Tiralová, Z. (2007a). Biocídy na ochranu dreva – aplikácia v interiéroch/Biocides for wood preservation – application in interiors/. Stavebné materiály, 3(6):52-54

[2] Yang V W and Clausen C A 2007 Inhibitory effect of essential oils on decay fungi and mold growth on wood *Proc. One Hundred Third Annual Meeting of the American Wood Protection Association* (St. Louis Missouri) vol 103 ed C McCown and B Williams (Birmingham, American wood protection association) pp 62-70

[3] Pánek M, Reinprect L and Hulla M 2014 Ten essential oils for beech wood protection-efficacy against wood-destroying fungi and moulds, and effect on wood discoloration *BioResources* **9 (3)** 5588-603

[4] Zyani M, Mortabit D, El Abed S, Remmal A and Koraichi S I 2011 Antifungal activity of five plant essential oils against wood decay fungi isolated from an old house at the medina of fez *International Research Journal of Microbiology* **2 (3)** 104-08

[5] Xie Y, Wang Z, Huang Q and Zhang D 2017 Antifungal activity of several essential oils and major components against wood-rot fungi *Industrial Crops and Products* **108** 278–85

[6] Voda K, Boh B, Vrtačnik M, and Pohleven F 2003 Effect of the antifungal activity of oxygenated aromatic essential oils compounds on the white-rot Trametes versicolor and the brown-rot Coniophora puteana *International Biodegradation and Biodeterioration* **51 (1)** 51-59

[7] Jarić S, Mitrović M and Pavlović P 2015 Review of ethnobotanical, phytochemical, and pharmacological study of Thymus serpyllum L *Hindawi Publishing Corporation Evidence-Based Complementary and Alternative Medicine* 10

[8] Wang S Y, Chen P F and Chang S T 2005 Antifungal activities of essential oils and their constituents from indigenous cinnamon (Cinnamomum osmophloeum) leaves against wood decay fungi. *Bioresource Technology* **96** 813–18

[9] Cheng S S, Liu J Y, Chang E H and Chang S T 2008 Antifungal activity of cinnamaldehyde and eugenol congeners against wood-rot fungi *Bioresource Technology* **99** 5145–49

[10] Axinte L, Cuzman O A, Feci F, Palanti S and Tiano P 2011 Cinnamaldehyde, a potential active agent for the conservation of wood and stone reliefs *European Journal of Science and Theology* **7 (1)** 25-34

[11] Yen T B and Chang St 2008 Synergetic effect of cinnamaldehyde in combination with eugenol against wood decay fungi *Bioresource Technology* **99** 232–36

[12] Jovanović A, Bugarski B, Šavikin K and Zdunić G 2013 Biological activity and ethnomedicinal use of Thymus vulgaris and Thymus serpyllum *Healing Raw Materials* **33** 3–17

[13] Vladić J 2017 Savremene metode ekstrakcije rtanjskog čaja (Satureja montana L.), hemijski sastav i biološka aktivnost odabranih ekstrakata *Doktorska disertacija* Univerzitet u Novom Sadu, Tehnološki fakultet. Novi Sad.

[14] Bezić N, Skočibušić M, Dunkić V 2005 Phytochemical composition and antimicrobial activity of *Satureja montana* L. and *Satureja cuneifolia* Ten. essential oils. Acta Bot. Croat. 64 (2), 313–322, ISSN 0365–0588

[15] Milosavljević PN, Veličković T D, Randelović V N 2000 Antibakterijska svojstva i hemijski sastav etarskih ulja čubra (*Satureja kitaibelii* Wierzb. ex Heuff.); Proceeding of 6th Symposium on Flora of the Southeastern Serbia, Sokobanja, 2000, 181-185

[16] Serrano, C., Matos, O., Teixeira, B., Ramos, C., Neng, N., Noguneira, J., Nunes, M.L., Marques, A., (2011). Antioxidant and antimicrobial activity of *Satureja montana* L. extracts. Science of Food and Agriculture**, 91 (9)**, 1554 – 1560

[17] Skočibušić, M., Bezić, N., (2004). Phytochemical analysis and *in vitro* antimicrobial activity of two *Satureja* species essential oils. Phytotherapy research, Vol. 18, Issue 12, 967 - 970

[18] Tzortzakis N G 2008 Impact of cinnamon oil-enrichment on microbial spoilage of fresh produce *Innovative food science and emerging technologies* **10 (1)** 97-102

[19] Ahmada S, Capoorb M and Khatoona F 2013 Phytochemical analysis and growth inhibiting effects of Cinnamomum cassia bark on some pathogenic fungal isolates *Journal of Chemical and Pharmaceutical Research* **5 (3)** 25-32

[20] Araar H 2009 *Cinnamon Plant Extracts: A Comprehensive Physico-Chemical and Biological Study for Its Potential Use as a Biopesticide* Thessis Istituto Agronomico Mediterraneo di Bari Master of Science Collection 562

[21] Alastruey-Izquierdo A, Gómez-López A, Arendrup C M, Lass-Florl C, Hope W W, Perlin S D, Rodriguez-Tudela L J and Cuenca-Estrellaa M 2012 Comparison of dimethyl sulfoxide and water as solvents for echinocandin susceptibility testing by the EUCAST methodology *Journal of Clinical Microbiology* **50 (7)** 2509–12

[22] Randhawa A M and Aljabre M S H M 2007 Dimethyl sulfoxide (DMSO) has an additive effect and alters minimal inhibitory concentrations of antifungal drugs *Journal of Rawalpindi Medical College* **11 (2)**

[23] Šućur, J 2015 *Biopesticidal Activity of Extracts of Selected Plant Species of the Lamiaceae Family* Doctoral dissertation University of Novi Sad Faculty of natural sciences Department of Chemistry, Biochemistry and Environmental Protection

[24] Amara S, Perrot T, Navarro D, Deroy A, Benkhelfallah A, Chalak A, Daou M, Chevret D, Faulds C B, Berrin J G, Morel-Rouhier M, Gelhaye E and Record E 2018 Enzyme activities of two recombinant heme-including peroxidases *Tv*DyP1 and *Tv*VP2 identified from the secretome of *Trametes versicolor Applied and Environmental Microbiology* American Society for Microbiology **84 (8)** 02826-17

Mathematical model of the influence of some parameters of solid wood machinability on the roughness of the processed surface Ra during planning

Atif Hodžić, Aldin Bijelić and Damir Hodžić

University of Bihać, Faculty of Technical Engineering, Bosnia and Herzegovina

E-mail: atif.hodzic@unbi.ba

Abstract. We consider the machinability of the material as a technological feature that expresses the ability of the material to remove the maximum number of shavings from its machined surface in the minimum time with satisfactory processing quality, with as little cutting force as possible and as little tool wear as possible. The aim of the experimental research in this work is to examine the significance of the influential kinematic parameters of the roughness of the machined surface, i.e. of wood density (ρ), feeding speed (s') and the number of spiral cutting knices (z) in the process of planning massive wood on the roughness of the newly created processing surface, which will vary in 14 trials, of which 6 are repetitions in the central point of the compositional plan, where the roughness parameter Ra is obtained as an output value, and the analysis of experimental data from the point of view of possible achievement of a better quality of the processed surface. The obtained mathematical model is essentially applicable and can be used to optimize the machinability parameters in the planning process of solid wood, and the experimental results can be used in further research into other parameters of the machinability of solid wood in the planning process.

1. Introduction

Machined wood surfaces are mostly rough to a greater or lesser extent and as such are not technologically processed enough for the gluing process or for surface treatment. Wood processing by planning is carried out with rotary tools, and as a result, on the planed surface we have traces of the cutting trajectory in the form of cyclotide sections, which can be seen as undulations on the machined surface. On the other hand, structural irregularities in the form of regularly shaped furrows define the roughness of the processed surface, and their amplitude and wavelength depend on the specific processing. The precision of the processing and the roughness of the processed surface directly affect the quality of the processing. When the deviation of the actual values of the processed surface is within the tolerance values, we can say that from a technological point of view the processing is correct, although often in direct production we have cases where the errors are within the allowed limit values and that the processed details differ from each other in terms of processing quality.

The quality of the machined surface is defined by different types of unevenness and their deviations from the ideal flat surface. Usually, when measuring the texture of a machined surface, it is the roughness that is analyzed, while waviness and shape deviation are the elements that are filtered out of the collected data. Roughness can occur as a result of processing conditions and regimes, or as a result of the internal structure of the wood's anatomical structure [1].

The basic numerical parameters of the roughness of the machined surfaces are determined according to the standard, starting from the middle line of the profile according to the "M" system, and the most commonly used in practice are:

Ra (arithmetic average deviation of the profile from the middle line)

Rz (mean height of bumps)

The mean arithmetic deviation Ra represents the mean arithmetic value of the distance of all points of the effective profile from the mean line and can be represented by the following mathematical expression [2]:

$$R_a = \frac{1}{l}\int_0^l |y|dx \tag{1}$$

where is:

Ra – mean arithmetic deviation of the profile (μm)

l – reference length (μm)

y (x) – the height of the roughness profile with regard to the reference point.

The paper examines the influence of wood density (ρ), feed speed (s') and the number of spiral cutting knives (z) on the roughness of the processed surface of the Ra parameter in the process of planing massive wood on a multi-sided planer. The experiment was carried out under real production conditions in doo "Interprocess" Tešanj, and the roughness parameter Ra was measured with a contact and laser measuring device in the laboratory of the Biotechnical faculty in Ljubljana. A stochastic mathematical model of the influence of wood density, feed rate, number of spiral cutting knives and their interactions on the roughness of the machined surface, i.e., the Ra parameter, was developed.

2. Experiment planning

The roughness of machined wood surfaces is almost impossible to reliably theoretically determine using analytical models, and the application of experimental analysis of the investigated influential parameters is of essential importance for improving the treatment regimen, while keeping the costs and quality of treatment at a satisfactory level.

The purpose of the experiment plan is to generate a mathematical model, that is, an equation (polynomial of the second degree) that describes the process. If the studied parameters in the experiment are really those that affect the process, and the data obtained by the experiment are of acceptable accuracy and precision, then it is possible to develop a model that reliably describes the process [3].

2.1. Resources needed to run an experiment

The experimental research related to this work is based on wood specimens with density in the range of 390.23 - 790.19 kg/m3, which were previously processed into clean blanks and then planed flat on the four-sided planer Profimat 23 Fortec, manufactured by Weining GmbH. with a change in feed rate and number of spiral cutting knives in three values of the same. The dimensions of the clean workpieces are 620 x 60 x 24 mm.

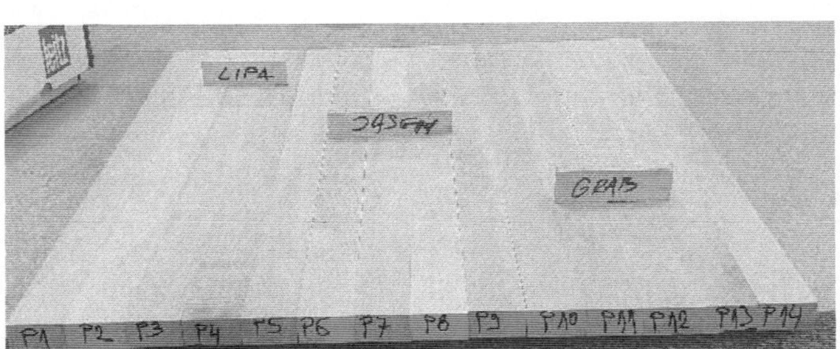

Figure 1. Samples of solid wood

The humidity of the samples was measured with an electric hygrometer, on the surface immediately before the technological planning operation. Wood density was determined by the volumetric method, where small samples were cut from large samples. The dimensions of the samples are measured with a sliding scale, and their weight with a precise scale.

Spiral cutting heads were used for the planning operation on four-sided planers. The number of cutting knives can be different, which primarily depends on the type of wood being processed and the parameters of the processing regime. In this case, the number of spiral erasers was 4; 6 and 8.

Figure 2. Cutting heads with spiral cutting knives

The Profimat 23 Fortec planer from the German manufacturer Weining GmbH has five working shafts, with a shaft diameter of 40 mm, on which cutting heads with knives (cutting knives) are mounted. This planer has an NC control system with electronic two-axis control, using a Siemens simatic OP7 programmer. The feeding speed and cutting speed, i.e., the number of revolutions of the working spindles are automatically adjusted by a potentiometer on the display of the machine's control table.

Figure 3. Four-sided planer Profimat 23 Fortec

The roughness parameters Ra were measured at three measuring points, at a distance of 15 cm from the ends, and in the middle of the sample using a contact profilimeter with a TMR 120 nedle. The characteristics of the profilimeter with a TMR 120 needle are: measurement parameters (µm): Ra; Rz; Rq and Rt; stroke length (mm): 6; sampling length (mm): 0.25; 0.80 and 2.50; estimated length (mm): 1.25; 10.0; measuring range (µm): 0.05-15.0 (Ra and Rq), respectively 0.1-100.0 (Rz and Rt); display error ± 15 %; display variation ≤ 12 %. At the Biotechnical faculty in Ljubljana, the roughness of the processed surface was measured with a LEXT 3D Measuring Laser Microscope OLS5000 laser profilimeter.

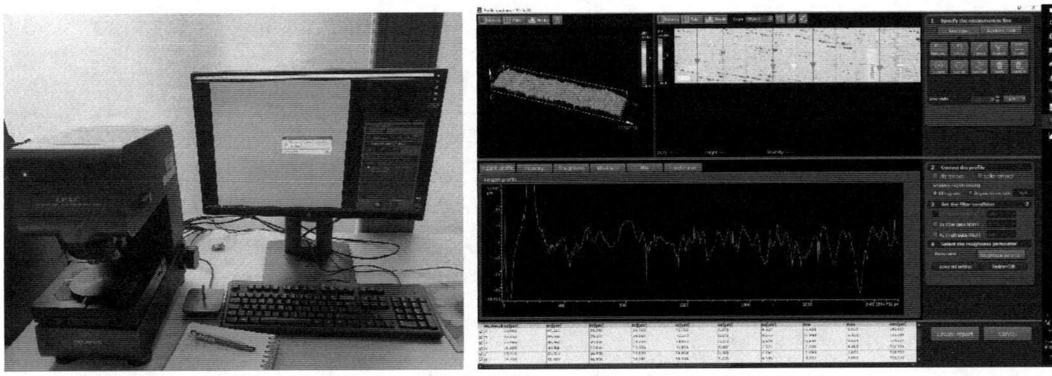

Figure 4. Roughness meter LEXT 3D Measuring Laser Microscope OLS5000

The roughness parameter Ra was measured on five measurement areas, and the mean values of all eight measurements for Ra were taken as relevant values.

When planning an experiment, there are several ways to process data. As it was shown in the work setting that we need to use three input factors that we varied on three levels, minimum, maximum and medium, the central composite plan (CCD – Central Composite Design) was used to create the model. It is a modified experiment plan used for greater precision of the experiment and to cover a larger experimental space [4].

2.2. *Design and results of the experiment*

When performing the planing operation on the Profimat 23 Fortec four-sided planer, processing was carried out by cutting with the separation of grain on three types of wood of different density, on samples with clean dimensions of 620 x 60 x 24 mm. Disturbing processing parameters during the experiment do not change and have constant values: cutting speed v; cutting height h; wood moisture u; cutter diameter D; maximum chip thickness hmax < 0.8 mm; cutting force Frez; cutting angle δ; displacement per one cutter e; dullness of the blade Sharpness Zoštr < 0.5 h of work and geometry of the cutter.

During the experiment, three variables of input processing parameters were examined: wood density ρ, feeding speed s' and number of spiral cutting knives z, and their influence on the dependently variable characteristics of roughness, as the output value of the mean arithmetic deviation of the profile Ra.

The total number of trials in this experiment was determined by three independent variables. The physical and coded values of the parameters of the process of planning solid wood on a multi-sided planer vary on three levels min-medium-max. Total number of trials [2]:

$$N = 2^k + n_0 = 2^3 + 6 = 14 \text{ experiments} \tag{2}$$

where is:

$k = 3$ – number of changing variables,

$n_0 = 6$ – the number of repetitions in the central point of the plan.

The formalized state, describing the real process of solid wood milling, is defined by the following quantities:

a) Input quantities:
- wood density ρ (gr/cm^3).
- feed rate s' (m/min),
- number of spiral cutting knives z,

b) Output quantities:
- arithmetic mean deviation Ra (μm),

c) Process state function: $Ra = f(\rho, s', z)$. (3)

IOP Conf. Series: Materials Science and Engineering 1298 (2023) 012010 doi:10.1088/1757-899X/1298/1/012010

The measured values of humidity, tests using an electric hygrometer, and the density of tests using the volumetric method are shown in the following Table 1.

Table 1. Measured values of humidity, mass and density of wood

The type of wood	Humidity % (average value)	Mass in kg (mean value)	Density in kg/m^3 (average value)
Linden	12,32	0,0609	390,23
Ash	11,96	0,0999	640,21
Hornbeam	12,54	0,1233	790,19

Table 2. Physical and coded values of Xji

Planing process variables			Encoded values		
			X_{j2}	X_{j3}	X_{j1}
		-	-1	0	1
Natural values	**Influential values**	ρ **(X1i)** (gr/cm^3)	0,39	0,64	0,79
		s' **(X2i)** (m/min)	13	23	33
		z **(X_{3i})** (number of cutting knives)	4	6	8

According to the determined plan of the experiment, the selected input parameters vary, while other factors are kept constant during the execution of the process. This formalized description of a real process with already given input quantities of a defined experiment, and in order to apply optimal technology in a given processing process, requires the selection of a sufficiently accurate and reliable mathematical model, because this is a condition for the existence of a set of multivariate solutions from which it is possible to optimally define it. The experimental measurement results for the roughness parameter Ra are given in the following table:

Table 3. Experimental measurement results for the Ra parameter

Ni	ρ	s'	z	*Ra*
	gr/cm3	m/min	number of cutting knives	µm
P.1	0,39	13	4	5,50
P.2	0,79	13	4	3,78
P.3	0,39	33	4	5,88
P.4	0.79	33	4	3,83
P.5	0,39	13	8	7,15
P.6	0,79	13	8	4,99
P.7	0,39	33	8	7,30
P.8	0,79	33	8	4,67
P.9	0,64	23	6	4,42
P.10	0,64	23	6	4,44
P.11	0,64	23	6	4,24
P.12	0,64	23	6	4,47
P.13	0,64	23	6	4,48
P.14	0,64	23	6	4,80

For the value of Ra, a linear model was created based on the obtained roughness data. For the mathematical model, the homogeneity of the dispersions was determined according to Cochran's criterion and the adequacy of the model based on the criterion of the multiple regression coefficient. Statistical data processing will be performed according to the variance analysis method and the significance of the model coefficients according to the p coefficient based on Fisher's criterion and the defined level of significance α.

3. Mathematical modeling

Mathematical modeling means the mathematical description of physical, technological, economic and other processes and systems, which result in an analytical, stochastic, numerical, graphical, statistical, genetic or other model [3]. In other words, it is a description of the legality of changes in process or system parameters in a certain time and space. On the basis of defined mathematical models, further optimization, simulation, revitalization and management of processes and systems are carried out.

For models, it is possible to improve the performance of the system in exploitation in such a way that selected parts of the system are modeled and based on the models thus set, analysis is carried out and conclusions are drawn about the activities that should be undertaken on the real system. They can be physical and abstract. Physical models are proportional copies of physical quantities in an experimental trial and as such are limited in application by high production costs. Due to their simplicity and relatively low costs of the experiment, abstract models are used far more in practice, which can be descriptive, analog and mathematical [4].

The results of stochastic modeling are mathematical models, and the main goal is to determine the legality between the input and output parameters. The input parameters determine the conditions of the process and the output results of the process. The dependence of the output parameters on the input parameters can be shown as based on the expression [5]:

$$y_i = f(x_i) \tag{4}$$

By replacing the input/output physical quantities with independently/dependently variable ones, and we present the process state function F = f(ρ, s', z) in polynomial, the formalized description turns into a mathematical one, based on the expression [3]:

$$Y = b_0 + b_1X_1 + b_2X_2 + b_3X_3 + b_{11}X_1^2 + b_{22}X_2^2 + b_{33}X_3^2 + b_{12}X_1X_2 + b_{13}X_1X_3 + b_{23}X_2X_3 + b_{123}X_1X_2X_3 \tag{5}$$

where are:

- Y=F- dependently variable quantity that corresponds to the physical, i.e. experimental, value of tensile fracture force,
- x_1, x_2, x_3, - independent variables that correspond to the physical values of the processing mode T,t,ρ
- b_0, b_1, b_2, b_3, b_{11}, b_{22}, b_{33} -. regression coefficients of the mathematical model

Formalized description F = f(ρ, s', z) turns into mathematical Y = f (x1, x2, x3), which besides the influence on both quantities x1, x2, x3 separately, also describes the influence of thei interactions x12, x11, x22 on the output value Y.

By introducing coded values according to the arrangement of the experiment plan and the position of the value of the coded quantity, a basic table is obtained for further statistical data processing.

Table 4. Experiment plan with coded quantities for Ra

Ni	*X0*	*X1*	*X2*	*X3*	*X1 X2*	*X1 X3*	*X2 X3*	*Ra*	*Ra*^{2}
1	1	-1	-1	-1	1	1	1	5,50	30,25
2	1	1	-1	-1	-1	-1	1	3,78	14,29
3	1	-1	1	-1	-1	1	-1	5,88	34,57
4	1	1	1	-1	1	-1	-1	3,83	14,67
5	1	-1	-1	1	1	-1	-1	7,15	51,12
6	1	1	-1	1	-1	1	-1	4,99	24,90
7	1	-1	1	1	-1	-1	1	7,30	53,29
8	1	1	1	1	1	1	1	4,67	21,81
9	1	0	0	0	0	0	0	4,42	19,53
10	1	0	0	0	0	0	0	4,44	19,71
11	1	0	0	0	0	0	0	4,24	17,97
12	1	0	0	0	0	0	0	4,47	19,98
13	1	0	0	0	0	0	0	4,48	20,07
14	1	0	0	0	0	0	0	4,80	23,04
							Σ	69,95	365,2

By multiplying the coded values with the experimental value of the roughness parameter Ra, we obtain the data with which we will define the regression coefficients of the model.

Table 5. Determining the regression coefficients of the Ra model

Ni	*X0· Ra*	*X1· Ra*	*X2· Ra*	*X3· Ra*	*X1 X2· Ra*	*X1 X3· Ra*	*X2 X3· Ra*
1	5,50	-5,50	-5,50	-5,50	5,50	5,50	5,50
2	3,78	3,78	-3,78	-3,78	-3,78	-3,78	3,78
3	5,88	-5,88	5,88	-5,88	-5,88	5,58	-5,88
4	3,83	3,83	3,83	-3,83	3,83	-3,83	-3,83
5	7,15	-7,15	-7,15	7,15	7,15	-7,15	-7,12
6	4,99	4,99	-4,99	4,99	-4,99	4,99	-4,99
7	7,30	-7,30	7,30	7,30	-7,30	-7,30	7,30
8	4,67	4,67	4,67	4,67	4,67	4,67	4,67
9	4,42	0	0	0	0	0	0
10	4,44	0	0	0	0	0	0
11	4,24	0	0	0	0	0	0
12	4,47	0	0	0	0	0	0
13	4,48	0	0	0	0	0	0
14	4,80	0	0	0	0	0	0
Σ	*69,95*	*-8,56*	*0,26*	*5,12*	*-0,8*	*-1,32*	*-0,57*
bi	*4,996*	*-1,07*	*0,00325*	*0,64*	*-0,1*	*-0,165*	*-0,0712*

The penultimate row in Table 5 represents the sum of the corresponding column and the last row represents the regression coefficient for the corresponding coded quantity. The regression coefficients bi are the mean values of individual columns, because for coded quantity X_0 we take the mean value of 14 repetitions of the experiment, and for other coded quantities and interactions between quantities, the mean value of 8 repetitions of the experiment, i.e. from the basic plan of the experiment. Therefore, the linear form of the model of the roughness parameter Ra can be written in the form:

$$Y = 4{,}996 - 1{,}07X_1 + 0{,}64X_3 - 0{,}1X_1X_2 - 0{,}165X_1X_3 - 0{,}0712X_2X_3 \tag{6}$$

In order to determine the significance of the regression coefficients for the linear model, we will perform an ANOVA analysis of variance for the given experiment. Variance analysis involves calculating the sum of squares, degrees of freedom and mean square value for given input factors, interactions between them, the overall model and the deviation from the model and the pure error of the model. When labeling, we will use the letters A, B and C for the input factors [6].

The sum of squares (SS – English Sum of Squares) of the input factors and the interaction between them is calculated using the equation:

$$SS = \frac{(\sum X_i \cdot R_a)^2}{N_o} \tag{7}$$

where is:
SS – sum of squares,
N0 – the number of basic repetitions of the experiment (in our case N0= 8)

Based on the formula (7) the sum of the squares is : SSA = 9,159; SSB = 0,008; SSC = 3,277, and for the interactions of factors A, B and C the sum of squares is SSAB = 0,08; SSAC = 0,2178; SSBC = 0,041;

The sum of squares of the model is obtained by adding the individual and interaction sums of squares:

$$SS_{Model} = SS_A + SS_B + SS_C + SS_{AB} + SS_{AC} + SS_{BC} = 12{,}78 \tag{8}$$

The total value of the sum of squares of the experiment is obtained using the equation:

$$SS_{total} = \sum R_a^2 - \frac{(\sum R_a)^2}{N} \tag{9}$$

where N = 14 and represents the total number of repetitions of the experiment.

The total value of the sum of squares of the experiment is:

$$SS_{total} = 365{,}2 - \frac{69{,}95^2}{14} = 15{,}699 \tag{10}$$

After calculating the total value of the sum of the squares of the experiment and the sum of the squares of the model, the sum of the squares of the remainder of the experiment can be calculated as follows:

$$SS_{remainder} = SS_{total} - SS_{model} = 15{,}699 - 12{,}780 = 2{,}9198 \tag{11}$$

The remainder of the analysis of variance can be divided into two parts. One part refers to the deviation from the model (Lack of Fit) and the other part represents a pure error (Pure Error). The pure error of the experiment refers to the central part of the experiment, i.e. the part where the input data is repeated, measurement 9, 10, 11, 12, 13 and 14. In order to calculate the sum of the squares of the pure error of the experiment, we will extract only the results of the experiment of the central part and based on table 6 arrive at the value of the sum of squares of the pure error..

Table 6. Data for calculating the sum of squares of the pure error of the experiment

Ni	R_a	$(R_{ai} - -R_a)^2$
9	4,42	0,06208
10	4,44	0,0122
11	4,24	0,0728
12	4,47	0,03125
13	4,48	0,0537
14	4,80	0,0223
	$-R_a = 4{,}475$	$\Sigma = 0{,}2544$

The sum from the last column of table 6 represents the sum of the squares of the pure error of the experiment, that is:

$$SS_{pure\ err} = 0{,}2544 \tag{12}$$

The sum of the squares of deviations from the model is obtained by subtracting the sum of the squares of the pure error from the sum of the squares of the experimental error.

$$SS_{Deviation} = SS_{Remainder} - SS_{pure\ err} = 2{,}9198 - 0{,}2544 = 2{,}6654 \tag{13}$$

The total degree of freedom of the experiment is obtained using the equation:

$$dF_{total} = N - 1 = 14 - 1 = 13 \tag{14}$$

where N is the total number of repetitions of the experiment.

The degrees of freedom of factors A, B and C and the interaction between the factors are determined based on the basic levels of the factors. Basic levels of input factors are mini-mum and maximum, so k = 2.

The degrees of freedom will be:

$$dF_i = k - 1 = 2 - 1 = 1 \tag{15}$$

The degree of freedom of the model is obtained as the sum of the degrees of freedom of the three input factors and the three interactions between the factors.

$$dF_{model} = dF_A + dF_B + dF_C + dF_{AB} + dF_{AC} + dF_{BC} = 6 \tag{16}$$

The degree of freedom of the remainder of the experiment is obtained by subtracting the degree of freedom of the model from the total degree of freedom.

$$dF_{Remainder} = dF_{Total} - dF_{model} = 13 - 6 = 7 \tag{17}$$

The degree of freedom of the rest of the experiment consists of two parts, one is the deviation from the model and the other is pure error.

The degree of freedom of pure error is obtained when the number of repetitions of the experiment in the central point (NC) subtract from 1. In this experiment NC = 6.

$$dF_{pure\ err} = N_C - 1 = 6 - 1 = 5 \tag{18}$$

The degree of freedom of deviation from the model is obtained when we subtract the degree of freedom of pure error from the degree of freedom of the remainder of the experiment.

$$dF_{Deviation} = dF_{Remainder} - dF_{Pure\ err} = 7 - 5 = 2 \tag{19}$$

After determining the sum of squares and the degrees of freedom, it is easy to calculate the mean value of the squares (MS - English Mean of Squares), which represents the quotient of the sum of the squares and the degrees of freedom.

$$MS = \frac{SS}{dF} \tag{20}$$

The total mean square value of the experiment is obtained according to the equation.

$$MS_{total} = \frac{SS_{total}}{dF_{total}} = \frac{15{,}699}{13} = 1{,}2076 \tag{21}$$

The mean values of the squares of the input factors and the interaction between the factors are obtained according to the equations:

$$MS_A = \frac{SS_i}{dFi} \tag{22}$$

and is: MSA = 9,159; MSB = 0,008; MSC = 3,27; MSAB = 0,08; MSAC = 0,2178 and MSBC = 0,041.

The mean square value of the model is obtained according to the expression:

$$MS_{model} = \frac{SS_{model}}{dF_{model}} = \frac{12{,}78}{6} = 2{,}13 \tag{23}$$

Finally, the mean squared values of the remainder of the experiment, the pure errors of the experiment and the deviations from the model are obtained according to the expressions:

$$MS_{Remainder} = \frac{SS_{Remainder}}{dF_{Remainder}} = \frac{2{,}918}{7} = 0{,}416 \tag{24}$$

$$MS_{Pure\ err} = \frac{SS_{Pure\ err}}{dF_{Pure\ err}} = \frac{0{,}2544}{5} = 0{,}05088 \tag{25}$$

$$MS_{Deviation} = \frac{SS_{Deviation}}{dF_{Deviation}} = \frac{2{,}6654}{2} = 1{,}3327 \tag{26}$$

After we have calculated the mean values of the squares, we can do the last step in the variance analysis, which is the determination of the calculated value of Fisher's F coefficient. In these cases, it is sufficient to determine the computational F coefficient of the model, input factors and interaction factors. The calculated value of the F coefficient is obtained as the quotient of the corresponding mean square value and the mean square value of the remainder of the experiment.

$$F_{model} = \frac{MS_{model}}{MS_{remainder}} = 5{,}1201 \tag{27}$$

Based on the formula (27), the values of the coefficients Fi for the input factors A, B, C and their mutual interactions are obtained: FA = 22,017; FB = 0,0192; FC = 7,8606; FAB =0,192; FAC =0,523 and FBC=0,09855.

After the Fisher's coefficients are determined, the statistical coefficient p is calculated, by which it can be seen whether and which factors are significant. The values of the statistical coefficient p as well as all other values obtained by analysis of variance are shown in table 7.

Table 7. Results of the analysis of variance for the linear model of the roughness parameter Ra

Source of variation	*SS*	*dF*	*MS*	*F*	*Significance*
A	9,159	1	9,159	22,017	significant
B	0,008	1	0,008	0,0192	not significant
C	3,277	1	3,277	7,8606	significant
AB	0,08	1	0,08	0,192	not significant
AC	0,2178	1	0,2178	0,523	not significant
BC	0,041	1	0,041	0,09855	not significant
Model	12,78	6	2,13	5,1201	significant
Remainder	2,9198	7	0,416		
Deviation from the model	2,6654	2	1,3327		
Pure error	0,2544	5	0,05088		
Total	15,699	13	1,206		

In this example, $\alpha = 0.05$ or 5% was taken. The parameter from Table 8 is significant if its statistical value p is less than $\alpha = 0.05$. This means that factors A and C will be significant for the linear model of the roughness parameter Ra. What is important is the final model. After determining the significance of the input parameters in equation (5), only those regression coefficients whose magnitudes are significant are taken into account, so expression (5) will be transformed into:

$$Y = 4{,}996 - 1{,}07X_1 + 0{,}64X_3 \tag{28}$$

Decoding X_1, X_2 and X_3 from the equation (46) is done according to the expression [2]:

$$X_i = \frac{f_i - \dfrac{(f_{imin} + f_{imax})}{2}}{\dfrac{(f_{imax} - f_{imin})}{2}} \tag{29}$$

where f_{imin} are f_{imax} are the lower and upper value of the level of the input factor, the values of the quantities are obtained X1 = 5ρ – 2,95; X2 = s' – 2,3 i X3 = 0,5 z – 3. By including them in equation (28), the final model of the influence of the input factors of planning processing on a multi-sided grinding machine on the roughness parameter of the processed surface Ra is obtained:

$$Ra = 6{,}179 - 5{,}35\rho + 0{,}32n \tag{30}$$

By including the given values for the wood density, displacement and cutting speed in equation (30), we can obtain the values of the surface roughness parameter Ra in [μm] according to the shown linear model. Table 9 shows the values of Ra according to the model and according to the experimental results.

Table 8. Comparative presentation of experimental and linear parameter model results Ra

Pon	*A[g/cm^3]*	*B[m/min]*	*C*	*Ra exp [μm]*	*Ra model [μm]*
1	0,39	13,0	4	5,50	5,37
2	0,79	13,0	4	3,78	3,23
3	0,39	33,0	4	5,88	5,37
4	0,79	33,0	4	3,83	3,23
5	0,39	13,0	8	7,15	6,65
6	0,79	13,0	8	4,99	4,51
7	0,39	33,0	8	7,30	6,65
8	0,79	33,0	8	4,67	4,51
9	0,64	23,0	6	4,42	4,68
10	0,64	23,0	6	4,44	4,68
11	0,64	23,0	6	4,24	4,68
12	0,64	23,0	6	4,47	4,68
13	0,64	23,0	6	4,48	4,68
14	0,64	23,0	6	4,80	4,68

As already emphasized earlier, after obtaining the model equation, it is useful to check the homogeneity of dispersions using Cochran's criterion and the adequacy of the model by determining the multiple regression coefficient.

Checking the homogeneity of dispersions according to the Cochran criterion is performed on the basis of the expression:

$$K_h = \frac{max\sigma_j^2}{\sum_{j=1}^{n} \sigma_j^2} \leq K_t(dF_j, n) \tag{31}$$

where is:
K_t- tabular value according to Cochran's criterion for degrees of freedom dFj and N,
dFj – degree of freedom (dFj = n-1)
n – number of repetitions
σ_j^2 – sample variance.

For this example, this means that repetitions at the central point will be taken and the magnitude of Kh will be determined and compared to the table. For this determination, we need the experimental and model values of the parameter Ra in the central points, table 9.

According to the data in Table 9, Cochran's coefficient is [2]:

$$K_h = \frac{0{,}1024}{0{,}232} = 0{,}441 \tag{32}$$

Table 9. Data for calculating Cochran's coefficient

Pon.	Ra	$\sigma_j^2 = (R_{ai} - R_a)^2$
9	4,42	0,032
10	4,44	0,036
11	4,24	0,0576
12	4,47	0
13	4,48	0
14	4,80	0,1024
	$\overline{R_a} = 4{,}47$	$\sum_{j=1}^{n} \sigma_j^2 = 0{,}2322$
		$\max\sigma_j^2 = 0{,}1024$

We will determine the table value Kt based on the degree of freedom dFj = n - 1 = 6 – 1 = 5 and n = 6 so Kt (5,6) =0,445. As $K_h > K_t$ we can conclude that the dispersion is not homogeneous [7].

Table 10. Data for determining the multiple regression coefficient R according to the linear model

Ni	*Experiment*	*Linear model*	$(Y_j^E - Y_j^M)^2$	$(Y_j^E - -Y^E)^2$
	Ra [μm] = Y_j^E	Ra [μm] = Y_j^M		
1	5,50	5,37	0,0169	0,254
2	3,78	3,23	0,3025	1,4786
3	5,88	5,37	0,2601	0,7815
4	3,83	3,23	0,3600	1,3596
5	7,15	6,65	0,2500	4,6397
6	4,99	4,51	0,2304	0
7	7,30	6,65	0,4225	5,3084
8	4,67	4,51	0,0256	0,1063
9	4,42	4,68	0,0676	0,3318
10	4,44	4,68	0,0576	0,3091
11	4,24	4,68	0,1936	0,5715
12	4,47	4,68	0,0441	0,2767
13	4,48	4,68	0,0400	0,2663
14	4,80	4,68	0,0144	0,0384
	$-Y^E = 4{,}996$		Σ = 2,2853	Σ =15,7219

Adequacy of the model based on the multiple regression coefficient is determined by calculating the value of R and calculated according to the equation:

$$R = \sqrt{1 - \frac{\sum_{j=1}^{n} \left(Y_j^E - Y_j^M\right)^2}{\sum_{j=1}^{n} \left(Y_j^E - -Y^E\right)^2}} \tag{33}$$

where are: Y_j^E – experimental value j – that repetition; Y_j^M– model value j – that repetition; $-Y^E$ – mean value of experimental replicates.

Therefore, the multiple regression coefficient of the logarithmic model can be calculated as:

$$R = \sqrt{1 - \frac{2{,}2853}{15{,}7219}} = 0{,}92446 \tag{34}$$

The value of the regression coefficient R=0.92446 means that the linear model describes the accuracy of the experiment results with 92.45%, which is a very good accuracy of the model [8].

4. Results and discussion

For the purposes of this work and to confirm the obtained statistical results, the Design Expert software package was used. It is a commercial software used for experiment planning and optimization of results, as well as statistical processing and visual display [9].

Table 11. Display of input of experimental results roughness Ra

		Factor 1	*Factor 2*	*Factor 3*	*Response 1*
Std	Run	A:Wood density	B:Shift	C:Number of spirals	Roughness Rz
1	3	-1	-1	-1	5,5
2	4	1	-1	-1	3,78
3	9	-1	1	-1	5,88
4	2	1	1	-1	3,83
5	11	-1	-1	1	7,15
6	12	1	-1	1	4,99
7	5	-1	1	1	7,3
8	1	1	1	1	4,67
9	6	0	0	0	4,42
10	13	0	0	0	4,44
11	8	0	0	0	4,24
12	7	0	0	0	4,47
13	10	0	0	0	4,48
14	14	0	0	0	4,8

Table 12. Results of the analysis of variance for the linear model of the roughness parameter Ra

ANOVA for Linear model
Response 1: R1

Source	***Sum of Squares***	***df***	***Mean Square***	***F-value***	***p-value***	
Model	12,44	3	4,15	12,66	0,0010	significant
A-A	9,16	1	9,16	27,95	0,0004	
B-B	0,0084	1	0,0084	0,0258	0,8756	
C-C	3,28	1	3,28	10,00	0,0101	
Residual	3,28	10	0,3277			
Lack of Fit	3,11	5	0,6225	18,85	0,0029	significant
Pure Error	0,1651	5	0,0330			
Cor Total	15,72	13				

The Model F-value of 12,26 implies the model is significant. There is only a 0,10% chance that an F-value this large could occur due to noise. P-values less than 0,0500 indicate model terms are significant. In this case A, C are significant model terms. Values greater than 0.1000 indicate the model terms are not significant. If there are many insignificant model terms (not counting those required to support hierarchy), model reduction may improve your model. Figure 5 shows the spatial position of experiment repetition points for the selected type of model with 14 experiment repetitions.

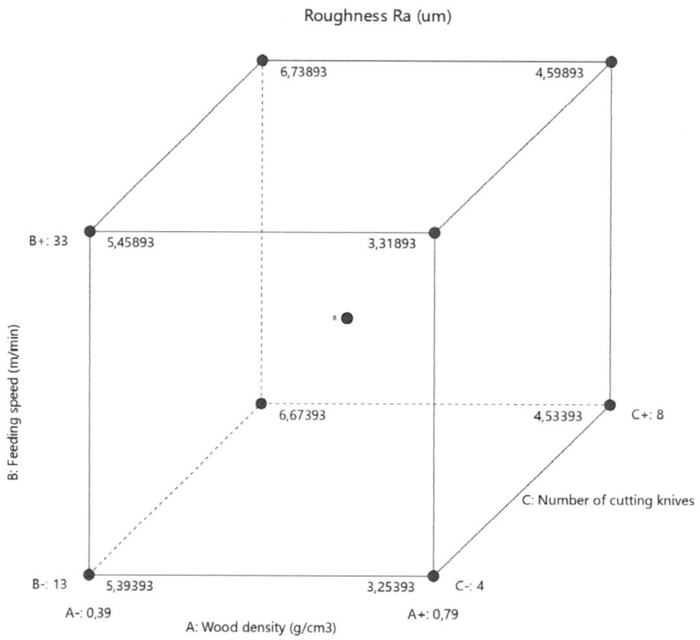

Figure 5. Spatial representation of the repetition of the experiment of the linear roughness model Ra

The arrangement of the results of the experiment and the type of distribution can be shown histogram, Figure 6.

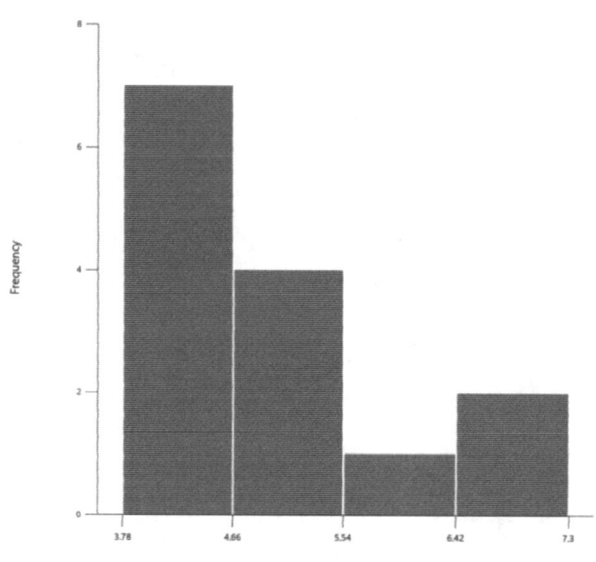

Figure 6. Histogram distribution of the results of the experiment of the linear roughness model Ra

IOP Conf. Series: Materials Science and Engineering 1298 (2023) 012010 doi:10.1088/1757-899X/1298/1/012010

In the theory of statistical data processing, there is the term FDS (Fraction of Design Space) diagram, that is, the diagram of the distribution of variance in the experimental space, Figure 7. The diagram shows the dependence of the distribution of the variance of the model and the error of the experiment.

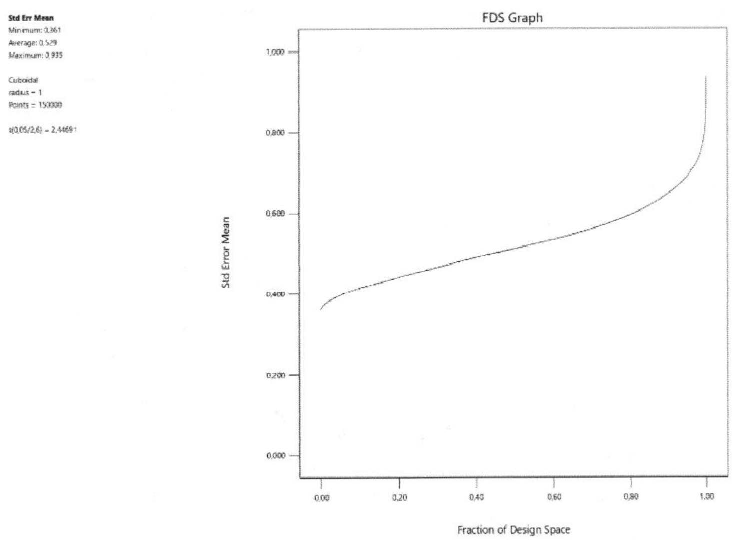

Figure 7. FDS diagram of the experiment of the linear roughness model Ra

By means of a disturbance diagram (eng. Perturbation Plot) the effects of all input factors at a certain point of the experimental space can be compared. Figure 8 shows the disturbance plot of the input factors of the model for the central point.

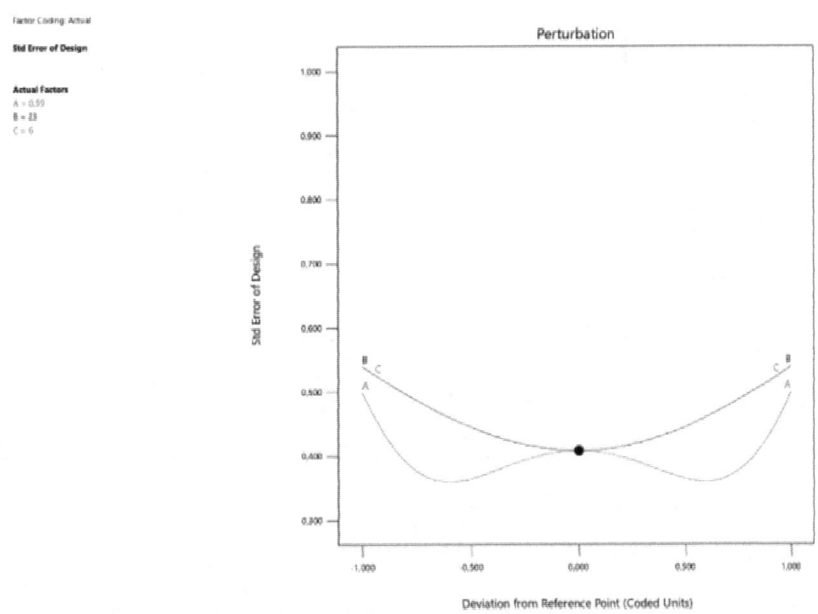

Figure 8. FDS diagram of the experiment of the linear roughness model Ra

The diagrams in Figure 9 show the dependence of the output value of the model, the rough-ness parameter Ra, on the input values, wood density, displacement, and number of spirals. On the abscissa of the input values, the minimum value is shown with "-1" and the maxi-mum value of the input value with "+1". It can be seen from the diagram that the value of the roughness parameter depends significantly on the density of the wood and the number of spirals of the tool. As the wood density value increases, Ra decreases, and as the number of tool spirals increases, Ra increases.

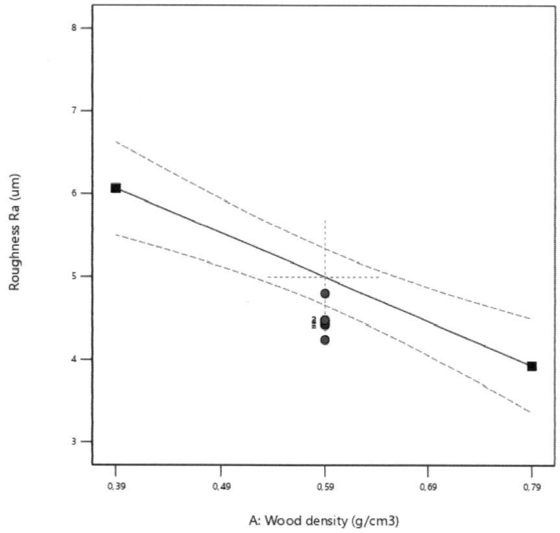

a)

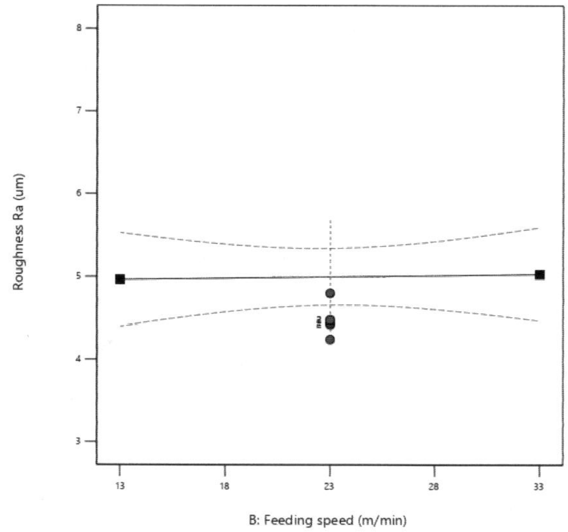

b)

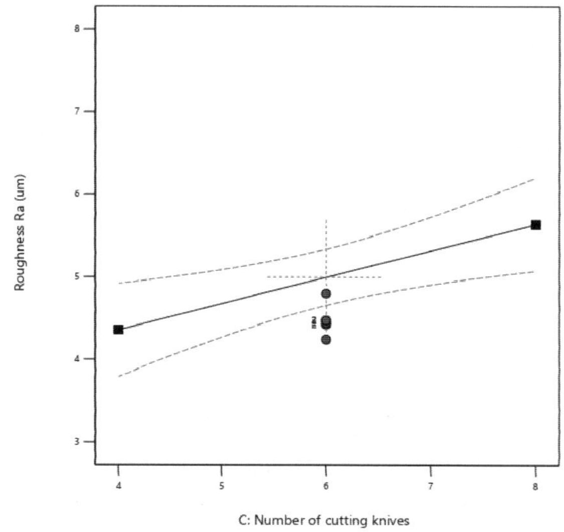

c)

Figure 9. Dependence of the output value of the roughness model Ra on the input parameters of the experiment a) wood density, b) displacement, c) number of spirals

It is also possible to graphically display the arrangement of experimental and model values of the roughness parameter Ra, Figure 10. This diagram helps to reveal the states of the experiment in which the output values do not match well with the experimental quantities. According to the theory, the data points should be evenly distributed in relation to the line, which should be at an angle of 45°. In the example of the linear model, the output values of the roughness parameter Ra on the horizontal axis are the experimental values and on the vertical axis the model values and the arrangement of the model values in relation to the experimental ones is good.

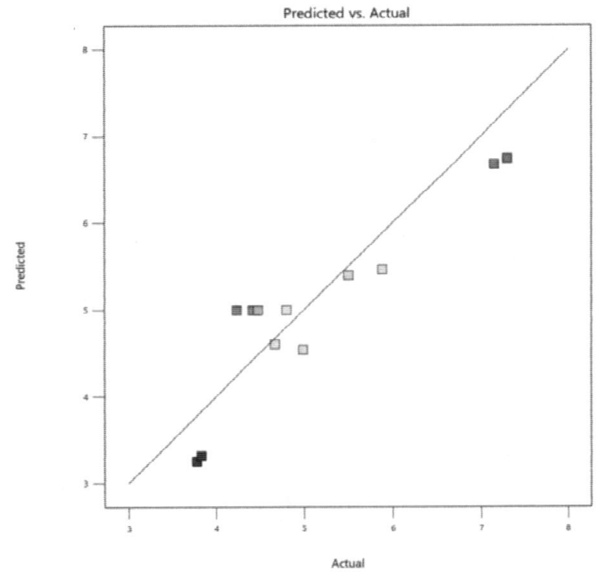

Figure 10. Schedule of experimental and model values of the roughness parameter Ra

Figure 11 shows the normal probability plot of the model showing whether the experimental remainders follow a normal distribution. In that case, the points follow a straight line, so it is to be expected that at certain points there will be a certain distance from the straight line, as in the case shown, which is a consequence of the deviation of the model values from the experimental values in some repetitions of the experiment.

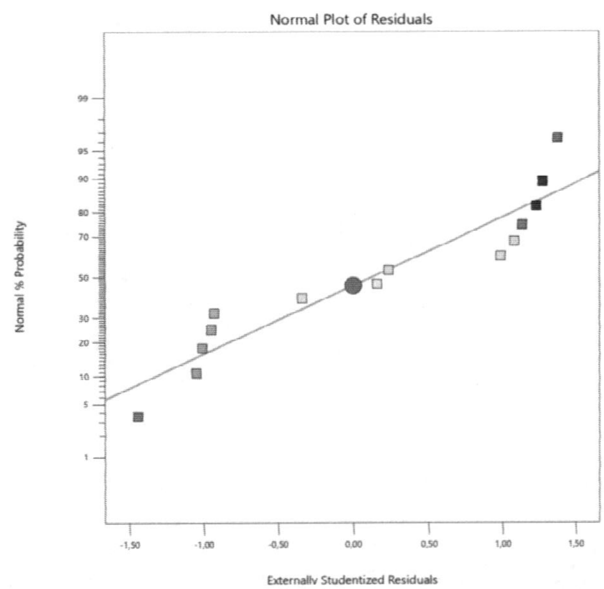

Figure 11. Normal distribution of experimental residuals

It is also possible to graphically display the relationship between the predicted model values of Ra and the rest of the experiment. This graph is shown in Figure 12.

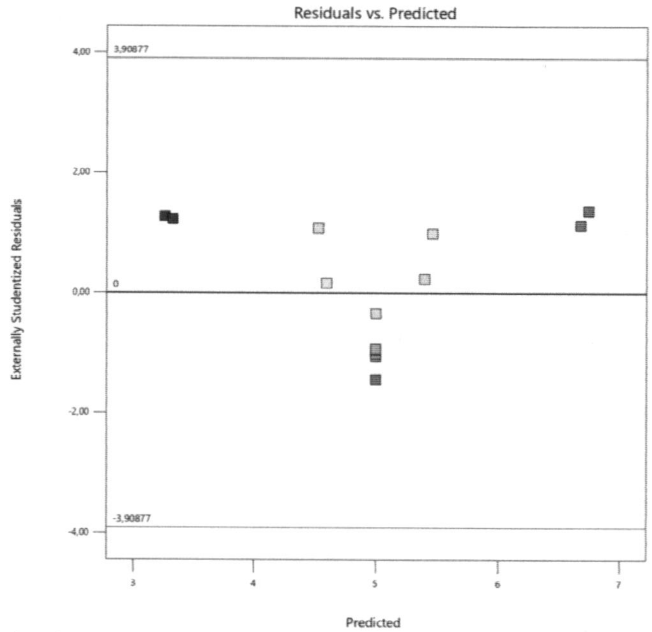

Figure 12. Relation between the model values of Ra and the remainders of the experiment

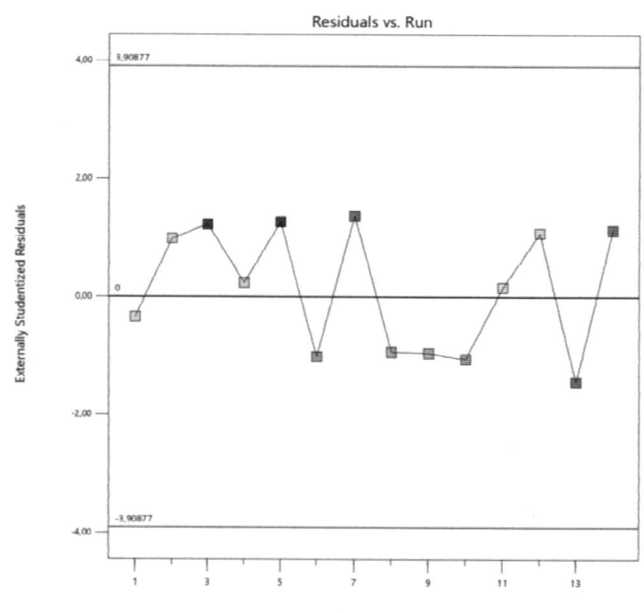

Figure 13. The ratio of the number of repetitions of the experiment to the remainders of the experiment

All diagrams from Figures 5-13. were generated in Design Expert software based on experimental and model results of the roughness parameter Ra. What can be seen from the diagram confirms the mathematically obtained results, and the basic conclusion is that the obtained model for the roughness

IOP Conf. Series: Materials Science and Engineering 1298 (2023) 012010 doi:10.1088/1757-899X/1298/1/012010

parameter Ra is adequate with high accuracy for the linear model. There are certain points in the experiment for which the model result deviates more from the experimental one, which is quite natural due to the process itself and possible measurement error. If the points where the model results deviate more from the experimental results were to be removed from the analysis, the accuracy of the model would be even higher, but this is not necessary. For each model, the most significant influence is the magnitude of the shift, whether it is a single input factor or an interaction between input factors.

Table 13 shows the experimental values and the values obtained using the model, as well as the remainder between the results for each repetition.

Table 13. Comparison of experimental and model results of parameter Ra

Rep.	*Exp. value*	*Model value*	*Remainder*
1	5,5	5,39	0,1061
2	3,78	3,25	0,5261
3	5,88	5,46	0,4211
4	3,83	3,32	0,5111
5	7,15	6,67	0,4761
6	4,99	4,53	0,4561
7	7,3	6,74	0,5611
8	4,67	4,60	0,0711
9	4,42	5	-0,5764
10	4,44	5	-0,5564
11	4,24	5	-0,7564
12	4,47	5	-0,5264
13	4,48	5	-0,5164
14	4,8	5	-0,1964

5. Conclusion

By planning and performing an experiment on three levels of measuring the quality of roughness, the obtained measurement results were used to create a mathematical model of the significance of machinability parameters on the quality of the processed surface of solid wood in the process of planning on a multi-sided planer. By analyzing the experimental results, it can be concluded:

- Mathematical modeling, with the application of experimental research on the real model, will confirm the legality of interlinking the input parameters of the machinability of solid wood with the output size of the quality of the machined surface, where the input parameters define the conditions of the process taking place, and the output parameter is the result of the modeled process.
- The F value for the model of the roughness parameter Ra of 12.26 implies that the model is significant. There is only a 0.10% chance that an F value of this magnitude could occur due to some other disorder. In this case, the input parameters wood density and number of cutting knives are significant model sizes.
- The results obtained from experimental testing can be used for modeling the rough-ness of the machined surface of all types of solid wood in the planning process on clean-sided planers.

A reliable mathematical model of the mean arithmetical deviation of the profile from the mean line of the profile Ra was obtained, which can serve technologists, depending on the process parameters, to start the production process in a short time without long-term testing, which today in practice represents a basic, long-term, and unreliable approach to the problem.

IOP Conf. Series: Materials Science and Engineering 1298 (2023) 012010 doi:10.1088/1757-899X/1298/1/012010

References

[1] Gürgen, A. et al. (2021) 'Optimization of CNC operating parameters to minimize surface roughness of pinus sylvestris using integrated artificial neural network and genetic algorithm', Maderas. *Ciencia y tecnología*, 24. doi:10.4067/s0718-221x2022000100401.

[2] Jones, B. and Montgomery, D, (2019) 'Design of Experiments - Modern Approach',*Wiley.*

[3] Jurković, M. (1999) 'Matematičko modeliranje inženjerskih procesa i sistema', *University of Bihać, Faculty of Mechanical Engineering.*

[4] Bjelić, A., and Hodžić, A. (2017) 'The influence of the choice of cutting tool on the quality of the machined surface of massive wood' ('Uticaj izbora reznog alata na kvalitet obrađene površine masivnog drveta') *11th International Scientific Conference on Production Engineering–RIM 2017.*

[5] Hazir, E. and Koc, K.H. (2016) 'Optimization of wood machining parameters in CNC routers: Optimization of wood surface machining parameters in CNC routers: Response surface methodology (RSM) approach', *International Journal of Scientific Research Engineering & Technology (IJSRET)*, 5(10), pp. 0–0. doi:10.4067/s0718-221x2019005000406.

[6] Hasanagić, R., Hodžić, A. and Jurković, M. 'Modelling and optimization of tensile break force of solid wood elements leng the nedbyfinger joint', *Journal of Adhesion Science and Technology, 2019, 33(11), 1013-1027. DOI:10.1080/01694243.2019.1690266*

[7] Škaljić, N. (2010) 'Analysis of influencing parameters on the roughness of the planed wood Surface', *Doctoral dissertation, Faculty of Mechanical Engineering.*

[8] Hasanagić, R., Hodžić, A. and Bjelić, A. "Mathematical modelling of surface roughness in orthogonal cutting of massive wood" Wood Science Journal, 2019, 16(2), 74-80. DOI: 10.1080/17480272.2019.1608297

[9] Ramić, D., Hodžić, D., Hodžić, A. "Modelling of Influential Machining parameters to the Surface Quality of Wood", Journal of Physics: Conference Series, Volume 2212, **DOI** 10.1088/1742-6596/2212/1/012006

IOP Conf. Series: Materials Science and Engineering 1298 (2023) 012011 doi:10.1088/1757-899X/1298/1/012011

Comparative Analysis of Bending Strength Properties between Glued and Un-Glued Finger Joints in Beech and Fir Wood Species

Umejr Šljivo[1], Redžo Hasanagić[1], Leila Fathi[2] and Mohsen Bahmani[2]
[1]Department of Wood Science and Technology, Faculty of Technical Engineering, University of Bihać, Bihać, Bosnia and Herzegovina
[2]Faculty of Natural Resources and Earth Science, Department of Industrial Engineering Furniture, Shahrekord University, Shahrekord, Iran

E-mail: redzo.hasanagic@gmail.com

Abstract. The aim of this study was to investigate the mechanical behavior of beech and fir finger joints under laboratory conditions. The samples were manufactured using a 9 mm finger joint with glued surfaces, in accordance with the EN 14 080 standard. Polyurethane adhesive of class D3, commonly used for the production of exterior wooden structures in Bosnia and Herzegovina, was applied to the samples. The specimens were subjected to destructive four-point bending tests according to the BAS EN 408 standard, and the achieved bending strength was statistically evaluated and compared to the results of unglued samples.

1. Introduction

Construction relies heavily on timber as a building material, but waste timber and discarded sawmill sections can create challenges for the industry. To address this, it is essential to explore ways to utilize these timber residues in construction and optimize resources while reducing waste. Additionally, glued-laminated timber (glulam) has emerged as a popular engineered wood product for structural applications. While traditionally produced using softwood species, there is a growing interest in incorporating hardwood species into glulam production. [1-3] The bonding strength of finger joints is a crucial technical property for longitudinally joined elements. Therefore, selecting the appropriate finger joint composition is of utmost importance, and considerable attention is dedicated to this aspect. [4] To enhance process efficiency and predictability, numerous authors [5-9] have investigated various aspects of finger joint strength using experimental research, mathematical modeling, and numerical simulation. Their research aims to achieve a single objective - improving process efficiency and control to enhance predictability. The bonding strength of solid wood finger joints depends on several key parameters, including height, tooth length, tip thickness, and slope, as discussed in the research paper [10]. When studying the effect of bonding PVAc adhesive to resistance classes D1, D2, and D3 in accordance with Standard EN 204:2001, researchers found that 10 mm long teeth and resistance class D3 of the PVAc adhesives exhibited the highest strength [11-14]. Timber is known to exhibit its maximum strength parallel to the grain. As a result, recent research in this field has focused on developing end joints capable of effectively transmitting a significant portion of this strength [15-23]. The studies referenced in papers [24-27] analyzed how the surface of the ring peak affects the compression strength of bonded parts, as well as the influence of finger parameters on the tensile strength of jointed fingers. The primary goal of this research is to provide a comprehensive analysis of the key factors that significantly affect the bending strength of joined wood, as well as the behavior of toothed joints under real loads, and to compare the results to those of solid wood. Through experimental research, we aim to gain a deeper understanding of the underlying principles that govern the strength and durability of joined wood and solid wood, especially in the context of modern manufacturing techniques and design int orientation, moisture content, and temperature, we aim to identify the optimal conditions for achieving maximum bending strength in joined wood and solid wood, according to the BAS EN 408 standard [28].The findings of our research have practical implications for the production and design of wooden products, such as furniture and carpentry. By leveraging the insights gained from our experiments, manufacturers can optimize the use of joined wood and solid wood in their manufacturing processes, leading to stronger and more durable products. This, in turn, can enhance the quality and longevity of finished wood products and contribute to the overall sustainability of the woodworking industry.

2. Materials and methods

2.1. Materials

In this experiment, we used solid wood joined with finger joints and massive four-point bending samples made from beech (Fagus sp.) and fir (Abies sp.) with average measured densities of 0.755 g/cm^3 and 0.400 g/cm^3, respectively. The rectangular samples, measuring 380 x 20 x 20 mm, were prepared according to BAS EN 408 standards [28] for the four-point bending process. To elongate the solid wood samples, we used a joint with a wedge and a mini tooth with a geometric ratio of 9 mm in length. The average humidity of the samples was in the range of 7-8%. We used Bostik's polyvinyl acetate (PVAC) D3 glue, which has water resistance certified as D4 BS-EN-204-D4, and a bonding time of 7 minutes. The wooden samples used in the experiment were prepared by Malak Janj-Donji Vakuf a company specializing in the production of solid panels consisting of segments connected with toothed joints in both length and width. The company handles the entire production process, from sawmill processing to wood expansion.

2.2. Wood Density Determination

The method for determining the density of the wood samples followed ISO standard 3131:1975 [29]. The wooden samples were cut into cubic shapes with dimensions of 50 x 25 x 15 (mm). After that, the sample is air-dried to a constant mass under conditions controlled according with the standard. Next, the dried sample was weighed on a digital balance with a minimum accuracy of 0.1 g. The density was then calculated by dividing the mass of the sample by its volume, expressed in grams per cubic centimeter (g/cm^3). The volume of the samples was determined by measuring their dimensions using a micrometer, and the volume was calculated using equation (1).

$$\rho = \frac{m}{V}\,[{}^{g}/_{cm^3}] \tag{1}$$

Where:
m= sample mass (g)
V= sample volume (cm^3)

2.3. Method for Determining Moisture Content of Wood Samples

Wooden samples with dimensions of 50 x 25 x 15 mm were taken and weighed on a digital balance with an accuracy of at least 0.1 g. The initial weight of each sample was recorded. Then, the samples were dried in a drying chamber for 24 hours at a temperature of 103±2°C, in accordance with ASTM D2016 standard [30]. After drying, the samples were weighed again and the final weight was recorded. The difference between the initial and final weights of the samples was used to calculate the moisture content using the following equation (2):

$$MC = [(W1 - W2) / W2]\, x\, 100\ \% \tag{2}$$

Where:

- MC - moisture content (%),
- W1 - initial weight of the sample (g),
- W2 - final weight of the sample (g).

2.4. Wood Bending Method According to BAS EN 408 Standard

The wood bending method according to BAS EN 408 standard is used to determine the bending characteristics of wooden elements, such as strength and stiffness. This method is applied to wood samples that have been processed by the dry or wet method, including laminated wood. The test was carried out in the following steps:

- Samples were made from wooden elements of standard dimensions and shapes, as specified in item 2.1.
- The samples were then placed on the SHIMTZU 50kN test machine at the Technical Faculty in Bihać (Figure 1). The testing machine enabled force to be applied to the samples in the direction of their length. The samples were attached to the testing machine so that the applied force was evenly distributed on the sample. The samples were gradually bent under pressure until they broke. The force was applied in accordance with the procedures prescribed by the BAS EN 408 standard [28]. After the samples were subjected to force and fracture occurred, the obtained results were analyzed. The analysis included the determination of the breaking point, maximum breaking force, and modulus of elasticity.

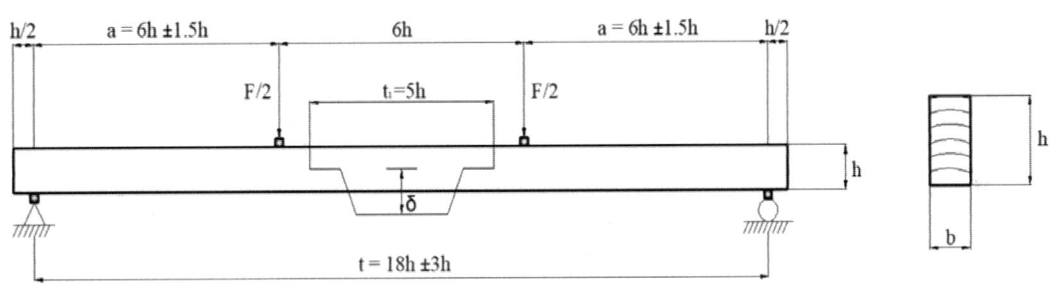

Figure 1. A schematic representation of the four-point bending test configuration [28]

Bending strength was calculated using the formula (3):

$$f_m = \frac{3 \cdot l_A}{(b \cdot h^2) \cdot F_{max}} \tag{3}$$

The modulus of elasticity was calculated using the formula (4):

$$Em, l = \frac{a \cdot l_1^2 (F_2 - F_1)}{16 \cdot I\, (\delta_2 - \delta_1)} \tag{4}$$

Where is:

- F_{max} (N) - maximum force,
- f_m (N/mm^2) - bending strenght,
- $E_{m,l}$ (N/mm^2) - modulus of elasticity,
- F (N) - load force,
- δ(mm) - displacement,
- h (mm) - sample thickness,

IOP Conf. Series: Materials Science and Engineering 1298 (2023) 012011 doi:10.1088/1757-899X/1298/1/012011

- b (mm) - sample width,
- F_2, F_1= load in newtons on the regression line with a correlation coefficient of 0.99
- a (mm) - distance between a loading position and the nearest support in a bending test in millimeters
- I (mm^4) - second moments of area, in millimetres to the fourth power
- l_l (mm) - gauge length for the determination of modulus of elasticity in millimetres.

3. Results and discussion

The studies were conducted to examine the effects of joints on solid wood of fir and beech compared to samples that were not joined. Table 1 shows the results of the four-point bending test, while Figure 2 shows the differences in maximum force on the bending strength between clean and finger-jointed wood species.

Table 1. Test results of fir and beech samples with finger joints and without.

Samples	Glue	Thickness h (mm)	Width b(mm)	ρ (g/cm³)	MC (%)	F max (N)	E (N/mm²)	f_m (N/mm²)
				No finger joint				
B1	D3	20.84	20.31	0.70	7.81	2959.82	14613.26	66.42
B2	D3	20.70	20.24	0.72	8.64	2977.90	16656.34	68.01
B3	D3	20.25	20.70	0.69	7.37	2866.93	18196.42	66.09
B4	D3	20.29	20.65	0.75	7.51	3195.31	16310.44	67.11
B5	D3	20.21	20.64	0.75	7.39	1543.83	10564.23	68.93
B6	D3	20.59	20.22	0.71	7.32	1638.82	10330.17	70.57
B7	D3	20.66	20.19	0.72	7.34	2439.58	15394.01	69.05
J1	D3	20.05	20.46	0.72	8.37	1543.83	10932.05	74.64
J2	D3	20.37	20.54	0.70	7.43	1638.82	11289.44	69.60
J3	D3	19.98	20.42	0.66	7.96	1906.31	12022.83	76.34
J4	D3	20.33	20.33	0.73	8.45	1434.72	10058.07	71.27
J5	D3	20.31	20.34	0.70	8.23	1773.53	11134.01	71.45
J6	D3	20.11	20.46	0.77	8.31	1552.38	10274.81	75.42
J7	D3	19.97	20.45	0.70	8.24	1610.88	10120.01	76.21
				With finger joint				
B-Z 1	D3	19.69	20.39	0.41	8.10	1157.86	1801.55	82.04
B-Z 2	D3	19.97	20.45	0.42	7.86	943.851	1756.89	77.55
B-Z 3	D3	19.82	20.31	0.48	8.23	880.265	1781.31	80.38
B-Z 4	D3	19.75	20.31	0.45	7.32	1140.04	1824.75	82.92
B-Z 5	D3	19.96	20.24	0.42	7.36	1049.96	1789.30	79.88
B-Z 6	D3	20.03	20.16	0.38	797	1208.01	1789.57	79.65
B-Z 7	D3	19.69	20.26	0.41	8.07	1135.09	1838.01	84.24
J-Z 1	D3	20.05	20.45	0.44	8.82	662.033	1779.60	77.93
J-Z 2	D3	20.12	20.47	0.44	8.51	974.981	1684.17	73.17
J-Z 3	D3	20.06	20.34	0.42	8.74	848.635	1978.92	87.04
J-Z 4	D3	20.15	20.44	0.50	8.40	1070.94	1676.71	72.73
J-Z 5	D3	20.00	20.47	0.39	8.88	850.956	1760.65	77.41
J-Z 6	D3	20.12	20.40	0.46	8.83	916.855	1766.06	76.99
J-Z 7	D3	19.97	20.47	0.49	9.56	1021.92	1719.36	75.82

IOP Conf. Series: Materials Science and Engineering 1298 (2023) 012011 doi:10.1088/1757-899X/1298/1/012011

Where is:

- B - beech samples without finger joint,
- J - fir samples without finger joint,
- BZ - beech samples with finger joint,
- JZ - fir samples with finger joint.

This study mainly focused on bending strength and modulus of elasticity in the experiment. When it comes to bending strength (f_m) of wooden samples, it refers to the maximum stress that the material can withstand when bent. This is an important property for materials used in construction and making objects that undergo bending or bending loading. On the other hand, the modulus of elasticity ($E_{m.l}$) model relates to the material's ability to deform under the action of a force without breaking. The measured samples consist of two types of wood, beech (B) and fir (J), and were measured in two variants: without finger joints (samples B and J), with finger joints (samples BZ and JZ), which in total gives 4 sample variants. Each sample variant was measured in relation to five parameters: density (ρ), moisture content (MC), models of elasticity ($E_{m,l}$), bending strength (fm), and maximum breaking force (F_{max}), shown in table 1.

When comparing models of elasticity ($E_{m,l}$) and bending strength (f_m) between samples with and without joints, it can be seen that samples without joints had higher elasticity and bending strength compared to samples with joints. This difference was most pronounced in samples of beech, while fir samples had less differences in elasticity and bending strength between samples with and without joints.

The percentage difference of maximum bending force at four-point bending test between beech samples and samples joined with a finger joint ranged from 26% to 60% on average, as shown in Figure 2. While the difference in maximum force for fir samples was on average 41%, indicating a decrease in maximum force for samples tested under laboratory conditions. Analysis of the percentage differences and maximum and minimum differences between samples with and without finger joints showed that the former exhibited significantly greater differences in strength and modulus of elasticity, with differences about 50% depending on the specific parameter measured. These findings suggest that the presence of finger joints can have a significant impact on the mechanical properties of the wood connection.

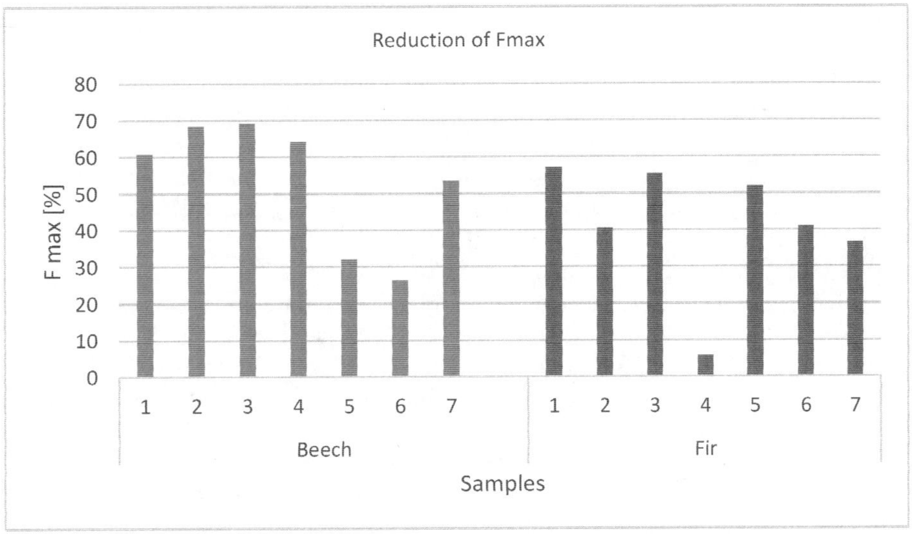

Figure 2. Percentage loss of force in the tested samples

Looking at the differences in moisture content between these variants, they are not significant. When it comes to bending strength, the differences between samples without and with finger joints are also insignificant. As for the differences in moisture content (MC), it is evident that the moisture content of all samples was around 8%, as can be seen from Table 1.

Regarding the differences in density (ρ) between samples with and without joints, it can be noticed that samples without joints had slightly higher density compared to samples with joints. This difference was most pronounced in samples of beech, while fir samples had less density differences between samples with and without joints. It is important for the moisture content and density to be consistent between the samples in a 4-point bending test, as significant differences can affect the results and make them less reliable. The fact that there were no significant differences in moisture content and only slight differences in density between samples with and without finger joints is a positive indication that the testing conditions were consistent and the results can be more reliable.

4. Conclusion

The experiment aimed to measure the bending strength (f_m) and modulus of elasticity ($E_{m,l}$) of two types of wood, beech (B) and fir (J), with and without finger joints. The study found significant differences between modules of elasticity and the samples in bending force with and without finger joints. However, samples with finger joints showed greater differences in strength compared to the samples without finger joints, indicating that the finger joints significantly affected the strength of the wood joint. Future research will focus on exposing the samples to outdoor conditions in exposure classes 1 and 2, with some samples being protected by with a coating for external use. The differences between samples exposed to outdoor conditions and those tested in laboratory conditions will be investigated.

References

[1] Ahmad Z, Lum W C, Lee S H, Razlan M A and Wan Mohamad W H 2017 Mechanical properties of finger jointed beams fabricated from eight Malaysian hardwood species *Construction and Building Materials* **145** 464–73 Online: http://dx.doi.org/10.1016/j.conbuildmat.2017.04.016

[2] Morin-Bernard A, Blanchet P, Dagenais C and Achim A 2021 Glued-laminated timber from northern hardwoods: Effect of finger-joint profile on lamellae tensile strength *Construction and Building Materials* **271** 121591 Online: http://dx.doi.org/10.1016/j.conbuildmat.2020.121591

[3] De Silva S and Liyanage V 2019 Suitability of finger jointed structural timber for construction *Journal of Structural Engineering & Applied Mechanics* **2** 131–42 Online: http://dx.doi.org/10.31462/jseam.2019.03131142

[4] Hodžić A and Hasanagić R 2017 Matematičko modeliranje sile loma na istezanje masivnog drveta produženog sa zupčastim vezom | Zbornik Međunarodnog kongresa o procesnoj industriji – Procesing *Matematičko modeliranje sile loma na istezanje masivnog drveta produženog sa zupčastim vezom | Zbornik Međunarodnog kongresa o procesnoj industriji – Procesing* Online: https://www.izdanja.smeits.rs/index.php/ptk/article/view/2344

[5] Hasanagić R, Fathi L, Kapić Z, Bahmani M, Crnkić A, Hrnjica B and Humar M and Jurković M 2023 Experimental and numerical determination of the longitudinal modulus of elasticity in wooden structures Drewno **65** *210* 1–13 doi: 10.12841/wood.1644-3985.421.10

[6] Abdul Hamid N H, Ahmad M, Suratman M N and Abu F 2015 Bending Strength of Finger Jointed Kelat Wood (*Syzygium* spp.) as Affected by Finger Length and Orientation *Advanced Materials Research* **1134** 138–42 Online: http://dx.doi.org/10.4028/www.scientific.net/amr.1134.138

[7] Pěnčík J 2015 Tests of Wooden Specimens from Scots Pine (Pinus sylvestris) with the Help of Anisotropic Plasticity Material Model *Drvna industrija* **66** 27–33 Online: http://dx.doi.org/10.5552/drind.2015.1362

[8] Hasanagić R, Hodžić A and Jurković M 2019 Modelling and optimization of tensile break force of solid wood elements lengthened by finger joint *Journal of Adhesion Science and Technology* 1–15 Online: http://dx.doi.org/10.1080/01694243.2019.1690266

[9] Hasanagić R 2018 Modeling and prediction of fracture force to tighten solid wood elements by genetic programming *Tehnika* **73** 653–7 Online: http://dx.doi.org/10.5937/tehnika1805653h
[10] Jokerst R W 1981 *Finger-Jointed Wood Products* (United States: United States Department of Agriculture Forest Service) Online: https://www.fpl.fs.usda.gov/documnts/fplrp/fplrp382.pdf
[11] Vassiliou V, Barboutis I and Karastergiou S 2006 Effect of PVAc bonding on finger-joint strength of steamed and unsteamed beech wood (Fagus sylvatica) *Journal of Applied Polymer Science* **103** 1664–9 Online: http://dx.doi.org/10.1002/app.25079
[12] Bašić N, Kulenović E, Hozdić E and Srce Ljepljenje lameliranih elemenata za proizvodnju građevinske stolarije *Ljepljenje lameliranih elemenata za proizvodnju građevinske stolarije* Online: https://hrcak.srce.hr/clanak/184409
[13] Klippel M, Frangi A and Hugi E 2014 Experimental Analysis of the Fire Behavior of Finger-Jointed Timber Members *Journal of Structural Engineering* **140** Online: http://dx.doi.org/10.1061/(asce)st.1943-541x.0000851
[14] Konnerth J, Valla A, Gindl W and Müller U 2006 Measurement of strain distribution in timber finger joints *Wood Science and Technology* **40** 631–6 Online: http://dx.doi.org/10.1007/s00226-006-0090-9
[15] Selbo M 1963 Effect of joint geometry on tensile strength of finger joints *Forest Products Journal* **13** 390–400
[16] Dutko P, Steller S and Kozelouh B 1982 Research into and Experience of the Use of Finger-Joints in Timber Structures in Czechoslovakia *Production, Marketing and Use of Finger-Jointed Sawnwood* 35–47 Online: http://dx.doi.org/10.1007/978-94-015-3859-6_6
[17] Frangi A, Bertocchi M, Clauß S and Niemz P 2011 Mechanical behaviour of finger joints at elevated temperatures *Wood Science and Technology* **46** 793–812 Online: http://dx.doi.org/10.1007/s00226-011-0444-9
[18] Smardzewski J 1996 Distribution of stresses in finger joints *Wood Science and Technology* **30** Online: http://dx.doi.org/10.1007/bf00244442
[19] Boyd S W, Dulieu-Barton J M and Rumsey L 2006 Stress analysis of finger joints in pultruded GRP materials *International Journal of Adhesion and Adhesives* **26** 498–510 Online: http://dx.doi.org/10.1016/j.ijadhadh.2005.07.003
[20] Özçifçi A and Yapıcı F 2008 Structural performance of the finger-jointed strength of some wood species with different joint configurations *Construction and Building Materials* **22** 1543–50 Online: http://dx.doi.org/10.1016/j.conbuildmat.2007.03.020
[21] Gong M, Delahunty S and Chui Y H 2014 Influence of number of finger joints per stud on mechanical performance of wood shearwalls *Construction and Building Materials* **50** 335–9 Online: http://dx.doi.org/10.1016/j.conbuildmat.2013.09.041
[22] Milner H R and Yeoh E 1991 Finite Element Analysis of Glued Timber Finger Joints *Journal of Structural Engineering* **117** 755–66 Online: http://dx.doi.org/10.1061/(asce)0733-9445(1991)117:3(755)
[23] Serrano E and Gustafsson P J 1999 Influence of bondline brittleness and defects on the strength of timber finger-joints *International Journal of Adhesion and Adhesives* **19** 9–17 Online: http://dx.doi.org/10.1016/s0143-7496(98)00048-7
[24] Kishan Kumar V S, Upreti N K and Khanduri A K 2010 Effect of finger tip area on the compression strength of finger jointed sections *Journal of the Indian Academy of Wood Science* **7** 25–9 Online: http://dx.doi.org/10.1007/s13196-010-0005-0
[25] Singh C P, Kumar K, Shukla S and Gupta S 2016 Maximum Tensile Stress of Eucalypt *Research Journal of Agriculture* **4** 10–3 Online: http://www.isca.me/AGRI_FORESTRY/Archive/v4/i8/2.ISCA-RJAFS-2016-024.pdf
[26] Hemmasi A H, Khademi-Eslam H, Roohnia M, Bazyar B and Yavari A 2014 Elastic Properties of Oak Wood Finger Joints with Polyvinyl Acetate and Isocyanate Adhesives | Hemmasi | BioResources *Elastic Properties of Oak Wood Finger Joints with Polyvinyl Acetate and Isocyanate Adhesives | Hemmasi | BioResources* Online:

https://jtatm.textiles.ncsu.edu/index.php/BioRes/article/view/BioRes_09_1_849_Hemmasi_Elastic_Properties_Oak_Wood/2524

[27] Amoah M, Kwarteng F O and Dadzie P K 2014 Bending properties and joint efficiency of some tropical hardwoods finger-jointed in green and dry states *Journal of the Indian Academy of Wood Science* **11** 57–64 Online: http://dx.doi.org/10.1007/s13196-014-0118-y

[28] Anon ISBIH *ISBIH* Online: https://isbih.gov.ba/standard/274682

[29] Anon 1975 ISO 3131:1975 *ISO* Online: https://www.iso.org/standard/8289.html

[30] Anon ASTM D2016 : Standard Test Methods for MOISTURE CONTENT OF WOOD *ASTM D2016 : Standard Test Methods for MOISTURE CONTENT OF WOOD* Online: https://global.ihs.com/doc_detail.cfm?document_name=ASTM%20D2016&item_s_key=00016225

Optimization of solid wood workability parameters in the planing process

[1]Husein Rošić, [1]Aldin Bjelić, [1]Atif Hodžić and [2]Miran Merhar

[1]University of Bihać, Faculty of Technical Engineering, Bosnia and Herzegovina
[2]University of Ljubljana, Biotechnical Faculty, Slovenia

husein.rosic@unbi.ba

Abstract. It is known from theory and practice that the workability of wood depends on structural parameters that are closely related to the physical, mechanical and chemical properties of the type of wood itself, and disturbance parameters that refer to the technological and geometric parameters determined by the specific processing regime. That machinability, in addition to the mechanical output sizes, is often expressed by the quality, that is, the roughness of the processed surface. By defining a mathematical model in the process of planing solid wood in which the input sizes are processing parameters: wood density (ρ), feed rate (m/min) and number of cutting spirals (z), and the spilled sizes are praamters of roughness of the treated surface (Ra and Rz), and by applying optimization methods, optimal solutions for the process of planing solid wood on planer machines will be determined, so that the obtained Yoptim model will aim to improve the workability of solid wood, specifically its roughness of the processed surface.

1. Introduction

We consider the machinability of the material as a technological feature that expresses the ability of the material to remove the maximum amount of chips from its treated surface in the minimum time with satisfactory processing quality, with as little cutting force and as little tool wear as possible. Due to its anatomical specificities and the fact that wood is an anisotropic, hygroscopic and orthotropic polymer material, defining the processing parameters of solid wood is much more complicated than for metal. The workability of wood depends on a large number of parameters that are caused by physical, mechanical, chemical and aesthetic properties of wood on the one hand, and parameters of the processing regime on the other hand, which can be technologically geometric. The mentioned parameters do not have the same importance in all production processes and technological procedures of wood processing. Therefore, it is necessary to rank the parameter for each individual case, according to the technical and technological requirements.

A better quality of the processed surface, therefore less roughness, means a higher quality product, and materials, for which a better processed surface is obtained with the same processing regimes, are also said to be more processable. Indicators of the quality of the treated surface refer to two groups of parameters:

- initial quality parameters,
- disruptive quality parameters

The initial processing quality parameters refer to the geometric and kinematic accuracy of the processed surface, while the disturbance parameters depend on the processing subsystem and can be caused by: thermal, static and mechanical deformations, friction and wear, and vibrations. In general, the inaccuracy of the processed surface is manifested through the inaccuracy of the dimensions and shape of the processed detail, and the roughness of the processed surface. Errors in the quality of the treated surface can be structural irregularities (caused by errors in the anatomical structure of the wood) and micro-irregularities (caused by the processing mode). Micro unevenness appears on the treated surface as undulations and surface roughness, and in the form of destructive unevenness. According to GOST, roughness during mechanical processing is characterized by parameters *Rz*, i.e. by the mean height of bumps and *Ra* - by the mean arithmetic deviation of the profile from the middle line of the profile.

Structural irregularities, i.e., their size and arrangement, depend on the botanical type of wood and cannot be avoided, which means that they follow the cutting of the wood. Micro unevenness is caused by cutting and refers to the roughness of the newly created cutting surface. This surface is not geometrically two-dimensional, because smaller or larger depressions or protrusions are created on the surface during cutting. The causes of the formation of micro bumps are as follows [2]:

- tool marks,
- bumps from residual deformations,
- traces of vibration of machines, tools and workpieces,
- cutting modes,
- state of sharpening of the blade,
- type of cutting,
- physical and mechanical properties of wood, especially volumetric mass and ageing.

Newly created processed surfaces are never ideally smooth, whether they are flat or curved, but should be viewed three-dimensionally in the form of roughness and undulations. They are the same consequence, on the one hand of the anatomical structure of the wood, and on the other hand, unevenness is caused by cutting. The quality of processing is influenced by factors such as cutting direction, blade geometry and state of sharpening, wood thickness, inaccuracy of tool sharpening, shape and type of chips, technological parameters (cutting speed, feed rate, cutting depth, etc.). The roughness of the treated surface is very important in the woodworking process, above all for the quality of the technological operations of gluing, pressing, sanding, application of surface treatment and assembly into the finished product.

The paper analysed the influence of certain parameters of the workability of solid wood on the quality of the treated surface during planing on a multi-sided planer, which were selected based on pre-research activities of the workability of wood species that are mainly exploited in these areas.

The experimental results were mathematically modelled with the aim of obtaining a mathematical model of the significance of the parameters in the planning process on a multi-sided planer through the roughness output values of the processed surface *Ra* and *Rz*.

Based on the experimental results that were carried out in real production in "Interprocess" Ltd. Tešanj, two mathematical models of the influence of the input parameters ρ, s' and z on the output parameters of the processing quality *Ra* and *Rz* in the process of planing solid wood on the Profimat 23 Fortec multi-sided planer of the German manufacturer were obtained. Weining GmbH:

$$y = \mathrm{f}\,(x_1, x_2, x_3) \tag{1}$$

where are:

y – roughness parameters of the processed surface *Ra* and *Rz* (μm)
x_1 - wood density ρ (g/cm3)
x_2 - feed rate s' (m/min)
x_3 - number of cutting spirals z.

In order for the models to be as acceptable as possible for the future selection of processing mode parameters in the process of planning on multi-sided planers, the results of this experimental research were optimized, with the aim of obtaining optimal values of the input parameters, where the minimum values of *Ra* and *Rz* are obtained. The optimization procedure is expressed mathematically as searching for the minimum or maximum of the objective function [8]. In general, the objective function is a scalar function F with one or more parameters *xi* depending on how many variables are observed during the model optimization procedure. The objective function is denoted by $F(x_1, x_2, ..., x_n)$ or $F(x)$. It is a unique function that must describe all the properties that we want to achieve by optimizing the model. The optimization method of mathematical programming was chosen, based on the assumption that the equation describing the process will be non-linear due to the number of variables and their interaction, in which the minimum values of *Ra* and *Rz* will be sought, with appropriate explicit limitations of the values of the input variables [7].

Optimizing the significant parameters of the solid wood planning process on multi-sided planers, in a model that is essentially applicable, optimal values of workability parameters were obtained for which the roughness of the wood, expressed by the values of *Ra* and *Rz*, is minimal.

2. Experimental Research

Experimental research related to application was carried out in "Interprocess" Ltd Tešanj on the Profimat 23 Fortec four-sided planer, manufactured by Weining GmbH. The dimensions of the clean workpieces were 620 x 80 x 24 mm. The specimens with three different densities in the range of 390.23 - 790.19 kg/m^3 were planed with various combinations of three different feeding speeds and three different number of cutting spirals. Then the roughness parameters *Ra* and *Rz* were measured with a contact profilimeter with a TRM 120 needle at three measuring points, at a distance of 15 cm from the ends, and in the middle of the sample (Figure 1).

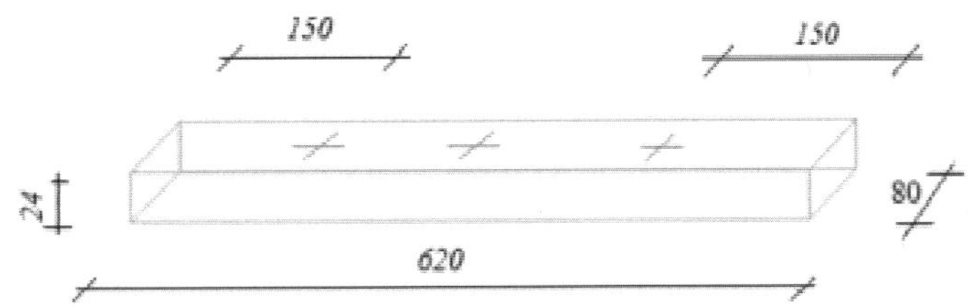

Figure 1. Schedule of measurements of parameters *Ra* and *Rz*

In addition, the roughness parameters *Ra* and *Rz* were measured on five measuring areas, with a laser roughness meter LEXT 3D Measuring Laser Microscope OLS5000 (Figure 2), and the mean values of all eight measurements for *Ra* and *Rz* were taken for the relevant sizes.

Figure 2. Laser roughness cutter LEXT 3D

Wood moisture was measured with an electric hygrometer and was 12 ± 0.5 % for all treated samples. Small samples measuring 0.25 x 0.25 x 0.24 mm were cut from the large samples, and the wood density values and their average were determined using the volumetric method (Table 1).

Table 1. Measured values of humidity, mass and density of wood

The type of wood	Humidity (%) (average value)	Mass (kg) (mean value)	Density (kg/m^3) (average value)
Linden (*Tilia platyphyllos*)	12.32	0.0609	390.23
Ash (*Fraxinus excelsior*)	11.96	0.0999	640.21
Hornbeam (*Carpinus betulus*)	12.54	0.1233	790.19

The total number of trials in this experiment was determined by three independent variables. The physical and coded values of the parameters of the process of planning solid wood on a multi-sided planer vary on three levels min-medium-max. Total number of trials [8]:

$$N = 2^k + n0 = 2^3 + 6 = 14 \text{ experiments} \quad (2)$$

where is:

$k = 3$ – number of changing variables,

$n_0 = 6$ – the number of repetitions in the central point of the plan.

The formalized state, describing the real process of solid wood milling, is defined by the following quantities:

a) Input sizes:
- wood density ρ (gr/cm^3).
- feeding speed s' (m/min),
- number of spiral cutters z,

b) Output sizes:
- mean arithmetic deviation of Ra (μm),
- average height of bumps Rz (μm)

c) Process state function: Ra (Rz) = f (ρ, s', z).

Table 2. Physical and coded values of Xji

PLANING PROCESS VARIABLES			ENCODED VALUES		
			Xj2	Xj3	Xj1
			-1	0	1
NATURAL VALUES	Influential values	ρ (X1i) (gr/cm^3)	0.39	0.64	0.79
		s' (X2i) (m/min)	13	23	33
		z (X_{3i}) (number of cutters)	4	6	8

The experimental measurement results for the parameters Ra and Rz are given in the following table 3:

Table 3. Experimental measurement results for the *Ra* and *Rz*

Number of experiments Ni	*Experimental measurement results*				
	ρ (gr/cm^3)	s' (m/min)	z (number of cutters)	*Ra* (µm)	*Rz* (µm)
P.1	0.39	13	4	5.50	48.24
P.2	0.79	13	4	3.78	40.34
P.3	0.39	33	4	5.88	51.15
P.4	0.79	33	4	3.83	48.67
P.5	0.39	13	8	7.15	50.23
P.6	0.79	13	8	4.99	41.42
P.7	0.39	33	8	7.30	56.30
P.8	0.79	33	8	4.67	45.48
P.9	0.64	23	6	4.42	42.83
P.10	0.64	23	6	4.44	42.48
P.11	0.64	23	6	4.24	42.66
P.12	0.64	23	6	4.47	42.71
P.13	0.64	23	6	4.48	42.71
P.14	0.64	23	6	4.80	42.35

A linear model was created for the roughness parameters *Ra* and *Rz* based on the obtained roughness data. For each of the models, the homogeneity of dispersions was determined according to Cochran's criterion and the adequacy of the model based on the criterion of the multiple regression coefficient.

3. Mathematical modeling

The basic goal of mathematical modeling is to determine the legality of mutual connections of input and output parameters, of which the input parameters define the conditions of the process, and the output parameters are the results of the modeled process [2].

In this work, stochastic modeling was applied, along with the application of active experimental research on a real object in the process of planing on a multi-spindle milling machine Profimat 23 Fortec, manufactured by Weining GmbH in "Interprocess" Ltd Tešanj. A regression analysis was used to determine the functional relationship between dependently variable quantities *yi* (parameters *Ra* and *Rz*) and independently variable quantities *xi* (wood density, feed rate and number of spiral cutters). The formalized description $\boldsymbol{F} = \mathbf{f}(\rho, s', z)$ turns into the mathematical $\boldsymbol{Y} = \mathbf{f}\,(x_1, x_2, x_3)$, which besides the influence on both quantities x_1, x_2, x_3 separately, also describes the influence of their interactions x_{12}, x_{11}, x_{22} to the output value Y. By introducing the coded values according to the layout of the experiment plan and the position of the value of the coded quantity, a basic table is obtained for further statistical data processing (Table 4).

Table 4. Experiment plan with coded quantities for *Rz*

Ni	*X0*	*X1*	*X2*	*X3*	*X1 X2*	*X1 X3*	*X2 X3*	R_z	R_z^2
1	1	-1	-1	-1	1	1	1	48.24	2327.097
2	1	1	-1	-1	-1	-1	1	40.34	1619.268
3	1	-1	1	-1	-1	1	-1	51.15	2616.320
4	1	1	1	-1	1	-1	-1	48.67	2368.767
5	1	-1	-1	1	1	-1	-1	50.23	2523.053
6	1	1	-1	1	-1	1	-1	41.42	1715.616

7	1	-1	1	1	-1	-1	1	56.30	3169.690
8	1	1	1	1	1	1	1	45.48	2068.430
9	1	0	0	0	0	0	0	42.83	1834.410
10	1	0	0	0	0	0	0	42.48	1804.551
11	1	0	0	0	0	0	0	42.66	1819.871
12	1	0	0	0	0	0	0	42.71	1824.140
13	1	0	0	0	0	0	0	42.71	1824.140
14	1	0	0	0	0	0	0	43.35	1879.221
							Σ	638.57	29394.58

By multiplying the coded values with the experimental value of the roughness parameter *Rz*, we obtain the data with which we will define the regression coefficients of the model (Table 5).

Table 5. Determining the regression coefficients of the linear model of the roughness parameter *Rz*

Ni	*X0· Rz*	*X1· Rz*	*X2· Rz*	*X3· Rz*	*X1 X2· Rz*	*X1 X3· Rz*	*X2 X3· Rz*
1	48.24	-48.24	-48.24	-48.24	48.24	48.24	48.24
2	40.34	40.34	-40.34	-40.34	-40.34	-40.34	40.34
3	51.15	-51.15	51.15	-51.15	-51.15	51.15	-51.15
4	48.67	48.67	48.67	-48.67	48.67	-48.67	-48.67
5	50.23	-50.23	-50.23	50.23	50.23	-50.23	-50.23
6	41.42	41.42	-41.42	41.42	-41.42	41.42	-41.42
7	56.30	-56.30	56.30	56.30	-56.30	-56.30	56.30
8	45.48	45.48	45.48	45.48	45.48	45.48	45.48
9	42.83	0	0	0	0	0	0
10	42.48	0	0	0	0	0	0
11	42.66	0	0	0	0	0	0
12	42.71	0	0	0	0	0	0
13	42.71	0	0	0	0	0	0
14	43.35	0	0	0	0	0	0
Σ	638.57	-30.01	21.37	5.03	3.41	-9.25	-1.11
bi	45.612	-3.751	2.671	0.629	0.426	-1.156	-0.138

The penultimate row in Table 5 represents the sum of the corresponding column and the last row represents the regression coefficient for the corresponding coded quantity. The regression coefficients would be the mean values of individual columns, because for coded quantity X_0 we have taken the mean value of 14 repetitions of the experiment, an d for other coded quantities and interactions between quantities, the mean value of 8 repetitions of the experiment, i.e., from the basic plan of the experiment.

Therefore, the linear form of the model of the roughness parameter *Rz* can be written in the form:

$$Y = 45.612 - 3.751X_1 + 2.671X_2 + 0.629X_3 + 0.426X_1X_2 - 1.156X_1X_3 - 0.138X_2X_3 \quad (3)$$

The sum of squares (SS) of the input factors and the interaction between them is calculated using the equation:

$$SS = \frac{(\sum X_i . R_i)^2}{N_0} \tag{4}$$

where is:
SS – sum of squares,
N_0 – number of basic repetitions of the experiment (in our case N_0= 8)

Table 6. Data for calculating the sum of squares of the pure error of the experiment

Ni	*Rz*	$(R_{zi} - R_z)^2$
9	42.83	0.0116
10	42.48	0.0961
11	42.66	0.0169
12	42.71	0.0064
13	42.71	0.0064
14	43.35	0.3136
	$R_z = 42.79$	$\Sigma = 0.451$

After the Fisher coefficients have been determined, we will calculate the statistical coefficient p, which can be used to see if and which factors are significant.
The values of the statistical coefficient p as well as all other values obtained by analysis of variance are shown in Table 7.

Table 7. The results of the analysis of variance for the linear model of the roughness parameter *Rz*

Source of variation	*SS*	*dF*	*MS*	*F*	Significance
A	112.575	1	112.575	0.42	significant
B	57.085	1	57.085	0.21	significant
C	3.163	1	3.163	0.012	not significant
AB	1.454	1	1.454	0.0052	not significant
AC	10.695	1	10.695	0.039	not significant
BC	0.154	1	0.154	0.006	not significant
Model	185.126	6	30.854	0.13	significant
Error	0.451	7	0.0902		
Deviation from the model	1914.035	2	957.02		
Pure error	0.451	5	0.0902		
Total	2099.61	13	160.893		

The significance of the coefficients is determined based on the significance level $\alpha = 0.05$, which means that for the linear model of the roughness parameter *Rz,* the first and second input factors *A* and *B* will be significant.

What is important is that we got a final model that is significant.
Once we have determined the significance of the input parameters, we can return to equation (4) and take into account only those regression coefficients whose magnitudes are significant, so expression (4) will be transformed:

$$Y = 45.612 - 3.751X_1 + 2.671X_2 \tag{5}$$

Decoding of quantities X_1, and X_2 from equation is performed according to expression.
According to equation (5) we will decode the sizes X_1, X_2 and X_3:

$$X_1 = \frac{\rho - \frac{(\rho_{lower} + \rho_{upper})}{2}}{\frac{(\rho_{upper} - \rho_{lower})}{2}} = \frac{\rho - \frac{0.39 + 0.79}{2}}{\frac{0.79 - 0.39}{2}} = \frac{\rho - 0.59}{0.2} = 5\rho - 2.95 \quad (6)$$

$$X_2 = \frac{s - \frac{(s_{lower} + s_{upper})}{2}}{\frac{(s_{upper} - s_{lower})}{2}} = \frac{s - \frac{13 + 33}{2}}{\frac{33 - 13}{2}} = \frac{s - 23}{10} = 0.1s - 2.3 \quad (7)$$

Inserting expressions (6) and (7) into equation (5) we get:

$$R_z = 50.534 - 18.755\rho + 0.2671s \quad (8)$$

For the linear model of the roughness parameter *Rz* from equation (8), we can determine the homogeneity of dispersions according to Cochran's criterion and the adequacy of the model based on the multiple regression coefficient.

Table 8. Data for calculating Cochran's coefficient

Ni	*Rz*	$\sigma_j^2 = (R_{zi} - R_z)^2$
9	42.83	0.0016
10	42.48	0.0961
11	42.66	0.1690
12	42.71	0.0640
13	42.71	0.0640
14	43.35	0.3136
	$R_z = 42.79$	$\sum_{j=1}^{n} \sigma_j^2 = 0.7083$
		$max\sigma_j^2 = 0.3136$

According to the data in Table 8, Cochran's coefficient is:

$$K_h = \frac{0.3136}{0.7083} = 0.442 \quad (9)$$

Based on the obtained data in the central points, the Cochran's coefficient for the logarithmic model is K_h=0.442 and the tabular K_t=0.445. As $K_h < K_t$ we can conclude that the dispersion is homogeneous.

Table 9. Data for determining the multiple regression coefficient *R* according to the model for *Rz*

Ni	Experiment	Linear model	$(Y_j^E - Y_j^M)^2$	$(Y_j^E - -Y^E)^2$
	Ra [μm] = Y_j^E	*Ra* [μm] = Y_j^M		
1	48.24	46.69	2.4025	6.9169
2	40.34	39.19	1.3225	27.772
3	51.15	52.03	0.7744	30.692
4	48.67	44.53	17.139	9.3636

5	50.23	46.69	12.5316	21.344
6	41.42	39.19	4.9729	17.556
7	56.30	52.03	18.233	114.276
8	45.48	44.53	0.9025	0.0169
9	42.83	44.67	3.3856	7.7284
10	42.48	44.67	4.7961	9.7969
11	42.66	44.67	4.0401	8.7025
12	42.71	44.67	3.8416	8.41
13	42.71	44.67	3.8416	8.41
14	43.35	44.67	1.7424	5.1076
	$Y^E = 45.61$		Σ = 79.9258	Σ =276.093

Adequacy of the model based on the multiple regression coefficient is determined by calculating the size of *R* and calculated according to equation (10). Therefore, the multiple regression coefficient of the logarithmic model can be calculated as:

$$R = \sqrt{1 - \frac{79.9258}{276.093}} = 0.843 \quad (10)$$

The value of the regression coefficient *R=0.843* means that the linear model describes the accuracy of the experiment results with *84.3%*, which is a good accuracy of the model.

In the same way, a mathematical model for the roughness parameter *Ra* was obtained by stochastic modeling.

Table 10. Experiment plan with coded quantities for Ra

Ni	*X0*	*X1*	*X2*	*X3*	*X1 X2*	*X1 X3*	*X2 X3*	*Ra*	Ra^2
1	1	-1	-1	-1	1	1	1	5.50	30.25
2	1	1	-1	-1	-1	-1	1	3.78	14.29
3	1	-1	1	-1	-1	1	-1	5.88	34.57
4	1	1	1	-1	1	-1	-1	3.83	14.67
5	1	-1	-1	1	1	-1	-1	7.15	51.12
6	1	1	-1	1	-1	1	-1	4.99	24.90
7	1	-1	1	1	-1	-1	1	7.30	53.29
8	1	1	1	1	1	1	1	4.67	21.81
9	1	0	0	0	0	0	0	4.42	19.53
10	1	0	0	0	0	0	0	4.44	19.71
11	1	0	0	0	0	0	0	4.24	17.97
12	1	0	0	0	0	0	0	4.47	19.98
13	1	0	0	0	0	0	0	4.48	20.07
14	1	0	0	0	0	0	0	4.80	23.04
							Σ	69.95	365.2

By multiplying the coded quantities with the experimental value of the roughness parameter *Ra*, we obtain the data with which we will define the regression coefficients of the model.

The regression coefficients bi are the mean values of each column because for coded quantity X_0 we take the mean value of 14 repetitions of the experiment and for other coded quantities and interactions between quantities the mean value of 8 repetitions of the experiment, i.e. from the basic plan of the experiment.

Table 11. Determining the regression coefficients of the Ra model

Ni	*X0· Ra*	*X1· Ra*	*X2· Ra*	*X3· Ra*	*X1 X2· Ra*	*X1 X3· Ra*	*X2 X3· Ra*
1	5.50	-5.50	-5.50	-5.50	5.50	5.50	5.50
2	3.78	3.78	-3.78	-3.78	-3.78	-3.78	3.78
3	5.88	-5.88	5.88	-5.88	-5.88	5.58	-5.88
4	3.83	3.83	3.83	-3.83	3.83	-3.83	-3.83
5	7.15	-7.15	-7.15	7.15	7.15	-7.15	-7.12
6	4.99	4.99	-4.99	4.99	-4.99	4.99	-4.99
7	7.30	-7.30	7.30	7.30	-7.30	-7.30	7.30
8	4.67	4.67	4.67	4.67	4.67	4.67	4.67
9	4.42	0	0	0	0	0	0
10	4.44	0	0	0	0	0	0
11	4.24	0	0	0	0	0	0
12	4.47	0	0	0	0	0	0
13	4.48	0	0	0	0	0	0
14	4.80	0	0	0	0	0	0
Σ	*69.95*	*-8.56*	*0.26*	*5.12*	*-0.8*	*-1.32*	*-0.57*
bi	*4.996*	*-1.07*	*0.00325*	*0.64*	*-0.1*	*-0.165*	*-0.0712*

Therefore, the linear form of the model of the roughness parameter Ra can be written in the form:

$$Y = 4.996 - 1.07X_1 + 0.64X_3 - 0.1X_1X_2 - 0.165X_1X_3 - 0.0712X_2X_3 \tag{11}$$

Table 12. Results of the analysis of variance for the linear model of the roughness parameter R

Source of variation	SS	dF	MS	F	Significance
A	9.159	1	9.159	22.017	significant
B	0.008	1	0.008	0.0192	not significant
C	3.277	1	3.277	7.8606	significant
AB	0.08	1	0.08	0.192	not significant
AC	0.2178	1	0.2178	0.523	not significant
BC	0.041	1	0.041	0.09855	not significant
Model	12.78	6	2.13	5.1201	significant
Remainder	2.9198	7	0.416		
Deviation from the model	2.6654	2	1.3327		
Pure Error	0.2544	5	0.05088		
Total	15.699	13	1.206		

In order to determine the significance (significance) of the regression coefficients for the linear model, we will perform an ANOVA analysis of variance for the given experiment. Variance analysis involves calculating the sum of squares, degrees of freedom and mean square value for given input factors, interactions between them, the overall model and the deviation from the model and the pure error of the model. When labeling, we will use the letters A, B and C for the input factors.

After Fisher's coefficients have been determined, the statistical coefficient p is calculated, which can be used to see if and which factors are significant. The values of the statistical coefficient p as well as all other quantities obtained by analysis of variance are shown in Table 12.

In this example, $\alpha = 0.05$ or 5% was taken. The parameter from Table 12 is significant if its statistical value p is less than $\alpha = 0.05$. This means that for the linear model of the roughness parameter *Ra*, the factors A and C will be significant. What is important is the final model. After determining the significance of the input parameters in equation (11), only those regression coefficients whose magnitudes are significant are taken into account, so expression (11) will be transformed into:

$$Y = 4.996 - 1.07X_1 + 0.64X_3 \tag{12}$$

The decoding of the quantities X_1, X_2 and X_3 from equation (13) is performed according to the expression:

$$X_i = \frac{f_i - \frac{(f_{imin} + f_{imax})}{2}}{\frac{(f_{imax} - f_{imin})}{2}} \tag{13}$$

where f_{imin} and f_{imax} are the lower and upper values of the level of the input factor, the values of the quantities $X_1 = 5\rho - 2.95$; $X_2 = s' - 2.3$ and $X_3 = 0.5\,z - 3$ are obtained. By including them in equation (13), the final model of the influence of the input factors of planning processing on a multi-faceted grinding machine on the roughness parameter of the processed surface *Ra* is obtained:

$$Ra = 6.179 - 5.35\rho + 0.32n \tag{14}$$

As already emphasized earlier, after obtaining the model equation, it is useful to check the homogeneity of dispersions using Cochran's criterion and the adequacy of the model by determining the multiple regression coefficient.

Checking the homogeneity of dispersions according to the Cochran criterion is performed on the basis of the expression:

$$K_h = \frac{max\sigma_j^2}{\sum_{j=1}^{n} \sigma_j^2} \leq K_t(dF_j, n) \tag{15}$$

where is:

K_t- tabular value according to Cochran's criterion for degrees of freedom *dFj* i *N*,
dFj – degree of freedom (*dFj = n-1*)
n – number of repetitions
σ_j^2 – sample variance.

For this example, this means that repetitions at the central point will be taken and the magnitude of *Kh* will be determined and compared to the table. For this determination, we need the experimental and model values of the parameter *Ra* in the central points, Table 9.
According to the data in Table 9, Cochran's coefficient is:

$$K_h = \frac{0.1024}{0.232} = 0.441 \tag{16}$$

Table 13. Data for calculating Cochran's coefficient

Ni	*Ra*	$\sigma_j^2 = (R_{ai} - -R_a)^2$
9	4.42	0.032
10	4.44	0.036
11	4.24	0.0576
12	4.47	0
13	4.48	0
14	4.80	0.1024
	$-R_a = 4.47$	$\sum_{j=1}^{n} \sigma_j^2 = 0.2322$
		$max\sigma_j^2 = 0,1024$

We will determine the table value Kt based on the degree of freedom $dFj = n - 1 = 6 - 1 = 5$ and $n = 6$ so Kt (5.6) =0.445. As $K_h > K_t$ we can conclude that the dispersion is not homogeneous.

Table 14. Data for determining the multiple regression coefficient R according to the linear model

Ni	*Experiment*	*Linear model*	$(Y_j^E - Y_j^M)^2$	$(Y_j^E - -Y^E)^2$
	Ra [μm] = Y_j^E	Ra [μm] = Y_j^M		
1	5.50	5.37	0.0169	0.254
2	3.78	3.23	0.3025	1.4786
3	5.88	5.37	0.2601	0.7815
4	3.83	3.23	0.3600	1.3596
5	7.15	6.65	0.2500	4.6397
6	4.99	4.51	0.2304	0
7	7.30	6.65	0.4225	5.3084
8	4.67	4.51	0.0256	0.1063
9	4.42	4.68	0.0676	0.3318
10	4.44	4.68	0.0576	0.3091
11	4.24	4.68	0.1936	0.5715
12	4.47	4.68	0.0441	0.2767
13	4.48	4.68	0.0400	0.2663
14	4.80	4.68	0.0144	0.0384
	$Y^E = 4.996$		Σ = 2.2853	Σ =15.7219

Adequacy of the model based on the multiple regression coefficient is determined by calculating the value of R and is calculated according to the equation:

$$R = \sqrt{1 - \frac{\sum_{j=1}^{n} \left(Y_j^E - Y_j^M\right)^2}{\sum_{j=1}^{n} \left(Y_j^E - -Y^E\right)^2}} \quad (17)$$

where are: Y_j^E – experimental value of j – of that repetition; Y_j^M– model value of j – of that repetition; -Y^E – mean value of experimental repetitions.

Therefore, the multiple regression coefficient of the logarithmic model can be calculated as:

$$R = \sqrt{1 - \frac{2.2853}{15.7219}} = 0.92446 \quad (18)$$

The value of the regression coefficient R=0.92446 means that the linear model describes the accuracy of the experiment results with 92.45%, which is a very good accuracy of the model.

By including the default values for the wood density, feed speed and number of cutters in the equation (13), that is (), we can obtain the values of the surface roughness parameters *Ra* and *Rz* in [μm] according to the linear model shown. Table 15 shows the values of the roughness parameters *Ra* and *Rz* according to the model and according to the experimental results.

Table 15. Comparative presentation of experimental and linear model results of parameters *Ra* and *Rz*

Ni	*A(g/cm³)*	*B(m/min)*	*C*	*Ra exp (μm)*	*Ra model (μm)*	*Rz exp (μm)*	*Rz model (μm)*
1	0.39	13.0	4	5.50	5.37	48.24	46.69
2	0.79	13.0	4	3.78	3.23	40.34	39.19
3	0.39	33.0	4	5.88	5.37	51.15	52.03
4	0.79	33.0	4	3.83	3.23	48.67	44.53
5	0.39	13.0	8	7.15	6.65	50.23	46.69
6	0.79	13.0	8	4.99	4.51	41.42	39.19
7	0.39	33.0	8	7.30	6.65	56.30	52.03
8	0.79	33.0	8	4.67	4.51	45.48	44.53
9	0.64	23.0	6	4.42	4.68	42.83	44.67
10	0.64	23.0	6	4.44	4.68	42.48	44.67
11	0.64	23.0	6	4.24	4.68	42.66	44.67
12	0.64	23.0	6	4.47	4.68	42.71	44.67
13	0.64	23.0	6	4.48	4.68	42.71	44.67
14	0.64	23.0	6	4.80	4.68	43.35	44.67

Based on the data from Table 15, we observe that the experimental results for the roughness parameters *Ra* and *Rz* are close to the values calculated by the stochastic model for *Ra* and *Rz*.

4. Optimization of processing parameters

Optimization can be understood as the process of defining a solution for given initial conditions from a set of possible solutions, where there are no restrictions when it comes to the choice of the object of optimization [9]. In order to be able to talk about optimization at all, it is first necessary to identify the goal, that is, the quantitative measure of the behavior of the system being studied. The process of defining the goal, variables and eventual limitations for a given problem is called the modeling process.

Designing a good model is the first step, and very often the most important, in the optimization process. The optimization procedure is expressed mathematically as searching for the minimum or maximum of the objective function [4]. In general, the objective function is a scalar function F with one or more parameters *xi* depending on how many variables are observed during the model optimization procedure. The objective function is denoted by F ($x_1, x_2, ..., x_n$) or as F (x). It is a unique function that must describe all properties that we want to achieve by optimizing the model. For the purposes of this work, the optimization was done in the Design expert software. We defined the minimum roughness values *Ra* and *Rz* and determined the calculation of optimal values for 100 repetitions (Table 16).

Table 16. Calculation of optimal values for 100 repetitions

Ni	*ρ*	*s'*	*n*	*Rz*	*Ra*	*Desirability*	
1	**0.789**	**15.145**	**7.465**	**40.179**	**4.378**	**1.000**	**Selected**
2	0.790	13.000	8.000	39.747	4.534	1.000	
3	0.790	13.000	4.000	38.489	3.254	1.000	
4	0.763	14.633	6.431	40.205	4.184	1.000	
5	0.772	13.682	4.780	39.246	3.600	1.000	
6	0.779	14.456	7.618	40.221	4.475	1.000	
7	0.765	13.408	7.746	40.243	4.587	1.000	
8	0.767	15.747	4.624	39.853	3.586	1.000	
9	0.784	13.437	4.188	38.769	3.345	1.000	
10	0.777	17.916	4.176	40.105	3.397	1.000	
11	0.752	15.518	4.930	40.168	3.763	1.000	
12	0.790	15.973	4.634	39.485	3.467	1.000	
13	0.770	14.323	6.122	39.889	4.046	1.000	
14	0.779	14.474	5.246	39.487	3.718	1.000	
15	0.762	14.733	4.277	39.570	3.500	1.000	
16	0.757	13.918	6.830	40.248	4.340	1.000	
17	0.776	17.654	4.242	40.074	3.422	1.000	
18	0.760	16.846	4.488	40.227	3.581	1.000	
19	0.788	19.145	4.201	40.229	3.348	1.000	
20	0.784	16.403	6.176	40.190	3.992	1.000	
21	0.787	13.756	6.988	39.680	4.227	1.000	
22	0.779	14.323	7.737	40.224	4.513	1.000	
23	0.766	13.234	4.245	39.072	3.460	1.000	
24	0.780	16.143	6.013	40.151	3.962	1.000	
25	0.784	13.569	7.175	39.758	4.305	1.000	
26	0.781	14.437	5.749	39.592	3.866	1.000	
27	0.780	14.772	7.768	40.335	4.519	1.000	
28	0.767	14.340	5.204	39.651	3.765	1.000	
29	0.773	13.381	7.144	39.905	4.354	1.000	
30	0.785	16.196	5.916	40.039	3.904	1.000	
31	0.754	13.506	4.063	39.327	3.470	1.000	
32	0.777	16.245	6.316	40.326	4.074	1.000	
33	0.785	16.162	6.674	40.277	4.149	1.000	
34	0.789	14.194	7.826	40.027	4.487	1.000	
35	0.780	14.120	5.732	39.521	3.865	1.000	
36	0.787	17.948	4.936	40.164	3.586	1.000	
37	0.770	14.423	5.902	39.847	3.976	1.000	
38	0.782	13.552	7.301	39.820	4.353	1.000	
39	0.728	13.355	4.694	39.969	3.810	1.000	
40	0.780	13.455	7.998	40.053	4.588	1.000	
41	0.743	13.707	6.237	40.269	4.225	1.000	
42	0.768	13.580	6.798	39.943	4.271	1.000	
43	0.782	14.664	7.687	40.247	4.483	1.000	
44	0.780	16.344	4.119	39.617	3.359	1.000	

45	0.765	16.493	5.077	40.236	3.745	1.000	
46	0.768	16.267	4.462	39.918	3.530	1.000	
47	0.775	18.438	4.120	40.257	3.389	1.000	
48	0.783	15.282	7.309	40.264	4.356	1.000	
49	0.789	17.805	5.574	40.293	3.780	1.000	
50	0.773	13.601	7.295	39.998	4.399	1.000	
51	0.789	14.518	5.644	39.422	3.788	1.000	
52	0.787	16.099	5.885	39.964	3.883	1.000	
53	0.760	13.457	7.127	40.165	4.419	1.000	
54	0.778	18.912	4.066	40.308	3.357	1.000	
55	0.762	13.042	7.917	40.256	4.657	1.000	
56	0.784	14.753	7.299	40.109	4.348	1.000	
57	0.782	14.819	7.534	40.231	4.432	1.000	
58	0.745	15.524	4.567	40.179	3.682	1.000	
59	0.788	18.537	4.690	40.230	3.505	1.000	
60	0.753	13.251	4.440	39.384	3.592	1.000	
61	0.788	15.096	7.954	40.328	4.536	1.000	
62	0.744	13.228	4.575	39.589	3.683	1.000	
63	0.783	17.301	4.389	39.896	3.431	1.000	
64	0.756	13.311	4.245	39.296	3.518	1.000	
65	0.772	13.363	6.440	39.687	4.131	1.000	
66	0.755	13.006	6.549	39.942	4.255	1.000	
67	0.783	14.451	7.940	40.241	4.555	1.000	
68	0.773	13.465	7.415	40.010	4.440	1.000	
69	0.771	17.754	4.068	40.133	3.392	1.000	
70	0.760	13.918	5.132	39.654	3.780	1.000	
71	0.714	13.383	4.126	40.050	3.700	1.000	
72	0.790	13.565	7.353	39.700	4.330	1.000	
73	0.742	13.433	6.503	40.291	4.313	1.000	
74	0.709	13.130	4.356	40.152	3.801	1.000	
75	0.775	14.840	7.279	40.293	4.390	1.000	
76	0.780	16.475	4.409	39.728	3.448	1.000	
77	0.777	15.738	6.582	40.272	4.157	1.000	
78	0.787	17.130	4.447	39.792	3.427	1.000	
79	0.786	13.001	7.928	39.800	4.532	1.000	
80	0.781	15.067	6.354	39.953	4.063	1.000	
81	0.749	15.218	5.370	40.275	3.917	1.000	
82	0.790	16.424	6.358	40.151	4.021	1.000	
83	0.762	13.621	6.373	39.922	4.164	1.000	
84	0.788	17.865	4.239	39.908	3.359	1.000	
85	0.771	13.900	6.875	39.999	4.281	1.000	
86	0.774	13.168	5.860	39.414	3.934	1.000	
87	0.783	16.760	5.166	39.995	3.678	1.000	
88	0.784	18.869	4.064	40.196	3.327	1.000	
89	0.769	13.461	7.931	40.238	4.625	1.000	
90	0.761	13.223	6.001	39.713	4.047	1.000	
91	0.785	15.475	6.861	40.139	4.203	1.000	
92	0.785	13.429	7.775	39.885	4.490	1.000	
93	0.746	13.094	6.876	40.237	4.408	1.000	
94	0.738	13.242	6.352	40.274	4.287	1.000	
95	0.780	17.855	4.278	40.058	3.411	1.000	
96	0.719	13.884	4.044	40.076	3.652	1.000	
97	0.789	15.740	4.320	39.334	3.369	1.000	
98	0.781	18.854	4.187	40.288	3.383	1.000	
99	0.763	13.269	6.614	39.885	4.235	1.000	
100	0.776	13.558	7.694	40.059	4.512	1.000	

Figure 3 shows 3D contour plot of input parameters and desirability of the optimization. There are regions of responses presented with colors. Desirability of 1, red color, means good optimization response.

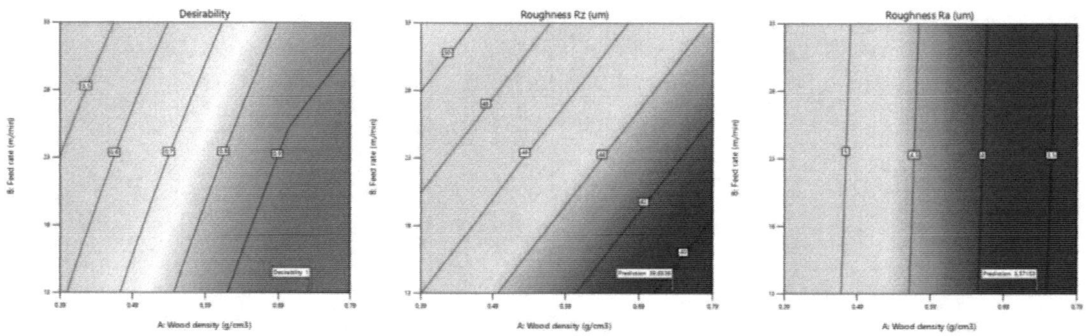

Figure 3. Contour plot of the optimization

Optimization can also be shown as a function of changing the value of the input parameters of the process, Figure 4.

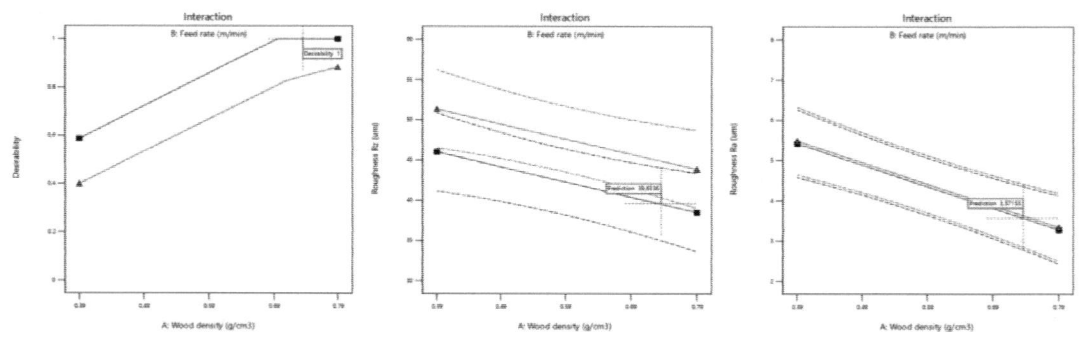

Figure 4. Optimization as function of input parameters values

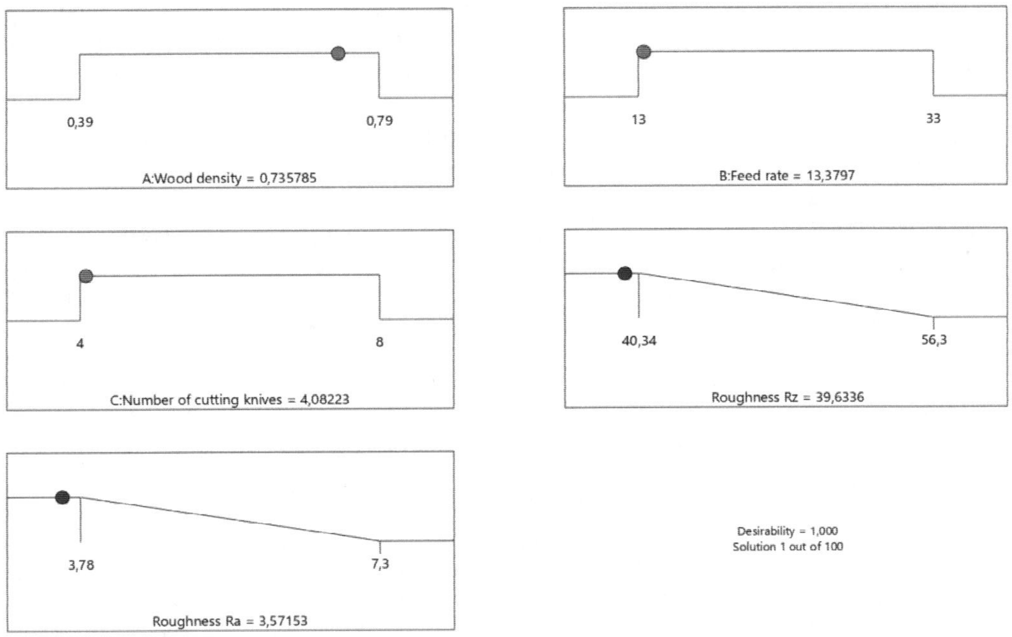

Figure 5. Ramps for optimal values

Figure 5 shows ramps for optimal values. Ramps are a graphical view of each optimal solution. Optimal factor settings are shown with red points and optimal response prediction values are displayed in blue.

5. Conclusion

The optimization of the influential parameters of the workability of solid wood in this work was carried out in order to achieve the maximum quality class of the treated surface, which is harmonized with the quality of the finished product. Using this method, it is possible to avoid exploitation tests and adjustments of input parameters in the process of planning massive wood on a planer machine, which are based on the subjective decision of the person who performs them. It is also possible to speed up the production process, because the time of realization of the production process will be shorter for the time spent on exploitation tests.

Based on the conducted research on the influence of wood density, feeding speed and number of spiral cutters on the process of planing solid wood on a multi-sided planer, it can be concluded:

- The results obtained from the experimental testing are essentially applicable and can be used for modeling the roughness parameters of the processed surface of all types of solid wood in the process of planing on planer machines.
- The results of stochastic modeling are reliable mathematical models of the mean arithmetic deviation of the profile from the mean line of the profile *Ra*, that is, the mean height of bumps *Rz* that can serve technologists, depending on the process parameters, to start the production process in a short time without long-term testing.
- The obtained mathematical models showed that the density of wood and the number of spiral cutters are far more significant input parameters on the roughness parameter *Ra*, while the input parameter feeding speed and the interaction of input parameters have no significant influence on *Ra*.
- The input parameters wood density and feeding speed are more influential factors for the roughness parameter *Rz,* while the number of spirals and interaction of the input parameters have no significant influence on the size of the roughness parameter *Rz*.

- Optimization as a function of changing input parameters gave optimal values of input parameters ρ = 0.7885 g/cm^3; s' = 14.1447 m/min and z = 7.4651 spiral cutters for optimal solutions of the output parameters of the roughness of the processed surface, which tend to the minimum values, i.e. Ra = 4.3776 μm and Rz = 40.1794 μm.
- By achieving the optimal parameters of the process of improving the quality of the processed surface of solid wood, before starting the production process, exploitation tests and increased costs of the processing process itself are avoided.
- The optimized model will be used in practical application to determine the optimal parameters of the machinability of solid wood in the process of planing on CNC machines in order to obtain the roughness of the machined surface of the appropriate roughness class according to the standard.

References

[1] Berni, R. and Burbui, M. (2014) 'Process optimization of a superfinishing machine through experimental design and mixed response surface models', Quality Engineering, 26(4), pp. 404–415. doi:10.1080/08982112.2013.872794.

[2] Bjelić, A. and Hodžić, A. (2017) 'The influence of the choice of cutting tool on the quality of the machined surface of massive wood' ('Uticaj izbora reznog alata na kvalitet obrađene površine masivnog drveta'), *11th International Scientific Conference on Production Engineering–RIM 2017* Sarajevo

[3] Dean, A., Voss, D. and Draguljić, D. (2017) 'Design and analysis of Experiments', *Springer Texts in Statistics* [Preprint]. doi:10.1007/978-3-319-52250-0.

[4] Ender, H. and Kucuk, H. K. (2016) 'Optimization of wood surface machining parameters in CNC routers: Response surface methodology (RSM) approach' *International Journal of Scientific Research Engineering & Technology (IJSRET), ISSN 2278 – 0882 Volume 5, Issue 10.*

[5] Jones, B. and Montgomery, D. (2019) 'Design of Experiments - Modern Approach',*Wiley*.

[6] Ratnasingam, J. and Scholz, F. (2005) 'Optimal surface roughness for high-quality finish on Rubberwood (Hevea brasiliensis)', *Holz als Roh- und Werkstoff,* 64(4), pp. 343–345. doi:10.1007/s00107-005-0068-6.

[7] Sofuoglu, D.S. (2016) 'Determination of optimal machining parameters of massive wooden edge glued panels which is made of Scots pine (PinussylvestrisL.) using Taguchi design method' *Eur. J. Wood Prod.*, doi 10.1007/s00107- 016-1028-z.

[8] Jurković, M. (1999) 'Mathematical modeling of engineering processes and systems' ('Matematičko modeliranje inženjerskih procesa i sistema'), Univerzitet u Bihaću, Mašinski fakultet.

Eco-friendly adhesives for wood-based composite production

Minka Ćehić, Hasan Talić and Salah Eldien Omer

University of Bihać, Technical Faculty Bihać, Dr. Irfana Ljubijankića bb

e-mail: minka.cehic@unbi.ba

Abstract. In the process of creating composites,especially if we talk about wood-based composites, great attention should be paid on a variety of factors which affect the final composite properties.One of the essential elements is proper adhesive selection. Most common used are synthetic formaldehyde adhesives. We are aware how dangerous emission of free formaldehyde can be; we tend to minimize its usage. Using adhesives which are safe for the environment and humans as well,we tend to find safer composites which are ecologically acceptable and recognized as a better option.The paper briefly describes the adhesives used so far and their dangers, and selects several types of adhesives with basic characteristics that can meet the requirements for the production of ecologically acceptable composite panels, with an emphasis on wood-based composites. Certain tests and achievements in the development of these adhesives are also presented.

1. Introduction

Wood-based composite materials are becoming more prevalent in a variety of applications, and this trend of increased development, production, and application of these materials has been evident for several decades.

A composite or composite material is any substance that is made up of two or more securely connected components that have diverse physical and chemical properties. These materials have distinct limits; they do not mix or lose their identity or qualities. On the contrary, when they are combined, they form a functional material that has been specialized to become stronger, lighter, more efficient, or more resistant, for example, to the action of electricity, stronger, and more durable, i.e. their properties have been significantly improved and changed in comparison to the properties of the original components.

The composite's structure is made up of a continuous phase called matrix, in which one or more discontinuous phases called the active filler or reinforcement are embedded.The discontinuous phase improves mechanical qualities and can affect other composite properties like as thermal, acoustic, electrical, magnetic, and optical properties. In wood composites, the filler is wood.

The wood is chopped into the shape and size of lamella, noodles - macrochips, chips, fibers, or grains of wood flour. A specific binder is used as a matrix, and various synthetic adhesives, various polymer binders, so-called plastics, and inorganic binders are most often used. Synthetic adhesives include adhesives based on resorcinol or phenolic resin and formaldehyde , melamine-formaldehyde and urea-formaldehyde resin-based adhesives, polyurethane adhesives, polyvinyl acetate adhesives, polyacrylic adhesives, and silicone-based or epoxy-based adhesives.

2. Conventional adhesives for the production of wood-based composites

Wood composites can be made in different shapes, often as panels or beam elements, flat or curved, but also in other shapes. Adhesives are used in the production of load-bearing elements of wooden structures, such as various supports and columns, purlins, rafters and roof trusses, and in the production of wood-based panel materials, such as plywood, chipboard, fiberboard, OSB boards, for the production of laminated products, carpentry, wooden floors and furniture.

Adhesives based on urea-formaldehyde resin, polyurethane, epoxy, polyester and polyvinyl acetate are often used for the production of the mentioned products. [1, 2, 3, 4]

The most commonly used glue is urea-formaldehyde (UF) glue (about 80% of wood-based panel composites are made using this glue). [5] The approximate worldwide annual consumption of urea-formaldehyde adhesives amounts to about 11 million tons. Urea-formaldehyde glue is obtained by a chemical reaction of polycondensation from urea and formaldehyde with the separation of side products such as water and formaldehyde.

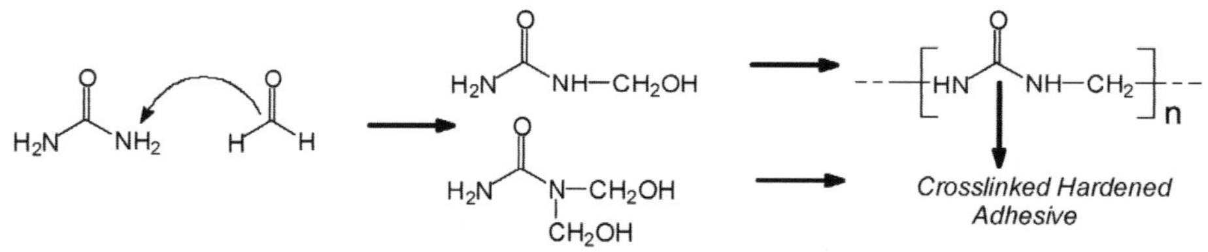

Figure 1. Scheme of the basic reaction between urea and formaldehyde in the process of preparing UF resin [3]

The hardening of these adhesives is enhanced by the application of suitable agents (hardeners) at normal temperature or by heating, i.e., under the influence of elevated temperature. In this way, the glue changes from a liquid state to a hardened, non-melting and insoluble state.The positive properties of the application of these glues are: the glue is light in colour, so that the cured joint is light or colourless, high adhesive ability with wood, fast hardening, high strength and stability of the quality of glued joints, rich sources and low price of raw materials, low production costs, and the possibility of producing relatively "harmless" glues.

Melamine is an expensive substance and that is why adhesive resins are more often produced as melamine-urea-formaldehyde (MUF) resins than of pure melamine-formaldehyde resins, especially because they both have the same effect.Two types of melamine resins are produced: the first MUF resins, with 2 - 5% melamine, i.e. improved urea-formaldehyde resins, that are applied for interior use, and the second MUF resins, with 30 to 40% melamine, which are applied for external and semi - external use.Figure 2 shows a schematic representation of these resins.

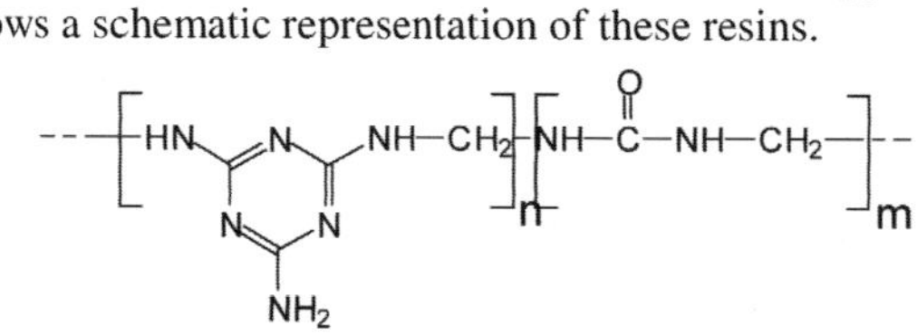

Figure 2. Schematic view of MUF resin [3].

The second most important adhesives for the production of wood composites are phenol-formaldehyde adhesives. The world consumption of these adhesives is about 3 million tons per year. A schematic view of PF glue is given in Figure 3.

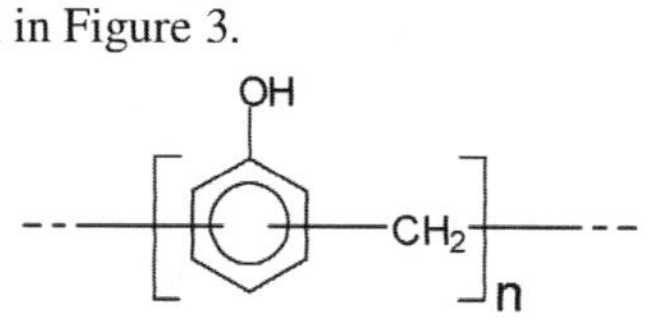

Figure 3. Schematic view of PF glue [3].

These adhesives are used in structures for external use.Phenol-resorcinol-formaldehyde (PRF) adhesives are cold-hardening. A schematic representation of these adhesives is given in the figure 4.

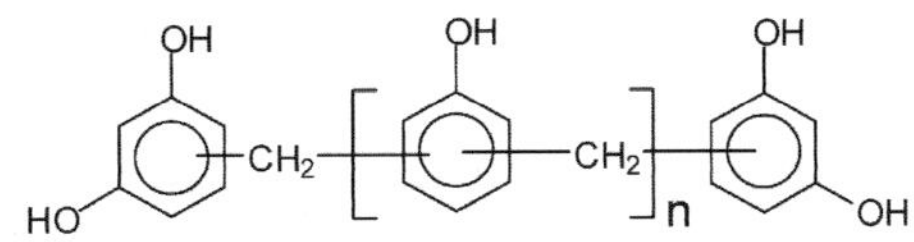

Figure 4. Schematic representation of RPF glue [3].

The cost of these adhesives is high (because resorcinol is expensive), yet they harden at room temperature exactly because of the presence of resorcinol.As a result, they are employed in the manufacture of glulam beams, LVL, and other similar goods used in outdoor temperature conditions and under heavy loads.Because of the high cost and the unique goods in which they are used, only a tiny amount (small volume) of these resins are produced and used (about 30 thousand tons per year). These, however, are resins with extraordinarily high property values.The emission of formaldehyde vapour and the toxicity of waste water during the production of adhesives represent a serious environmental problem with all of the above-mentioned adhesives, and this was also the cause of the development of new adhesives. The earliest adhesives utilized were isocyanate polymer adhesives and one-component polyurethane adhesives. In the production of wood composites, only polymer 4,4'-diphenyl methane diisocyanate (pMDI) is used. A schematic view of this glue is given in Figure 5.

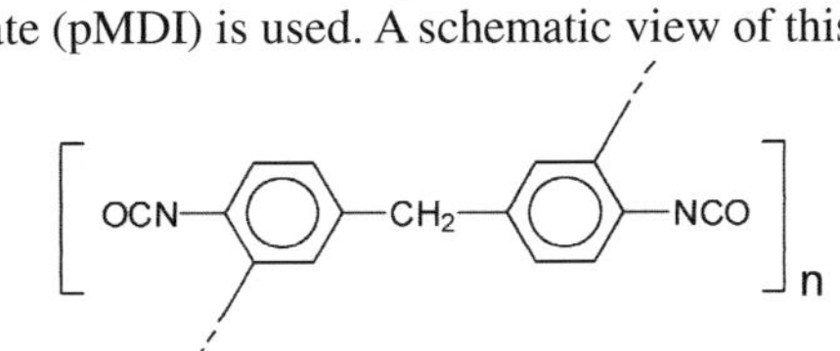

Figure 5. Schematic view of pMDI [3].

pMDI adhesive can be considered a type of polyurethane adhesive because pMDI reacts with OH during curing, forming urethane type crosslinks.

pMDI is an excellent adhesive, and when gluing wood composites, significantly smaller amounts of this adhesive are needed compared to the amount of applied UF, MUF and PF adhesive for the same purpose. But pMDI glue is also relatively more expensive glue. However, even this glue does not meet the strict environmental regulations, if we take into account the emission of steam and the relative toxicity of the glue.The main competition for RPF adhesives are one-component PUR adhesives for the same application area. Figure 6 shows a one-component PUR adhesive.

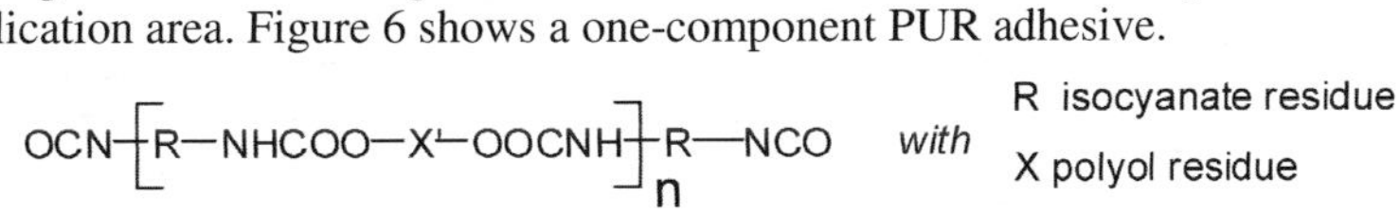

Figure 6. Schematic representation of PUR [3].

The main advantage of PUR glue is easier handling because it does not need to be mixed with a hardener or catalyst. However, precisely because of their structure, these adhesives are prone to yielding bonded joints subject which are subject to both creep and temperature-dependent creep. This is the main disadvantage of PUR adhesives as well as their relatively high price.

3. Emission of volatile substances and requirements of environmental acceptability and sustainability

One of the main sources of volatile organic compounds (VOC) in the air in closed spaces is the adhesives used for the production of wood composites intended for use in construction or interior decoration. Over 500 different compounds have been identified in indoor air including formaldehyde, phenol, resorcinol, vinyl acetate, methyl acetate, ethyl acetate, ethanol, methanol, acetone, benzene, toluene, trimethylbenzene, phenoxyethanol, dibutyl phthalate, 2-butanone,methylpentane, dimethylpentane, methyl hexane, cyclohexane and methyl cyclohexane.

These organic compounds can come from paints, varnishes, glues, floor coverings, wallpapers and other materials used. Air testing in indoor industrial plants in China showed that 27% of the total VOC

IOP Conf. Series: Materials Science and Engineering 1298 (2023) 012013 doi:10.1088/1757-899X/1298/1/012013

in the air originates from industrial plants that use paint, while 14% of the total VOC in the air comes from industrial plants that use glue.[6]

3.1. Ecological consequences of harmful substances

Various materials are used in the construction of structures, the interior decoration of various buildings, and the production of products that we use every day, including wood-based materials as well as non-wood materials that can be painted, varnished, glued, or otherwise treated during the production process. These materials can have an impact on human health and the environment, either positively or negatively. The discharge of a particular amount of dangerous compounds might have a detrimental influence, which varies depending on the end of life of the product utilized. Also, the technological process of production of materials or finished products can have a positive as well as a negative impact on the environment at a certain stage. A special problem is the disposal and environmentally friendly disposal of products that are no longer used (waste products).

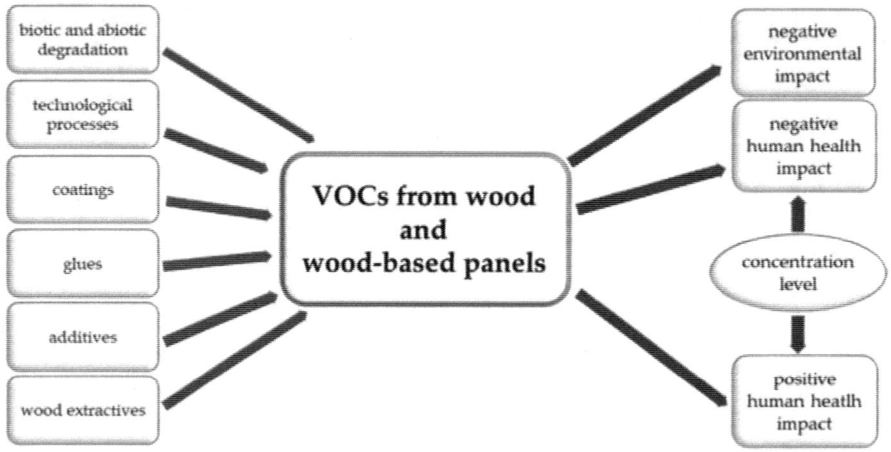

Figure 7. Sources of volatile organic compounds (VOC) from wood and wood-based panels and their impacts [6, 7].

In the production of conventional composite materials or finished products based on wood, significant amounts of adhesives are used, such as adhesives based on urea-formaldehyde resin, polyurethane, polyvinyl acetate, phenol-resorcinol formaldehyde, epoxy and polyester. The process of production - synthesis of these adhesives also has an impact on the human health and environment.

Volatile organic compounds from adhesive have a negative effect on the health of people and animals if they are inhaled or these compounds come into contact with the skin. They cause allergic reactions, irritation of the nasal passages, eyes, mouth and lungs, cause respiratory problems, migraines, and have carcinogenic and mutagenic properties and effects. [8] McDermott et al. have established in 2007 that small doses of toluene, hexane and methyl ethyl ketone damage the membranes of human neurons in just five days. [5] The waste water produced by the production - synthesis of adhesive resins contain a large amount of polluting substances and have high concentrations of compounds that are toxic to aquatic biocenoses. [9,10,11]

Formaldehyde is a gas with a characteristic pungent smell and is colourless.The chemical reactivity of formaldehyde is very high, and it is unstable. If the air contains formaldehyde in a concentration of the 1 ppm (1.23 mg/m3), it will strongly irritate eyes and mucous membranes. A concentration of formaldehyde in the air of 60 mg/m3 damages lung tissue. Even smaller concentrations of formaldehyde in the air cause damage and, with longer exposure, serious illnesses.Workers who work for more than five years in chipboard factories whose air contains 0.47 mg/m3 of formaldehyde fall ill with conical obstructive disease (COPD).The International Agency for Research on Cancer (IARC) has classified formaldehyde in Group 1 as a known human carcinogen. It is associated with the development of brain cancer, cancer of the nasal sinuses,cancer of the nnasopharynx,prostate cancer, and the development of leukaemia.Formaldehyde also negatively affectsreproductive health, causing an increase in menstrual disorders, increases the incidence of inflammatory diseases of the reproductive organs, increases the risk of spontaneous abortions during pregnancy, and the development of adverse effects on the fetus, primarily a decrease in birth weight.The lethal dose of formaldehyde (LD50) for humans, if swallowed, is 500 mg/kg.

Phenol is a solid crystalline substance of colourless to white colour, which is sensitive to light and air (in air it gradually changes colour to red) and dissolves in water. Commercial phenol is a liquid with a characteristic smell (sweet, similar to tar), which evaporates more slowly than water. Phenol can be found in the air and in water after leaving the production, use and disposal of products containing phenol. Phenol decomposes in the air within a day and a half, but it can remain in water for up to a week or more. Humans and animals can be exposed to phenol if they inhale or come into direct skin contact with contaminated air, water or a product containing phenol. In this way, phenol is absorbed in the body, causing certain complaints (health problems) depending on the length of exposure. Long-term exposure to phenols in the workplace is associated with cardiovascular disease, but workers are also exposed to other chemicals at the same time. They also have a harmful effect on the central nervous system. Ingestion of liquid products containing concentrated phenols can lead to serious gastrointestinal damage and even death. The lethal dose for humans, if swallowed, is 167 mg/kg of weight. [10] 70-80% of phenol vapour is retained in the lungs. Short-term exposure can cause irritation of the respiratory tract and muscle spasms in animals, and long-term exposure to high concentrations of phenol can cause damage to the heart, kidneys, liver and lungs.

Benzene is the simplest aromatic hydrocarbon. It is a colourless, volatile liquid with a characteristic smell that boils at 80 °C. It is absorbed in the human or animal body through the respiratory system and through the skin. Benzene is toxic, especially its vapours, which are retained in the lungs (80 %retention). Benzene accumulates in fatty tissue, has a narcotic effect, damages the bone marrow and weakens the immune system. Benzene is a carcinogenic substance that causes blood cancer.

Vinyl chloride (VC) is a gas with a characteristic smell of chloroform, which easily polymerizes into polyvinyl chloride and is difficult to dissolve in water. It is absorbed by the respiratory system and by contact with the skin. Vinyl chloride has a narcotic effect, irritates the skin and mucous membranes, damages the liver and can cause lung, brain and blood cancer.

Styrene is a colourless oily liquid that has an irritating sweet smell and is poorly soluble in water. It is absorbed by the respiratory system and has neurotoxic properties, and can cause depression in humans. It is considered mutagenic because its oxides form adducts with DNA and proteins.During the adhesive synthesis process, as an integral part of waste water, or from materials and/or from a finished product that is no longer used (waste product), adhesive compounds come into contact with water streams and enter the composition of this water.

Table 1. Some industrial processes that generatewastewater containing formaldehyde. [28].

Process	Chemical oxygen demand (g/l)	Formaldehyde concentration [g/l]
Dimethyl terephthalateproduction	30	3.0 – 5.0
Phenol-formaldehyde resinssynthesis	1.8	0.1
Urea-formaldehyde adhesives synthesis	0.46 – 3.0	0.2 – 4.0

Certain adhesive compounds are ecotoxic. The ecotoxicity of compounds on aquatic organisms is determined based on the rate of mortality, ontogeny and growth inhibition.The characteristic values of the constituent substances of wastewater from the synthesis of urea-formaldehyde adhesives are shown in table 2. [11]

Table 2. Characteristics of wastewater produced during the production of urea-formaldehyde adhesives.

Parameter	Characteristic value in [g/l]	
	Range	Average value
Chemicaloxygendemand	0.46 – 3.9	2.3
Formaldehyde concentration	0.22 – 4.0	0.95

Total Kjeldahl Nitrogen	0.12 – 0.81	0.35
NH4 (Ammonia)	0.003 – 0.018	0.008
Total suspended solids	0.019 – 0.15	0.06
Volatile suspended solids	0.016 – 0.14	0.05

Formaldehyde is a moderately ecotoxic compound. For example, formaldehyde concentration in water (LC(EC)50) of 50-60 mg/dm^3 is already harmful to crustacean (*Daphnia magna*), while for fish the minimum concentration is 2.3 mg/dm^3 (for *Tanichthys albonubes*), and upwards.

The ecotoxicity of phenol is different, so the LC50 value is in the range of 0.1-500 mg/dm^3 for fish, or 0.005-175 mg/dm^3 for crustaceans.Benzene is considered a mildly or moderately toxic compound for aquatic organisms. Toxic LC50 concentrations are in the range of 9-425 mg/dm^3 for fish, in the range of 10-200 mg/dm^3 for insects and in the range of 18-373 mg/dm^3 for the crustacean *Daphnia magna*).

3.2 Environmental acceptability and sustainability requirements

The dangers of the aforementioned volatile substances have been extensively publicized for a long time (decades). However, little information on the toxicity of adhesives and additives is available in the literature. Acrylamide, 2,5-hexanedione, terpenes and aldehydes, and coniferous resins were examined and evaluated.Some countries recognized the dangers of these drugs to human health a long time ago, in the 1980s, and established the first restrictions and guidelines. For example, the US Agency for Housing and Urban Development (HUD) has established formaldehyde emission limitations of 0.30 ppm for chipboard and 0.20 ppm for plywood if they are to be used in the US.

These limit values are high by today's standards, but it is significant that the process of regulating formaldehyde emissions from wood-based panels has continued since then, and with regulation ATCM 93120, issued by the Californian Air Resource Board (CARB) and whose application is fully has been since 2012, and with the adoption of the Law on Formaldehyde Standards for Composite Wood Products (adopted by the Environmental Protection Agency EPA), the maximum values of formaldehyde emissions are prescribe

At the same time (during the 1980s), West Germany, Denmark and the Netherlands were the first European countries to regulate formaldehyde concentrations in the air in closed spaces. The West German classification of wood-based boards according to formaldehyde potential expressed as a preforator value had the greatest influence.[12] Three emission classes were adopted, marked with the letter "E" and a number representing the formaldehyde emission level. Class E1 denoted boards whose formaldehyde emission according to the perforator method is less than 10 mg/100 g of board. Formaldehyde emission in E2 class boards was 10-30 mg/100 g board, and in E3 class boards formaldehyde emission was 30-40 mg/100 g board. Panels of class E2 and E3 required additional processing (surface processing or processing of surfaces and edges of panels).

The European Committee for Standardization (CEN) defines formaldehyde emission limits in technical standards for each type of wood-based panels.

For example, the limits of formaldehyde emission in panels are prescribed: for chipboard - norm EN 312, for fiber boards - norm EN 622-1, for OSB boards - norm EN 300, for plywood - norm EN 636 and for all boards used in construction - norm EN 13986.[13, 14, 15, 16, 17]

According to these norms and the limitation of formaldehyde emission, the products are divided into two classes: E1 and E2.E1 class wood-based panels should have a concentration of formaldehyde in the air $\leq$ 0.124 mg/m^3, if determined by the chamber method (EN 717-1; ISO 12460-1), or approximately 0.1 ppm. [18, 19] In many European countries, it is allowed to use wood-based panels only of class E1.In many European countries, only wood-based boards of class E1 are allowed to be used.

Table 3. The formaldehyde emission classes for some wood-based panels in accordance with EN norms and California Air Resources Board (CARB) final regulation 93120.

Emission class	Type of wood-basedpanel	Upper emission limit	Test method
E1	All	0.124mg/m^3	Chamber method

		≈0.10 ppm	EN 717-1
CARB phase 2	Hardwood plywood	0,05 ppm	Large chamber method ASTM E 1333
	Particleboard	0,09 ppm	
	Medium density fiberboard(MDF)	0,11 ppm	
	thin MDF	0,13 ppm	

Many methods are used to investigate VOCs. The most commonly used test methods are: chamber method (test chamber), gas analysis method, extraction method (called the perforator method), desiccator method. These methods are standardized in Europe and the USA by appropriate standards and norms: ISO 12460-1 (EN 717-1), ISO 12460-3 (EN 717-2), EN ISO 12460-5 (EN 120), ASTM E 1333 , ASTM D 6007, ASTM D 5562.[18 - 26]

4. Production of environmentally acceptable adhesives for wood composites

When compared to previous periods, the process of inventing and producing new materials or upgrading the qualities of current materials is changing today. The circular bio economy, environmental protection, and ecological acceptability and sustainability of production are becoming increasingly important in all stages of the manufacturing process, as well as in application.

Given this, and the fact that a significant amount of wood composites are produced using formaldehyde-based glue, tests and research into the possibility of developing new composites, optimization of the production process of conventional and new types of composites, optimization of the used raw materials, and the application of new types of raw materials from alternative natural and renewable sources have been conducted in recent decades. all with the aim of reducing the emission of harmful formaldehyde. [12, 27, 31]

Many factors can affect the emission of formaldehyde and other VOCs - in the air: the type of resin, the molar ratio between formaldehyde and other compounds in the glue (such as urea); the degree of polycondensation; the type of wood; the water content, the type and quality of additives; conditioning methods, and the application of external wood coatings. Operating conditions and factors (parameters), such as temperature and humidity, efficiency of the air conditioning system, air conditioning system efficiency, room usage, or any interior renovation or equipment repair, can also have a significant impact on VOC emissions. [9]

The production of wood-based panels with reduced or no formaldehyde emissions can be achieved in several ways:

1. By reducing the molar ratio of formaldehyde and another substance, e.g., urea, which participates in the production of these adhesives,
2. Application of a formaldehyde cleaner,
3. Replacement of these adhesives, e.g., urea-formaldehyde adhesives, with other synthetic adhesives that are produced without formaldehyde (and thus no formaldehyde emission),
4. Replacement of formaldehyde-based glues, e.g., urea-formaldehyde glues, with bioadhesives,
5. Use of alternative reactants instead of formaldehyde.

Attempts to reduce formaldehyde emissions from wood-based panels and to bring emissions within acceptable limits (panels class E1 and better) are achieved by optimizing the synthesis and formulation of adhesives, primarily of urea-formaldehyde adhesives as the most commonly used adhesives.The ratio of formaldehyde to urea in the new adhesives is 1.02. A further decrease in the ratio of formaldehyde to urea would diminish the effect of formaldehyde as the main reactant in the polycondensationand cross linking reactions of urea-formaldehyde adhesives.

Application of formaldehyde cleaner is an old idea and has long been used to lowering formaldehyde emissions from wood-based panels bonded with urea-formaldehyde adhesive. Formaldehyde cleaners can be applied before the pressing process, as a pre-treatment when formaldehyde cleaners are mixed with urea-formaldehyde adhesive or used as an impregnation agent that is sprayed on the surface of the board, or applied by a post-treatment process by spraying on the surface of the finished board.

Urea itself (for subsequent addition to the prepared UF glue) and ammonia and ammonia salts can be used as formaldehyde cleaners; (although good results are achieved, their use is limited due to health problems. In order to reduce formaldehyde emissions, fillers in the glue can and are used, such as wheat flour and clay (sepiolite), which reduce the amount of applied UF resin. In this way, certain properties of the composite material are improved, but formaldehyde emissions are also reduced.

UF glue, as well as other glues in the synthesis of which formaldehyde participates, are being replaced by new types of glues such as isocyanateglues (pMDI glue) and one-component polyurethane glues. Also, for special application conditions (increased ambient humidity, etc.), MUF and PF glue are used. These adhesives are applied in smaller quantities, and it is also possible to modify them by applying certain biosubstances, such as lignin, which affects the emission of formaldehyde.

Reduction of formaldehyde emissions is also achieved by using alternative reactants. Knowing that urea is a non-toxic substance, and formaldehyde is toxic, in order to produce environmentally friendly adhesives, the use of other reactants instead of formaldehyde began to be considered and researched. So far, the greatest effect has been achieved with the use of glyoxal. [27] It is a non-volatile and non-toxic aldehyde, which is colourless and relatively cheap, but it is also less reactive than formaldehyde.

4.1 *Bio-based adhesives for the production of wood composites*

Bio-based adhesives include adhesives derived from natural, non-mineral and organic sources. Various biomass resources such as lignin, starch, plant proteins, tannin, bark and vegetable oils have been used as renewable raw materials to synthesize bio-based adhesives.

The most researched biomaterial for adhesives is lignin. Lignin is present in the lignocelluloses of wood, grasses, residues of agricultural crops and other plants. It is an aromatic polymer of high molecular weight based on phenylpropane units in a densely cross-linked structure. Lignin-based adhesives can be divided into two groups: phenol-formaldehyde lignin-based adhesives and formaldehyde-free lignin-based adhesives. In the first group, lignin is used as a partial substitute for phenol, while in the second, lignin reacts with non formaldehide chemicals to create bio-adhesives that are less toxic. [3, 9, 28, 29]

The amount of replacement of phenol with lignin in phenol formaldehyde adhesives is below 50%. The presence of lignin decreases the reactivity of the resin, resulting in a longer reaction time. The use of a suitable crosslinker is critical for the creation of lignin-based adhesives that can match the reaction speed and adhesion strength criteria of the wood panel industry. [2, 29, 30]

Lignin-based adhesives can be divided into two groups: phenol-formaldehyde lignin-based adhesives and formaldehyde-free lignin-based adhesives. In the first group, lignin is used as a partial substitute for phenol, while in the second, lignin reacts with chemicals other than formaldehyde to create bio-adhesives with less toxicity. The amount of phenol replacement in phenol formaldehyde adhesives is below 50%. The addition of lignin reduces the reactivity of the resin, which leads to an increase in reaction time. The use of an appropriate crosslinker is an important factor for the development of lignin-based adhesives that can meet the requirements of reaction speed and adhesion strength in the wood panel industry.

The development of the production process of formaldehyde-free glue based on lignin is in several directions.

The innovative thermoreactive resin was synthesized from a combination of polyfurfuryl alcohol (PFA) and lignin with two different lignin contents of 20 and 30% by weight. In the lignin-furfural resin system, furfural and lignin were used to replace formaldehyde and phenol in phenol-formaldehyde resins, respectively.Another direction in the development of formaldehyde-free adhesives based on lignin is characterized by the replacement of formaldehyde with glycosal in an alkaline state. The modified lignin was mixed with pMDI and applied as awood glue (or more precisely for the production of chipboards).

A formaldehyde-free, lignin-based wood glue can also be produced using Kraft lignin and polyethyleneimine (PEI). If during the synthesis of the glue the weight ratio of lignin/PEI is 2:1, the glue that can be used for the production of plywood will be produced. Dimethylated Kraft lignin (DKL) and polyethyleneimine (PEI) can be used for the production of glue in a weight ratio of 1:4 to 4:1. In this way, the obtaining of certain characteristics of the glued joint is influenced. The best shear strength results were achieved when the DKL/PEI weight ratio was 1:1.

Ammonium lignosulfonate (AL) oxidized in the presence of H_2O_2 in an alkaline state and then mixed with polyethyleneimine(PEI) can also be used as a green binder.

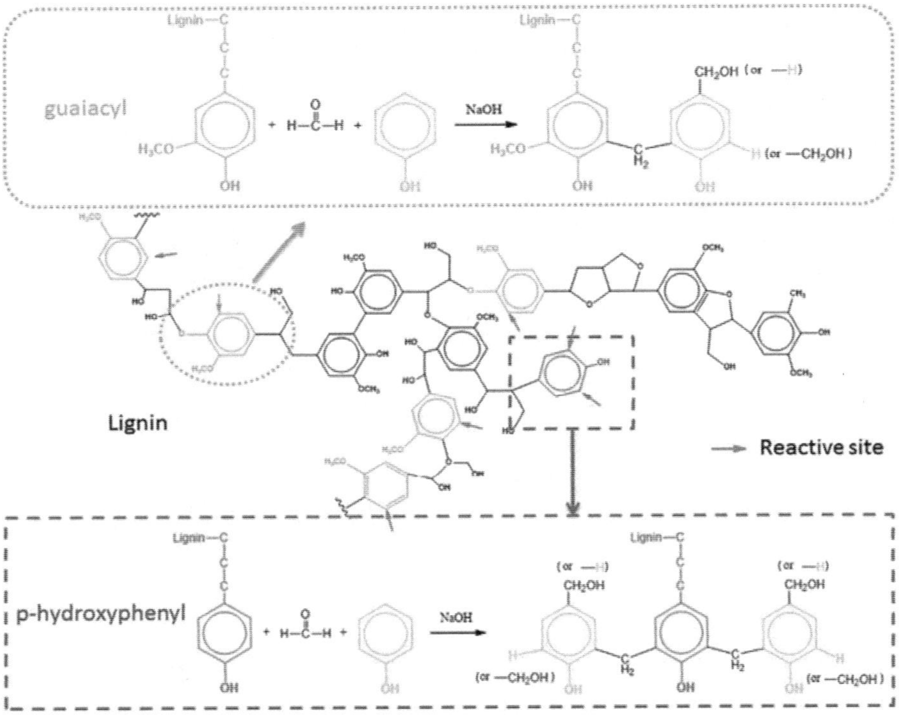

Figure 8. The reaction equation for the formation of lignin-based glue. [5]

Starch is a natural polymer, which has a wide range of appearance and application.Starch is also used, among other things, for the production of environmentally friendly and sustainable adhesives, because of its renewability, availability, good adhesion and low price.

Starch is a mixture of two different polysaccharide fractions - amylose and amylopectin. The proportion of amylose and amylopectin varies depending on the botanical origin and affects the properties of the wood glue.

In the process of synthesis of starch-based adhesives, hydrogen bonding forces play a significant role. Starch creates hydrogen bonds with water molecules, which leads to poor water resistance of the glue itself and the glued joint. A higher bond strength value and better water resistance is achieved by combining starch with another component, for example polyvinyl alcohol, formaldehyde, isocyanates and tannins.

A typical chemical modification of starch is achieved by esterification. Esterification of corn starch was achieved by reaction with maleic anhydride and this starch was cross-linked with polyisocyanate prepolymer. The optimal amount of prepolymer is 10% by weight.

Another way to produce a modified starch-based adhesive is by adding a blocked isocyanate and using agents. The ecologically acceptable starch-based adhesive was synthesized by grafting polymerization of vinyl acetate monomer onto corn starch and crosslinking polymerization with N-methylol acrylamide. [30, 31, 32]

An environmentally friendly glue that can be used for the production of wood composites is epoxy glue. This glue is used in the synthesis processes of starch-epoxy glue and tannin-epoxy glue.

Tannins are natural polyphenols. There are two types of tannins: hydrolysable tannins and condensed tannins. Tannins can be found in the leaves, seeds, roots and stems of various plant species, but only a few plants have high enough concentrations to make their extraction worthwhile.

Significant amounts of tannin can be extracted from the following types of wood: chestnut, oak, pine, maple, birch, willow, wattle, eucalyptus, myrtle, etc.

The possibility of partially replacing phenol (up to 50%) with hydrolysable tannins has been known for about fifty years. However, the naturally low molecular structure, low level of phenol substitution, limited production and relatively high price have influenced the lesser use of this type of tannin. About 200,000 tons of condensed tannin are produced in the world, which accounts for 90% of the world's commercial tannin production.

The possibility of using tannin as an adhesive for the production of wood-based panels depends on the content of reactive polyphenols and the reactivity of these components towards formaldehyde. Tannins can be used alone as adhesives (with the formaldehyde component as a crosslinker) or in combination with aminoplastic or phenolic resins. Thus, for example, an MDF board produced with tannins that replace parts of phenol in phenol-ureaformaldehyde resins or even with 100% tannin resin can meet the quality requirements for internal use, but not for external use. [30, 33, 34, 35]

Soy-based adhesives are synthesized by partial hydrolysis of soy protein. The effectiveness of soy-based adhesive depends on the morphology, viscosity, particle size and nature of the bond with the substrate. The proteins in the glue can be modified to improve water resistance and bond strength. Common methods for improving water resistance and bond strength include alkaline modification of soy protein, urea modification of soy protein, enzyme modification, surfactant, cross-linking and grafting modification. [6, 36, 37]

Adhesives derived from soy protein are available in three different forms: an ineffective crude extract-based adhesive, a protein-rich concentrate, or an effective adhesive-based protein isolate. A good adhesive for chipboards and plywood is synthesized by the co-reaction of condensed and hydrolysed tannins and soy protein isolate (SPI).).[33, 38, 39] The co-reaction is caused by the reaction of amino groups of protein side chains that replace some of the phenolic hydroxyl groups of tannins.[40, 41]

Carbohydrate adhesives are adhesives in which sugar molecules are used as modifiers for formaldehyde adhesives. Typically, the modification process involves the creation of degradation compounds, which are ultimately used as a source for adhesive formation. Certain carbohydrate compounds are reduced in an acidic environment, leading to the formation of furfur aldehyde and furfuryl alcohol, which are ultimately used to produce wood-based panel adhesives .[1, 31, 32, 42]

Modification of bio-based adhesives with different dispersing agents and crosslinkers is being done to improve their strength and water resistance properties so that these adhesives can compete with synthetic adhesives. Mixing proteins and carbohydrates can give good bio-adhesives with acceptable adhesive properties.

As we mentioned earlier, conventional adhesives, in the synthesis of which formaldehyde participates, have a whole series of advantages that we have already written about. However, the main problem is the harmful emissions of formaldehyde and waste water produced during the production of glue.

New types of adhesives with significantly reduced or no formaldehyde emissions are still in the development phase and efforts are being made to ensure that these adhesives have properties that are characteristic of conventional adhesives. In certain cases, the observed defects of the bio-adhesive can be improved by different combinations. For example, if you want to increase the water resistance of the glue based on soy protein, a certain amount of lignin is added. Adding a small amount of lignin, soy protein/lignin = 10: 2, concentration 12%, not only increases water resistance, but also adhesion strength and shear strength.

5. Conclusion

Significant research has been conducted in recent decades to develop ecologically acceptable adhesives for use in the manufacture of wood composites. The reason for this is that more and more attention is being paid to the harmful effects of adhesive substances on human and animal health, environmental pollution, and the need for wood composites to be as environmentally friendly and sustainable as possible, in accordance with circular economy principles.

Bio-based adhesives are a more environmentally friendly and long-lasting alternative to traditional adhesives. All of the bioadhesives indicated can greatly reduce the negative impact of hazardous emissions of formaldehyde and volatile organic compounds from wood composites, especially wood-based panels, on the environment and both human and animal health. Although these adhesives have a significant ecological advantage over conventional adhesives, they still have significant flaws or significant problems during synthesis, so that when conventional adhesives are replaced in the production process of wood composites by environmentally friendly adhesives, many more tests are required to determine the factors for formulating adhesives with optimal properties and to expand the application.

References

[1] Nordqvist P and Thedjil D and Khosravi S and LawtherM and Malmstrom E and Khabbaz F 2012 Wheat Gluten Fractions as Wood Adhesives-Glutenins Versus Gliadins *J. Appl. Polym. Sci.* 123 1530–1538 doi:10.1002/app.34312
[2] Pizzi A 2006 Recent developments in eco-efficient bio-based adhesives for wood bonding: Opportunities and issues *J. Adhes. Sci. Technol.* 20 829–846 doi:10.1163/156856106777638635
[3] Pizzi A and Papadopoulos A.N and Policardi F 2020 Wood Composites and Their Polymer Binders *Polymers* 12(5) 1115https://doi.org/10.3390/polym12051115
[4] Papadopoulos A.N 2021 Advances in Wood Composites III *Polymers* 13 163 doi:10.3390/polym13010163
[5] Luo J and Luo J and Yuan C and Zhang W and Li J and Gao Q and Chen H 2015 An eco-friendly wood adhesive from soy protein and lignin: Performance properties *RSC Advances* 122 doi https://doi.org/10.1039/C5RA19232C
[6] Papadopoulos A.N 2020 Advances in Wood Composites II *Polymers* 12 1552 https://doi.org/10.3390/polym12071552
[7] Adamova T and Hradecky J and Panek M 2020 Volatile Organic Compounds (VOCs) from Wood and Wood-Based Panels: Methods for Evaluation, Potential Health Risks, and Mitigation, *Polymers* 12 2289 doi:10.3390/polym12102289
[8] McDermott C and Allshire A and van Pelt F and Heffron J2007 Sub-chronic toxicity of low concentrations of industrial volatile organic pollutants in vitro *Toxicol Appl. Pharm.* 219 85-94 doi: 10.1016/j.taap.2006.12.004
[9] Łebkowska M and Załęska-Radziwiłł M and Tabernacka A 2017 Adhesives based on formaldehyde – environmental problems, *BioTechnologia*vol. 98(1) C pp. 53-65 doi: https://doi.org/10.5114/bta.2017.66617
[10] Lu J and Zhuo Q and Ren X and Qiu Y and Li Y and Chen Z and Huang K 2022 Treatment of wastewater from adhesive-producing industries by electrocoagulation and electrochemical oxidation *Process Safety and Environmental Protection* 157 27-536
[11] Vidala G and Jiangb Z.P and Omilc F and Thalassod F and Mendezc R and Lemac J.M 1999 Continuous anaerobic treatment of wastewaters containing formaldehyde and urea, *Bioresource Technology* 70 3 283-291
[12] Popović M and Điporovic-Momčilović M and Gavrilović-Grmuša I 2020 New standards and regulations on formaldehyde emission from wood-based composite panels *Zastita Materijala 61* (2) pp 152 – 160 doi: 10.5937/zasmat2002152P
[13] BAS EN 312:2012 Particleboards - Specifications
[14] BAS EN 622-1:2005Vlaknaste ploče - Specifikacije - Dio 1: Opšti zahtjevi
[15] BAS EN 300:2009Oriented Strand Boards (OSB) - Definitions, classification and specifications
[16] BAS EN 636+A1:2016Plywood - Specifications
[17] BAS EN 13986+A1:2016Wood-based panels for use in construction - Characteristics, evaluation of conformity and marking
[18] EN 717-1:2004 Wood-based panels - Determination of formaldehyde release - Part 1: Formaldehyde emission by the chamber method
[19] ISO 12460-1:2007 (en) Wood-based panels — Determination of formaldehyde release — Part 1: Formaldehyde emission by the 1-cubic-metre chamber method
[20] BAS EN ISO 12460-3:2022Wood-based panels — Determination of formaldehyde release — Part 3: Gas analysis method
[21] BAS EN 717-2:2002Wood-based panels - Determination of formaldehyde release - Part 2: Formaldehyde release by the gas analysis method

[22] EN ISO 12460-5:2015Wood-based panels - Determination of formaldehyde release - Part 5: Extraction method (called the perforator method) (ISO 12460-5:2015)

[23] EN 120 - Wood-based panels - Determination of formaldehyde release - Extraction method (called the perforator method)

[24] ASTM E1333-22 Standard Test Method for Determining Formaldehyde Concentrations in Air and Emission Rates from Wood Products Using a Large Chamber

[25] ASTM D6007-22 Standard Test Method for Determining Formaldehyde Concentrations in Air from Wood Products Using a Small-Scale Chamber

[26] ASTM D 5562 -95 Standard Test Method for Determination of the Acidity as Free Fatty Acids/Acid Number in the Presence of Ammonium or Triethanolamine Soaps / Note: Approved 2011, 2006, 2001

[27] Antov P and Savov V and Neykov N 2020 Sustainable Bio-Based Adhesives for Eco-Friendly Wood Composites A review *Wood Research* 65 (1) doi:10.37763/wr.1336-4561/65.1.051062

[28] Jawaid M and Tanveer A and Nasir M and Asim M 2021 Eco-Friendly Adhesives for Wood and Natural Fiber Composites: Characterization, Fabrication and Applications *Composites Science and Technology* 1st ed. 2021 EditionSpringer doi:10.1007/978-981-33-4749-6

[29] Klasnja B and Kopitovic S 1992 Lignin-phenol-formaldehyde resins as adhesive in the production of plywood *Holz als Roh- und Werkst* 50282–285 https://doi.org/10.1007/BF02615352

[30] Papadopoulos A.N 2020 Advances in Wood Composites*Polymers* 12 48 https://doi.org/10.3390/polym12010048

[31] Wang P and Cheng L and Gu Z and Li Z and Hong Y 2015 Assessment of starch-based wood adhesive, quality by confocal Raman Microscopic detection of reaction homogeneity*Carbohydr. Polym.* 131 75–79DOI: 10.1016/j.carbpol.2015.05.044

[32] LiZ and WangJ and Li C and Gu Z and Cheng L and Hong Y 2015 Effects of montmorillonite addition on the performance of starch-based wood adhesive *Carbohydr. Polym.* 115 394–400 doi: 10.1016/j.carbpol.2014.08.106

[33] Trosa A and Pizzi A 2001 A no-aldehyde emission hardener for tannin-based wood adhesives for exterior panels *Holz als Roh- und Werkst* 59 266–271 https://doi.org/10.1007/s001070100200

[34] FengS and Yuan Z and Leitch M and Shui H and XuC 2016 Effects of bark extraction before liquefaction and liquid oil fractionation after liquefaction on bark-based phenol formaldehyde resoles*Ind. Crop. Prod.* 84 330–336 doi:10.1016/j.indcrop.2016.02.022

[35] Feng S and Yuan Z and Leitch M and Xu C 2015 Adhesives formulated from bark bio-crude and phenol formaldehyde resole *Ind. Crop. Prod.* 76 258–268 doi:10.1016/j.indcrop.2015.06.056

[36] Pan Z and CathcartA and Wang D 2006 Properties of particleboard bond with rice bran and polymeric methylene diphenyl diisocyanate adhesives*Ind. Crop. Prod.* 23 40–45 doi:10.1016/j.indcrop.2005.03.004

[37] Savov V and Antov P and Zhou Y and Bekhta P 2023 Eco-FriendlyWood Composites: Design, Characterization and Applications *Polymers* 15 892 https://doi.org/10.3390/polym15040892

[38] Santoni I and Pizzo B 2013 Evaluation of alternative vegetable proteins as wood adhesives *Ind. Crop. Prod.* 45148–154DOI:10.1016/j.indcrop.2012.12.016

[39] Xu Y and Han Y and Li J and Luo J and Sheldon Q.S and Li J and Gao Q and Mao A 2022 Research Progress of Soybean Protein Adhesive: A Review *Journal of Renewable Materials* 10(10) 2519-2541

[40] Ghahri S and Pizzi A and Hajihassani R 2022 A Study of Concept to Prepare Totally Biosourced Wood Adhesives from Only Soy Protein and Tannin *Polymers* 14(6) 1150 https://doi.org/10.3390/polym14061150

[41] Li H and Wang Y and Xie W and Tang Y and Yang F and Gong C and Wang C and Li X. and Li C 2023 Preparation and Characterization of Soybean Protein Adhesives Modified with an Environmental-Friendly Tannin-Based Resin *Polymers* 15(10) 2289 https://doi.org/10.3390/polym15102289

[42] Ferdosian F and Pan Z and Gao G and Zhao B and2017Bio-Based Adhesives and Evaluation for Wood Composites Application *Polymers* 9(2) 70-79 https://doi.org/10.3390/polym9020070

IOP Conf. Series: Materials Science and Engineering 1298 (2023) 012014 doi:10.1088/1757-899X/1298/1/012014

Performance of Coconut Wood in Timber Structures: A Review of Its Properties and Applications

Leila Fathi[1], Redžo Hasanagić[2], Aldin Bjelić[2] and Mohsen Bahmani[1]
[1]Faculty of Natural Resources and Earth Science, Department of Industrial Engineering Furniture, Shahrekord University, Shahrekord, Iran
[2] Department of Wood Science and Technology, Faculty of Technical Engineering, University of Bihać, Bihać, Bosnia and Herzegovina

E-mail: leila.fathi@sku.ac.ir

Abstract. Timber structures have been a popular choice for construction due to their natural and aesthetic appeal. However, with the increasing focus on sustainability and eco-friendliness, alternative building materials are gaining popularity. One such material that has gained attention is coconut wood. Coconut wood is a by-product of the coconut industry and has several unique properties that make it an excellent choice for timber structures. This paper reviews the properties and applications of coconut wood in timber structures and discusses its advantages, limitations, and challenges. We discussed the physical and chemical properties and durability of coconut wood. The average density of coconut palm wood ranges from 0.41-1.11g/cm^3, while its moisture content ranges from 50% to 400%. Coconut wood has low shrinkage and swelling rates, reducing the risk of cracking or warping. The holocellulose content is about 67% while the lignin content is approximately 25%. Chemical and natural products, are effective in protecting coconut wood against decay and insect attack. Understanding such characteristics of coconut wood is critical for its optimal utilization in various industries. By employing appropriate preservation techniques and utilizing this versatile and sustainable resource, coconut wood can continue to provide significant benefits for communities and industries around the world.

1. Introduction

Timber structures have been used for centuries due to their natural and aesthetic appeal. However, in recent years, the focus has shifted towards sustainable and eco-friendly building materials. One such material that has gained attention is coconut wood. Coconut wood is a by-product of the coconut industry and is an attractive alternative to traditional timber due to its fast-growing nature and superior properties [1]. The coconut palm tree (Cocos nucifera) is a monocotyledonous species that belongs to the Arecaceae family. Coconut wood comes from the trunk of coconut palm trees, which are primarily grown in tropical regions around the world [2]. The coconut palm tree is native to the tropical regions of Southeast Asia and Melanesia, but it has been widely cultivated and naturalized in many other parts of the world. Today, coconut palms can be found growing in tropical and subtropical regions throughout the world, including Africa, the Caribbean, South America, and the Pacific Islands. In some areas, coconut palms are an important cash crop, providing income and livelihoods for local communities. They are particularly well-suited to coastal regions with sandy soils and high levels of rainfall, as they require ample water and good drainage [3]. However, coconut palm trees are also vulnerable to a number of pests and diseases, which can have devastating impacts on local populations [4]. For example, the coconut rhinoceros beetle, a destructive pest that feeds on the sap of young palm trees, has caused significant damage to coconut plantations in parts of the Pacific region [5]. Overall, the distribution of coconut palm trees reflects their adaptability and value as a crop, as well as the complex ecological and economic factors that influence the growth and cultivation of agricultural crops around the world [6]. Indonesia, the Philippines, and India are also major producers of coconut wood and timber. Additionally, there has been a growing trend in recent years towards using sustainable materials such as coconut wood, which may lead to an increase in production as demand for eco-friendly products continues to grow [7]. It estimated that approximately 20 million cubic meters of coconut lumber is produced each year from harvested mature tree [8]. Figure 1 indicate transverse section of a coconut palm stem.

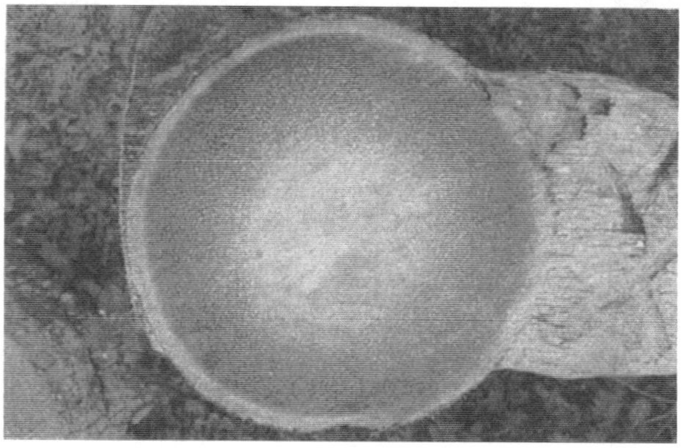

Figure 1. Transverse section of a coconut palm stem [9]

Coconut timber has many applications as both a structural and interior design material. The harder, high-density timber is suitable for general structural purposes such as pillars, trusses, rafting, furniture, window and door frames, floors, decking and floor joists [10, 11]. The construction industry provides the bulk of the demand for coconut lumber. In the construction of buildings, coconut lumber is used in large volumes as scaffolding and as form lumber. Selected and graded coconut lumber is also used as house posts, girders, trusses, door jambs and sidings (Fao.org). In research [12] states that coconut wood is strong and can be used to make furniture or decorative items like picture frames and sculptures. In a study by [12], coconut wood is used as the main material for furniture design based on local culture.

2. Physical properties

The distribution of bundles, as well as the proportion of fibers to other cell types in vascular bundles and thickness of fiber and ground parenchyma cell walls, are responsible for variations in wood density [13]. These variations in density can have a significant impact on the mechanical properties of the wood [1]. The density of coconut wood is an important property that affects its elasto-mechanic properties per unit area. The concentration of vascular bundles, which contain the structural fibers responsible for providing strength and stiffness, also contributes significantly to the mechanical properties of the wood. These properties, in turn, are essential for visual or mechanical grading of the wood [14].

2.1. Density

Wood density is an important factor to consider when choosing a type of wood for a particular application. It can affect the strength, durability, and workability of the wood. Density, for example, helps to determine the physical and mechanical properties and characterize different kinds of wood and woody materials for their intended use [1]. The average air dry and oven dry density of coconut stem was 400 and 460 kg/m3, respectively [15]. The density distribution of coconut palm wood varies depending on the location and age of the tree [2]. In general, the trunk of a mature coconut palm tree is divided into three main parts: the outer shell, the inner shell, and the core. The outer shell, also known as the "husk," is a fibrous layer that surrounds the entire trunk of the tree. It is lightweight and has a low density, typically ranging from 0.2 to 0.4 g/cm3 [17]. This layer provides some protection to the inner layers of the trunk, but it is not used for any commercial purposes. The inner shell, or "mesocarp," is a denser layer of the trunk that surrounds the core. It is composed of lignified fibers that are arranged in a helical pattern, which gives the wood its characteristic strength and flexibility. The density of this layer ranges from 0.6 to 1.2 g/cm3, depending on the age of the tree and other environmental factors [17,18]. The core, or "endocarp," is the densest part of the trunk and is located at the center of the tree. It is composed of hard, lignified fibers that provide structural support to the entire tree. The density of this layer can reach up to 1.5 g/cm3, making it one of the densest woods available [19]. Overall, the density distribution of coconut palm wood reflects the unique structure and composition of the tree, as well as its adaptation to the tropical environment. While the outer shell is not typically used for commercial purposes, the inner shell and core are highly valued for their strength, durability, and versatility in a wide range of applications, from construction to furniture-making. The main reason for density variation is seen in (1) the number and distribution of vascular bundles, (2) the dimension (diameter) as well as thickness of the cell walls of the bundles and (3) the cell wall thickness of the ground parenchyma. The density of wood and the concentration of vascular bundles per unit area are important properties for visual and mechanical grading procedures [14]. Figure 1 gives a qualitative impression of the density distribution over the stem from 80 year old Philippine palms.

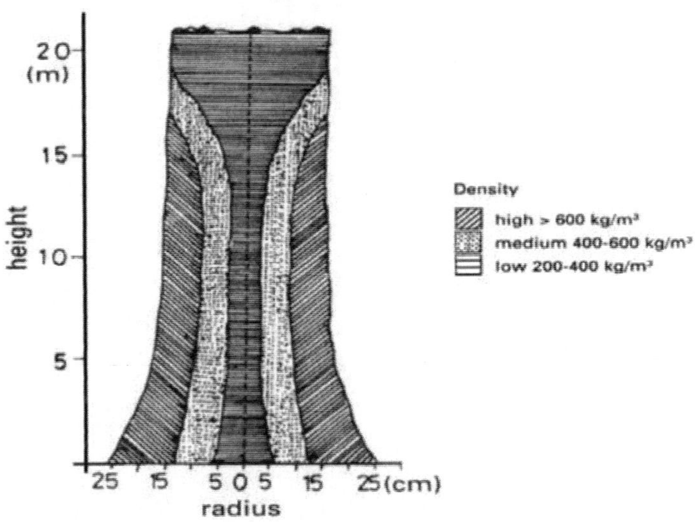

Figure 2. Density distribution in a mature coconut palm stem [20]

Frühwald et al. [21] investigated the mechanical properties such as bending strength and MOE, compression strength, shear strength and impact bending of Indonesian coconut wood. They concluded that strength properties of coconut wood are within the range of common traditional timbers, though generally about 20-30% lower when compared on the basis of matching densities. This drawback can be compensated by either larger cross sectional dimensions (for structural members under static or dynamic load) or by selecting wood in a higher density range. Density and related strength properties were strongly correlated with tree age.

2.2 Moisture content

The moisture content of coconut wood can vary depending on various factors such as the location within the tree, age of the tree, and environmental conditions. The moisture content of coconut stem ranged from 50% to 400% (Figure 3) [21]. In general, coconut wood has a relatively high moisture content when compared to many other hardwoods.

IOP Conf. Series: Materials Science and Engineering 1298 (2023) 012014 doi:10.1088/1757-899X/1298/1/012014

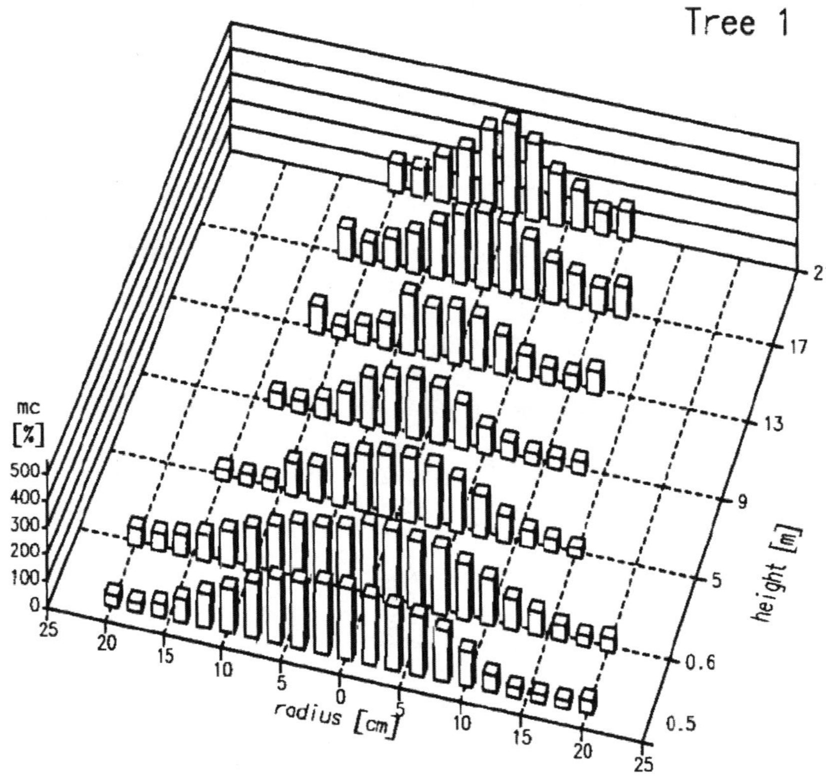

Figure 3. Moisture distribution across and along the stem of tree [21]

According to [22], the moisture content of coconut wood can vary widely. It generally increases rapidly with increased stem height and is particularly high in the central cylinder, with a marked increase from the peripheral zone towards the center. In research [23] found that moisture content is closely related to basic specific gravity. The density of coconut wood is also an important factor affecting its moisture content. Low-density wood typically contains less vascular bundles and more parenchyma cells, resulting in higher water storage capacity or greater moisture content. This relationship between density and moisture content is particularly relevant for coconut wood, which is known for having relatively low density compared to many other hardwoods. According to [9], there is a close relationship between the moisture content and the oven-dry density of coconut wood. Hard timber, with a density of 600 kg/m^3 or more, tends to have a low moisture content, while soft timber has a higher moisture content. This suggests that coconut lumber should be kiln-dried in accordance with its density group rather than being mixed together. However, little information is available regarding the sorption behavior of coconut wood, including its equilibrium moisture content (EMC). Frühwald et al. in paper [21] conducted tests on EMC at various relative humidities and found that it is similar to that of common timber species. This similarity is due to the comparable composition of the main constituents of coconut wood and other hardwoods, such as holocelluloses and lignin.

Overall, the moisture content of coconut wood is an important factor to consider when evaluating its suitability for different applications. High moisture content can affect the strength and durability of the wood, as well as its dimensional stability, so it is essential to properly dry coconut wood before using it in construction or other applications.

2.3 Shrinkage

The dimensional stability of wood is influenced by several factors, including the anisotropy of shrinkage and swelling as well as the orientation of the wood fibers within the trunk [24]. The orientation of the fibers, commonly referred to as the fiber angle or grain angle, plays a significant role in how the wood responds to changes in moisture content. Shrinkage and swelling are significant factors that can lead to defects such as checks and splits in conventional wood. However, when it comes to coconut palm wood, the tangential and radial shrinkage values in these two directions are not significantly different, which makes the wood more dimensionally stable compared to conventional wood. This means that coconut palm wood is less prone to deformations such as warping or cracking due to changes in moisture content [25]. This unique property of coconut palm wood makes it a valuable material for various applications, including construction and furniture making [1]. Shrinkage and swelling values in wood are often related to density, which generally increases from the center of the tree to the bark. In addition, these values can also increase up to a certain height and then decline. This is because the density and structure of the wood at different locations within the tree can affect its response to changes in moisture content. Moreover, shrinkage is closely associated with the number of vascular bundles (VBs) per unit area, as VBs are a significant contributor to the density of the wood. Studies have shown that the volumetric shrinkage of coconut wood can vary depending on the location within the tree. For example, [26] reported a volumetric shrinkage of 10% for wood from the outer portion of the tree, while the intermediate and center portions exhibited a lower range of 5-7%. In research [27] also observed similar results with a 10% volumetric shrinkage at the periphery but up to 22% for low-density wood found in the core. Frühwald and Peek [28] suggested that this high level of shrinkage in low-density wood may be due not only to the shrinkage of the cell wall substance but also to cell collapse caused by excessively severe drying conditions. Additionally, it is important to note that longitudinal shrinkage in coconut wood is very low, typically less than 1%. It is also worth mentioning that low-density wood, defined as having a density of less than 200 kg/m^3, shows a marked tendency towards cell collapse. All of these factors contribute to the dimensional stability of coconut wood and highlight its unique properties as a building material.

3. Chemical properties

Coconut palm wood is known for its unique chemical properties that make it a highly desirable material for various industrial and commercial applications. These properties are largely attributed to the chemical composition of the wood. Coconut wood shares similar main wood constituents with other timber species [1]. Coconut wood is comparable to Philippine hardwood/softwood and bamboo as far as holocellulose, lignin and pentosan content are concerned. Nevertheless, it is known to contain a higher percentage of ash content [25]. The proximate chemical composition of coconut wood consists of inorganic pure ash (0.75% with a range of 0.25-2.4%), silica (0.07% with a range of 0.01-0.2%), holocellulose (66.7%), lignin (25.1%), pentosans (22.9%), and starch (ranging from 4.3% to 4.6%). The pH-value of coconut wood falls at approximately 6.2. These values have been reported by various studies including [29, 30, 31]. Additionally, the presence of inorganic pure ash and silica may have implications for certain industrial application. Understanding the chemical composition of coconut wood is important for its optimal utilization in various industries. As mentioned above, Coconut wood contains different chemical constituents in varying proportions, depending on its location within the tree. The outer zone of coconut wood has the highest holocellulose content. While the holocellulose content decreases from the outer to the inner zone, there is no significant difference between the middle and inner zones [32]. Cellulose is the primary source of the mechanical properties of wood, comprising about 42% of coconut wood, which is similar to many other wood species. Its cellulose content consistently decreases from the outer to the inner zone. Coconut wood's cellulose content is within a range suitable for the paper and pulp industry. The outer zone of coconut wood also has the highest lignin content, which is a significant contributor to the higher strength properties of this zone. The lignin amount of 25.1% places coconut wood at the end of the normal range of 11-27% reported for non-woody biomass [33] and closely resembles the range reported for softwoods (24-37%) and hardwoods (17-

30%) [34, 35] Additionally, coconut wood contains minor constituents such as resins and tannins. Overall, understanding the chemical composition of coconut wood is crucial for determining its suitability for various industrial applications. While coconut wood shares similarities with other timber species, it also possesses unique chemical characteristics that make it valuable for specific uses, such as in paper and pulp production. Ash refers to inorganic substances such as silicates, sulfates, carbonates, or metal ions [36]. Compared to common woods such as aspen, yellow polar, and white oak, coconut wood has a significantly higher ash content ranging from 0.25% to 0.75% on average, up to a maximum of 2.4%. However, the ash content of coconut wood is generally lower when compared to the bark of most wood species, such as white oak bark (1.64%) and Douglas-fir bark (1.82%). Despite this, the higher ash content in coconut wood can negatively impact processing [32]. It is worth noting that palm woods, including coconut wood, have high silica (hydrated silicon dioxide) content, which poses a significant challenge to saw-millers and other industries seeking to utilize them. This high silica content is a major obstacle to the use of coconut wood and was highlighted during the Coconut Stem Utilization Seminar held in 1977. Understanding the chemical composition of coconut wood is essential for determining its suitability for various applications. While its high ash and silica content may present challenges for certain industries, coconut wood's unique chemical properties make it a valuable resource for other uses.

4. Durability and protection of Coconut wood

Coconut wood, like other types of wood, is susceptible to decay and insect attack (Bahmani et al. 2014). To improve its durability and protect it from these threats, various organic and inorganic preservatives have been tested [25, 37, 38]. Protecting coconut wood through the application of preservatives is essential for extending its lifespan and ensuring its long-term usability. Coconut palm wood is a popular choice for furniture and other applications due to its durability and resistance to decay [37]. However, it is still important to protect coconut wood by using preservatives to prevent it from being damaged by insects and other pathogen

One study by [32] investigated the effectiveness of four different organic preservatives (copper amine complex, borax, boric acid, and propiconazole) on coconut wood. The results showed that all four treatments provided significant improvements in resistance to termite attack and decay compared to untreated wood. However, copper amine complex and borax were found to be the most effective preservatives for improving the durability of coconut wood. In addition to organic preservatives, inorganic preservatives such as copper chrome arsenate (CCA) have also been tested on coconut wood. A study by [39] evaluated the efficacy of CCA-treated coconut wood against fungal decay and found that it significantly improved the wood's resistance to fungi compared to untreated wood. Another approach to protecting coconut wood is through the use of natural extracts. A study [40] examined the effectiveness of a natural extract derived from neem leaves on coconut wood. The results showed that the extract significantly inhibited the growth of fungi and termites, indicating its potential as a natural preservative for coconut wood. Overall, both organic and inorganic preservatives have demonstrated effectiveness in improving the durability of coconut wood. The choice of preservative will depend on the specific application and environmental conditions, as well as considerations of sustainability and safety.

5. Challenges and Limitations

While coconut wood has several advantages, it also has some limitations. One significant challenge is the limited availability of the material, as coconut trees are primarily grown for their fruit and not for timber production. Additionally, coconut wood requires specialized processing techniques, which may increase production costs. Finally, there is a lack of standardization in grading and testing coconut wood, which may affect its acceptance in the construction industry [3].

6. Conclusion

Coconut wood has significant potential as a sustainable and eco-friendly building material for timber structures. Its unique properties make it an attractive alternative to traditional timber, particularly in coastal areas where traditional timber is susceptible to damage. However, there are some limitations and challenges that need to be addressed before coconut wood can be widely accepted in the construction industry. Coconut wood has several unique properties that make it an excellent material for timber structures. Firstly, it has a high strength-to-weight ratio, which means it is lighter than many other types of wood but can still withstand heavy loads. Additionally, it is highly durable and resistant to decay, insects, and fungi, making it an ideal choice for outdoor applications. Furthermore, coconut wood has low shrinkage and swelling rates, reducing the risk of cracking or warping. With further research and development, coconut wood could become a viable option for sustainable timber structures.

References

[1] Fathi L, Fruehwald K and Fruehwald A 2014 *The potential use of timber from palm trees for building purposes Proceeding from the world conference on timber Engineering, Quebec,* Canada. Pp172

[2] Fathi L. 2014 *Title of PhD thesis*. University of Hamburg, Germany, pp. 248.

[3] Kumar B M and Kunhamu T K 2022 Nature-based solutions in agriculture: A review of the coconut (Cocos nucifera L.)-based farming systems in Kerala, "the Land of Coconut Trees" Nature-Based Solutions 2 100012 Online: http://dx.doi.org/10.1016/j.nbsj.2022.100012

[4] Schmidt O, Bahmani M, Koch G, Potsch T and Brandt K 2016 Study of the fungal decay of oil palm wood using TEM and UV techniques International Biodeterioration & Biodegradation 111 37–44 Online: http://dx.doi.org/10.1016/j.ibiod.2016.04.014

[5] Etebari K, Hereward J, Sailo A and Ahoafi E M 2021 *The coconut rhinoceros beetle, Oryctes rhinoceros (Coleoptera: Scarabaeidae): current status and future perspectives,* Curr Res Insect Sci. 1:100015

[6] Nampoothiri K U K, Krishnakumar V, Thampan P and Achuthan Nair M 2018 *The Coconut Palm (Cocos nucifera L.) - Research and Development Perspectives*

[7] Srivaro S, Lim H, Li M 2022 Pasztory Z. Struct, Vol. 40, pp. 237-246. supplement the reference

[8] Srivaro S, Pásztory Z, Le Duong H A, Lim H, Jantawee S and Tomad J 2021 Physical, mechanical and thermal properties of cross laminated timber made with coconut wood *European Journal of Wood and Wood Products* **79** 1519–29 Online: http://dx.doi.org/10.1007/s00107-021-01741-y

[9] Killmann W 1983 Some physical properties of the coconut palm stem Wood Science and Technology 17 167–85 Online: http://dx.doi.org/10.1007/bf00372314

[10] Moreno M L, Kuwornu J K M and Szabo S 2020 Overview and Constraints of the Coconut Supply Chain in the Philippines International Journal of Fruit Science 20 S524–41 Online: http://dx.doi.org/10.1080/15538362.2020.1746727

[11] Srivaro S, Tomad J, Shi J and Cai J 2020 Characterization of coconut (Cocos nucifera) trunk's properties and evaluation of its suitability to be used as raw material for cross laminated timber

production Construction and Building Materials 254 119291 Online: http://dx.doi.org/10.1016/j.conbuildmat.2020.119291

[12] Suprobo F P and Santosa A 2017 Models of furniture design using coconut wood based on local culture for global and domestic markets *Creative Industries Journal* **10** 89–99 Online: http://dx.doi.org/10.1080/17510694.2017.1282304

[13] Butterfield B and Meylan B 1979 Distribution of polyphenols in coconut stem wood: Histochemical observations Int Assoc Wood Anatomists Bull. **2** 35-36.

[14] Sulc V 1979 Grading coconut wood *Proceedings Coconut Wood-1979* 22–71

[15] Rana M, Das A and Ashaduzzaman M 2015 Physical and mechanical properties of coconut palm (Cocos nucifera) stem Bangladesh Journal of Scientific and Industrial Research **50** 39–46 Online: http://dx.doi.org/10.3329/bjsir.v50i1.23808

[16] Kuttankulangara Manoharan S, Megalingam RK 2019 *title* In: Information, Communication and Computing Technology: Third International Conference, ICICCT 2018, New Delhi, India, May , Revised Selected Papers 3. Springer Singapore, pp. 207-222.

[17] Nguyen X T, Hou S, Liu T and Han X 2016 A potential natural energy absorption material – Coconut mesocarp: Part A: Experimental investigations on mechanical properties International Journal of Mechanical Sciences 115–116 564–73 Online: http://dx.doi.org/10.1016/j.ijmecsci.2016.07.017

[18] Ghosh A, Razzino C do A, Dasgupta A, Fujisawa K, Vieira L H S, Subramanian S, Costa R S, Lobo A O, Ferreira O P, Robinson J, Terrones M, Terrones H and Viana B C 2019 Structural and electrochemical properties of babassu coconut mesocarp-generated activated carbon and few-layer graphene Carbon **145** 175–86 Online: http://dx.doi.org/10.1016/j.carbon.2018.12.114

[19] Schmier S, Hosoda N and Speck T 2020 Hierarchical Structure of the Cocos nucifera (Coconut) Endocarp: Functional Morphology and its Influence on Fracture Toughness Molecules **25** 223 Online: http://dx.doi.org/10.3390/molecules25010223

[20] Killmann and Fink D 1996 Coconut palm stem processing Protrade Deutsche Gesellschaft

[21] Frühwald A, Peek RD, Schulte M 1992 Coconut wood utilization in Asia and the Pacific. GTZ, Hamburg pp. 352.

[22] Nicholson JM, Swarup R 1976 Coconut stem utilization seminar proceedings. Tonga, pp. 65-102.

[23] Kloot 1952. This reference is indispensable, add it

[24] Kocaefe D, Huang X, Kocaefe Y 2015 Leaching of organic compounds from coconut coir fibers and its impact on termite resistance of particleboards. Curr Forestry Rep. Vol.1, pp.151-161.

[25] Arancon and Apfsos J 2009 he situation and prospects for the utilization of coconut wood in Asia and the Pacific Asia-Pacific Forestry Sector Outlook Study II, Working Paper NAPFSOS II/WP/2009/15

[26] Richolson (1975) This reference is indispensable, add it

[27] McConchie DL 1975 New Zealand Forest Service, Forest Research Institute, Forest Product

[28] Frühwald and Peek (1992) This reference is indispensable, add it

[29] Gapasin, D P 1986 Coconut research and development in the Philippines Cord **2** 34

[30] Arancon in 1997 This reference is indispensable, add it

[31] Poulter R, Hopewell G 2010 Secondary cocowood products. Potting mix. (DEEDI), Brisbane, Australia.

[32] Anon Physical, Chemical, And Mechanical Properties of Bamboo and Its Utilization Potential for Fiberboard Manufacturing - ProQuest Physical, Chemical, And Mechanical Properties of Bamboo and Its Utilization Potential for Fiberboard Manufacturing - ProQuest Online: https://www.proquest.com/openview/ddface4f432d2e0036fec86af25188f7/1?pq-origsite=gscholar&cbl=18750&diss=y

[33] Bagby MO 1971 The effect of extractives on paper properties. Tappi. Vol. **54** 11.

[34] Fengel D and Wegener G 2011 Wood: chemistry, ultrastructure, reactions (Walter de Gruyter)

[35] Dence 1992 This reference is indispensable, add it

[36] Rydholm S 1965 *Pulping processes*, Interscience publishers, John wiley and sons, New York, London, Sydney, pp. 635-644.

[37] Anoop et al. 2018 This reference is indispensable, add it

[38] Jeeshma V J, Anoop E V, Dhamodaran T K, Vidyasagaran K, Kunhamu T K and Gopakumar S 2018 Diffusion treatment of coconut palm wood using organic preservatives Journal of Tropical Agriculture **56**

[39] Imken A A P, Brischke C, Kögel S, Krause K C and Mai C 2020 Resistance of different wood-based materials against mould fungi: a comparison of methods European Journal of Wood and Wood Products **78** 661–71 Online: http://dx.doi.org/10.1007/s00107-020-01554-5

[40] Ney FP, Malco DCL, Senoro DB 2019 Catajay-Mani M. Sustain Environ Res. Vol. 29, 1, pp. 1-13

IOP Conf. Series: Materials Science and Engineering 1298 (2023) 012015 doi:10.1088/1757-899X/1298/1/012015

Numerical analysis of wood beam bending with verification of breaking force

Fatka Kulenović[1], Nedim Hurem[2] and Amra Talić-Čikmiš[3]

[1] Technical Faculty of Bihac, University of Bihać, B&H
[2] HARA d.o.o., Visoko, B&H
[3] Faculty of Mechanical Engineering, University of Zenica, B&H
E-mail: amra.talic.cikmis@unze.ba

Abstract. The work represents the continuation of the research of the maximum bending stress force of a wooden beam up to the moment of failure, which is defined experimentally and with a corresponding mathematical model. The goal was to confirm the accuracy of the mathematical model of the breaking force using numerical simulation and the SolidWorks software. Five different types of wood and five different beam thicknesses were used as input parameters whose influence was monitored trough the simulation. The stress deformation state and deflection for the calculated fracture forces were analysed separately for each variant. The breaking force model, which is defined depending on the density of the wood and the thickness of the board, proved to be adequate for defining the maximum stresses and deflections when bending beams.

1. Introduction

As wood being an organic material, which is created by a natural biological process, various irregularities (defects) are possible that more or less affect the possibility of its use in constructions. There is a difference between structural changes, which are already formed in living wood during growth, and changes that are the consequences of physical, mechanical and chemical influences that may occur during growth or during processing and use.

Also, there are errors that are not the result of wood properties, but machine or human errors that occur during wood processing.

Laboratory tests [1] have established that when the fibers are beveled 1:8, only about 53% from a total value of the bending strength remains, and for beveling 1:15, about 76%. 100% bending strength of the wood remains if the bevel of the fibers is 1:20, or for example, the compressive strength at a fiber bevel of 1:15 remains 100%, and for a fiber bevel of 1:6 it remains somewhere about 56%. 100% bending strength of the wood remains if the bevel of the fibers is 1:20.

In the case of tensile stress, the skewness of the fibers leads to tension perpendicular to fibers, where the strength of the wood is the lowest.

Mechanical properties of the beam are determined by the standard procedure, on test samples made of wood with a humidity of up to 12%. Table 1 shows the tensile (ft), compressive (fc) and shear (fv) values of different types of wood given in the literature[2].

Table 1. Mechanical characteristics of different types of wood [2].

Species	$f_{t,90}$[MPa]	$f_{t,0}$[MPa]	$f_{c,90}$[MPa]	$f_{c,0}$[MPa]	F_v[MPa]
Spruce	3	90	6	30	7
Pine	4	10	4	30	10
Larch	2	100	8	35	9
Beech	7	140	9	40	11
Oak	4	90	8	40	11

When wood is tested to fracture by axial tension, the stress-strain relationship is fairly linear until the ultimate load, and the fracture is brittle. Under axial pressure, wood is a much more ductile material, with a linear stress-expansion relationship up to the limit of proportionality, followed by plastic yielding until the limit expansion is reached. Variability of materials, non-linear behavior under pressure and the presence of defects make it impossible to determine the bending behavior of wood directly on the basis of tension and pressure tests. Therefore, the current standards consider flexural strength as an independent characteristic.

Table 2 [3] explains the main forms of wood breakage during pure bending depending on the ratio between tension and compression strength in the direction of the fibers (η = ft/fc). In technical practice, given forms of failure have different probability of occurrence.

Table 2. The main forms of wood breakage during pure bending [3].

	Description of the break	Condition	Diagram moment curve
1	Brittle fracture in the tension zone with a linear relationship between the bending moment and the curvature until fracture.	$\eta<1$	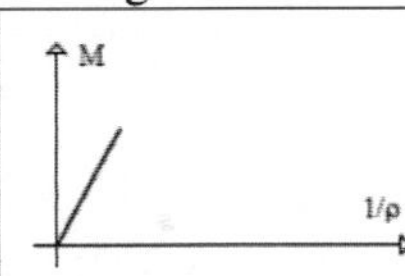
2	Fracture in the tension zone followed by plasticization of the pressure zone and lowering of the neutral axis. The moment-curve relationship is different from a linear trend.	$\eta\geq1$	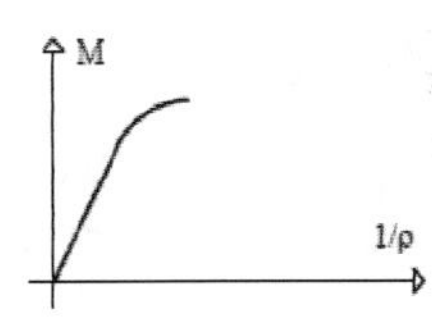
3	Fracture in the tension zone with high ductility due to plasticization of the cross section. The moment-curve diagram shows a small downward trend.	$\eta>1$	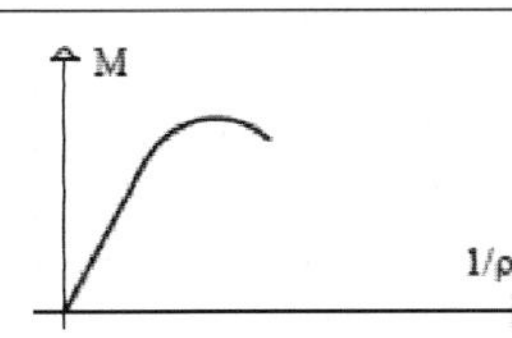
4	Fracture in the pressure zone with high ductility due to considerable plasticization of the cross section. The moment-curve diagram shows an evident downward trend.	$\eta>>1$	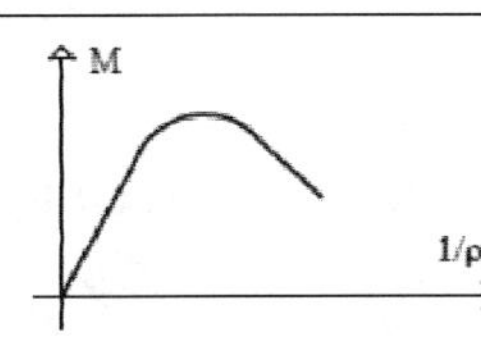

The main goal of the numerical analysis and the bending simulation of the beams made of different types of wood is to determine the correctness of the mathematical model of the breaking force and the results obtained by numerical and experimental procedures [6]. Numerical analysis was performed using the finite element method (FEM) with the commercial software SolidWorks.

2. Numerical modeling

The model that was created for numerical analysis was created in such a way as to represent the experimental model and its properties as accurately as possible (Figure 1).

The mathematical model of the breaking force was determined (defined) by the experimental test and method of the planned experiment [4, 5, 6], as follows:

$$F_{\max} = -21{,}97 + 0{,}029\rho + 0{,}53b \tag{1}$$

where ρ is density of the wood, b is thickness of the board.

After the creation of a 3D model of a wooden beam and adding the appropriate material and its characteristics, a new study is created within the construction tree - Simulation. It is a static simulation, so all other parameters are left predefined. The setup of boundary conditions comes down to setting fixed supports and setting the load action. When it comes to fixed supports, since it is a beam that is placed evenly on the supports, end edges of the beam are defined as fixed supports. The placement of fixed supports is important for the implementation of the simulation, because in this way we limit our 3D model to a certain level that we need to perform the analysis and make it possible for the analysis, and ensure that our model does not "slip" with the impact of an external load. We apply the load in a form of a maximum force that is previously calculated for each beam individually, and we set it to act perpendicular to the beam in the very canter of the beam.

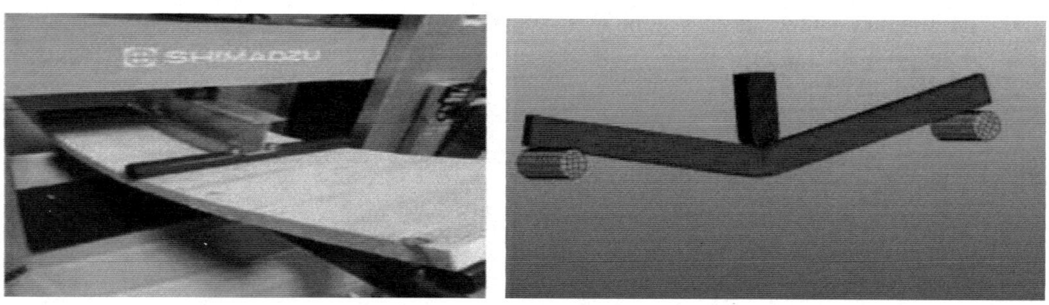

Figure 1. Experimental and numerical models

In this way, all the necessary boundary conditions have been entered, and we can start creating the network. Network generation in SolidWorks can generally represent a very complex problem, especially when it comes to complex 3D models, where in most cases it will be necessary to improve the network to some final result. However, since the 3D model of our wooden beam is not complex, it is possible to automatically create a standard mesh with the recommended size of finite elements (Figure 2.). We will change the parameters to improve the mesh in the sense that it will not be a coarser mesh but a finer one, in order to obtain more reliable results. Due to the load concentration in the very canter of the beam, the mesh will be smaller/finer at that point, and by clicking on "Details" we can see all the necessary details of the mesh itself. If the mesh is well placed, good results can be expected. By setting up the network, the 3D model is fully prepared for starting the analysis and analysing the obtained results.

According to the results of the experimental test, the fracture criteria are based on the state of maximum stress, that is, when the tensile stress in the longitudinal direction, in a certain step of the movement, reaches the limit value.

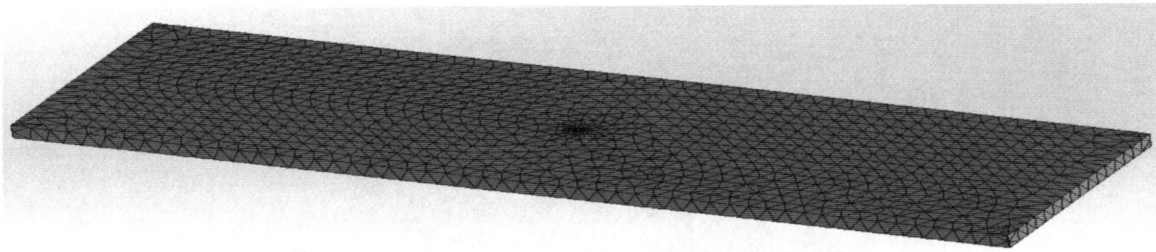

Figure 2. Initial discretization 3D slab model

2.1. Material characteristics

Defining an appropriate model for the material is essential to achieve an accurate prediction of the mechanical characteristics of the support. In unstrengthened glulam beams, failure usually occurs in the tension zone, before any plasticization occurs in the compression zone, so the linear-elastic model is quite appropriate. A linear-elastic ideal-plastic material model was used for the behavior of wood under pressure, while a linear-elastic (brittle) model was used for wood exposed to tension.

Wood is an anisotropic material, which means that it has different material properties in different directions. During the numerical simulation, it is necessary to enter the corresponding material coefficients for the material, such as the modulus of elasticity, the shear modulus and the Poisson's coefficients (Table 3). In this case, the coefficients are taken from the literature [4].

Table 3. Overview of the characteristics of wood materials.

Wood			
Modulus of elasticity E [MPa]		Yield stress (MPa)	
E_1	11080	$\bar{\sigma}_{11}$	36.3
E_2	886	$\bar{\sigma}_{22}$	5
E_3	554	$\bar{\sigma}_{12}$	6.1
Poisson's coefficient n [-]		$\bar{\sigma}_{13}$	6.1
ν_1	0.37	$\bar{\sigma}_{23}$	3
ν_2	0.42	$\bar{\sigma}^0$	36.3
ν_3	0.47		
Shear modulus [MPa]			
G_{12}		791	
G_{13}		744	
G_{23}		79	

When setting up the simulation pre-processing, it was necessary for different types of wood, with different densities and different board thicknesses to calculate the breaking force with which the beam will be loaded during the simulation.

2.2. Results of numerical analysis

Numerical analysis, on previously described models formed by finite elements, was performed for an unreinforced support made of one type of wood. The models were loaded incrementally until the limit load was reached. The values of the bending deflection determined by numerical calculation with average values are shown in Figure 3.

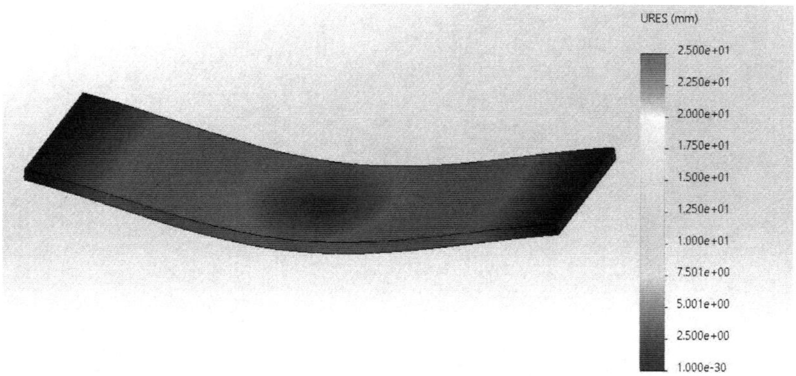

a)

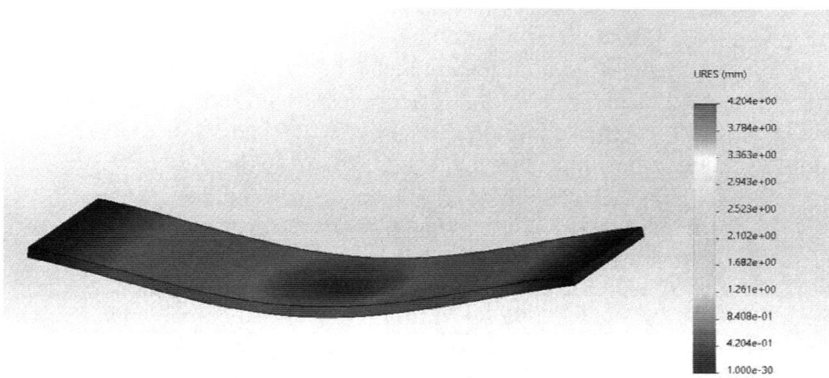

b)

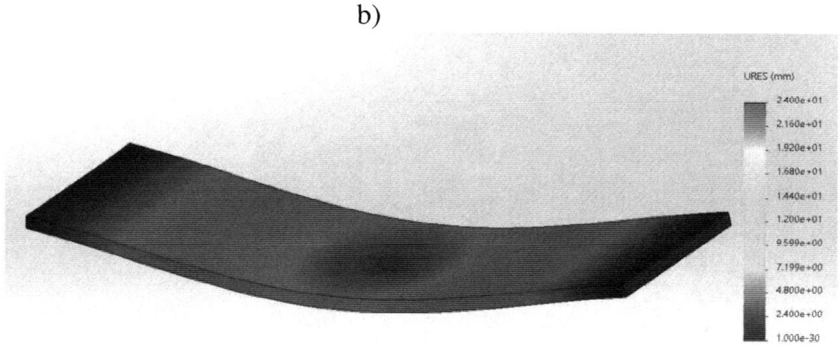

c)

Figure 3. Values of vertical displacement (mm) in the MKE model,

a) beech 18 mm, b) poplar 18 mm, c) oak 22 mm

Figure 4 shows the stress deformation state for three types of wood and three thicknesses at the limit load of the calculated force. Analysis of the stress state shows the fracture zone and the formation of cracks in the wood structure itself.

The results of the stress state distribution and displacement are shown (Figure 3 and Figure 4) for beech with a thickness of 18 mm, poplar 18 mm and oak with a thickness of 22 mm. Other obtained results obtained by numerical simulation of different types of wood and thickness are given in Table 4.

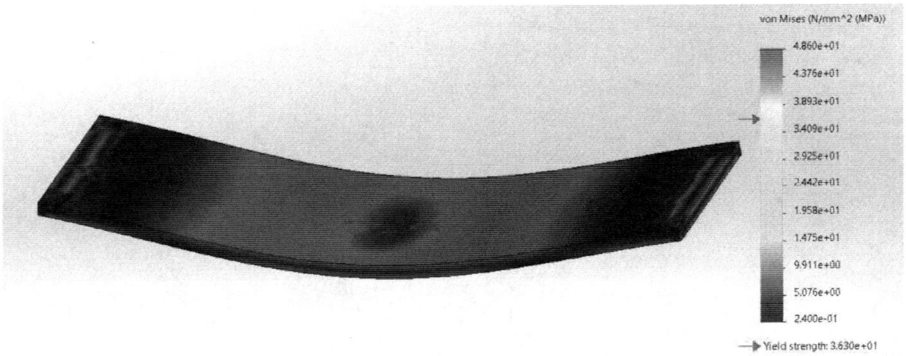

a)

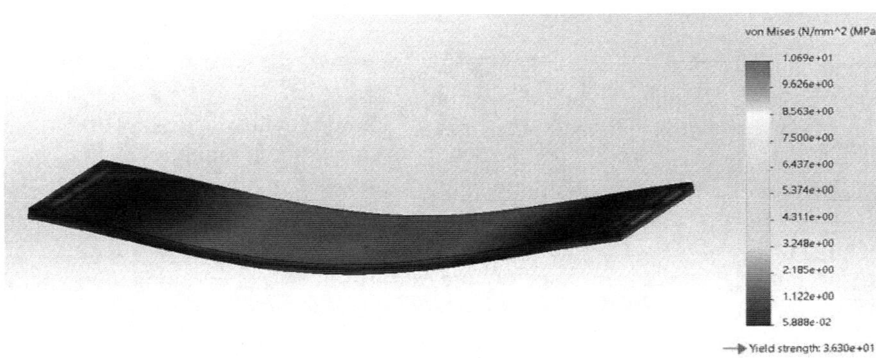

b)

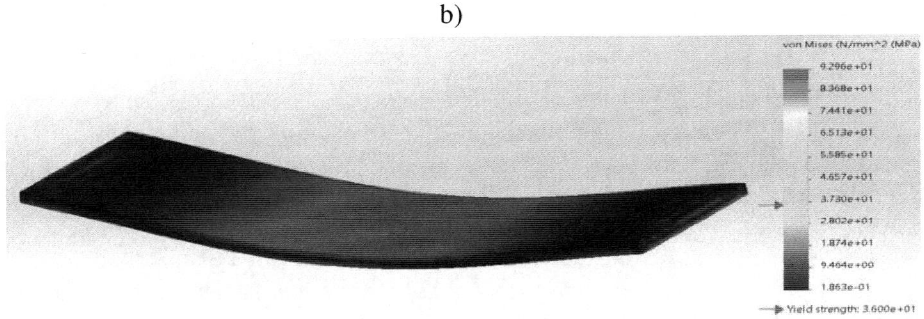

c)

Figure 4. Stress distribution in layers on the section of the MKE model, a) beech 18 mm, b) poplar 18 mm, c) oak 22 mm

It can be noticed that the fracture occurred as a result of exceeding the bending stress. Also, the indicator of the usability of the numerical method applied on beam bending is in the obtained results if the displacement is analysed for different types of wood. The relationship between strength, density and thickness of the material, i.e. different wood, is noticeable. From this comes the usefulness and usability of the method itself, especially in the initial phase of determining the characteristics and behaviour of wooden beams.

Table 4. Display of input parameters and SW simulation results.

No.	Type of wood	Thickness [mm]	Density [kg/m^3]	Breaking force of the plate [kN]	Max. deflection [mm]	Max stress [MPa]
1	beech	18	560	3.81	25.84	39.56
2	beech	20	560	4.87	25	48.6
3	beech	24	560	6.99	22.76	65.56
4	poplar	18	450	0.62	4.2	10.69
5	spruce	20	450	1.68	6.42	34.51
6	ash	20	450	1.68	6.87	31.16
7	oak	22	560	5.93	24	92.96

3. Conclusion

The conducted numerical simulation showed that the used mathematical model of the breaking force gives us good results when applied to the process of wooden beam bending. Calibration of the parameters of the material model was performed using literature data within the Solid Works software. For the given load of the calculated breaking force, simulation and analysis of the stress state and flexural strength-displacement were performed. The obtained results are satisfactory even with small deviations considering the conducted experimental research. It can be concluded that on a simple model satisfactory results can be obtained and the behaviour of the wood until its breaking point can be predicted. For a more detailed analysis of material characteristics, taking into account a larger number of influential parameters, the application of numerical simulation as a method for analysing the behaviour of wood under load would be a more complete method.

Further tests should be carried out on more complex constructions in order to optimize and dimensionally define some structural elements.

REFERENCES

[1] Gojković M and Stojić D 1996 *Drvene konstrukcije* (Građevinski fakultet, Beograd)

[2] Glišović I R 2013 Teoretska i eksperimentalna analiza nosača od lepljenog lameriranog drveta ojačanih karbonskim trakama, doktorska disertacija, Beograd

[3] CNR Advisory Committee on Technical Recommendations for Construction, *Guidelines for the design and construction of externally bonded FRP systems for strengthening existing structures. Timber structures*, CNR-DT 201/2005

[4] Hurem N, Talić-Čikmiš A and Imamović A 2022 Exsperimental analisis of distribution of bending stresses in layers of wood with numerical verification, 11^{8h}. Conf. MTSM 2022, Split, September 22th-23th, pp 42-50

[5] Hurem N, Talić-Čikmiš A and Imamović A 2023 Analysis of stress distribution during bending in wood layers using numerical simulation, 13^{8h} Conf. Quality 2023, Neum, June 19th-21th, pp 459-464

[6] Hurem N and Hodžić D 2020 Planiranje eksperimenta i matematički model čvrstoće na savijanje ploče od masivnog drveta, časopis *Mašinstvo*, Vol. 17, No. 3-4, pp. 93-102

Use of the Lagrange interpolation polynomial to estimate the bending strength value of plywood

Hasan Talić, Melisa Haurdić and Fatka Kulenović

University of Bihać, Technical Faculty Bihać, dr. Irfan Ljubijankić bb

hasan.talic@unbi.ba

Abstract. In order to determine the bending strength of wood or wood-based panels, an experimental method is typically used to test samples of different sizes that involves bending or compression. However, such tests are expensive and time-consuming, so mathematical models are often applied that allow predicting the bending strength based on a smaller number of obtained test results. One such model is the Lagrange interpolation polynomial. The study discusses Lagrange's interpolation model. Using the results of tests done on plywood of various thicknesses, approximates the bending strength of plywood for values that could not be tested directly and provides an estimate of strength for particular values.

1. Introduction

Veneer board - plywood, is a board that has found application in the construction industry, but also in other wood products, due to its good properties such as high static load capacity, shaping ability, high fire resistance, good moisture resistance, aesthetic properties, and own weight of the structure. When it comes to the properties of wood or wood-based panels such as plywood, strength is a critical factor. It determines the utilization of those plates, which is done through experimental methods such as 3- or 4-point bending tests. Testing plywood using this approach takes time and money, hence alternative procedures are used to evaluate the strength of the boards. Non-destructive testing is a procedure for identifying physical and mechanical qualities without affecting the look or usability of a product. The advantage of this procedure is that it helps to preserve the material. In accordance with this, mathematical models are frequently used to forecast strength based on a lesser number of obtained test data. One such model, Lagrange's interpolation model, is described in the paper, where an approximation of the bending strength of plywood for values that were not directly measured, as well as an estimate of strength for specific values, was derived based on the bending strength results obtained by testing plywood of different thicknesses.

2. Lagrange polynomial

In the application of mathematics, it is often necessary to solve problems where the values of some function $y = f(x)$, are known, where $x_i \in [a, b]$, $i \in \mathbb{N}_0$, but the analytical form of the function $f(x)$ is not known or the analytical form is known function $f(x)$, but it is very complicated. In both cases, it is necessary to calculate the approximate value of the function $f(x)$ for the value of the argument x, which are different from the given values x_i.[6]

If you need to calculate the value of the function $f(x)$ for some x, $x_0 < x < x_n$, $x \neq x_i$, then that task is called interpolation (*interpoler* = insert, French). Therefore, the word interpolation means the procedure of finding the value of the function $f(x)$ for $x \in [a,b]$, $x_i < x < x_{i+1}, i = 0,1,...,n-1$, on basis of the table [5] :

Table 1. Value of function $f(x)$ at points x_i, $i \in [0,n]$

x	x_0	x_1	...	x_i	...	x_n
$f(x)$	$f(x_0)$	$f(x_1)$	...	$f(x_i)$	...	$f(x_n)$

The interpolation task consists in the following: instead of the known function $y = f(x)$, which is given by an analytical expression or a table, a simpler function $F(x; a_0, a_1, ..., a_n)$ should be constructed, where $a_i \in \mathbb{N}_0$ are some parameters using which the values of the function $f(x)$ can be found more easily. These values are found approximately, but with the required accuracy. In doing so, the condition is set that the function $F(x; a_0, a_1, ..., a_n)$ coincides with the function $f(x)$ at the points $x_i, i \in \mathbb{N}_0$, and that at the other points $x \in [a,b], x \neq x_i, i \in \mathbb{N}_0$ approximates the function $f(x)$ well, so let it be:

$$|f(x) - F(x;\ a_0, a_1, \ldots, a_n)| \leq \varepsilon, \tag{1}$$

Where $\varepsilon > 0$ is the permissible approximation error. If ε is smaller, then the approximation is more accurate, better.

It is also natural to require that the function F (which we call the interpolation function) is simple to calculate. Algebraic polynomials are most often used in practice for easier calculations. So, we are looking for the interpolation function in the form of a polynomial. The interpolation polynomial exists and is unique, but there are many forms of writing that polynomial. Each of these forms has a special name, and one of them is Lagrange's interpolation polynomial. The Lagrange interpolation polynomial is a mathematical method used to construct a smooth function that passes through certain points-nodes on a graph.

Let there be $n+1$ different nodes on the segment $[a,b]$: $x_0, x_1, ..., x_n$ and let $y_i = f(x_i), i \in \mathbb{N}_0$ be the values of the function $y = f(x)$ in those nodes. It is necessary to construct an interpolation polynomial $L_n(x)$, of degree no greater than n, such that:

$$L_n(x_i) = y_i, \quad i \in \mathbb{N}_0. \tag{2}$$

That is, we need to construct a polynomial $L_n(x)$ that passes through the points $M_i = (x_i, y_i), i \in \mathbb{N}_0$.

First, we need to construct an auxiliary polynomial $p_i(x)$ of the nth degree, which is equal to zero for $x = x_j, j \neq i$, and equal to one for $x = x_i$.

$$p_i(x_j) = \begin{matrix} 0, j \neq i \\ 1, j = i. \end{matrix} \tag{3}$$

As the polynomial is defined by its zeros up to a constant factor, that is:

$$p_i(x) = C_i(x - x_0)(x - x_1)\cdots(x - x_{i-1})(x - x_{i+1})\cdots(x - x_n), \tag{4}$$

where C_i is a constant. If we replace x with x_i in the last equality, we get:

$$1 = p_i(x) = C_i(x_i - x_0)(x_i - x_1)\cdots(x_i - x_{i-1})(x_i - x_{i+1})\cdots(x_i - x_n), \tag{5}$$

That is:

$$C_i = \frac{1}{(x_i - x_0)(x_i - x_1)\cdots(x_i - x_{i-1})(x_i - x_{i+1})\cdots(x_i - x_n)}, \tag{6}$$

IOP Conf. Series: Materials Science and Engineering 1298 (2023) 012016 doi:10.1088/1757-899X/1298/1/012016

so it is:

$$p_i(x) = \frac{(x-x_0)(x-x_1)\cdots(x-x_{i-1})(x-x_{i+1})\cdots(x-x_n)}{(x_i-x_0)(x_i-x_1)\cdots(x_i-x_{i-1})(x_i-x_{i+1})\cdots(x_i-x_n)}. \quad (7)$$

The required interpolation polynomial has the form:

$$L_n(x) = \sum_{i=0}^{n} p_i(x) \cdot y_i, \quad (8)$$

That is:

$$L_n(x) = \frac{(x-x_0)(x-x_1)\cdots(x-x_{i-1})(x-x_{i+1})\cdots(x-x_n)}{(x_i-x_0)(x_i-x_1)\cdots(x_i-x_{i-1})(x_i-x_{i+1})\cdots(x_i-x_n)} \cdot y_i. \quad (9)$$

This interpolation polynomial is called the Lagrange interpolation polynomial.[5]

3. Veneer panel – plywood

Given its advantageous features, wood is one of the most popular building materials. Although solid wood has been used as a construction material throughout history, with the advancement of technology in wood processing and processing, we now have an increasing number of wood products that compete as an excellent building material. Thus, we have glued laminated supports of various sorts (LVL, LLD, LSL, CLT), wood-based panels, and composite materials containing wood. Due to their good properties, such as high static load capacity, the possibility of various shaping, high fire resistance, good moisture resistance, improvement of aesthetic properties, large spans of supports, own weight of the structure, they have found application in the construction of buildings in construction for walls, ceilings, roofs and are they can be used for multi-story buildings as well as other wood products.

Veneer panel - plywood, are panels made of at least 3 sheets of veneer that are joined together with glue. The veneer sheets in the panel should be stacked in such a way that the wood fibers of the adjacent sheets are at a certain angle, mostly 90°, but they can be 60°, 45° and even 30°, depending on the type of panel. The adhesives used must meet the requirements of durability and environmental protection.[1]

Figure 1. Veneer panel – plywood

When we talk about veneer panels - plywood, we are referring to one of the most versatile wooden panels, which is utilized in practically every industry. As a result of its low weight, high strength, and stability, it is commonly employed in construction as partition walls, wall coverings, floor slabs, and other parts where these properties are required. They are utilized as table tops, sofas, kitchen counters, door and window sashes, and other straight and curved pieces in the furniture sector. Due to their low weight in relation to the carrying capacity, they are particularly interesting in the fields of aircraft construction, shipbuilding and automotive.[2]

All building materials have mechanical, physical, acoustic, thermal, aesthetic, ecological, and other vital attributes that greatly influence their location and technique of application. In Europe, all wood-

IOP Conf. Series: Materials Science and Engineering 1298 (2023) 012016 doi:10.1088/1757-899X/1298/1/012016

based panels used in building must meet the standard BAS EN 13 986 (Wood-based panels for use in construction - Characteristics, conformity evaluation, and labeling). This standard stipulates that wood-based boards (plywood, chipboard, fibreboard, OSB boards, LVL boards, etc.) should have the following properties: bending strength, modulus of elasticity (bending stiffness), tensile strength, quality of gluing, thickness swelling, formaldehyde emissions , moisture resistance, water resistance, water vapor permeability, fire resistance, sound insulation (sound absorption), thermal conductivity, biological durability, specific strengths, impact resistance, etc. Considering the place of use of the selected wood-based panels, all or some of the properties are checked, that is, whether they are used in dry, wet or outdoor climate conditions.

3.1. The strength of plywood

The strength of plywood can vary based on a variety of factors, including the type of plywood used, its thickness, the type of glue used, the number of sheets, the thickness of the sheets, the way of orientation of the sheets, the method of pressing, and the quality of the manufacturing itself. Plywood is well-known for its comparatively high strength and stability. Thicker plywood is typically stronger than thinner plywood because it has more layers of wood that are glued together. Furthermore, plywood with more layers tends to be stronger than plywood with fewer layers. However, it is important to note that plywood is not as strong as solid wood. Also, the strength of plywood can vary depending on the direction of the load. For example, plywood can be stronger when loaded along the grain of the layers than when loaded cross-grain.

The bending strength and modulus of elasticity of wood-based panels are determined in accordance with the BAS EN 310 standard. The measurement is made by applying force to the sample's center (middle), which is supported by two supports. The modulus of elasticity is calculated using the ratio of the linear area of the force curve for the load test; however, the value produced is merely the current value and does not represent the true modulus; this approach incorporates torsion and twisting. The bending strength of each sample is determined by determining the twist value M at the greatest reading of the force F_{max}. The shredder on which the test is performed must contain the following:

- Two parallel, cylindrical roller bearings (which serve as supports), the length of which exceeds the width of the test and the diameter of 15±0,5 mm. The distance between the supports must be adjustable,
- Cylindrical force head, of equal length and diameter 30±0,5 mm, placed parallel to the supports at the same distance,
- A suitable instrument that measures the deflection of the sample during measurement with a precision of 0,1 mm,
- A suitable measuring instrument that can calculate the force on the test with a permissible deviation of 1% from the measured value. [3]

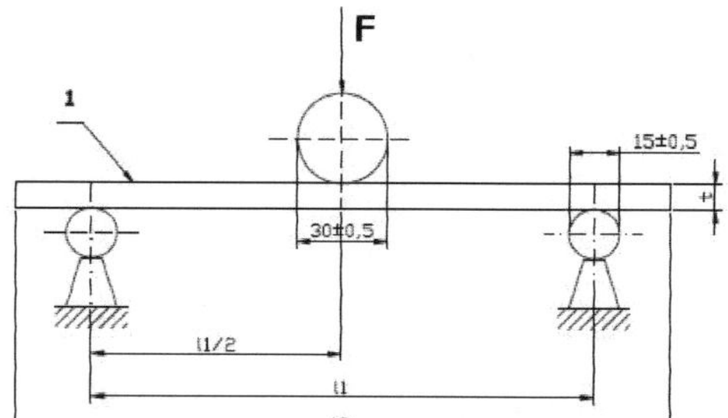

Figure 2. Test and equipment for testing the bending strength properties of panels

IOP Conf. Series: Materials Science and Engineering 1298 (2023) 012016 doi:10.1088/1757-899X/1298/1/012016

The dimensions of the samples are as follows: *l*- sample, width [*b*] is 50±1 mm, *t* - sample thickness, length [l_2] should be 20 times greater than the basic thickness plus 50 mm, maximum length of 1050 mm, and minimum length of 150 mm. Sampling and cutting of samples should be performed according to the standard BAS EN 326-1. Bending strength is calculated based on the formula:

$$f_m = \frac{3F_{max}l_1}{2bt^2} \tag{10}$$

where is: F_{max} – maximum force [*N*], l_1 – distance between supports [mm], *b* – sample width [mm], *t* – sample thickness [mm].[3]

Today, there are various methods of testing plywood that are used in order to obtain data on its characteristics. A division can be made between destructive and non-destructive testing of plywood. Non-destructive testing is a procedure by which some physical and mechanical properties can be identified or errors in a material can be observed without disturbing or changing its appearance or usability. The advantage of these test methods in relation to destructive ones is that they contribute to the preservation of materials, which achieves better quality control. In these tests, the experimental samples are of smaller dimensions, and it is possible to determine several different mechanical properties on the same samples. This group also includes computer programs that provide the data needed to assess the suitability of the materials from which the product is made. [4]

4. Application of the model

In the following, the application of the Lagrange polynomial to determine the strength of plywood is given. Given that the strength of plywood depends on a number of factors, strengths were taken for different thicknesses and different directions of force action (parallel and vertical). Based on this, results were obtained by calculation for thicknesses that are within the thickness interval [9,24] millimeters, and for which results were not obtained experimentally. The strength data used were taken from the product declaration (Declaration of performance, no.2.6/2017, Harmonized standard EN 13986:2004+A1:2015, Producer: PAGED SKLEJKA S.A. Poland).

On the basis of the data listed in table 2, we constructed the Lagrange polynomial, and predicted the values for the plywood thicknesses within the board thickness segment [9,24], which are not directly listed in the table. Where *x* - the nominal thickness of the plywood panel in mm, and the value *y* - characteristic value bending strength ecc. ($f_{m\square}$ - according to standard EN 789 in N/mm²).

Table 2. Plywood strength values ($f_{m\square}$) for different thicknesses

X (mm)	9	12	15	18	21	24
Y (N/mm²)	26,7	28	28	25,2	26,8	27,8

Using the formula[5] :

$$L_n(x) = \sum_{i=0}^{n} p_i(x) \cdot y_i \, , \tag{11}$$

and by including the data from the table, we get the Lagrange interpolation polynomial:

$$L_5(x) = \frac{(x-12)(x-15)(x-18)(x-21)(x-24)}{(9-12)(9-15)(9-18)(9-21)(9-24)} \cdot 26 + \frac{(x-9)(x-15)(x-18)(x-21)(x-24)}{(12-9)(12-15)(12-18)(12-21)(12-24)} \cdot 28 + \frac{(x-9)(x-12)(x-18)(x-21)(x-24)}{(15-9)(15-12)(15-18)(15-21)(15-24)} \cdot 28 + \frac{(x-9)(x-12)(x-15)(x-21)(x-24)}{(18-9)(18-12)(18-15)(18-21)(18-24)} \cdot 25{,}2 + \frac{(x-9)(x-12)(x-15)(x-18)(x-24)}{(21-9)(21-12)(21-15)(21-18)(21-24)} \cdot 26{,}8 + \frac{(x-9)(x-12)(x-15)(x-18)(x-21)}{(24-9)(24-12)(24-15)(24-18)(24-21)} \cdot 27{,}6 = -0{,}00072359x^5 + 0{,}058745x^4 - 1{,}84643x^3 + 28{,}01x^2 - 204{,}71x + 603{,}3$$

The given polynomial can also be represented by a graph (Figure 3), where the relationship between the strength and thickness of the plywood is visible.

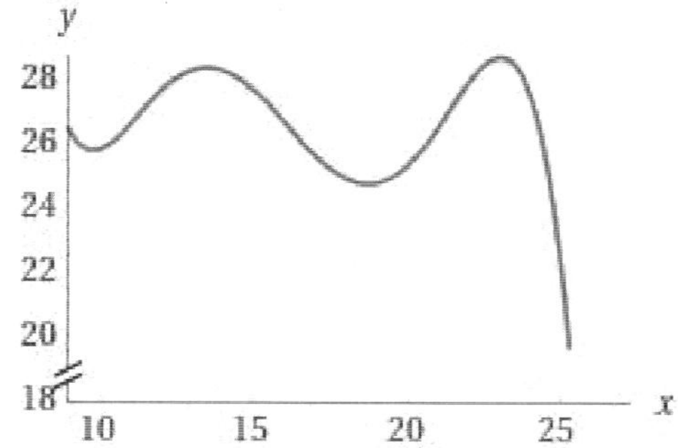

Figure 3. Graphic representation of the relationship between strength and thickness of plywood

Now, with the help of the Lagrange interpolation polynomial, we obtained strength data for thicknesses of: 10 mm, 17 mm, 20 mm, 22 mm and 23 mm, which were not part of the experimental testing of the plywood panel for which we took the data.

$$L_5(10) = 26{,}161$$
$$L_5(17) = 25{,}956$$
$$L_5(20) = 25{,}672$$
$$L_5(22) = 28{,}286$$
$$L_5(23) = 29{,}032$$

In this way, through calculations, the stated values of plywood strength (N/mm^2) were obtained.

Considering that the strength of the slab depends on the direction of stretching of the fibers during the test, the difference is whether the action of the mortar is parallel or perpendicular to the fibers. For this reason, a calculation was made for the same thickness of plywood - x, and vertical characteristic value bending strength ecc. ($f_{m\perp}$ - according to standard EN 789 in N/mm^2).

Table 3. Plywood strength values ($f_{m\perp}$) for different thicknesses

X (mm)	9	12	15	18	21	24
Y (N/mm^2)	11,5	15,1	12,2	18,8	17,4	17

The Lagrange polynomial, now has the form:

$$L_5(x) = \frac{(x-12)(x-15)(x-18)(x-21)(x-24)}{(9-12)(9-15)(9-18)(9-21)(9-24)} \cdot 11{,}5$$
$$+ \frac{(x-9)(x-15)(x-18)(x-21)(x-24)}{(12-9)(12-15)(12-18)(12-21)(12-24)} \cdot 15{,}1$$
$$+ \frac{(x-9)(x-12)(x-18)(x-21)(x-24)}{(15-9)(15-12)(15-18)(15-21)(15-24)} \cdot 12{,}2$$
$$+ \frac{(x-9)(x-12)(x-15)(x-21)(x-24)}{(18-9)(18-12)(18-15)(18-21)(18-24)} \cdot 18{,}8$$
$$+ \frac{(x-9)(x-12)(x-15)(x-18)(x-24)}{(21-9)(21-12)(21-15)(21-18)(21-24)} \cdot 17{,}4$$
$$+ \frac{(x-9)(x-12)(x-15)(x-18)(x-21)}{(24-9)(24-12)(24-15)(24-18)(24-21)} \cdot 17$$
$$= 0{,}00205761x^5 - 0{,}171553x^4 + 5{,}56636x^3 - 87{,}6505x^2 + 668{,}69x - 1960{,}8$$

The graph of this polynomial (Figure 4) describes the relationship between the vertical strength and the thickness of the plywood.

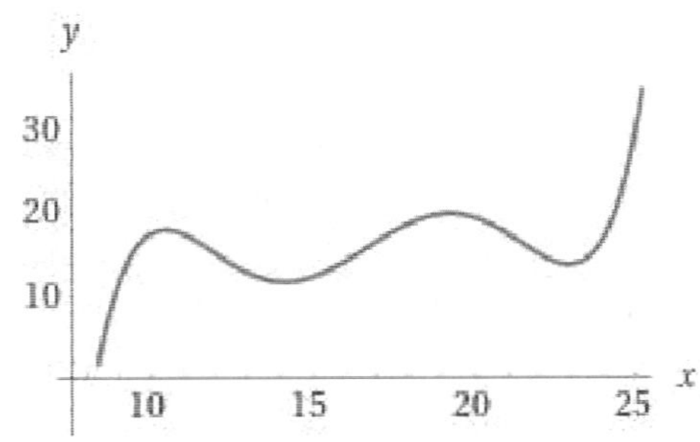

Figure 4. Graphic representation of the relationship between strength (vertical) and thickness of plywood

Analogous to the assessment of the strength of the plywood in the parallel direction, it was also done in the vertical direction for the same thickness of the board and the following values were obtained (N/mm^2):

$$L_5(10) = 17{,}641$$
$$L_5(17) = 16{,}696$$
$$L_5(20) = 19{,}552$$
$$L_5(22) = 14{,}984$$
$$L_5(23) = 13{,}778$$

The above values, for selected thicknesses, are obtained through calculation and the use of the Lagrange polynomial.

5. Conclusion

Many scientific domains rely heavily on mathematical models for prediction and analysis, optimization, comprehending complicated processes, testing and verification, and the production of new theories and knowledge.

The Lagrange interpolation model is presented in this study, which is used to approximate an unknown function using an interpolation polynomial. The objective of interpolation is to build a polynomial via a group of data points that passes through those points, allowing the value of the function between those points to be predicted. The bending strength values f_m parallel and perpendicular to the fibers were taken for plywood of various thicknesses (9, 12, 15, 18, 21, 24), and a model was developed based on them, with the help of which values for plywood thicknesses that were not tested experimentally were calculated (10, 17, 20, 22, 23). This significantly improves efficiency, safety, sustainability, and innovation in manufacturing and construction of wooden materials and products. The application of this model is possible for solid wood and other wood-based materials in predicting different characteristics.

In short, the application of mathematical models provides quantitative data to support inferences that lead to new theories and knowledge. They are a key tool for progress in many areas of science and technology. Thus, in the wood industry, where non-destructive testing methods and the application of various mathematical models are increasingly applied in order to reduce the time and costs of testing as well as protect the wood itself, The combination of mathematical models, experimental data, and engineering knowledge enables a better understanding of the mechanical properties of wood and a safer and more efficient design and construction of wooden products.

References

[1] Ćehić M and Omer S E 2018 Pločasti materijali na bazi drveta *University of Bihać*
[2] Šernek M and Jošt M 2004 Konstrukcijski kompozitni les *Les (Ljubljana)*, letnik 56, number 7/8, str. 230-235. URN:NBN:SI:DOC-MFDGUWCP from http://www.dlib.si
[3] BAS EN 310 2001 *Wood-based panels – Determination of modulus of elasticity in bending and of bending strength*
[4] Brezović M and Pervan S and Petrak J and Prekrat S 2018 Method of Estimation of Laminated Wood Properties *Drvena Industrija* 69 (1) 49-54 (2018) UDK: 630*832.28; 630* 812.7 doi:10.5552/drind.2018.1738
[5] Scitovski R 2004 Numerička matematika *University of Osijek,* UDK: 517(075.8)
[6] Zolić A 2008 Numerička matematika *Faculty of Mathematics*, Belgrade

IOP Conf. Series: Materials Science and Engineering 1298 (2023) 012017 doi:10.1088/1757-899X/1298/1/012017

Design and FPGA Implementation of Compensator for Sharpening CIC Filter

Gordana Jovanovic Dolecek and Gabriel Alejandro Martinez Novelo

Department of Electronics, Institute INAOE, Puebla Mexico

gordana@ieee.org, gabriel.martinez@inaoe.mx

Abstract. This paper presents a novel compensator design for sharpened CIC (cascade-integrator-comb) proposed in the literature. Sharpened CIC provides higher aliasing attenuation than the CIC filter. However, its passband droop is higher than the corresponding CIC filter and must be compensated. Our motivation was to design a decimator with better compensation than the one proposed in the literature. The proposed decimation filter has two coefficients and six adders. The coefficients are determined using particle swarm optimization (PSO) in MATLAB. Two designs are presented. The first one has two multipliers and six adders. The second is a multiplierless design obtained by presenting optimal coefficients in the signed power-of-two (SPT) form. The proposed design is compared with the design from the literature. The designed compensator is implemented in a field-programmable gate array (FPGA). Details of the implementation are described in the paper.

1. Introduction

Decreasing the sampling rate in the digital domain has many applications in communications, software-defined radio, and oversampled analog-digital converters, among others [1-4]. Decreasing the sampling rate introduces aliasing, which must be eliminated by the digital decimation or antialiasing filter [5]. The cascade-integrator-comb (CIC) filter is the simplest decimation multiplierless filter. The CIC filter must attenuate the aliasing in the bands around the filter's zeros, called folding bands. As such, attenuation in the folding bands should be high. However, the attenuation provided by the CIC filter is insufficient, especially in the first folding band. Many authors proposed different methods to increase CIC aliasing rejection while trying to keep as much as possible its low complexity [6-11]. Additionally, since the CIC frequency characteristic in the passband of interest is not flat should be compensated by a filter called a compensator. The design of CIC compensators is presented, for example, in works [12-14]. The authors in [6] presented sharpening CIC filters to improve aliasing rejection and the compensators to decrease a high droop in the passband of interest. Different sharpened polynomials are considered, and the compensator designs for several passbands of interest. However, the compensators in [6] do not provide flexibility in design, and the compensator magnitude characteristic in dB is not zero for the zero frequency and should be normalized.

The work proposed here aims to improve the compensator design proposed in [6]. The first goal is to provide flexibility in the design by trading off the number of compensator adders and the compensation quality expressed in the absolute value of the maximum passband deviation in the passband of interest.

The next goal is to get a naturally normalized compensator characteristic.

IOP Conf. Series: Materials Science and Engineering 1298 (2023) 012017 doi:10.1088/1757-899X/1298/1/012017

The rest of the paper is organized as described in the following. Section 2 briefly presents the minimax sharpened CIC filters and designed optimum coefficient compensators from [6]. Section 3 elaborates on the proposed compensator design and is illustrated with two examples. The comparison of the proposed design and the design from [6] is elaborated in Section 4. Section 5 presents the results of the FPGA design. Finally, the last section presents concluding remarks.

2. Minimax sharpened CIC filters and optimum compensators

This section briefly introduces the minimax sharpened CIC (SHCIC) filters and the designed corresponding optimum narrowband multiplierless compensators proposed in [6]. Four sharpening polynomials and the corresponding compensators for several passbands of interest are presented.

The coefficients of compensators are given in SPT (signed power-of-two forms). The number of adders varies from 4 to 8, and the corresponding values of the passband deviation vary from 0.03 dB to 0.24 dB, depending on the passband width of interest defined by the passband edge ω_p. The method is illustrated in the following example.

Example 1. The sharpened polynomial $p(x)=-2^{-6}x^2+x^4$, where x is the CIC filter. The passband of interest is defined by $\omega_p=\pi/4$. The compensator coefficients reported in [6] are given as, $g=[-1-2^3, 2+2^6, -1-2^3]$. Observe that the compensator coefficients should be normalized by the coefficient $K= -1-2^3+2+2^6-1-2^3=-2^4+2^6$.

The magnitude response of compensated SHCIC filter, denoted as $|H_{CG}(e^{j\omega})|$, is given as:

$$|H_{CG}(e^{j\omega})| = |H_{SH}(e^{j\omega})| \times |G(e^{j\omega})|, \tag{1}$$

where $|H_{SH}(e^{j\omega})|$ and $|G(e^{j\omega})|$ are magnitude responses of the SHCIC filter and the compensator [6], respectively.

Figure 1 compares the magnitude responses of the no compensated and compensated SHCIC filters in the passband, taking the normalized compensator in [6].

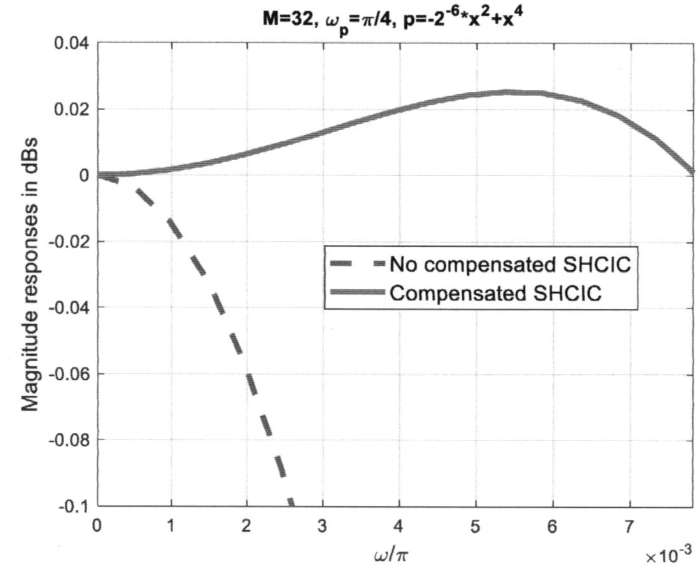

Figure 1. Magnitude responses of no compensated and compensated SHCIC filters in [6].

3. Proposed compensator for minimax sharpened CIC

The goal is to introduce a novel normalized compensator that exhibits a trade-off between the number of adders and the maximum absolute value of passband deviation δ in dB. We propose adapting the CIC compensator proposed in [14] to compensate for the SHCIC filter.

The magnitude characteristic of the compensator $C(e^{j\omega M})$ is in the sinusoidal form,

$$|C(e^{j\omega M})| = 1 + A_1 sin^4\left(\frac{\omega M}{2}\right) + A_2 sin^2\left(\frac{\omega M}{2}\right), \tag{2}$$

where M is the decimation factor, and A_1 and A_2 are the design parameters.

The system function of the compensator (2) has six adders and two coefficients, C_0 and C_1, [14]

$$C(z) = C_0(1 + z^{-4} - 2z^{-2}) + C_1(z^{-1} + z^{-3} + 2z^{-2}) + z^{-2}. \tag{3}$$

The compensator coefficients C_0 and C_1 are obtained from the parameters A_1 and A_2 in (2) as [14]:

$$C_0 = A_1 \times 2^{-4}. \quad C_1 = (A_1 + A_2) \times 2^{-2}. \tag{4}$$

The number of compensator adders from (3) and (4) is

$$N=6+N_0+N_1, \tag{5}$$

where N_0 and N_1 are the numbers of adders in the compensator coefficients C_0 and C_1, respectively. The magnitude response of compensated SHCIC filter, denoted as $|H_{CC}(e^{j\omega})|$, is given as:

$$|H_{CC}(e^{j\omega})| = |H_{SH}(e^{j\omega})| \times |C(e^{j\omega M})|, \tag{6}$$

Using particle swarm optimization (PSO) in MATLAB [15] we get the optimum parameters A_{1opt} and A_{2opt} which give the optimum absolute deviation of (6) in the passband defined by the passband edge ω_p. From (4) the optimum coefficients of the compensator are :

$$C_{0opt} = A_{1opt} \times 2^{-4}. \quad C_{1opt} = (A_{1opt} + A_{2opt}) \times 2^{-2}. \tag{7}$$

The compensator has six adders and two multipliers. To get a multiplierless design the optimum parameters A_{1opt} and A_{2opt} are presented in SPT (signed power-of-two) forms. The method is illustrated in the following example.

Example 2. We consider the same SHCIC as in Example 1. The optimal values of parameters are obtained as $A_{1opt} = 0.5546$, A_{2opt}=0.6674. The optimum value of the passband deviation is 0.0004 dB. Table 1 presents multiplierless design varying the number of adders from nine to eleven.

Table 1. SPT design parameters, compensator coefficients, number of adders N, and passband deviation δ in dB.

SPT form of A	SPT form of B	C_0	C_1	N	δ in dB
$2^{-1}+2^{-3}-2^{-5}$	$2^{-1}+2^{-3}+2^{-5}$	$2^{-5}+2^{-7}-2^{-9}$	$2^{-2}+2^{-4}$	9	0.0067
$2^{-1}+2^{-3}-2^{-5}-2^{-7}$	$2^{-1}+2^{-3}+2^{-5}+2^{-7}$	$2^{-5}+2^{-7}-2^{-9}-2^{-11}$	$2^{-2}+2^{-4}$	10	0.0011
$2^{-1}+2^{-4}-2^{-6}+2^{-8}$	$2^{0}-2^{-2}-2^{-4}-2^{-6}-2^{-8}$	$2^{-3}+2^{-6}-2^{-8}+2^{-10}$	$2^{-3}+2^{2}+2^{4}$	11	0.00056071

Magnitude responses of the compensated SHCIC filter with the optimum compensator and compensators from Table 1 are contrasted in Figure. 2.

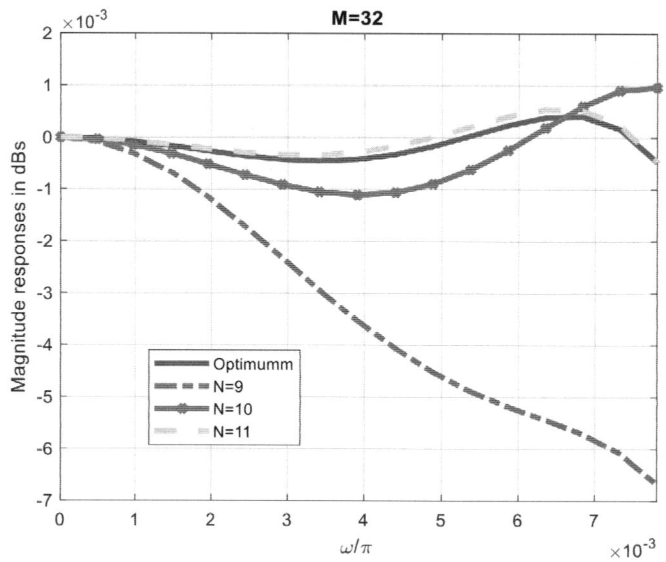

Figure 2. Magnitude responses in the passband of the compensated SHCIC using the designs from Table 1.

Example 3. We design the compensator for the sharpening polynomial $p(x)=2^{-10}x^2-2^{-4}x^4+x^6$, for $\omega_p=\pi/3$. The optimum values of the parameters of design are $A_{1opt} = 1.2459$, A_{2opt}=0.9919, while δ_{opt}=0.0055 dB. Table 2 presents the results of the multiplierless design, varying the number of adders from eight to ten. We can observe that the choice of N=10 equals practically to the optimum design.

Table 2. SPT design parameters, compensator coefficients, number of adders N, and passband deviation in dB.

SPT form of A	SPT form of B	C_0	C_1	N	δ in dB
2^0+2^{-2}	2^0	$2^{-4}+2^{-6}$	$2^{-1}+2^{-4}$	8	0.0166
2^0+2^{-2}	2^0-2^{-7}	$2^{-4}+2^{-6}$	$2^{-1}+2^{-4}-2^{-9}$	9	0.0062
2^0+2^{-2}	$2^0-2^{-7}-2^{-11}$	$2^{-4}+2^{-6}$	$2^{-1}+2^{-4}-2^{-9}-2^{-13}$	10	0.0056

Figure 3 contrasts the magnitude responses of the compensated SHCIC filters in the passband.

4. Comparisons with method in [6]

In this section we compare the proposed design with the method in [6] in the following examples.

Example 4. We compare the proposed design with the design in [6] given in Example 1. We take the design parameters from the first row of the Table 1. The values of the maximum passband deviations in [6] and the proposed design are 0.03 dB and 0.0067 dB, respectively, while the required numbers of adders are four, and nine, respectively.

The corresponding magnitude responses in the passband are contrasted in Figure 4. It can be observed that the proposed design achieves much better characteristic in the passband at an expense of the increased number of adders.

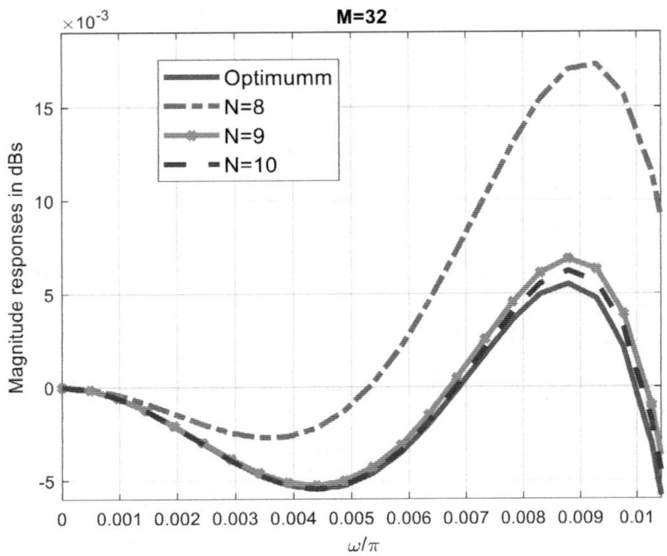

Figure 3. Magnitude responses in the passband of the compensated SHCIC using the designs from Table 2.

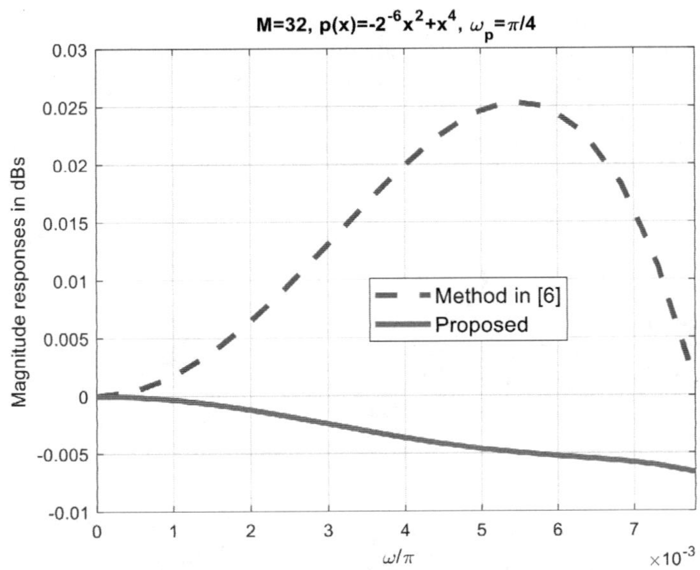

Figure 4. Comparison with the compensated SHCIC [6] and the proposed compensated SHCIC from Table 2.

Example 5. We compare the proposed design from the first row in Table 2 with the second design in [6]. The values of the maximum absolute values of the passband deviations, in [6] and the proposed design, are 0.0285 dB and 0.0166 dB, while the corresponding numbers of adders are five and eight. The magnitude responses in passbands are contrasted in Figure 5.

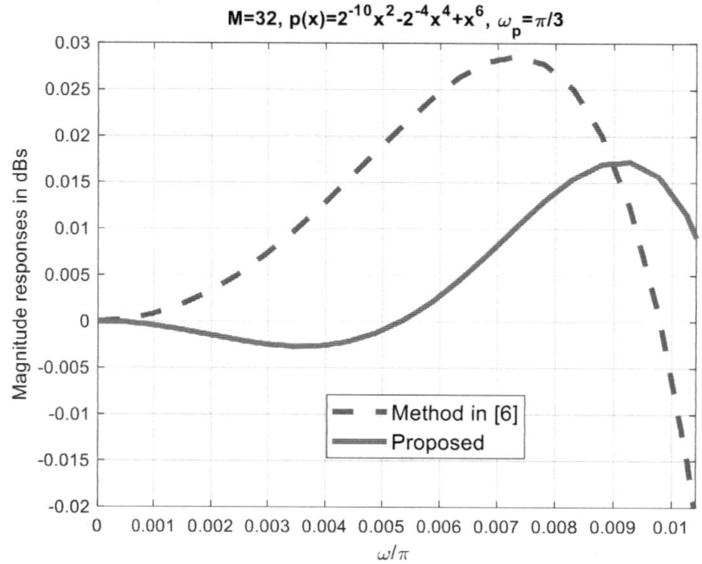

Figure 5. Comparison with the compensated SHCIC [6] and the proposed compensated SHCIC from Table 2.

5. FPGA implementation

The compensator was implemented in FPGA on an EP235F672 card using Quartus II software, designed to observe the impulse response, which impulse response has been generated from the FPGA itself.

Figure 6a shows the block diagram that was proposed to carry out the multiplication with C_o in equation (3), denoted as M_0, while Figure 6b shows the diagram provided by the Quartus II software. In a very similar way, block diagrams of multiplication with C_1, denoted M_1, in equation (3) are presented in Figure 6c. Similarly, Figure 6d shows the diagram provided by the Quartus II software.

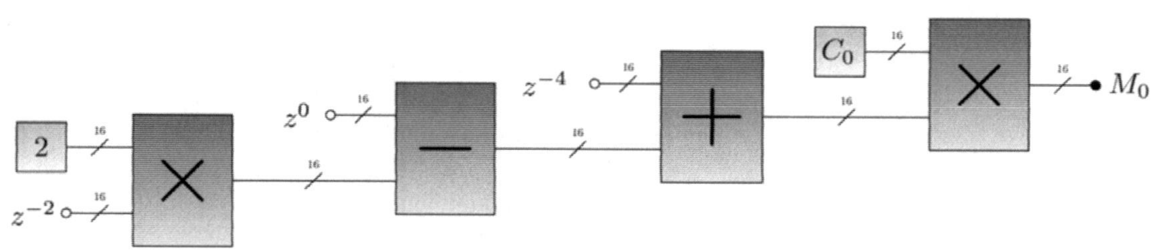

(a)

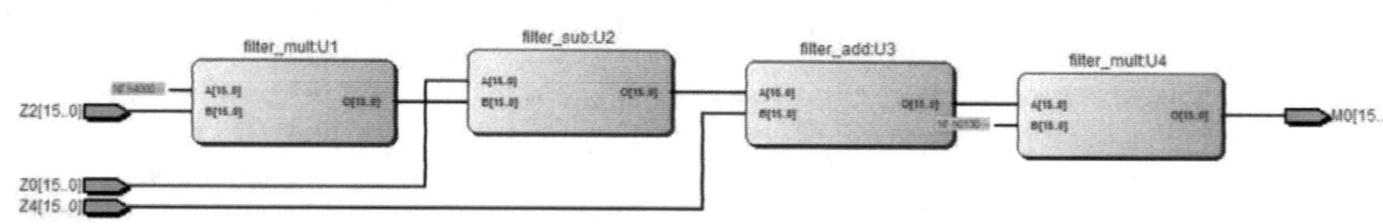

(b)

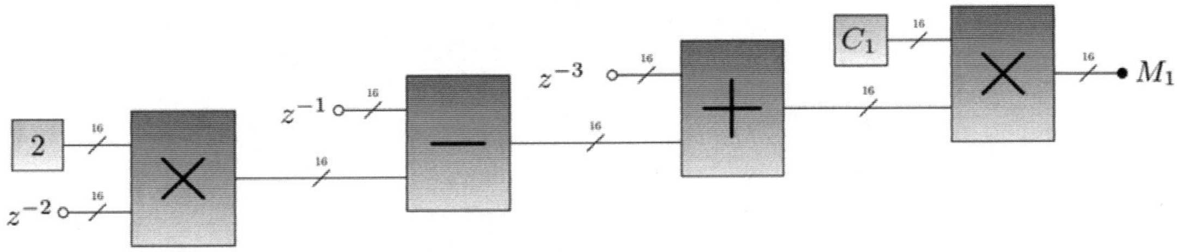

(c)

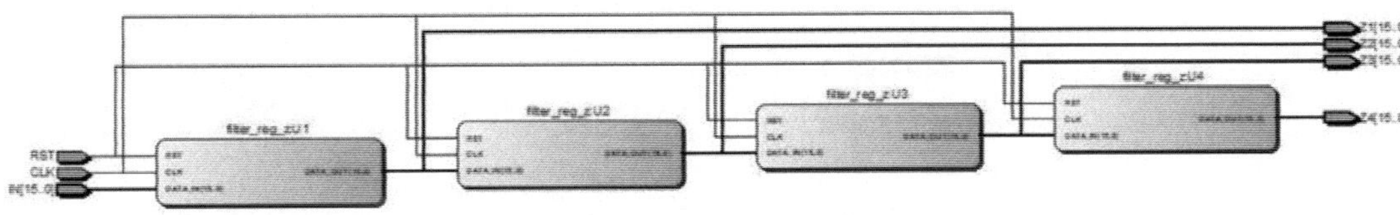

(d)

Figure. 6. Implementation of system function given in equation (3).

Figure 7 shows the highest hierarchy RTL diagram of the compensator implemented in the FPGA. The circuit is made up of 6 blocks. The first block "filter pulse:V4" generates a unit pulse as input to the system. The "filter_regs:V1" block are the shift registers, to delay the signal; at the output we have the four delayed signals, from one to four samples to use them as inputs to the operator blocks, "filter_c0:V2" and "filter_c1: V3" that represent the implementation of M_0 and M_1 respectively. The subtraction is made in the "filter_sub:V5" block, and finally the result is added with the signal Z_2 to obtain the output of the compensator.

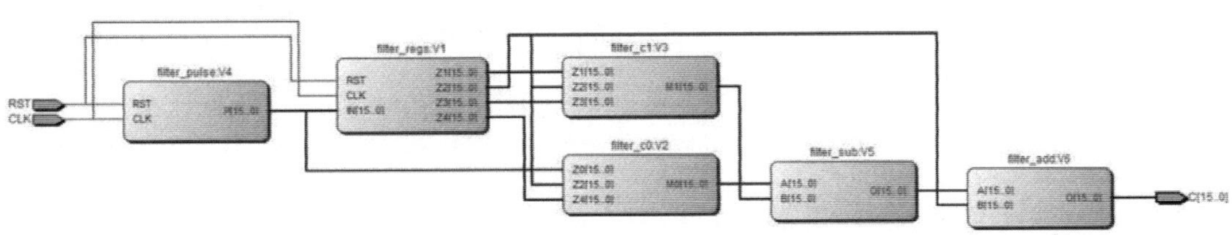

Figure 7. RTL diagram of the implemented compensator.

Summary of power and resource consumption is given in Figure 8.

PowerPlay Power Analyzer Summary	
PowerPlay Power Analyzer Status	Successful - Thu Jun 22 21:22:35 2023
Quartus II 64-Bit Version	13.0.1 Build 232 06/12/2013 SP 1 SJ Web Edition
Revision Name	filter_decimator
Top-level Entity Name	filter_decimator
Family	Cyclone II
Device	EP2C35F672C6
Power Models	Final
Total Thermal Power Dissipation	111.93 mW
Core Dynamic Thermal Power Dissipation	0.00 mW
Core Static Thermal Power Dissipation	79.93 mW
I/O Thermal Power Dissipation	32.00 mW
Power Estimation Confidence	Low: user provided insufficient toggle rate data

(a) Power consumption.

Flow Summary	
Flow Status	Successful - Thu Jun 22 21:22:35 2023
Quartus II 64-Bit Version	13.0.1 Build 232 06/12/2013 SP 1 SJ Web Edition
Revision Name	filter_decimator
Top-level Entity Name	filter_decimator
Family	Cyclone II
Device	EP2C35F672C6
Timing Models	Final
Total logic elements	49 / 33,216 (< 1 %)
Total combinational functions	49 / 33,216 (< 1 %)
Dedicated logic registers	7 / 33,216 (< 1 %)
Total registers	7
Total pins	18 / 475 (4 %)
Total virtual pins	0
Total memory bits	0 / 483,840 (0 %)
Embedded Multiplier 9-bit elements	0 / 70 (0 %)
Total PLLs	0 / 4 (0 %)

(b) Resource consumption.

Figure 8. Summary of power and resource consumption.

6. Conclusion

We present the design and FPGA implementation of the narrowband compensator for SHCIC filter proposed in the literature. The proposed compensator presents better passband compensation and is naturally normalized in the comparison with the designs given in [6].

The proposed design generally requires higher number of adders. However, since the compensator works at low rate, the complexity of the compensator is unsignificantly increased.

Additionally, the proposed design exhibits the flexibility expressed in the trade-off between the number of adders and maximum value of the deviation in the passband.

References

[1] Jovanovic Dolecek G 2018 *Advances in Multirate Systems* (NY: Springer) p 57.

[2] Pavan S, Schreier R, and Temes G C 1918 *Understanding Delta-Sigma Data Converters* (New Wiley-IEEE Press) p 517.

[3] Gan C, and Le X 2021 Improved CIC Decimation Filter on Software Defined Radio Proc. of *9th International Conference on Communications and Broadband Networking*, (Shanghai: ACM Digital Library) pp 232-8.

[4] Abinaya A, Maheswari M, and Alqahtani A S 2021 Heuristic analysis of CIC filter design for next-generation wireless applications *Arabian Journal for Science and Engineering* **46** pp.1257–68.

[5] Jovanovic Dolecek G 2002 *Multirate Systems: Design and Applications* (Hershey: IGP) pp 1-27.

[6] Dudarin A, Molnar G, and Vucic M 2018 Optimum multiplierless compensators for sharpened cascaded- integrator-comb decimation filters *Electronics Letters* **54** pp 971-2.

[7] Dudarin A, Vucic M, and Molnar G 2022 Decimation filters with minimum number of additions per output sample *El Letters* 58 pp. 246-8.

[8] Jovanovic Dolecek G 2022 Design of decimation filter with improved magnitude characteristic and low complexity *IEEE Access* **10** pp 63455 – 65.

[9] Aggarwal S, and Meher P K 2017 Enhanced Sharpening of CIC decimation filters, implementation and applications *Circ Syst and Sig Proc* **36** pp 2031–49.

[10] Gautam D, Khare K, and Shrivastava B P 2021 A novel approach for optimal design of sample rate conversion *IEEE Access* **9** pp 44436-41.

[11] Jovanovic Dolecek G 2022 Comb decimator design based on symmetric polynomials with roots on the unit circle: Two-stage multiplierless design and improved magnitude characteristic *Int J of Circ Th and Appl* **50** pp 2210-27.

[12] Jovanovic Dolecek G, Baez G R, Molina Salgado G, and de la Rosa J M 2017 Novel multiplierless wideband comb compensator with high compensation capability *Circ Syst and Sig Proc* **36** pp. 2031–49.

[13] Xu L, Yang L W, and Tian H 2019 Design of wideband CIC compensator based on particle swarm optimization *Circ Syst and Sig Proc* **38** pp. 1833-46.
[14] Jovanovic Dolecek G, and de la Rosa J M 2021 Design of wideband comb compensator based on magnitude response using two sinusoidals and particle swarm optimization *AEU-Int J of El and Comm* **130** 153570.
[15] https://www.mathworks.com/help/gads/particleswarm.html.

IOP Conf. Series: Materials Science and Engineering 1298 (2023) 012018 doi:10.1088/1757-899X/1298/1/012018

Averaging data on the unit hypersphere

Aladin Crnkić, Zinaid Kapić, Šejla Jusić and Melisa Haurdić

University of Bihać, Faculty of Technical Engineering, Bosnia and Herzegovina

E-mail: aladin.crnkic@unbi.ba

Abstract. Averaging data on the unit sphere S^d (also called a unit hypersphere) is a common problem in computer vision, robotics and other fields, with applications ranging from motion planning to DNA modelling. In this paper, we introduce a new method for averaging data represented as points on the unit sphere S^{d-1} using the d-dimensional generalized Kuramoto model. Our method is verified on a range of benchmark data sets and compared with common data averaging algorithms. Also, we showcase the applicability of this method for solving rotation averaging problem.

1. Introduction

Averaging of different forms of data is widely used across many scientific and engineering fields, including pattern recognition [1], computer vision [2], and camera calibrations [3]. This paper focuses on the averaging of a directional data set, in which data points are constrained to the surface of the $(d-1)$-dimensional unit sphere S^{d-1} in $\mathbb{R}^d$ vector space. The main difficulty in finding the average of unit vectors is their spherical nature since this data does not belong to Euclidean space. Most of the papers on averaging data on spheres are based on finding averages in $\mathbb{R}^d$ and renormalizing the result on the sphere, causing undesirable effects [4, 5]. Clark and Thompson compute averages using a stereographic projection [6]. This approach suffers from distortion, particularly for data points far from the center of the projection. The paper [7] uses Shoemake's approach introduced in defining spherical linear interpolation to allow computing averages of such data points.

In this paper, we introduce a new dynamical model for finding weighted and unweighted averages on the unit hypersphere that considers the spherical nature of the data. In other words, our approach takes into account a differential geometrical structure of a manifold such as the unit hypersphere.

For our method, we will use a model that is a generalization to higher dimensions of the Kuramoto model that mathematically describes the synchronization phenomenon. This phenomenon can be easily explained in the trivial example where two or more oscillating particles can align their oscillations in a phase or anti-phase with each other. It can be seen in different biological, physical, and social systems.

The Kuramoto model has been established in 1975 by Y. Kuramoto, and it describes the dynamics of a population of N phase coupled oscillators [8]

$$\dot{\varphi}_j(t) = \omega_j + \frac{K}{N}\sum_{i=1}^{N}(\varphi_j - \varphi_i), j = 1, \dots, N, \tag{1}$$

where $\varphi_j \in [0,2\pi)$ and ω_j are a phase and an intrinsic frequency of oscillator j respectively, and K is global coupling strength. The model (1) and its generalizations have found wide applications in studying synchronization phenomena across various systems, including complex networks, power grids, and interpolation problems [9-12].

In this paper, we use a dynamical model where the state of a d-dimensional generalized oscillator is described by a unit vector $Q \in \mathbb{R}^d$, or a point that lies on the surface of the unit sphere S^{d-1}. The dynamics of a given oscillator can be represented mathematically in the form of the following differential equation

$$\dot{Q} = WQ,$$

where W is the frequency of the generalized oscillator given in the form of the anti-symmetric $d \times d$ matrix. On the other hand, a system of coupled generalized oscillators on the unit sphere S^{d-1} is described by the following dynamical system

$$\dot{Q}_j = \frac{K}{N}\sum_{i=1}^{N}(Q_i - \langle Q_j, Q_i\rangle Q_j) + W_j Q_j, \qquad j = \overline{1,N}, \tag{2}$$

where K represents a global coupling strength and W_j is the frequency of the generalized oscillator j. The notation $\langle \cdot,\cdot \rangle$ stands for the standard inner product (dot-product) in $\mathbb{R}^d$. Model (2) is an extension of the classical Kuramoto model (1) to higher dimensions, i.e. model (1) has been extended from the unit circle S^1 to the unit hypersphere S^n of any dimension. In the following, we will use a variation of the model (2) for the purposes of our method.

The rest of the paper is structured as follows. In the second section, we introduce our method for solving average of weighted and unweighted data points on the unit hypersphere. In the third section, we will present simulation results that illustrate our approach on real-world and random data sets and showcase its applicability in solving rotation averaging problems. Finally, we will give a conclusion and outlook of our research.

2. Algorithms

In order to introduce dynamical model for averaging points $P_j, j = 1, \ldots, N$, that lie on the surface of the unit sphere S^{d-1}, let us assume that the frequencies of the generalized oscillator W_j is equal to zero for all j. Also, we set that $K = 1$ without loss of generality. According to the above, we can transform (2) into

$$\dot{Q}_j = \frac{1}{N}\sum_{i=1}^{N}\left(Q_i - \langle Q_j, Q_i\rangle Q_j\right), \qquad Q_j(0) = P_j, \qquad j = 1, \ldots, N. \tag{3}$$

If we take closer look at the previous model, its dynamics will tend to the mean field that corresponds to the alignment (synchronization) of the population where $Q_1 = Q_2 = \cdots = Q_N$. In this setup, the obtained equilibrium configuration represents the average of the considering points. Simultaneously, model (3) preserves the unit sphere S^{d-1}, i.e. if $Q_j(0) \in S^{d-1}$ implies that $Q_j(t) \in S^{d-1}$ for all t.

Analogously, we can easily introduce model for averaging of weighted points $P_j \in S^{d-1}, j = 1, \ldots, N$

$$\dot{Q}_j = \frac{1}{N}\sum_{i=1}^{N} w_i(Q_i - \langle Q_j, Q_i\rangle Q_j), \qquad Q_j(0) = P_j, \qquad j = 1, \ldots, N, \tag{4}$$

where w_i denotes weights of the point P_j for all j.

Furthermore, we will explain our method for averaging hyperspherical data in more detail through the algorithms.

The algorithm for unweighted data is as follows:

1: Enter N and $P_j, j = 1, \ldots, N$

2: Set tolerance ϵ, step δ, and $T = 0$

3: Solve (3) with $Q_j(0) = P_j$

IOP Conf. Series: Materials Science and Engineering 1298 (2023) 012018 doi:10.1088/1757-899X/1298/1/012018

4: Compute $\hat{Q}(t) = \frac{1}{N}\sum_{j=1}^{N} Q_j(t)$, for $t > 0$

4: loop

5: if $1 - \|\hat{Q}(T+\delta)\| < \epsilon$ then

return $A(P_1, \dots, P_N) = Q_j(T)$, for any j

6: else

$T = T + \delta$

7: end if

8: end loop

Analogously, the algorithm for weighted data is as follows:

1: Enter N, P_j and w_j for $j = 1, \dots, N$

2: Set tolerance ϵ, step δ, and $T = 0$

3: Solve (4) with $Q_j(0) = P_j$ and w_j

4: Compute $\hat{Q}(t) = \frac{1}{N}\sum_{j=1}^{N} Q_j(t)$, for $t > 0$

4: loop

5: if $1 - \|\hat{Q}(T+\delta)\| < \epsilon$ then

return $WA(P_1, \dots, P_N; w_j) = Q_j(T)$, for any j

6: else

$T = T + \delta$

7: end if

8: end loop

3. Simulations

For the first simulation we have generated 100 data points that belong to the unit sphere S^2. Data points are generated using von Mises-Fischer distribution on S^2 with the mean direction $\mu = (1,0,0)$ and concentration parameter $\kappa = 0.01$. Simulation results that illustrate our method are presented in Figure 1. Unweighted and weighted average is obtained at a moment $T = 6.09$ and $T = 12.58$, respectively, with tolerance $\epsilon = 10^{-5}$ and step $\delta = 0.01$.

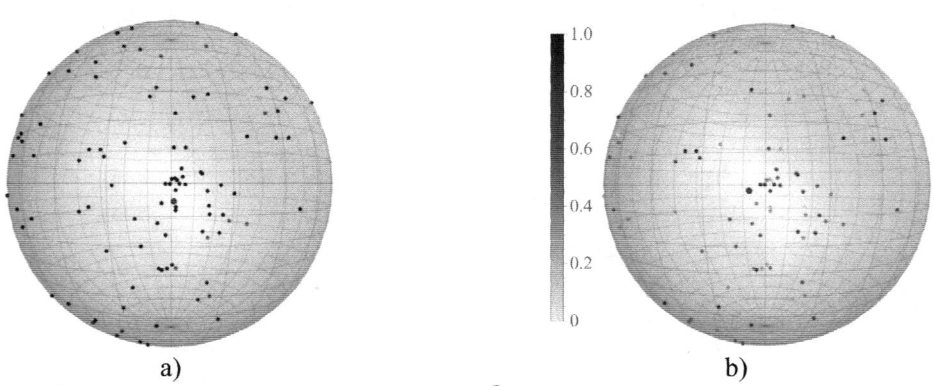

a) b)

Figure 1. Averages of data points on the unit sphere S^2: a) unweighted average; b) weighted average.

Figure 1a) shows the unweighted average of the data points, while Figure 1b) visualizes the weighted average for the same data points with corresponding weights generated randomly in the range [0,1]. As can be seen, weighted and unweighted averages have different values (see Table 1) since the weighted average considers the importance or contribution of each data point.

Table 1. Unweighted and weighted averages obtained with the data set with 100 points on S^2.

Averaging algorithm	Moment	Values
Unweighted average	$T = 6.68$	$[-0.159448, 0.987164, 0.00916746]$
Weighted average	$T = 13.93$	$[0.9113, 0.408656, -0.0503307]$

Since data points on the unit sphere S^3 can be represented with unit quaternions, this method is also applicable to the rotations. Unit quaternions are widely used tools for representing rotations of objects in three-dimensional space. Their mathematical notation is of the form $q = a + b\mathbf{i} + c\mathbf{j} + d\mathbf{k}$, where a, b, c, and d are real numbers for which $a^2 + b^2 + c^2 + d^2 = 1$, and $\mathbf{i}^2 = \mathbf{j}^2 = \mathbf{k}^2 = -1$ [13].

The rotation averaging has gained significant popularity with the rise of the computer vision and computer graphics [14]. Govindu [15-17] introduces rotation averaging in the field of structure from motion. Gramkow [18] compares different rotation metrics and methods for rotation averaging. Rotation averaging is also present in the 3D modelling [19]. This paper examines the single rotation averaging problem, defined by Moakher in paper [20]. Following the analogy in Euclidean space, the mean rotation of a given sequence of rotations is a minimization of the sum of squared distances from the given rotations. If we use Euclidean distance, the mean is called projected arithmetic mean. Geometric mean is obtained by using Riemannian metric and geodesic distance between rotations [13].

Let us assume that we have a given set of N points P_n, $n = 1, \dots, N$ in $\mathbb{R}^d$, the arithmetic mean $\bar{P}$ is calculated as [13]

$$\bar{P} = \underset{P\in\mathbb{R}^d}{\arg\min} \sum_{n=1}^{N} d_e(P, P_n)^2, \tag{5}$$

where $d_e(\cdot,\cdot)$ is the Euclidean distance in $\mathbb{R}^d$. This approach is commonly used to calculate the mean of points in Euclidean space. When it comes to finding the mean of rotations, the Karcher mean, also known as the geometric mean, is the preferred method [13]

$$\bar{P} = \underset{P\in\mathbb{R}^d}{\arg\min} \sum_{n=1}^{N} d_{geod}(P, P_n)^2, \tag{6}$$

where $d_{geod} = |\log x - \log y|$ represents a geodesic distance between points. The paper [13] also addresses a problem of finding weighted mean of points on the unit hypersphere where each data point is associated with a corresponding weight. By incorporating weights into the averaging process, the resulting mean represents a more robust and representative summary of the data, where the contributions of individual points are appropriately considered.

Our method is evaluated with one real-world and two randomly generated data sets of rotations. Simulation results obtained from our algorithm are compared to the arithmetic and geometric averages. Since a point on the unit sphere S^3 is impossible to visualize, we will use function known as Hopf's fibration [21] that transforms a point on the unit sphere S^3 into a point on the unit sphere S^2

$$H: S^3 \to S^2 \Rightarrow H(a,b,c,d) = \left(a^2 + b^2 - c^2 - d^2, 2(ad + bc), 2(bd - ad)\right).$$

Second data set is a real-world data set named Drill. Drill data set was obtained to assess human movement variations in performing specific tasks [22]. This data set has 720 observations of eight subjects drilling into a metal plate while monitored by infrared cameras. Rancourt used quaternions to represent the orientation of subject's wrist, elbow and shoulder in one of six positions. For simulation purposes we have used 30 rows of first subject's wrist movements.

Simulation results for this data set are presented in Figure 2 and Table 2.

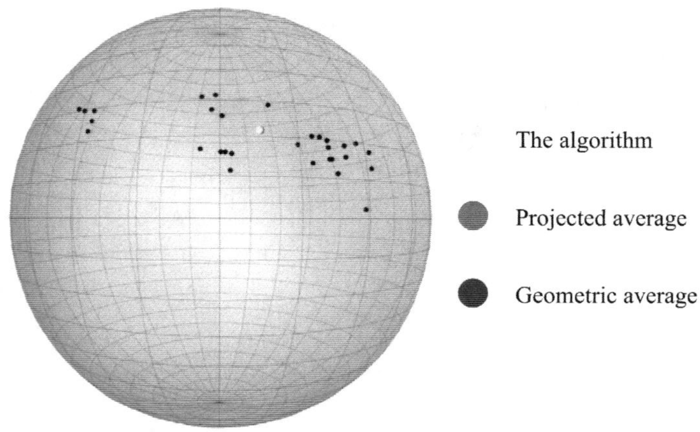

Figure 2. Averages obtained with a real-world data set Drill.

As is seen in Figure 2, two common (projected and geometric) averages and average obtained with our method overlap each other. Numerical values can be seen in Table 2.

Table 2. Numerical values of averages obtained with a real-world data set Drill.

Averaging algorithm	Values at a moment T=4.66
Arithmetic	$[0.9872858, -0.0698275, -0.1341495, 0.0489356]$
Geometric	$[0.9873313, -0.0697236, -0.1344122, 0.0474225]$
The algorithm	$[0.9873313, -0.0697236, -0.1344123, 0.0474227]$

The third data set contains 250 rotations generated using von Mises-Fischer distribution on S^3 with the mean direction $\mu = \left(\frac{1}{2}, \frac{1}{2}, \frac{1}{2}, \frac{1}{2}\right)$ and concentration parameter $\kappa = 0.01$. Each data point is assigned with corresponding weights generated randomly in the range [0,1]. Figure 3 and Table 3 present simulation results for this data set.

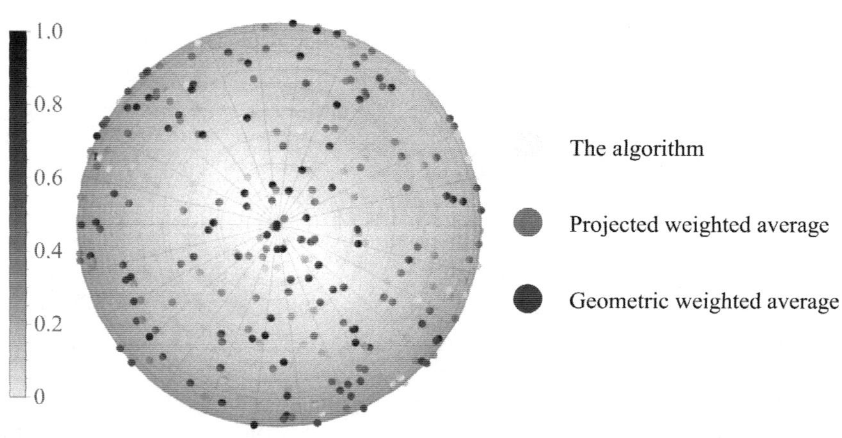

Figure 3. Weighted averages obtained with 250 (weighted) points randomly generated with von Mises-Fischer distribution on S^3.

Table 3. Numerical values of weighted averages obtained using different algorithms.

Averaging algorithm	Values at a moment T=13.68
Arithmetic weighted average	[0.997087, 0.029580,0.068253,0.017408]
Geometric weighted average	[0.991697, 0.083612,0.072090,0.065940]
The algorithm	[0.995508,0.068452,0.055995, 0.033789]

As is seen in Table 3, the geometric weighted average and weighted average introduced in this paper are almost identical, with values differing only in second or third decimal positions.

4. Conclusion

In this paper, we introduced new algorithms for finding unweighted and weighted averages on the unit hypersphere. These algorithms are based on the d-dimensional generalized Kuramoto model, which is an extension of the classical model to the unit hypersphere. We have shown the applicability of our method in solving rotation averaging problems. Simulations and numerical results of averages indicate that our algorithms give approximately the same results as geometric average. This method enables flexibility and applicability across numerous different situations and dimensions, making these algorithms valuable for a wide range of applications, especially in the field of computer vision, robotics, and 3D modelling. In future work, we will focus on expanding the evaluation on larger data sets and improving the computational time, as well as exploring the performance on other non-Euclidean manifolds. This model can be also adapted for cluster analysis on the unit hypersphere, which could be our potential future direction of research.

References

[1] Corbett J, Fischer J and Whitney D 2010 Averaging independent estimates improves pattern recognition *Journal of Vision* **9** 819–819 doi: 10.1167/9.8.819
[2] Xiaoyi Jiang, Munger A and Bunke H 2001 An median graphs: Properties, algorithms, and applications *IEEE Transactions on Pattern Analysis and Machine Intelligence* **23** 1144–51 doi: 10.1109/34.954604
[3] Dai Y, Trumpf J, Li H, Barnes N and Hartley R 2010 Rotation averaging with application to camera-rig calibration *Computer Vision – ACCV 2009* 335–46 doi: 10.1007/978-3-642-12304-7_32
[4] Alfeld P, Neamtu M and Schumaker L L 1996 Bernstein-Bézier polynomials on spheres and sphere-like surfaces *Computer Aided Geometric Design* **13** 333–49 doi: 10.1016/0167-8396(95)00030-5
[5] Wagner G 1990 On means of distances on the surface of a sphere (lower bounds) *Pacific Journal of Mathematics* **144** 389–98
[6] Clark R M and Thompson R 1984 Statistical comparison of palaeomagnetic directional records from Lake Sediments *Geophysical Journal International* **76** 337–68 doi: 10.1111/j.1365-246X.1984.tb05050.x
[7] Buss S R and Fillmore J P 2001 Spherical averages and applications to spherical splines and interpolation *ACM Transactions on Graphics* **20** 95–126 doi: 10.1145/502122.502124
[8] Kuramoto Y 1975 Self-entrainment of a population of coupled non-linear oscillators *International Symposium on Mathematical Problems in Theoretical Physics* 420–2 doi: 10.1007/bfb0013365
[9] Novikov A V and Benderskaya E N 2014 Oscillatory neural networks based on the Kuramoto model for Cluster Analysis *Pattern Recognition and Image Analysis* **24** 365–71 doi: 10.1134/S1054661814030146
[10] Guo Y, Zhang D, Li Z, Wang Q and Yu D 2021 Overviews on the applications of the Kuramoto

model in modern power system analysis *International Journal of Electrical Power & Energy Systems* **129** 106804 doi: 10.1016/j.ijepes.2021.106804

[11] Crnkić A and Kapić Z 2021 Interpolation on the special orthogonal group with high-dimensional Kuramoto model *IOP Conference Series: Materials Science and Engineering* **1208** 012037 doi: 10.1088/1757-899X/1208/1/012037

[12] Kapić Z and Crnkić A 2021 Interpolating rotations with non-abelian Kuramoto model on the 3-sphere *Lecture Notes in Networks and Systems* 610–6 doi: 10.1007/978-3-030-90055-7_48

[13] Hartley R, Trumpf J, Dai Y and Li H 2013 Rotation averaging *International Journal of Computer Vision* **103** 267–305 doi: 10.1007/s11263-012-0601-0

[14] Kapic Z, Crnkic A, Jacimovic V and Mijajlovic N 2021 A new dynamical model for solving rotation averaging problem 2021 20th International Symposium INFOTEH-JAHORINA (INFOTEH) doi: 10.1109/INFOTEH51037.2021.9400663

[15] Govindu V M 2001 Combining two-view constraints for motion estimation *Proceedings of the 2001 IEEE Computer Society Conference on Computer Vision and Pattern Recognition. CVPR 2001* doi: 10.1109/CVPR.2001.990963

[16] Govindu V M 2004 Lie-algebraic averaging for globally consistent motion estimation *Proceedings of the 2004 IEEE Computer Society Conference on Computer Vision and Pattern Recognition, 2004. CVPR 2004.* doi: 10.1109/CVPR.2004.147

[17] Govindu V M 2006 Robustness in motion averaging *Computer Vision – ACCV 2006* 457–66 doi: 10.1007/11612704_46

[18] Gramkow C 2001 On Averaging Rotations *Journal of Mathematical Imaging and Vision* **15** doi: 7–16 10.1023/A:1011217513455

[19] Chen Y, Zhao J and Kneip L 2021 Hybrid rotation averaging: A fast and robust rotation averaging approach *2021 IEEE/CVF Conference on Computer Vision and Pattern Recognition (CVPR)* doi: 10.1109/cvpr46437.2021.01022

[20] Moakher M 2002 Means and averaging in the group of *rotations SIAM Journal on Matrix Analysis and Applications* **24** 1–16 doi: 10.1137/S0895479801383877

[21] Lyons D W 2003 An elementary introduction to the Hopf fibration *Mathematics Magazine* **76** 87–98 doi: 10.1080/0025570X.2003.11953158

[22] Rancourt D, Rivest L-P and Asselin J 2000 Using orientation statistics to investigate variations in human kinematics *Journal of the Royal Statistical Society Series C: Applied Statistics* **49** 81–94 doi: 10.1111/1467-9876.00180

Power losses in the MOSFET transistors of switching-mode converters

Georgi Kunov[1], Tihomir Brusev[2] and Elissaveta Gadjeva[1]

[1]Faculty of Electronic Engineering and Technologies, Technical University of Sofia, Kl. Ohridski 8, 1797 Sofia, Bulgaria
[2]Faculty of Telecommunications, Technical University of Sofia, Kl. Ohridski 8, 1797 Sofia, Bulgaria

E-mail: brusev@ecad.tu-sofia.bg

Abstract. The power losses in the MOSFET transistors of switching-mode converters are investigated and analyzed. The results presented in this paper are achieved using software Cadence OrCAD. The simulation model suitable for estimation of energy dissipations in MOSFET transistors is proposed. The power losses as a function of different circuit's parameters, like switching frequency, power supply voltage and output current are evaluated. The PSpice model of MOSFET transistor "*BSC020N03MS*" of the company Infineon is used in the presented analysis. Control circuit, which realize the adaptive deadtime (ADT) control technique depending on the load current, is applied to reduce the switching power losses in the MOSFET transistor. Thus, the efficiency of the switching-mode converters can be increased.

1. Introduction

Nowadays the switching-mode converters are widespread used in the Internet of things (IoT), automotive industry and telecommunication applications [1], [2], [3]. They have to ensure desire energy to the loads and at the same time they have to be highly efficient circuits. Therefore, the power losses in the converters should be as minimum as possible. The designers have to optimize the operating circuit's parameters in order to increase the efficiency of switching-mode converters.

The major function in the converters is played by the switches [4]. On the other hand, they are one of the most energy consuming components in the switching-mode converters [5]. The output power transistors have the big impact over the efficiency of the whole converter's system. The power losses in the output MOSFET transistors depend on the electrical circuit's parameters. High efficiency of the switching-mode converters could be achieved if optimal circuit's parameters are chosen from the designers.

In this paper is proposed simulation model suitable for estimation of energy dissipation in MOSFET transistors in switching-mode converters. The power losses are evaluated and analyzed as a function of switching frequency, power supply voltage and output current of the circuit. The PSpice model of MOSFET transistor "*BSC020N03MS*" of the company Infineon is used in the presented analysis. The investigations of power losses in MOSFET transistors are realized using software Cadence OrCAD. The schematic of synchronous buck dc-dc converter is presented in Figure 1. The diode in the in the basic circuit's topology is replaced by MOSFET transistor Q_2. Thus, by eliminating the diode voltage drop, the efficiency of the converter can be increased.

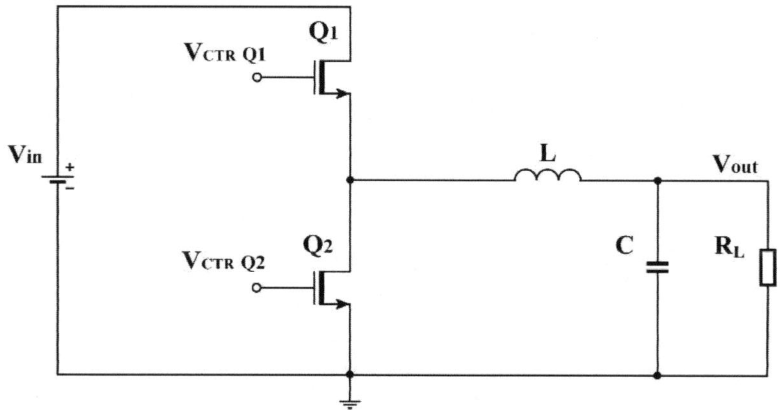

Figure 1. Schematic of synchronous buck dc-dc converter.

The output transistors of converters usually are controlled with Pulse-Width Modulation (PWM) technique. With development of the modern technologies is reduced the switching transition times of the power MOSFET transistors [4]. This allows power converters to operate at high switching frequency f_s [6]. The effect of increasing of f_s over the energy consumption of output switches is investigated and analyzed.

2. Power losses in MOSFET transistors investigated with Cadence OrCAD

The power losses in MOSFET transistors in switching-mode converters are divided into switching and conduction power losses [7]:

$$P_{losses,MOSFET} = P_{sw} + P_{cond}, \tag{1}$$

The switching losses in the hard-switched buck dc-dc converter can be expressed by formula [8]:

$$P_{sw(conv)} = P_{turn-on} + P_{turn-off} = \left(C_p\, V_{in}^2 + V_{in}\, I_{out}\, t_r\right) f_s, \tag{2}$$

where $P_{turn-on}$ and $P_{turn-off}$ are power switching losses in the output MOSFET transistors, during their turning-on and turning-off times respectively; C_p is parasitic junction capacitance of transistors when they are turning-on; t_r is a rise time of the transistor. The losses $P_{turn-off}$ are due to the V-I overlapping when the power switches are turning-off. As can be seen from the formula (2), the power losses in MOSFET transistors go up with increasing of the switching frequency f_s of the converter.

The conduction power losses P_{cond} of the MOSFET transistor are proportional to $R_{DS(on)}$ [7]. The total energy consumption of switching-mode converters can be decreased if output transistors have low drain-to-source on resistance.

The efficiency is one of the most important parameters of switching-mode converters and can be expressed by formula [9]:

$$\eta = \frac{P_{out}}{P_{in}} = \frac{P_{out}}{P_{out} + P_{losses}}, \tag{3}$$

where P_{out} and P_{in} are the output and the input power, respectively; P_{losses} are the overall power losses in the circuit. On the other hand, the total power losses in the output MOSFET transistors are dominating losses in the switching-mode converters [10], [11]. Therefore, the efficiency of the converters can be improved if the total power losses in the MOSFET transistors are decreased.

The total power losses in MOSFET transistor "*BSC020N03MS*" of the company Infineon are investigated with software Cadence OrCAD. The analyzed model is presented in Figure 2. The

MOSFET transistors Q_1 and Q_2 are regulated with PWM technique, which is widespread used control method for switching-mode converters [4]. In the first half period the transistor Q_1 is switched-on and the inductor L is connected to the power supply. During the second half period Q_1 is switched-off, while Q_2 is switched-on and the current is flowing through the transistor Q_2, because of the energy stored in the inductor L.

The passive snubbers are used to decrease the power losses in MOSFET transistors Q_1 and Q_2 [4]. They are formed by capacitors C_{sn1} and C_{sn2}, resistors R_{sn1} and R_{sn2} and diodes D_{sn1}, D_{sn2}, respectively for both switches. The snubbers have to reduce the switching stress spike of the transistors Q_1 and Q_2 [4]. In order to be regulated a stable inductor current with the desire value, the current feedback is used in the presented model. Thus, the power losses in the MOSFET transistors can be investigated at the different inductor current levels.

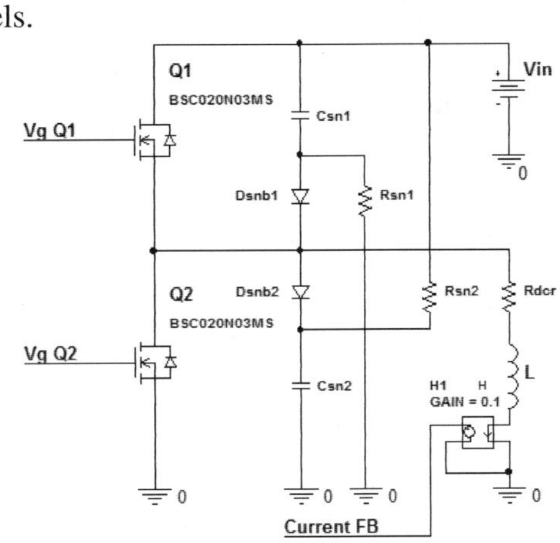

Figure 2. The investigated model with software Cadence OrCAD.

The circuit, which realizes the PWM control technique of the Q_1 and Q_2, ensures the adaptive deadtime (ADT) depending on the output current. Therefore, the switching power losses in the MOSFET transistors can be reduced. The block diagram of the investigated model with ADT is presented in Figure 3 [12].

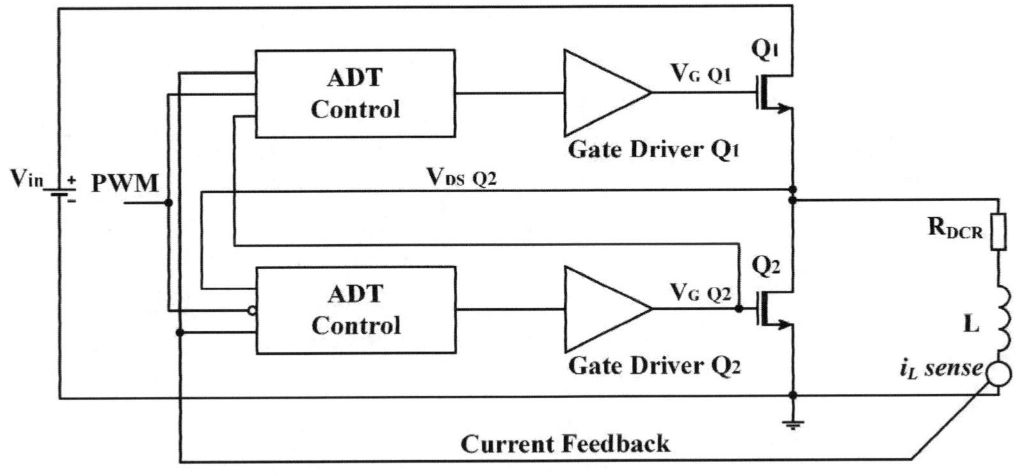

Figure 3. Block diagram of the investigated model with ADT control.

The ADT control used in this investigation, helps to be reduced the switching power losses $P_{turn\text{-}off}$ of the transistor Q_1 and $P_{turn\text{-}on}$ of the transistor Q_2. The parasitic capacitance C_{ds1} and C_{ds2}, of the both output MOSFET transistors Q_1 and Q_2 respectively, are discharged during the deadtime intervals. The ADT control circuit operates in such way, that when the voltage drop over the capacitance C_{ds2} became equal to zero, enable signal is generated and Q_2 is switched-on [12]. Therefore, the overall power losses of the MOSFET transistors are minimized.

The current feedback delivers the signal proportional to inductor current value I_L to the ADT control blocks shown in the Figure 3. The delay time between voltages V_{GQ1} and V_{GQ2}, which regulate the mode of operation of MOSFET transistors Q_1 and Q_2, depend on the I_L. The waveforms of the control signals V_{GQ1} and V_{GQ2} are presented in Figure 4 (a). As can be seen from the picture, the short-circuit power losses, which can appear if both transistors Q_1 and Q_2 are simultaneously switched-on, are prevented.

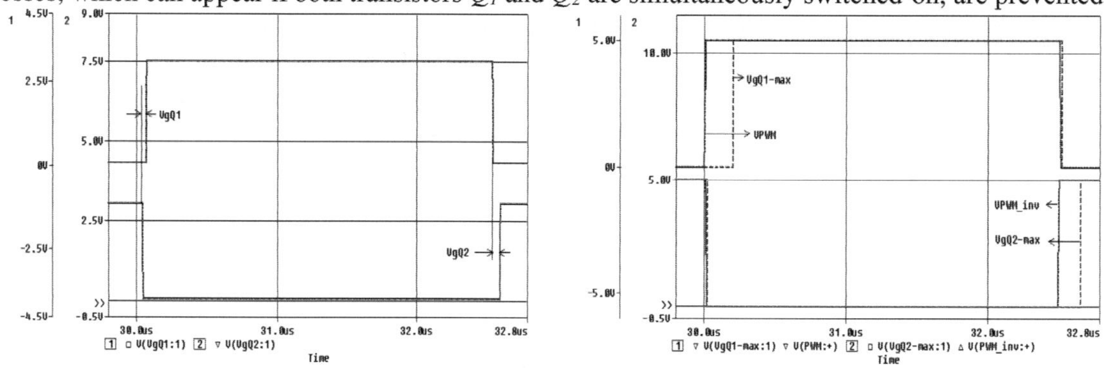

Figure 4. (a) The waveforms of the ADT control signals of the MOSFET transistors Q_1 and Q_2; (b) PWM control signals and the control signals of the MOSFET transistors Q_1 and Q_2 in the initial starting moment of the circuit's simulation.

In the initial moment, when the simulation of the investigated model starts, the maximum delay time between the control signals of the MOSFET transistors Q_1 and Q_2 is ensured. The waveforms of the PWM control signals, the voltages V_{GQ1} and V_{GQ2} in this case are presented in Figure 4 (b). After the initial starting moment, the adaptive deadtime control has priority and the control signals shown in Figure 4 (a) are applied to Q_1 and Q_2.

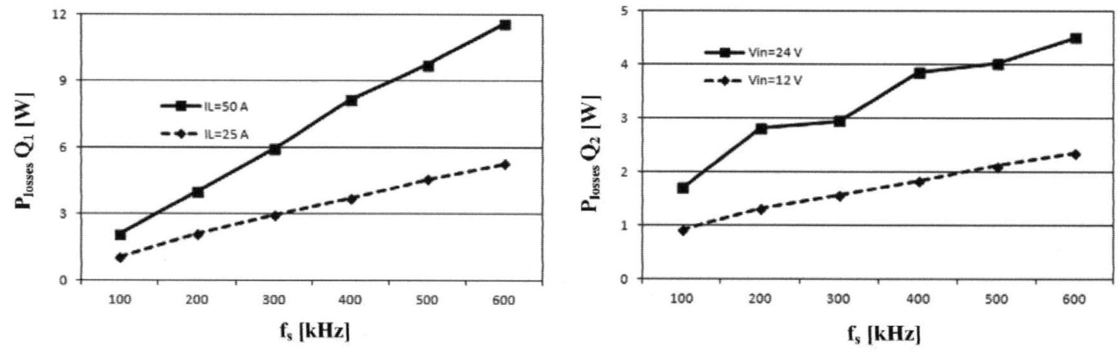

Figure 5. (a) Power losses of MOSFET transistor Q_1 (P_{losses} Q_1) at V_{in}=12 V and (b) power losses of MOSFET transistor Q_2 (P_{losses} Q_2) at I_L=25 A, as a function of switching frequency f_s.

The power losses in MOSFET transistor "*BSC020N03MS*" of the company Infineon are investigated as a function of switching frequency f_s. They are evaluated with Cadence OrCAD, using the model shown in Figure 2 and adaptive deadtime control technique illustrated in Figure 3. The values of f_s are changed between 100 kHz and 600 kHz, which is typical frequency range of most of switching-mode converter's controllers [13], [14].

The investigations are realized at different values of inductor current I_L and power supply voltage V_{in}. The power losses of MOSFET transistor Q_1 (P_{losses} Q_1) at V_{in}=12 V, I_L=25 A and I_L=50 A, are graphically presented in Figure 5 (a). As can be seen from the picture, they are proportional to the switching frequency and the inductor current values.

The power losses of MOSFET transistor Q_2 (P_{losses} Q_2) at I_L=25 A, V_{in}=12 V and V_{in}=24 V are illustrated in Figure 5 (b). The simulation results presented in Figure 5 (b) prove that energy consumptions in Q_2 go up, when power supply voltage is increasing.

The power losses in MOSFET transistor "*BSC020N03MS*" are investigated using two different regulation methods. They are the conventional hard-switching PWM control and the ADT control technique. In the conventional hard-switching PWM control the dead time, when both transistors Q_1 and Q_2 are switched-off, is fixed. In the ADT control technique this dead time varies and depends on the load current. The received results of energy consumption as a function of switching frequency are presented in Table 1. In this particular case the input voltage V_{in} is12 V, while the inductor current I_L is stabilized to equal to 25 A.

Table 1. Power losses in MOSFET transistor Q_1 and Q_2 as a function of f_s at V_{in}=12 V and I_L=25 A.

f_s [kHz]	P_{losses} Q_1 [W] hard-switching	P_{losses} Q_1 [W] ADT control	P_{losses} Q_2 [W] hard-switching	P_{losses} Q_2 [W] ADT control
100	1.75	1.05	1.56	0.92
200	2.47	2.09	1.65	1.32
300	3.98	2.95	2.34	1.56
400	4.8	3.7	2.72	1.84
500	5.73	4.58	2.86	2.12
600	6.58	5.27	3.1	2.35

The power losses of MOSFET transistor Q_2 (P_{losses} Q_2), when hard-switching PWM control and ADT control technique is used, are graphically presented in Figure 6.

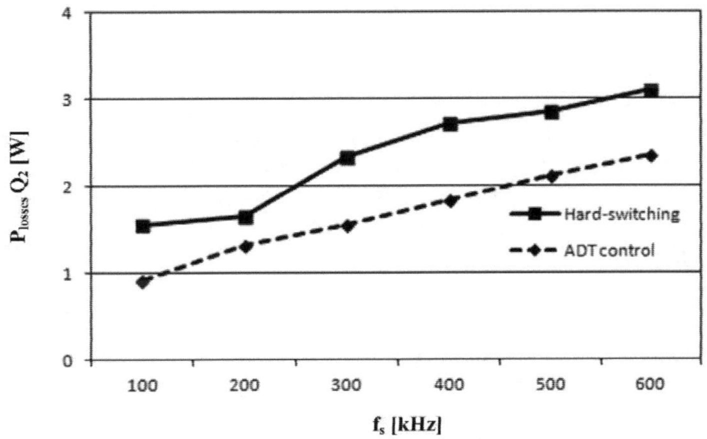

Figure 6. Power losses of MOSFET transistor Q_2 as a function of f_s.

As can be seen from the results presented in Table 1 and Figure 6, the total power losses in the output MOSFET transistors of switching-mode converters are minimized, when ADT control is used. Therefore, the efficiency of those circuits can be improved utilizing this regulation method. The analysis of the achieved results shows that power losses in the MOSFET transistors is decreased with average 38%, if ADT control instead of hard-switching PWM control method is used.

3. Conclusion

In this paper is presented the simulation model suitable for estimation of power losses in MOSFET transistors. They are evaluated with software Cadence OrCAD using PSpice model of transistor "*BSC020N03MS*" of the company Infineon. The received results prove that energy dissipations in MOSFET transistors go up, when the power supply voltage, output current and switching frequency of the converters are increasing. The total power losses in MOSFET transistors are investigated with the conventional hard-switching PWM method and the adaptive deadtime control technique depending on the load current. The analysis of the achieved results shows that they are decreased with average 38%, if ADT control is used. Therefore, the efficiency of the switching-mode converters can be increased.

References

[1] Q. Yu, Z. Jiang, Y. Liu and G. Long, *Research of Power Energy Management Control Strategy with IOT in NZEB*, IEEE International Conference on Smart Internet of Things (SmartIoT), pp. 319-322, Beijing, China 2020.

[2] O. Pelan, O. Cornea, N. Muntean, F. Blaabjerg, *Comparative Efficiency Evaluation of Buck and Hybrid Buck DC-DC Converters for Automotive Applications*, International Conference on Optimization of Electrical and Electronic Equipment (OPTIM), pp. 582-587, 2014.

[3] J. Sankman, M. K. Song, and D. Ma, *A 40-MHz 85.8%-Peak-Efficiency Switching-Converter-Only Dual-Phase Envelope Modulator for 2-W 10-MHz LTE Power Amplifier*, Symposium on VLSI Circuits Digest of Technical Papers, June 2014, Honolulu, pp. 1-2.

[4] H. Yu, *Driver Based Soft Switch for Pulse-Width-Modulated Power Converters*, Blacksburg, Virginia, 2005.

[5] J. Ham, H. Jung, H. Kim, W. Lim, D. Heo and Y. Yang, *A CMOS Envelope Tracking Power Amplifier for LTE Mobile Applications*, Journal of Semiconductor Technology and Science, vol. 14, no. 2, April 2014, pp. 235-245.

[6] N. Le Gallou, D. Sardin, C. Delepaut, M. Campovecchio, S. Rochette, *Over 10 MHz bandwidth envelope-tracking DC/DC converter for flexible high power GaN amplifiers*, IEEE MTT-S International Microwave Symposium Digest (MTT), pp.1-4, 5-10 June 2011.

[7] F. J. Ham, H. Jung, H. Kim, W. Lim, D. Heo and Y. Yang, *A CMOS Envelope Tracking Power Amplifier for LTE Mobile Applications*, Journal of Semiconductor Technology and Science, vol.14, No.2, pp. 235-245, April 2014.

[8] S. Chen, O. Trescases and W. T. Ng, *Fast Dead-Time Locked Loops for a High-Efficiency Microprocessor-Load ZVS-QSW DC/DC Converter*, 2003 IEEE Conference on Electron Devices and Solid-State Circuits, pp. 391-394, December 2003.

[9] R. W. Erickson, D. Maksimovic, *Fundamentals of Power Electronics*, Kluwer Academic Publishers, 2001.

[10] J. W. Kwak and D. B. Ma, *Comparative Topology and Power Loss Analysis on 48V-to-1V Direct Step-Down Non-Isolated DC-DC Switched-Mode Power Converters*, 2020 IEEE Energy Conversion Congress and Exposition (ECCE), pp. 943-949, October 2020.

[11] Y. Xiong, S. Sun, H. Jia, P. Shea, Z. J. Shen, *New Physical Insights on Power MOSFET Switching Losses*, 2009 IEEE Transactions on Power Electronics, pp. 525-531, February 2009.

[12] www.ti.com/product/TPS5210

[13] https://www.ti.com/power-management/non-isolated-dc-dc-switching-regulators/overview.html

[14] https://www.analog.com/en/product-category/switching-regulators.html

Software system for automatic reading, storing, and evaluating scanned paper Evaluation Sheets for questions with the choice of one correct answer from several offered

Đulaga Hadžić[1], Enes Saletović[1] and Zinaid Kapić[2]
[1]Polytechnic Faculty, University of Zenica, Zenica, Bosnia and Herzegovina
[2]Faculty of Technical Engineering, University of Bihać, Bihać, Bosnia and Herzegovina

E-mail: djulaga.hadzic@unze.ba

Abstract. Although the technology to automatically score multiple-choice tests has been around for several decades, it is still not as widely available or affordable, especially for paper-based test processing. The main reasons that hinder these processes are the lack of software systems capable of working with contents that are not filled optimally and do not require expensive scanners or other costly equipment. In this study, we present a software system for the automatic reading, storing, and evaluation of scanned assessment sheets. This software system offers a solution that only requires the usual scanned Evaluation Sheets in the form of a not high-resolution image and with the help of specific markers, the system performs reading, storage, and scoring. The user interface is designed to read and display the data from the Evaluation Sheets in detail so that they are very understandable to the user and allow him to quickly spot any errors. The tool has been validated over six years of use and has been continuously improved throughout that period. Thus, this software system achieves a high level of reliability and sensitivity to different levels of the quality of filling out Evaluation Sheets by students.

1. Introduction

A traditional multiple-choice question (MCQ) is a type of question in which a student selects one answer from several choices provided [1]. However, implementing online computer-assisted assessment (CAA) can be challenging due to the high cost of setup and maintenance. Multiple-choice tests are used in important examinations to determine students' academic excellence [2]. Multiple-choice questions (MCQ) tests are considered a direct form of written assessment [3]. Despite the adaptability of other alternatives to assessment techniques, such as modified essay questions, the MCQ is still considered the most reliable technique to avoid any misinterpretation of answers [4, 5].

In Bosnia and Herzegovina, specifically in Tuzla Canton, at the end of the school year, the graduation exam is held for all students of both primary and secondary schools. As part of the matriculation examination, the Integral test is taken, which contains a certain number of questions with four choices for each question, from which they can select only one answer. The Integral test contains questions over several areas (subjects). Grading or evaluation of completed tests is done percentage-ally (from 0 to 100%) for the entire test as well as for individual areas (subjects). In further work, as a case study, we will analyze the Integral test for elementary school students, which is conducted in paper form. Considering the large number of students taking the Integral test at the same time (approximately 4000 students), the need to automate the review and evaluation of completed tests arose. The integral test for elementary school students contains 60 questions, which means that it would be necessary in a very short time to manually review over 24,000 (60 questions x 4,000 students) questions and determine whether the students answered each of them correctly or incorrectly. Of course, this is only one part of

IOP Conf. Series: Materials Science and Engineering 1298 (2023) 012020 doi:10.1088/1757-899X/1298/1/012020

reviewing the tests. Additionally, it is necessary to calculate the overall test result for each student as well as the test results for individual areas (subjects) within that test.

At the beginning of the matriculation exam, the review of tests was conducted manually, which posed significant challenges due to the large number of participating students [6]. Thus, the need to implement a specific system that could automate the process of reviewing and evaluating tests without introducing any expensive additional IT equipment quickly emerged. Therefore, an Evaluation Sheet was added to the Integral Test, enabling students to transfer their answers from the test onto this sheet. In the following sections of this article, we will introduce an in-house software system specifically designed to process scanned evaluation sheets. This system efficiently analyzes the sheets and creates comprehensive results for students' tests from the Integral test. These results include both the overall test score and the individual area (subject) scores, presented in a percentage format.

2. Integral Test and Evaluation Sheet

As already mentioned, the Integral Test for elementary school students contains 60 questions from 14 different areas (subjects). The number of questions by specific areas (subjects) is shown in Table 1.

Table 1. Structure of the Integral Test

	Area (Subject)	Number of questions
1	Bosanski/Hrvatski/Srpski jezik i književnost (eng. Bosnian/Croatian/Serbian language and literatures)	5 (1-5)
2	Prvi strani jezik (eng. First foreign language)	5 (6-10)
3	Matematika (eng. Mathematics)	5 (11-15)
4	Fizika (eng. Physics)	5 (16-20)
5	Informatika (eng. Informatics)	5 (21-25)
6	Biologija (eng. Biology)	5 (26-30)
7	Hemija (eng. Chemistry)	5 (31-35)
8	Historija (eng. History)	5 (36-40)
9	Geografija (eng. Geography)	5 (41-45)
10	Tehnička kultura (eng. Technical culture)	3 (46-48)
11	Likovna kultura (eng. Art culture)	3 (49-51)
12	Muzička kultura (eng. Music culture)	3 (52-54)
13	Tjelesni i zdravstveni odgoj (eng. Physical and health education)	3 (55-57)
14	Građansko obrazovanje (eng. Citizenship education)	3 (58-60)

In the Integral test, each question is accompanied by four possible answers marked with A, B, C and D. Figure 1 shows the structure of the first question from the Integral Test held at the External Matura for elementary schools at the end of the 2021/2022 academic year.

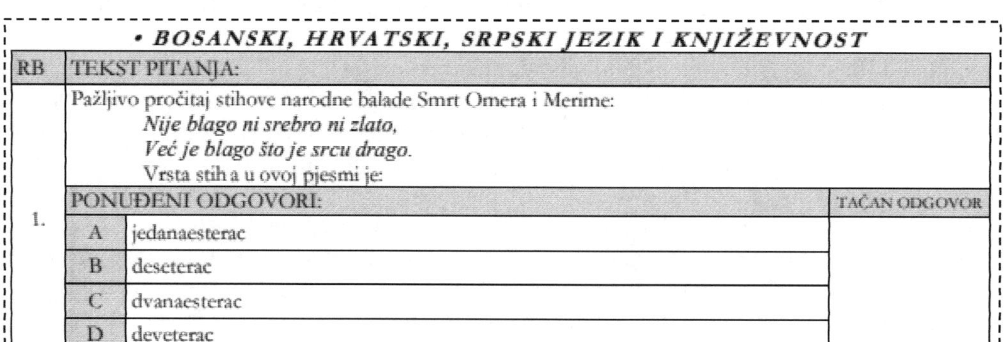
• *BOSANSKI, HRVATSKI, SRPSKI JEZIK I KNJIŽEVNOST*

RB	TEKST PITANJA:		
1.	Pažljivo pročitaj stihove narodne balade Smrt Omera i Merime: *Nije blago ni srebro ni zlato,* *Već je blago što je srcu drago.* Vrsta stiha u ovoj pjesmi je:		
	PONUĐENI ODGOVORI:		TAČAN ODGOVOR
	A	jedanaesterac	
	B	deseterac	
	C	dvanaesterac	
	D	deveterac	

Figure 1. Structure of questions in the Integral Test.

In Figure 1, the TAČAN ODGOVOR (CORRECT ANSWER) field is where the student marks the corresponding letter (A, B, C, or D) for the answer they believe is correct for each question. There are

60 questions with this structure in the Integral Test. An example of an Integral Test can be viewed at the following link.[1]

Once the student has answered all the questions or the ones they believe they can answer correctly, they proceed to transfer their marked answers to the Evaluation Sheet. Figure 2 provides an example of the Evaluation Sheet.

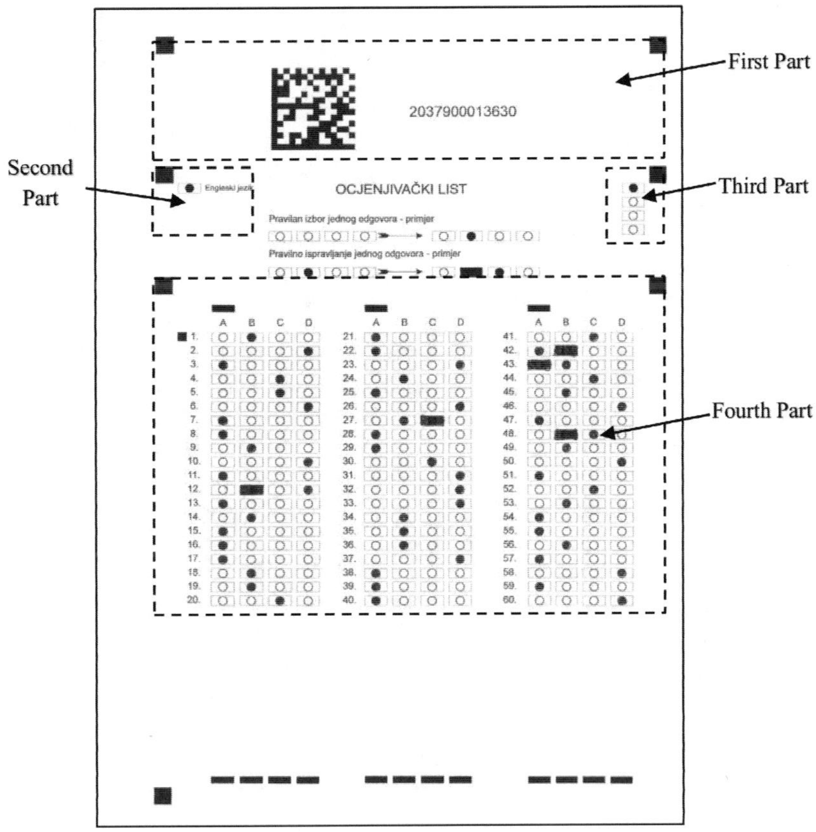

Figure 2. An example of a filled-in Evaluation Sheet.

The Evaluation Sheet consists of the following parts (Figure 2):

- The First Part contains the student's ID
- The Second Part contains the ID for the first foreign language
- The Third Part contains the ID for the test combination
- The Fourth Part is intended for the transfer of marked answers for questions (from 1 to 60) from the Integral Test to the Evaluation Sheet

The student's ID is 13 digits, the first 9 of which indicate the elementary school the student attends, and the last 4 digits represent the student's unique identifier in the entire system (it can appear only once and belong to one student). The student's ID is printed on the Evaluation Sheet in numerical form and in the form of a 2D barcode. An example of the student's ID written in the Evaluation Sheet in the previously mentioned two ways is shown in the First Part of Figure 3.

The student ID is printed in numerical form for the reason that the numerical form is comprehensible to humans and the student ID in the form of a 2D barcode is comprehensible to the software system that will process (evaluate) the Evaluation Sheets.

[1] http://cms1.pztz.ba/userfiles/malamatura/files/Testovi/Matura2022/MMatura_test_april_2022_en.pdf

IOP Conf. Series: Materials Science and Engineering 1298 (2023) 012020 doi:10.1088/1757-899X/1298/1/012020

In the second part of the Evaluation Sheet, there is an identification field for the student's first foreign language, which corresponds to the language they study in the final grades of elementary school. The software system can recognize two options for the first foreign language: English (Engleski jezik) and German (Njemački jezik). Figure 3 illustrates the available options that the software system can identify for this field.

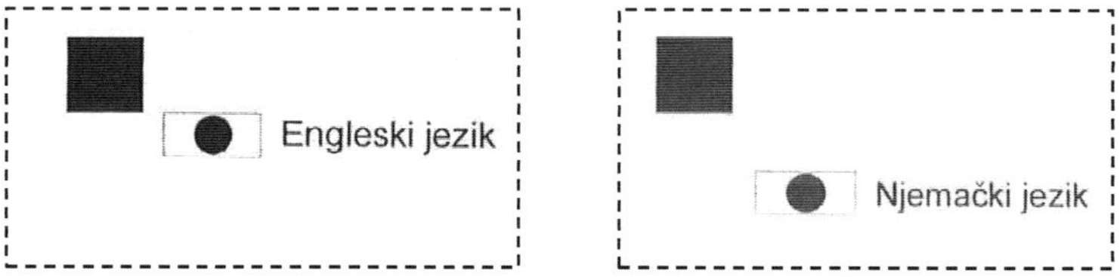

a. b.
Figure 3. Examples for First Foreign Language ID.

In Figure 3, two Evaluation Sheets are shown. Under **a.** only English is shown as the first foreign language and on the second Evaluation Sheet under **b.** only German language is displayed as the first foreign language, but in different positions (coordinates). Based on these positions (coordinates), the software system determines a student's first foreign language and records it. On the basis of the first foreign language, at the end of the Integral Test, the corresponding Evaluation Sheet with the appropriate ID for the foreign language is printed.

The ID of the test combination is recorded in the Third Part of the Evaluation Sheet. Figure 2 shows an example of a combination label for one test. The number of possible combinations of the Integral Test is four. Based on the example shown in Figure 2, the ID of the combination of the Integral test is 1. This combination of the Integral Test is important from the aspect of the combination of correct answers for the questions that will be used when evaluating (reviewing) the completed test, as well as the Evaluation Sheet can be linked with the appropriate combination of the Integral Test.

In the Fourth Part of the Evaluation Sheet, there are the transferred selected answers from the Integral Test. Figure 2 shows the Fourth Part of the Evaluation Sheet, which consists of 3 columns with 20 IDs, one for each question. Each question ID is preceded by its ordinal number (RB). Also, the ID for one question contains four offered selection options that represent the transmitted answer from the Integral test for that question.

In Figure 4, below the title of the Evaluation Sheet (OCJENJIVAČKI LIST), there is a short instruction, i.e. examples of how to fill in (transmit) an answer to a specific question in the Evaluation Sheet and an example of how to cancel that answer and fill in (transmit) another answer for the same question.

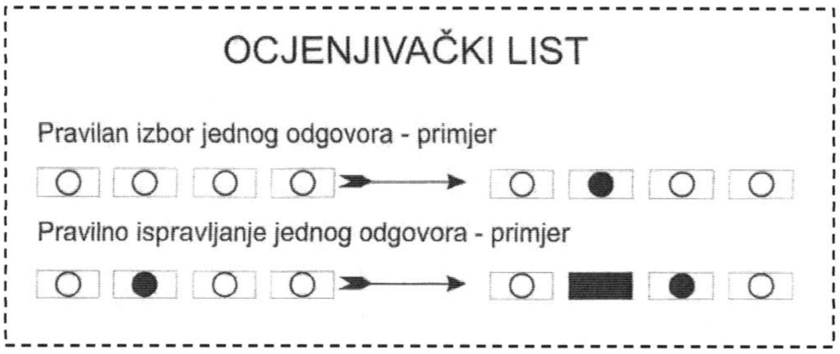

Figure 4. An example of the correct selection of an answer, with an example of how that answer is canceled and another is selected.

For example, in Figure 4, in the first example, the answer to a specific question is marked under B, and in the second example, answer B is canceled and a new answer is marked (transferred) under C for the same question.

IOP Conf. Series: Materials Science and Engineering 1298 (2023) 012020 doi:10.1088/1757-899X/1298/1/012020

3. Software system with main functionalities

As part of this project, a software system for the automatic reading, storing, and evaluation of scanned Evaluation Sheets was developed. The software system was developed using .NET technology in the C# programming language as a desktop application. This system uses already externally scanned Evaluation Sheets as input parameters, and their reading and recording in the database can be done in two ways:

- Semi-automatic - reading of Evaluation Sheets is done one by one, where after one Evaluation Sheet has been read, it is stored in the database by clicking on the **Postavi** (Set) button.
- Automatic - reads and writes to the database successively a large number of Evaluation Sheets placed in one folder, which usually represents all Evaluation Sheets for students of one elementary school.

3.1. Semi-automatic reading

Below is a detailed description of the semi-automatic reading and recording of data from the Evaluation Sheets through the mentioned software system. Figure 5 shows the interface of the part of the software system through which the Semi-automatic reading of Evaluation Sheets is performed.

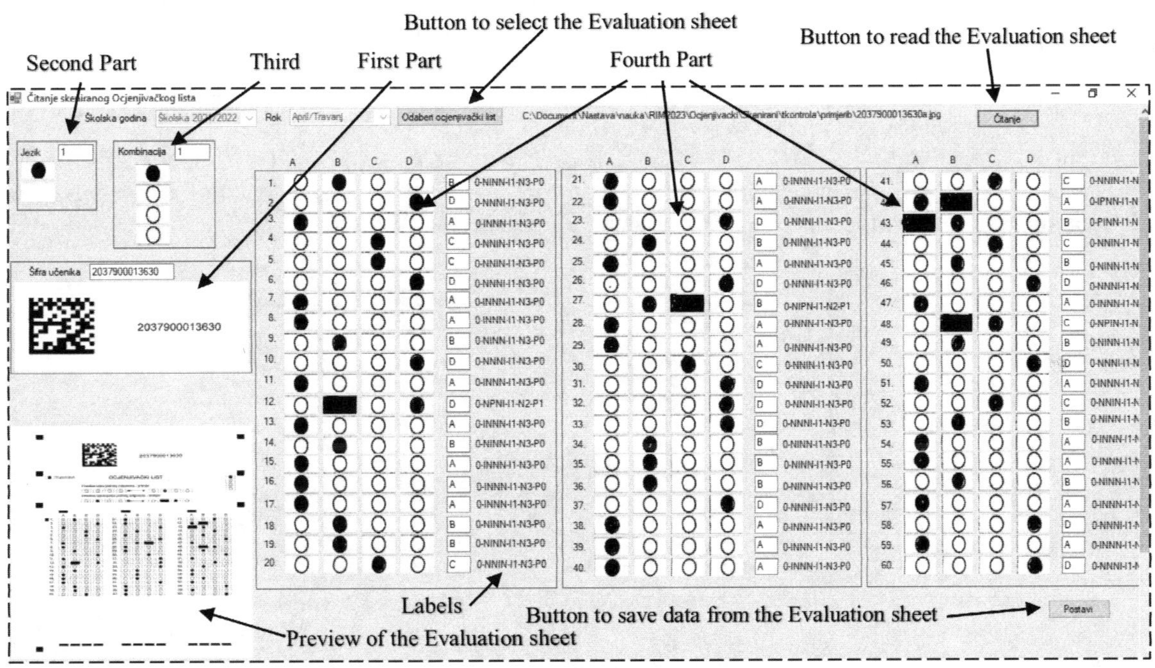

Figure 5. Interface for semi-automatic selection, reading, and storage of one scanned Evaluation Sheet

The user process a specific scanned Evaluation Sheet by clicking the "Odaberi ocjenjivački list" (Select evaluation sheet) button. Once selected, the Evaluation Sheet is read by clicking the "Čitaj" (Read) button and its content is populated in the corresponding fields of the interface shown in Figure 5. The interface design in Figure 5 aligns with the different parts of the Evaluation Sheet as shown in Figure 2. Figure 6 shows a more detailed perspective of the first, second, and third components of the Evaluation Sheet as seen in Figure 5.

Figure 6. Read the first, second, and third parts of the Evaluation Sheet from Figure 2

Next, we will provide an overview of how the software system reads the Evaluation Sheet. Different ways of processing this type of question and answer are presented in the literature. One approach involves creating a template or mask that matches the layout of the Evaluation Sheet. This mask is then compared against the scanned student grade sheet, allowing the system to extract the answers (true or false) to the questions based on the detected markings.

Optical Mark Reading (OMR) is a technique used to automatically read user input in surveys, questionnaires, interviews, and testing systems [7]. OMR uses sensors to evaluate student responses by determining whether a predefined position is empty or already marked [8].

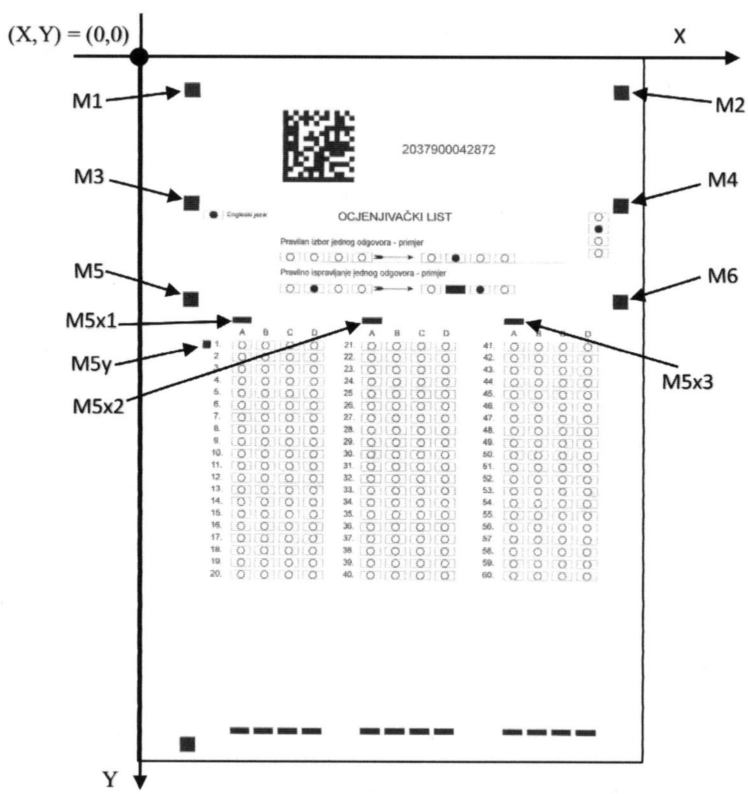

Figure 7. Markers and coordinate system for "movement" according to the Evaluation Sheet

The software system presented in this paper uses an optical mark reading approach. It "moves" around the Evaluation Sheet and reads individual contents. To move around the Evaluation Sheet, the software system uses the markers that are marked as $M1$ to $M7$ in Figure 7. Movement along the Evaluation Sheet starts from the point with coordinates $X = 0$ and $Y = 0$, located in the upper left corner of the Evaluation Sheet (Figure 7). The search is based on a function that works on the principle of comparing the colors of individual pixels. Hence, the software system will search all the pixels from point (0,0) to the first pixel that has a different color from the previous ones. In this case it will have a black color based on the previous pixels that are white. That pixel represents the top left corner of the $M1$ marker. The X and Y coordinates of the previously identified pixel are recorded for further calculations. The same methodology is used for all other markers in the Evaluation Sheet shown in Figure 7.

In the first step, after selecting the file with the Evaluation Sheet, the (X, Y) coordinates of the markers $M1$ and $M2$ are determined, and based on their positions, it is determined whether the Evaluation Sheet is "slanted", i.e. whether $Y_{M1} - Y_{M2} <> 0$, and if so, then the Evaluation Sheet is corrected (it is rotated by a certain angle to the left or right depending on whether the difference $Y_{M1} - Y_{M2}$ is less or greater than zero). Also, based on the distance between the markers $M1$ and $M2$, zooming (increased or reduced) is carried out according to the already defined standard dimensions of the Evaluation Sheet. After the previous corrections have been made, the Evaluation Sheet is saved in a separate file with a .jpeg extension for further processing.

In the second step, based on certain positions of the markers $M1\ (X_{M1}, Y_{M1})\ and\ M2\ (X_{M2}, Y_{M2})$, the searching of the **First part** in the Evaluation Sheet begins (Figure 2). After the 2D barcode is found, it is read using the defined function, and the read value is placed in the corresponding field in the interface shown in Figure 5 (First part).

In the third step, the "movement" along the Evaluation Sheet continues and the $M3$ marker with coordinates (X_{M3}, Y_{M3}) is found, and based on it, the **Second Part** in the Evaluation Sheet from Figure 2 is searched and the ID of the first foreign language is determined through the corresponding function for reading this type of mark. and records in the appropriate field in the interface shown in Figure 5 (Second Part).

In the fourth step, the $M4$ marker with coordinates (X_{M4}, Y_{M4}) is found, and based on it, the **Third Part** is searched in the Evaluation Sheet from Figure 2 and through the appropriate function for reading this type of marker, the ID of the test combination is determined and recorded in the corresponding field in the interface shown on Figure 5 (Third Part).

In the fifth step, marker $M5$ with coordinates (X_{M5}, Y_{M5}), is found, and then from that point marker $M5y$ with coordinates (X_{M5y}, Y_{M5y}) is found. The point with coordinates (X_{M5y}, Y_{M5y}) contains the Y coordinate of the first mark in the first column of the Fourth Part of the Evaluation Sheet. By searching from marker $M5$ with coordinates (X_{M5}, Y_{M5}), marker $M5x1$ with coordinates (X_{5x1}, Y_{5y1}) is found, then marker $M5x2$ with coordinates (X_{5x2}, Y_{5y2}) and finally marker $M5x3$ with coordinates (X_{5x3}, Y_{5y3}). Recorded coordinates X_{5x1}, X_{5x2} and X_{5x3} from markers $M5x1$, $M5x2$ and $M5x3$ represent the value of X coordinates of the first mark in the first column, the first mark in the second column and the first mark in the third column, respectively. This means that the coordinates of the upper left corner of the listed first markers by column were found and recorded, namely (X_{M5x1}, Y_{M5y}) for the first, (X_{M5x2}, Y_{M5y}) for the second and (X_{M5x3}, Y_{M5y}) for the third column.

After the previous three points have been identified and recorded, and given that the width and height of one mark, the distance between the marks and the distance between the lines (questions) of the marks are already known, it is possible to separate each of the marks through certain loops, i.e. 240 of them (60 questions x 4 marks). For each question we have 4 marks representing 4 possible answers, where the first mark represents the answer under A, the second mark represents the answer under B, the third mark represents the answer under C and the fourth mark represents the answer under D.

The following describes how the software system reads the transmitted answers from the Evaluation Sheet. The software system defined three possible label states within a question, which are shown in Figure 8.

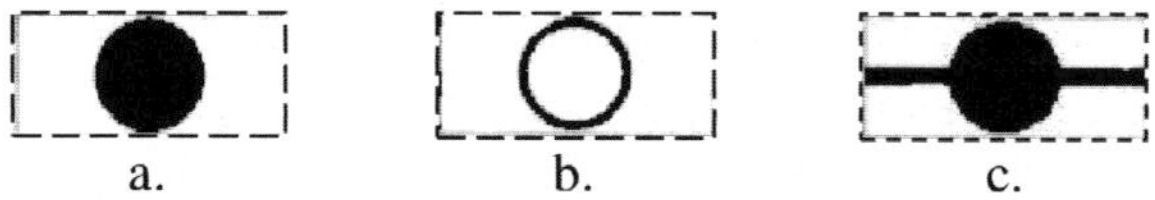

Figure 8. Three possible states of marks within one question: a. Izabran (Selected), b. Nije izabran (Not selected), and c. Poništen (Cancelled)

When analyzing the states of marks within a question, each of the four marks is compared with marks with three predefined states (Figure 8), and their similarities are calculated via the appropriate function.

The analyzed mark will represent the state with which it has the greatest similarity. The stated states under a., b., and c., in Figure 8 are recorded as I, N or P, respectively, in the Label located next to a specific question in the interface shown in Figure 5. In this way, in addition to recording the selected answer, it also records the answers which were selected and canceled as well as whether the student did not try to answer the given question at all or if he chose more than one offered answer. Based on the records stored on the label next to the marks for one question, the software system records in the corresponding field the answer chosen by the student (A, B, C, or D) or records the value 0 if the student did not offer an answer at all, or offered more than one answer.

Here it is necessary to further clarify the appearance of the mark for canceling the selected answer in Figure 8 since this mark does not have the same appearance as the mark for canceling the transferred answer shown in Figure 4. This was done for practical reasons, because the student will not always accurately cancel the marked answer, i.e. completely black out the answer mark, but it can be only partially. With that in mind, the response cancellation tag has the layout shown in Figure 8. Tag state under c. will always have the maximum similarity with the mark for canceling the answer, even in the case when it will not be completely filled in black.

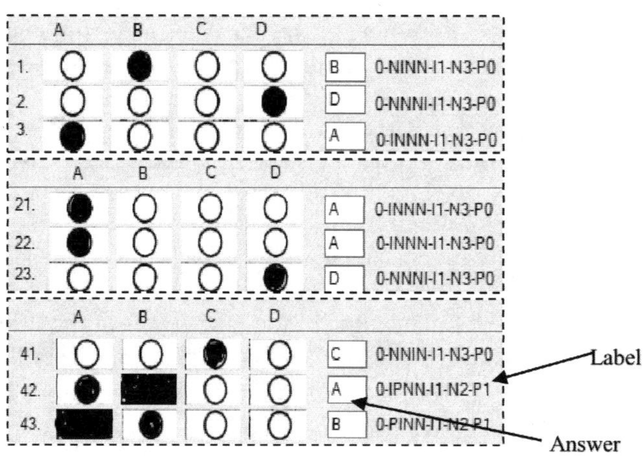

Figure 9. Examples of reading and recording the selected answer for the questions

Figure 9 shows examples of reading labels for individual questions and how the answer was recorded in the interface shown in Figure 5.

Let's analyze question ordinal number 42 from Figure 9. The Label for this question is filled with the following content 0-IPNN-I1-N2-P1 which means the following:

IPNN: I – Selected, P – Cancelled, N – Not selected, and N – Not selected.

I1 – 1 mark is selected

N2 – 2 marks not selected

P1 –1 mark canceled

In the field for recording the selected answer for question 42, the mark A is written, which represents the correct reading of the answer transferred to the Evaluation Sheet.

IOP Conf. Series: Materials Science and Engineering 1298 (2023) 012020 doi:10.1088/1757-899X/1298/1/012020

After all the data from the Evaluation Sheet have been read and recorded in the interface shown in Figure 5, by clicking the **Postavi** (Set) button, the read data is stored in the database. With this, one Evaluation Sheet was read and stored in the database. Also, by clicking on the button, the file, in which the corrected evaluation sheet is stored, is saved, named student ID with the extension .jpeg.

3.2. Problems when reading the Evaluation Sheets

There are several challenges that can arise during the reading of Evaluation Sheets, resulting in incomplete or inaccurate data extraction. This is especially true for failing to read the Student ID or First Foreign Language ID or Test Combination ID. The causes can be various, but the most common are:

- Adding unnecessary marks to the Evaluation Sheet by the student
- Greasy fingers from breakfast or food stains on the Evaluation Sheets.
- Transferring makeup or other substances onto the Evaluation Sheets, creating prints or stains.
- Excessively slanted or reverse scanning (rating sheet turned 180 degrees during scanning) of grading sheets.

The first reason could be solved by better preparation of the students for the matriculation exam, i.e. by filling in the evaluation sheets with the note that nothing else should be added to the Evaluation Sheet except to transfer their answers from the Integral Test to the Evaluation Sheet.

The second and third reasons could be prevented by better preparation of the teachers on duty who conduct the External Matura in a certain elementary school.

The fourth reason can be prevented by better preparation of Evaluation Sheets before the start of their scanning, as well as by purchasing better quality paper, at least for Evaluation Sheets, as well as by purchasing better quality scanners. In the previously mentioned cases when the Evaluation Sheet is not read in its entirety, then we have two cases:

- If it is a semi-automatic reading, then the user can manually enter the missing data by visual inspection of the Evaluation Sheet (Figure 5 - Review of the Evaluation Sheet) and then store the data from that Evaluation Sheet in the database via the button **Postavi** (Set).
- If it is an automatic reading, then for the Evaluation Sheets that are not read completely in the automatic reading interface, a message that the evaluation sheet was not read is printed with a description of the log, and it continues to read the next Evaluation Sheet. After the auto reader has finished its process of reading all the mark sheets e.g. for one school, then semi-automatic reading is applied to mark sheets that have not been successfully read.

In addition to the aforementioned controls, a certain number of logical controls were added to the software system in the sense that the student ID must comply with certain rules such as length, range in which the student ID can move, etc. Overall, the system has very good filters so that it is almost impossible for a grade sheet to enter the system with a student ID that does not belong to any student taking the matriculation exam, or that the first foreign language as well as the ID of the test combination has not been identified.

3.3 Scoring of Evaluation Sheets

Scoring of the Evaluation Sheets is done in such a way that the stored answers from the Evaluation Sheet are compared with the correct answers in the database and one point is added for each correct answer. After scoring all the questions (60), the total number of points is obtained. Based on the total number of points, a percentage of the maximum number of points is calculated, which represents the total result of the matriculation exam (testing) for one student.

When scoring the answers to the questions, the software system, based on the number and structure of questions in certain areas (subjects) shown in Table 1, calculates the total number of points per area (subjects), and then based on that, the percentage of the maximum number of points in a certain area (subject) is calculated field, which represents the result of taking the matriculation exam in a particular subject.

Figure 10 shows the interface for evaluating (scoring) stored evaluation sheets.

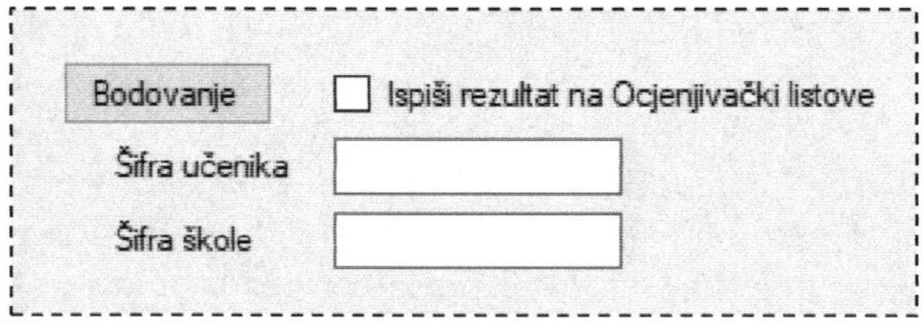

Figure 10. Interface for scoring (evaluation) of stored Evaluation Sheets

Scoring of Evaluation Sheets can be done in two ways:

- Individually for one student by entering the Student ID in the Student Code (Šifra učenika) field and clicking the **Bodovanje** (Scoring) button
- For all students of one school by entering the school ID in the School Code (Šifra škole) field and clicking on the **Bodovanje** (Scoring) button

Within the scoring interface (Figure 10) there is a **chk** field with the label "Ispiši rezultat na ocjenjivačke listove" (Print the result on the Evaluation Sheets), when included during scoring in the student's Evaluation Sheet, prints the overall test result, as well as the results by areas (subjects). Figure 11 shows an Evaluation Sheet that was scored in the previously described manner. All scoring results of the Evaluation Sheets are stored in the database.

Figure 11. Scored Evaluation Sheet for one student

IOP Conf. Series: Materials Science and Engineering 1298 (2023) 012020 doi:10.1088/1757-899X/1298/1/012020

Figure 12 shows how the software system printed the test results by area (subjects) as well as the total test result for one student (surname and first name of the student are covered) on the Evaluation Sheet.

1. Bosanski, hrvatski, srpski jezik i književnost (1-5): bodova:5 (100,00%)
2. Engleski jezik - prvi strani jezik (6-10): bodova:5 (100,00%)
3. Matematika (11-15): bodova:5 (100,00%)
4. Fizika (16-20): bodova:5 (100,00%)
5. Informatika (21-25): bodova:4 - 80,00%
6. Biologija (26-30): bodova:5 (100,00%)
7. Hemija/Kemija (31-35): bodova:5 (100,00%)
8. Historija/Povijest (36-40): bodova:5 (100,00%)
9. Geografija/Zemljopis (41-45): bodova:5 (100,00%)
10. Tehnička kultura (46-48): bodova:2 (66,67%)
11. Likovna kultura (49-51): bodova:3 (100,00%)
12. Muzicka kultura (52-54): bodova:3 (100,00%)
13. Tjelesni i zdravstveni odgoj (55-57): bodova:3 (100,00%)
14. Građansko obrazovanje (58-60): bodova:2 (66,67%)

Ukupno bodova: 57 Procenat: 95,00%

Figure 12. Printout of test results by areas (subjects) and in total

Likewise, next to each question, the software system puts a mark on whether the transmitted answer is correct or not, and along with that mark, it also puts the correct answer to that question for that test combination as it is shown in Figure 13.

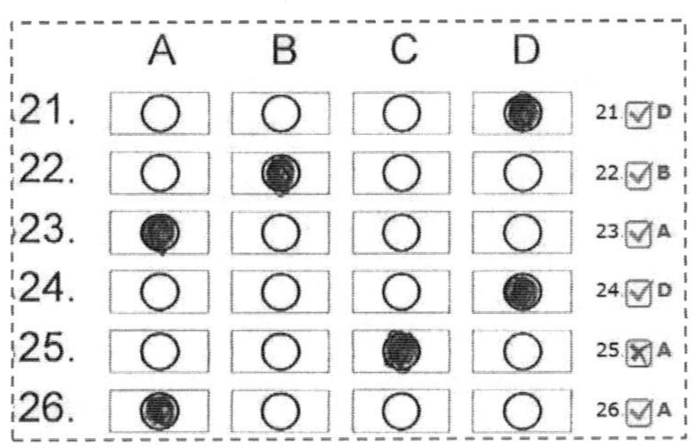

Figure 13. Print marks for correct or incorrect answers to questions

In Figure 13, it can be seen that all the answers were correctly read and evaluated, so only the 25th question was incorrectly answered with the mark C, and the correct answer was under A.

4. Conclusion

The implementation of the Software System for automatic reading, storing, and evaluation of scanned Evaluation Sheets in the field of education has been successfully developed with the use of various techniques of reading and recording marked answers, as well as other test data. In the last school year, about 4,000 assessment sheets were examined and evaluated within the framework of matriculation examinations in elementary schools of Tuzla Canton with the help of the described software system. Following the publication of test results, the scoring Evaluation Sheets, along with the corresponding test combination, were electronically provided to the students for review. While the system generally yielded satisfactory results, there was one student complaint regarding the interpretation of transferred answers on the Evaluation Sheet. The student had misunderstood the logic for canceling answers, marking the desired answer as 'Cancelled' and the incorrect answer as 'Correct'. Such instances provided insights into potential improvements for the system. In the initial years of application, a semi-automatic model of reading and scoring Evaluation Sheets was more used, through which the possibilities of improving the system itself were recognized. In recent years, as confidence in the results of the review of Evaluation Sheets has grown, the automatic model has been used more. Although the presented system was created for questions with the choice of one correct answer from several offered answers, it can be upgraded with very little effort for questions with the choice of more than one correct answer from several offered answers. Given that the developed system was not created for commercial use, but for internal use in the process of passing the matriculation exam for elementary school students, not much work was done on the appearance of the software interface itself. Of course, in future upgrades of this system, work will be done on improving the user interface as well as adding new features that will aim to increase the flexibility and capabilities of the software system.

References

[1] Awny Abbas A 2009 An automatic system to grade multiple choice questions paper based exams Journal of University of Anbar for Pure Science **3** 174–81 doi: 10.37652/juaps.2009.15510

[2] Nicol D 2007 E-assessment by design: Using multiple-choice tests to good effect Journal of Further and Higher Education **31** 53–64 doi: 10.1080/03098770601167922

[3] Afifi M and Hussain K F 2019 The achievement of higher flexibility in multiple-choice-based tests using image classification techniques International Journal on Document Analysis and Recognition (IJDAR) **22** 127–42 doi: 10.1007/s10032-019-00322-3

[4] Gronlund N E 1998 Assessment of student achievement. Sixth edition

[5] McCoubrie P 2004 Improving the fairness of multiple-choice questions: A literature review *Medical Teacher* **26** 709–12 doi: 10.1080/01421590400013495

[6] Kommey B, Keelson E, Samuel F, Twum-Asare S and Akuffo K K 2022 Automatic Multiple Choice Examination questions marking and grade generator software IPTEK The Journal for Technology and Science **33** 175 doi: 10.12962/j20882033.v33i3.14522

[7] Hafeez Q, Aslam W, Ikramullah Lali M, Ahmad S, Alqahtani M and Shafiq M 2023 Fault tolerant optical mark recognition *Computers, Materials & Continua* **74** 3829–47 doi: 10.32604/cmc.2023.026422

[8] Nguyen T D, Hoang Manh Q, Bui Minh P, Nguyen Thanh L and Hoang T M 2011 Efficient and reliable camera based multiple-choice test grading system The 2011 International Conference on Advanced Technologies for Communications (ATC 2011) doi: 10.1109/atc.2011.6027482

A method for maintaining a stable ISM band point-to-point wireless communication link in the presence of interference

Enes Saletović, Elmir Babović, Đulaga Hadžić and Nevzudin Buzađija

University of Zenica, Polytechnic Faculty, Department of software engineering, Fakultetska 1, 72000 Zenica, Bosnia and Herzegovina

E-mail: enes.saletovic@unze.ba

Abstract. This paper presents a new method that enables automatic channel change in case of intense interference on the active channel of a wireless communication point-to-point link. Channel change algorithms on the master and slave side of the communication link are presented, which enable the automatic selection of a new communication channel from a predefined set of available communication channels. By applying this method, it is possible to maintain a stable communication link in case of intense unpredictable interference on the transmitter or receiver side. The method applies to two-way master-slave communication links, regardless of the frequency spectrum and the number of used communication channels.

1. Introduction

ISM (Industrial, Scientific, Medical) radio frequency bands are used for communication in the field of industry, science and medicine and are regulated by the international standard of the International Telecommunication Union (ITU). By national legal solutions, ISM bands are more precisely defined. Due to the expansion in the field of Wi-Fi networks, Internet of Things (IoT) and built-in smart devices, more and more non-ISM devices use the free ISM band for data transmission, remote control, wireless telephony and other purposes. National laws limit the broadcast power to limit the use of non-ISM devices to a smaller area and reduce interference among users of the ISM band. For the use of the ISM band, by the defined legal solutions, no license is usually required, so some ISM bands are intensively used to establish point-to-point links for wireless communication. For example, sensor networks and remote control systems [1] most often use the ISM band 2.4 – 2.5 GHz.

However, the same ISM band is used by widely used Wi-Fi networks, microwave ovens, blue-tooth devices and numerous other devices, so signal interference is present [2]. In practice, two-way point-to-point links are most often used to enable control of data transmission and correction of errors in data transmission. When the transmitter (Point A) sends data to the receiver (Point B), using the appropriate communication protocol, the receiver sends an acknowledgment (ACK) in case of receiving the sent data packet (Figure 1). Depending on the receiver's response, the transmitter can continue sending a new data packet or resend the same packet in case of unsuccessful sending (no ACK confirmation). Each ISM frequency range is divided into several channels of the same width (Figure 1), so different channels can be used for data transmission [3].

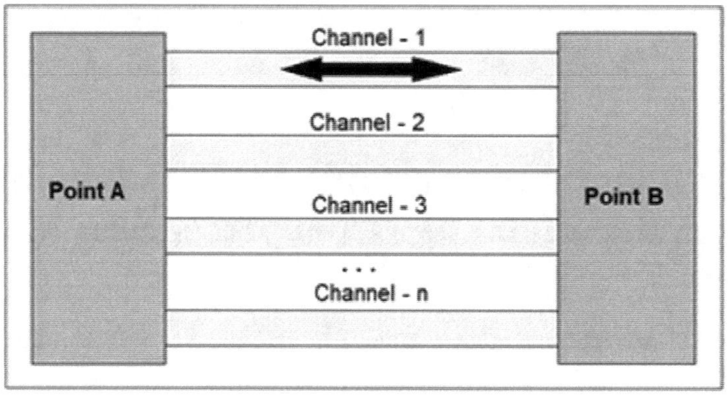

Figure 1. Point-to-point link.

2. Problem formulation

In general, data transmission via a point-to-point master-slave communication link is subject to the influence of stochastic interference due to signal interference from other devices at the same or close frequencies and due to other electromagnetic interference emitted by other devices in the environment [4]. Since interference is unpredictable, it is not possible to ensure uninterrupted data transmission. Interference intensity is usually higher in urban areas due to the larger number of ISM devices. Because other devices in the environment use different channels of the same ISM band, the load on individual channels can be different, so the problem of the temporary stoppage of data transmission can be solved by a synchronized change of channels on both sides of the communication link [5]. The proposed method contains an algorithm for automatically changing the communication channel in case of interruption of communication on the active channel, to select a more favorable channel.

3. Algorithm of the method

There are several methods for maintaining the ISM communication link, based on monitoring the load of the available communication channels [6]-[8]. This method implies the use of predefined channels for data transmission over a point-to-point master–slave link. Figure 2 (a) shows the algorithm for selecting the communication channel on the transmitter side.

Legend:
T_{TCW} – transmission waiting time for a channel
T_{LW} – transmission waiting time for the link
T_{CD} – channel downtime
T_{LD} – link downtime
ch – channel ordinal number
NoC – the number of channels used
C(NoC) – the array of channels used
ACK – confirmation of successful data transmission
T_{LS} – the time of the last successful transmission on the link
T_{CS} – channel's last successful transmission time
Millis() - time reading function

When the transmitter is turned on, the values of the necessary constants T_{CW}, T_{LW}, NoC, and C(ch) are loaded, and the values of the variables: ACK, ch, T_{LS} and T_{CS} are initialized. The transmitter sends the actual measured values of the used variables and waits for an acknowledgment from the remote receiver (ACK). If the reception is confirmed (ACK=1) from the receiver in the given time, the values of the variables containing the time of the last successful data transmission (T_{LS} and T_{CS}) are updated. Since the sending was realized within the given time T_{CW}, the process of sending the data repeats as long as the receiver confirms the data reception. If the transmitter does not receive an ACK confirmation within the given time T_{CW}, the serial number of channel ch is incremented and the data transmission is repeated on the next channel. If the variable ch becomes greater than the maximum sequence number of channel NoC, the value is reset to the initial value (ch=1). After changing the channel, a new T_{CS} time value is set, based on which the T_{CD} channel downtime is calculated. The link downtime T_{LD} is refreshed, which is used to interrupt the operation of the transmitter if no connection is established on any available channel, that is, when the condition ($T_{LD} > T_{TCW}$ NoC=T_{LW}) is met. Until that condition is met, the transmitter tries to send data on the selected channel. When a data transfer is successfully performed on a channel, that channel is used until another data transfer interruption occurs, after which a new channel is selected. The algorithm is terminated when it is determined that transmission is not possible on any available channel. In the described way, the transmitter automatically skips to the next available channel until data transmission succeeds, or until it is determined that data transmission is not possible on any available channel. A reduced set of used channels enables faster connection establishment because a smaller number of channels are searched on the receiver side after each interruption of data transmission on an active channel. A similar algorithm with appropriate time parameters is implemented on the receiver side. Figure 3 (b) shows the algorithm for selecting the communication channel on the receiver side.

Legend:
T_{RCW} – waiting time for a successful reception on the channel
T_{LW} – waiting time for a successful reception on the link
T_{CD} – channel downtime
T_{LD} – link downtime
ch – channel ordinal number
NoC – the number of channels used
C(NoC) – the array of channels used
ACK – confirmation of a successful data reception
T_{LR} – the time of the last successful reception for the link
T_{CR}– the time of the last successful reception for the channel
Millis() – time reading function

After switching on, the receiver initializes the necessary variables and sets the initial channel C(1). If a packet is received (ACK=1), the time values of the last packet reception T_{CR} and T_{LR} are refreshed. The received data is processed and the procedure is repeated until reception is interrupted on the selected channel (ACK=0) in the period $T_{CD}>T_{RCW}$. Since the interruption of packet reception is longer than the allowed time T_{RCW}, the receiver switches to the next channel (ch=ch+1), and the time of the last reception T_{CR} is re-initialized. If a packet is received on the new channel, communication continues on that channel. If no packet is received on the selected channel, after the allowed T_{RCW} waiting time, the receiver switches to the next channel (ch=ch+1). The procedure is repeated until communication with the transmitter is established or until the set time for establishing the link T_{LW} expires. The waiting time for a successful reception on the link T_{LW} is the same on the transmitter and receiver side. To enable the establishment of communication between the transmitter and the receiver after a temporary disconnection, the allowed channel connection recovery waiting time on the transmitter side must be much longer than the waiting time on the receiver side, that is, the relation applies:

$$T_{TCW} \geq T_{RCW}\, NoC \tag{1}$$

In this way, the transmitter will wait on each channel for the receiver to check each channel on its side to find the channel on which the transmitter is sending data. By programming the number of consecutive data packet transmission attempts and the waiting time between consecutive attempts on the transmitter side, the establishment of a connection on a new active channel is ensured. On both sides, the same channel array containing an identical set of channels is used. The channel selection can be arbitrarily defined in sequence C(NoC) where ($1 \leq$ ch $\leq$ NoC). In the case of higher communication speed, it is necessary to use a smaller set of mutually distant channels due to the interference of close channels.

IOP Conf. Series: Materials Science and Engineering 1298 (2023) 012021 doi:10.1088/1757-899X/1298/1/012021

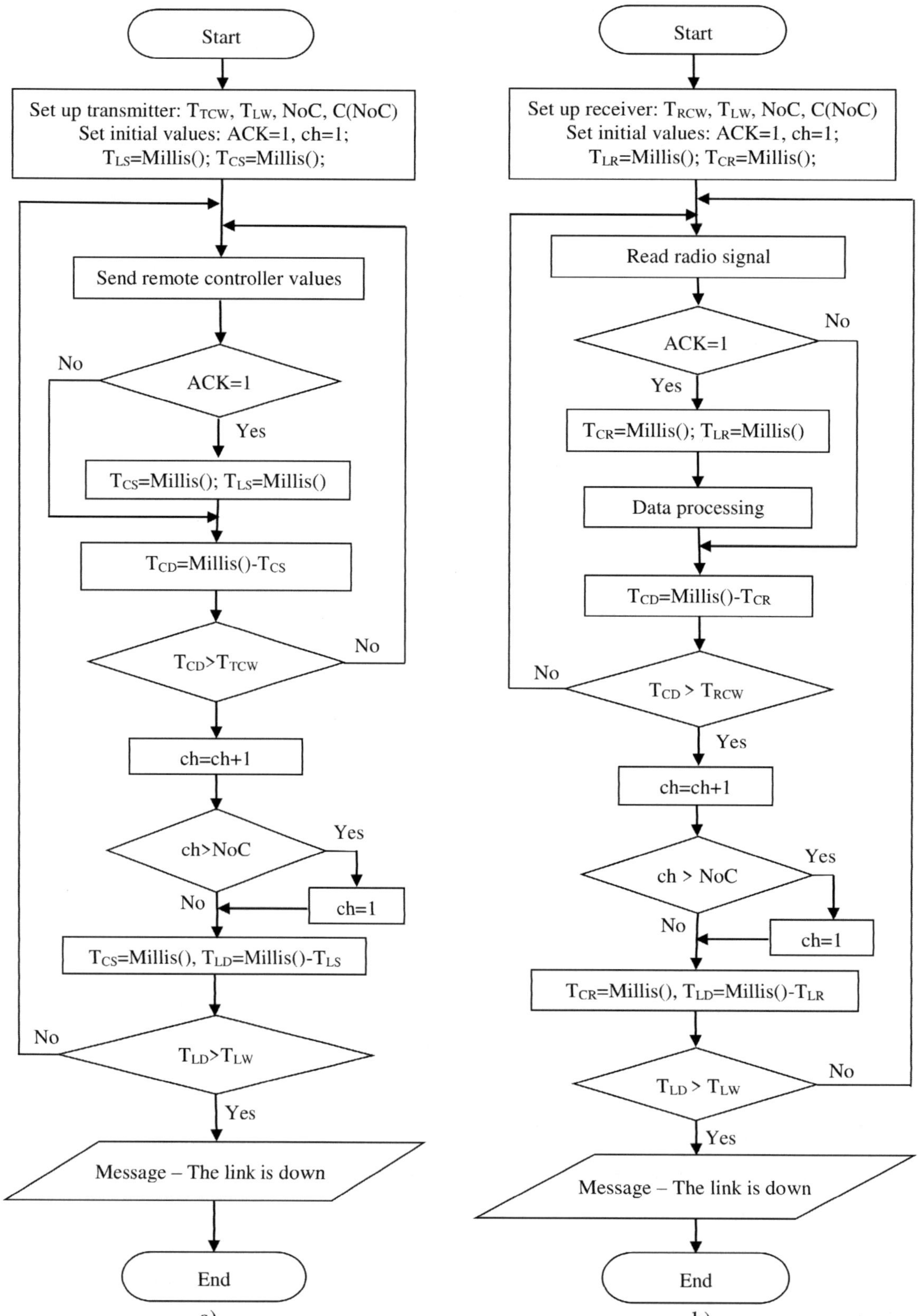

Figure 2. Transmitter (a) and receiver (b) algorithm flow diagram.

4. Algorithm testing with experimental results

The created algorithm was experimentally tested by simulating 4 different scenarios using Arduino development environment, 32-bit STM32F103F6C8 microcontroller and nRF24L01+ PA/LNA transceiver [1]. The transceivers used enable half-duplex wireless communication at distances of up to 1000 m, and operate on the ISM band 2.4 - 2.5 GHz. Testing was performed in an urban environment with the presence of various interferences emitted by Wi-Fi networks and other wireless devices. Wi-Fi networks use the ISM frequency band ranging from 2402 MHz to 2482 MHz. For testing purposes, three communication channels were selected (Table 1). Channel 90 uses a frequency of 2490 MHz, so it is not burdened by interference from surrounding Wi-Fi networks. Within the test software, with the appropriate libraries included, the transceiver parameters are set on both sides of the connection.

Table 1. Channels used for the method simulation.

Channel ordinal Number	Channel ch	Frequency (MHz)
1	10	2410
2	50	2450
3	90	2490

The following communication link failure scenarios are simulated:

Scenario 1: Transmitter does not receive an acknowledgment

The transmitter sends data on channel 10, but no confirmation arrives from the receiver. This scenario occurs when the receiver is turned off or when there is intense interference on the receiver or transmitter side, so the link is disabled on all channels. Due to the absence of confirmation of data reception from the receiver, working channels were periodically changed on the transmitter side until the maximum allowed waiting time on the transmitter side T_{LW} expired, when the transmitter reported the message: "The link is down". Figure 3 shows the experimental results for this scenario.

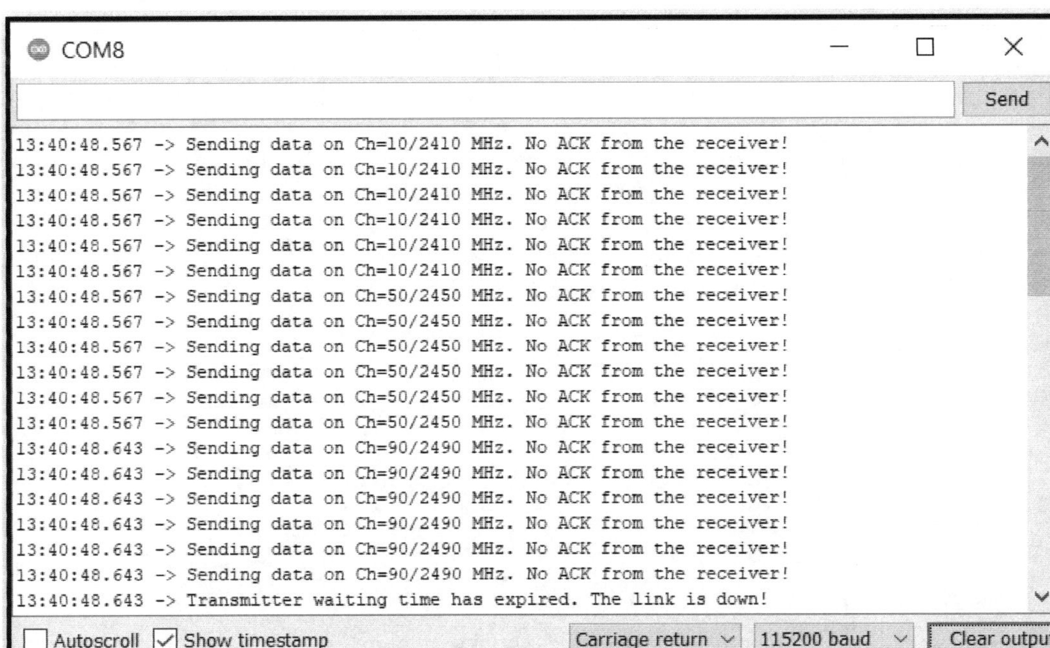

Figure 3. Experimental results for scenario 1 monitored by the Arduino serial monitor.

IOP Conf. Series: Materials Science and Engineering 1298 (2023) 012021 doi:10.1088/1757-899X/1298/1/012021

Scenario 2: The receiver does not receive data

This scenario occurs when the transmitter is turned off or when, due to intense interference on all channels, the communication link is disabled. The receiver listens for data on channel 10 until the T_{RCW} waiting time expires, after which it switches to channel 50. After the T_{RCW} time expires, the receiver switches to channel 90 due to the absence of a signal from the transmitter. After no data packet has been received on the third channel, the link waiting time T_{LW} expires and the receiver prints the message "Link is down". Figure 4 shows the experimental results for this scenario.

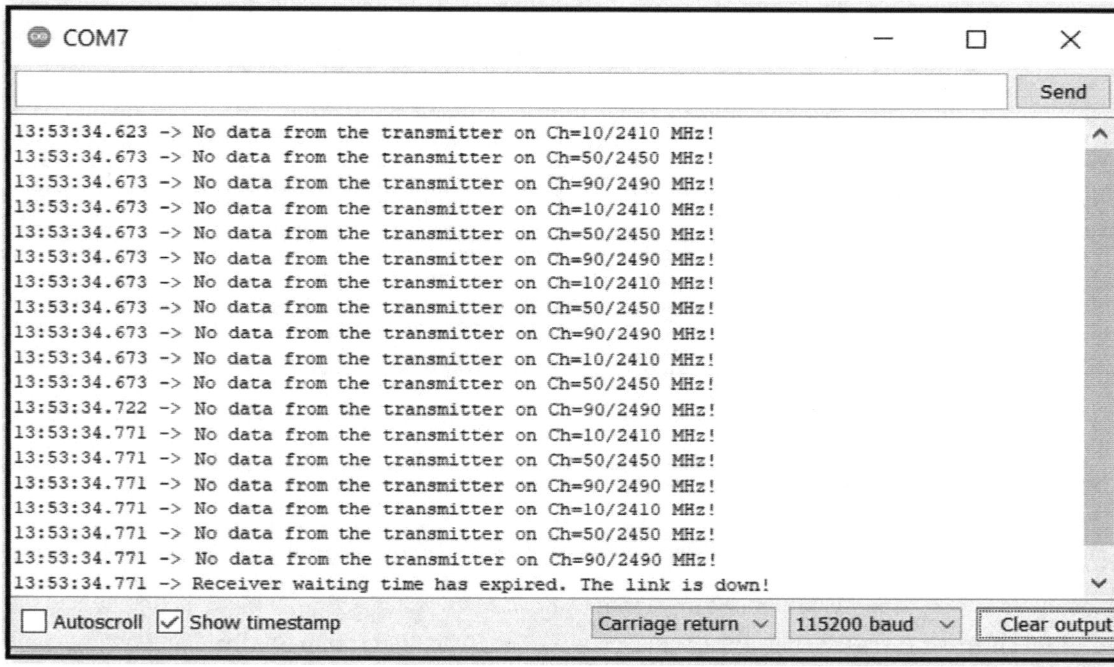

Figure 4. Experimental results for scenario 2 monitored by the Arduino serial monitor.

Scenario 3: Intense interference on the transmitter side

In this scenario, intense interference is simulated on the transmitter side, on channels 10 and 50, so sending data was only possible on channel number 90. When the transmitter is turned on, it starts sending data on the default channel. When the receiver is turned on, it listens to the default channel 10 first. Since the transmitter did not receive a data reception confirmation (ACK) from the receiver, after the T_{TCW} time expired, the transmitter tried to send data on channel 50. Since no data reception confirmation was received on this channel either, after the channel waiting time T_{TCW}, the transmitter switched to channel 90 where an ACK signal from the receiver was received and a communication link was established. Figure 5 shows the experimental results for this scenario.

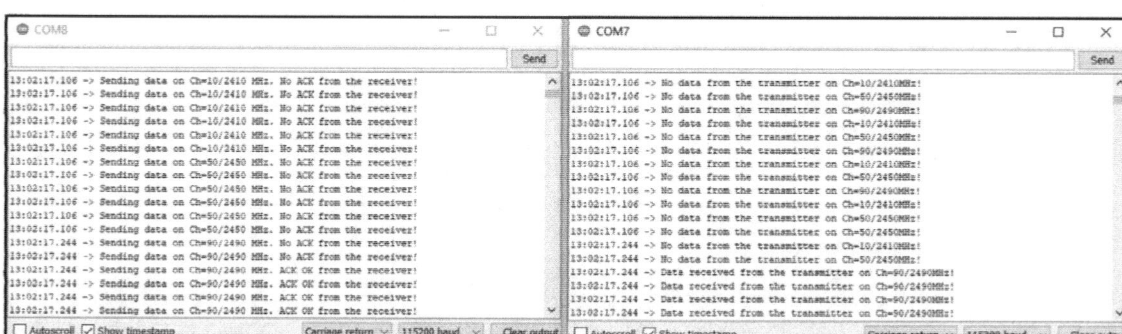

Figure 5. Experimental results for scenario 3 monitored by the Arduino serial monitor.

IOP Conf. Series: Materials Science and Engineering 1298 (2023) 012021 doi:10.1088/1757-899X/1298/1/012021

Scenario 4: Intense interference on the receiver side

In this scenario, intense interference on the receiver side is simulated, which makes communication on channels 10 and 50 impossible. Connecting on channels 10 and 50 is disabled by software, setting the receiver to search for channels 11 and 51 instead of channels 10 and 50. The transmitter started transmitting data on the default channel number after switching on. Due to the simulated interference on the receiver side, the link was not established, so the transmitter switched to the next channel after the T_{TCW} timeout. Since the link was not established on this channel either, due to the simulated interference on the receiver side, the transmitter changed the channel and started data transmission on channel 90 after the T_{TCW} time. Since the receiver confirmed the reception of data, the link was established on channel 90. Figure 6 shows the experimental results for this scenario.

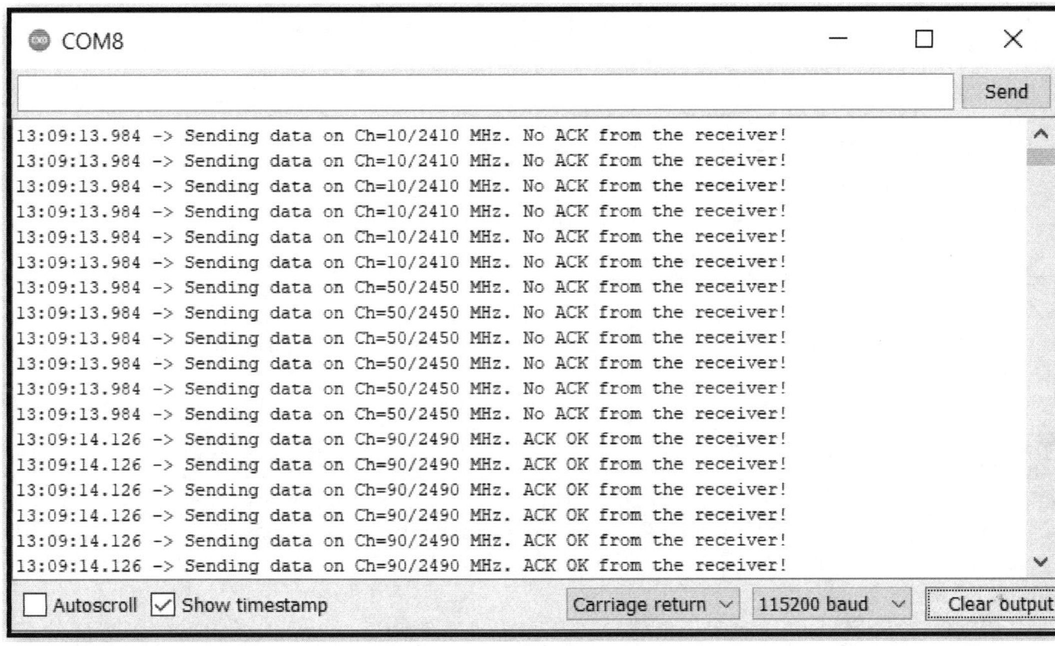

Figure 6. Experimental results for scenario 4 monitored by the Arduino serial monitor.

The presented experimental results confirm the validity of the created method for maintaining a stable point-to-point wireless communication link in the presence of interference. The speed and stability of the link based on the proposed method significantly depend on the choice of working channels as well as on the number of working channels. The optimal choice of working channels can be made based on a detailed analysis of interference on the used ISM frequency spectrum. The selected working channels should be evenly distributed so that they cover as much of the used frequency range as possible. Fewer channels enable faster link recovery because fewer channels are searched on both sides of the link. On the other hand, a larger number of evenly spaced working channels allows for greater link reliability because the probability of interference on all used channels is reduced. The choice of working channels represents a compromise between the required link recovery speed and link reliability. In urban areas, the most interference comes from Wi-Fi networks and IoT devices.

Standard transceivers provide adjustment of the packet-sending attempts on the working channel to overcome problems due to intermittent interference. Some modern transceivers in their firmware software have implemented similar algorithms for automatic channel switching in case of disconnection due to intense interference. Compared to other channel switching methods [7][8], the proposed method is simpler and more suitable for software implementation in the case of point-to-point master-slave systems such as remote control or data acquisition systems.

5. Conclusion

The presented method enables maintaining a stable ISM point-to-point master-slave link in the presence of interference and other types of disturbances. In real systems, point-to-point links are established along selected channels, where the distribution of interference is random. Therefore, by changing the active channel, a stable link can be maintained even in the presence of intense interference on multiple channels. The speed of channel synchronization and link establishment on a new frequency depends on the number of used channels. Depending on the required channel synchronization speed, fewer or more different channels spaced apart may be used to reduce the likelihood of simultaneous interference. By selecting the channel search step, a set of searched channels can be defined. For example, three equally spaced channels can be used from the total set of available channels that can be preset and defined on the transmitter and receiver sides.

The advantages of this method are:

- Simple implementation
- Applicability to point-to-point master-slave links
- Independence from the frequency range of the link
- Flexible choice of used channels
- Automatic adaptation of the link to new conditions

The disadvantages of this method are:

- Not applicable to unidirectional point-to-point links
- It is necessary to implement software algorithms on both sides of the link, on the transmitter (master) and the receiver (slave) side

References

[1] Saletović E, Babović E and Hadžić Đ 2023 Maintaining a stable point-to-point remote control link based on transceivers nRF24L01+ PA/LNA in presence of interferences *2023 22nd International Symposium INFOTEH-JAHORINA (INFOTEH)*, East Sarajevo, Bosnia and Herzegovina pp 1-5

[2] Wituski J and Dietl G 2020 Interference Investigation of Wireless ISM Band Devices in Rescue Helicopters *IEEE 5th International Symposium on Smart and Wireless Systems within the Conferences on Intelligent Data Acquisition and Advanced Computing Systems (IDAACS-SWS)* pp 1-6

[3] Watteyne T, Mehta A and Pister K 2009 Reliability Through Frequency Diversity: Why Channel Hopping Makes Sense in PE-WASUN *ACM* pp 116–123

[4] Sesham S and Sabat S L 2015 Spectrum Sensing for Cognitive Radio Networks *White Space Communication* p 117

[5] Kotsiou V, Papadopoulos G, Chatzimisios P and Theoleyre F 2019 Whitelisting Without Collisions for Centralized Scheduling in Wireless Industrial Networks *IEEE Internet of Things Journal* IEEE 6 (3) pp 5713-5721

[6] Zeng B, Dong Y and Lua D 2012 Point-to-Point Interference Measurement Approach for Large-Scale Wireless Sensor Networks *International Journal of Distributed Sensor Networks* vol 2012 pp 1-9

[7] Shih C F, Xhafa A E and Zhou J 2015 Practical frequency hopping sequence design for interference avoidance in 802.15.4e TSCH networks *International Conference on Communications (ICC)* London UK pp 6494-6499

[8] Figueiredo A, Luís M and Zúquete A 2022 A Fair Channel Hopping Scheme for LoRa Networks with Multiple Single-Channel Gateways *Sensors (Basel)* 22 (5260) pp 1-19

Interpolation on the unit hypersphere using the n-dimensional generalized Kuramoto model

Zinaid Kapić and Aladin Crnkić

University of Bihać, Faculty of Technical Engineering, Bosnia and Herzegovina

E-mail: zinaid.kapic@unbi.ba

Abstract. Computer graphics, robotics, and physics are one of the many domains where interpolation on the unit sphere S^n (often called a unit hypersphere or unit n-sphere) plays a crucial role. In this paper, we introduce a novel approach for achieving smooth and precise interpolation on the unit sphere S^{n-1} using the n-dimensional generalized Kuramoto model. The proposed algorithm finds the shortest and most direct path between two points on that non-Euclidean manifold. Our simulation results demonstrate that it achieves performance comparable to that of a Spherical Linear Interpolation algorithm. Also, the paper proposes the application of our algorithm in the interpolation of rotations that are presented in the form of four-dimensional data.

1. Introduction

A fundamental problem in several disciplines, including physics, computer graphics, and robotics, is interpolation on the unit sphere S^n [1]. Many applications, such as virtual reality simulations [2], robotic arm path planning [3], and 3D model animation [4], depend on the ability to interpolate smoothly and precisely between two points lying on that non-Euclidean manifold. In other words, interpolation involves finding a smooth and continuous path between that two points.

Utilizing Spherical Linear Interpolation (SLERP), which establishes a spherical path between two points on the sphere, is one typical way to accomplish this [5]. However, SLERP is susceptible to problems like singularities and non-uniformity of the interpolated path, especially when the two points on the sphere are far apart.

For our purposes, we will use a model that is an extension to higher dimensions of the well-known dynamical model that descibes the phenomenon of synchronization, the Kuramoto model. The ability of two or more oscillating systems to align their oscillations so that they are in phase with one another is referred to as a synchronization phenomenon. This can happen in many different systems, including biological, physical, and social systems. The Kuramoto model has been introduced by the Japanese physicist Yoshiki Kuramoto in 1975, and it describes an ensemble consisting N phase coupled oscillators [6]

$$\dot{\theta}_j(t) = w_j + \frac{K}{N}\sum_{i=1}^{N}(\theta_j - \theta_i), j = 1, \dots, N. \quad (1)$$

In this model, oscillators are represented by a set of coupled differential equations that describe the evolution of the phase θ_j of each oscillator over time. The coupling between the oscillators is controlled

IOP Conf. Series: Materials Science and Engineering 1298 (2023) 012022 doi:10.1088/1757-899X/1298/1/012022

by a parameter called the coupling strength K, which determines the degree to which the oscillators are able to synchronize with one another. The notion w_j stands for an intrinsic frequency of the oscillator j.

The Kuramoto model and its variations have been used to study synchronization phenomena in a variety of systems [7-9]. It has also been used to understand synchronization in complex networks [10], power grids [11], clustering of static and stream data [12,13], as well for solving rotation averaging and interpolation problems [14-16].

The rest of this paper is structured as follows. In the next section, we introduce the n-dimensional generalized Kuramoto model and provide the way how it can be applied to interpolate between two points on the unit sphere S^{n-1} in the n-dimensional Euclidean space. In the section 3, we present the results of simulations conducted on the two-dimensional unit sphere S^2 using the proposed algorithm. We also visualize the simulation results of data on the unit sphere S^3 using the Hopf fibration [17], a powerful tool for representing the geometry of the sphere. This section will also include a set of figures to demonstrate object rotation interpolation. Finally, in the Conclusion, we summarize the key contributions of our paper and discuss the possible upgrades of this method, and further research.

2. Algorithm

Each oscillator in (1) is described as a point on the unit circle S^1. Since the circle is the simplest example of non-Euclidean manifold, the Kuramoto model provides a simple, but highly non-trivial example of a dynamical system on non-Euclidean manifold. Consequently, Kuramoto model has been extended from the circle to the higher-dimensions of non-Euclidean manifold, like matrix groups [18-20], homogeneous spaces [21], and spheres [22, 23]. In this paper, we focus on the model where the state of a n-dimensional generalized oscillator is represented by a unit vector $X \in \mathbb{R}^n$, or a point on the $n-1$ dimensional unit sphere S^{n-1}. Its dynamics is given by the following ordinary differential equation

$$\dot{X} = WX,$$

where W is an anti-symmetric $n \times n$ matrix represented a frequency of the generalized oscillator. If we consider a system of coupled oscillators then their states on the unit sphere S^{n-1} are given by the following dynamical system

$$\dot{X}_j = F - \langle X_j, F \rangle X_j + W_j X_j, \qquad j = \overline{1,n} \tag{2}$$

where $F = F(X_1, X_2, \dots, X_n)$ is a global coupling function and the symbol $\langle \cdot,\cdot \rangle$ stands for standard inner product (dot-product) in $\mathbb{R}^n$. If the coupling function is given by equation

$$F = \frac{K}{N} \sum_{i=1}^{N} X_i,$$

then by substituting in (2), we obtain the n-dimensional generalized Kuramoto model

$$\dot{X}_j = \frac{K}{N} \sum_{i=1}^{N} \left(X_i - \langle X_j, X_i \rangle X_j \right) + W_j X_j\,. \tag{3}$$

In the abovementioned model, K represents a global coupling strength and W_j is a frequency of the generalized oscillator with index j.

In order to introduce dynamical model for interpolation between two points on the unit sphere S^{n-1}, let us assume that the starting point is given by a unit vector $Q \in \mathbb{R}^n$ and the final point by $R \in \mathbb{R}^n$. If we set that the coupling function is equal to final point (i.e. $F = R$), and for the initial conditions we take the starting point Q, we can transform (2) into

$$\dot{X}(t) = R - \langle X(t), R \rangle X(t), \qquad X(0) = Q. \tag{4}$$

In the following, we will use the model (4) for interpolation between two points Q and R on the unit sphere S^{n-1}.

Furthermore, consider the unit vectors in the form $X = [x_1, x_2, x_3]^T$, $R = [r_1, r_2, r_3]^T$ and $Q = [q_1, q_2, q_3]^T$ as points on the two-dimensional unit sphere S^2. Substitution into (4) yields the following system of ordinary differential equations

IOP Conf. Series: Materials Science and Engineering 1298 (2023) 012022 doi:10.1088/1757-899X/1298/1/012022

$$\begin{cases} \dot{x}_1(t) = r_1 - x_1(t)\langle X(t), R\rangle \\ \dot{x}_2(t) = r_2 - x_2(t)\langle X(t), R\rangle, \\ \dot{x}_3(t) = r_3 - x_3(t)\langle X(t), R\rangle \end{cases} \tag{5}$$

where $x_1(0) = q_1$, $x_2(0) = q_2$ and $x_3(0) = q_3$. If we parametrize the unit sphere S^2 with the implicit equation $x^2 + y^2 + z^2 = 1$ using spherical coordinates

$$x = \sin\theta\cos\phi\,, y = \sin\theta\sin\phi,\ z = \cos\theta\,, 0 \le \phi \le 2\pi, 0 \le \theta \le \pi,$$

then model (4) can be written as

$$\begin{cases} \theta'^{(t)} = \cos(\theta(t))\cos(\phi_r - \phi(t))\sin(\theta_r) - \cos(\theta_r)\sin(\theta(t)) \\ \phi'^{(t)} = \dfrac{\sin(\theta_r)\sin(\phi_r - \phi(t))}{\sin(\theta(t))} \end{cases}, \tag{6}$$

where $\theta(0) = \theta_q$ and $\phi(0) = \phi_q$. In model (6), starting and final point are $Q(\theta_q, \phi_q)$ and $R(\theta_r, \phi_r)$ respectively.

Furthermore, if we regard the unit sphere S^3, given by $x^2 + y^2 + z^2 + u^2 = 1$, in terms of Hopf coordinates

$$x = \sin\theta\cos\phi\,, y = \sin\theta\sin\phi,\ z = \cos\theta\cos\eta\,, u = \cos\theta\sin\eta,$$

where the angles take values $0 \le \theta \le \frac{\pi}{2}, 0 \le \phi \le 2\pi, 0 \le \phi \le 2\pi$. Then, substituting into the model (4), we obtain

$$\begin{cases} \theta'(t) = \cos\left((\theta(t))\right)\cos(\phi_r - \phi_t)\sin(\theta_r) - \cos(\theta_r)\cos(\eta_r - \eta(t))\sin(\theta(t)) \\ \phi'(t) = \dfrac{\sin(\theta_r)\sin(\phi_r - \phi(t))}{\sin(\theta(t))} \\ \eta'(t) = \dfrac{\cos(\theta_r)\sin(\eta_r - \eta(t))}{\sin(\theta(t))} \end{cases}, \tag{7}$$

where initial conditions are $\theta(0) = \theta_q, \phi(0) = \phi_q, \eta(0) = \eta_q$. In the abovementioned model, starting and final point are $Q(\theta_q, \phi_q, \eta_q)$ and $R(\theta_r, \phi_r, \eta_r)$ respectively. The substitutions in model (6) and (7) are valid for $\theta \ne 0$ and $\theta \ne \frac{\pi}{2}$.

Furthermore, we will explain our method for interpolating between two points on the unit sphere S^{n-1} in more detail through the algorithm.

The algorithm is as follows:

1: Enter Q and R

2: Set tolerance ϵ, step δ, and $T = 0$

3: Solve (4) with $X(0) = Q$

4: loop

5: if $||R - Q(t)|| < \epsilon$ then

return $X(t)$, for $t \in [0, T]$

6: else

$T = T + \delta$

7: end if

8: end loop

3. Simulations

In this section, we present the simulation results of our method. First simulation represents an interpolation curve in S^2 between a starting point given in spherical coordinates as $Q\left(\frac{\pi}{3},\frac{\pi}{6}\right)$ and ending point $R\left(\frac{\pi}{5},\frac{5\pi}{6}\right)$. These points can be converted to equivalent form as unit vectors: a starting point $Q = [0.75,0.433,0.5]$ and ending point $R = [-0.509, 0.294,0.809]$.

Figure 1 presents an interpolation curve between abovementioned points. This curve lies on the unit sphere S^2. Using the model (6) and setting the tolerance $\epsilon = 10^{-5}$, as well as the step $\delta = 0.01$, interpolation curve reaches the ending point in a moment $T = 12.06$.

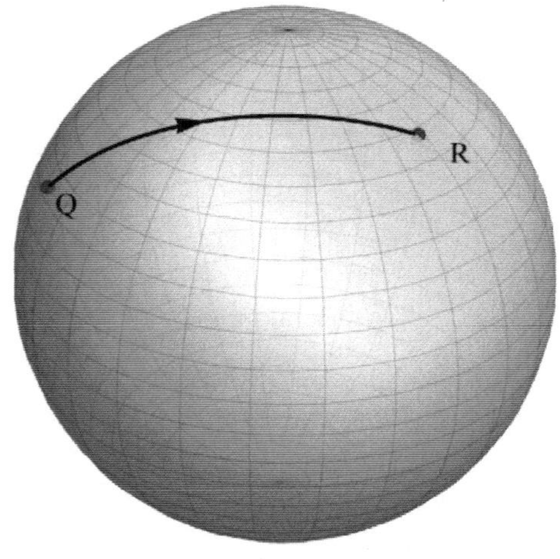

Figure 1. Interpolation between points Q and R.

For the second simulation we represent an interpolation curve in S^3 between three data points given in Hopf coordinates as $Q\left(0,\frac{\pi}{2},0\right)$, $P\left(\frac{\pi}{6},\frac{\pi}{2},0\right)$ and final position as $R\left(\frac{\pi}{6},\frac{2\pi}{3},0\right)$. In addition to four-dimensional unit vectors, these points can also be represented as unit quaternions.

With rotational matrices and Euler angles, unit quaternions are one of the most famous rotation representation metrics. Quaternions represent a number system defined as: $q = x + y\mathbf{i} + z\mathbf{j} + u\mathbf{k}$, where x, y, z, and u are real numbers and $\mathbf{i}^2 = \mathbf{j}^2 = \mathbf{k}^2 = -1$. Special group of quaternions are unit quaternions that represent rotation on unit sphere S^3 in four-dimensional space. Unit quaternions have Euclidean norm equal to 1. Their popularity rose with computer graphics and aircraft simulations, especially as a rotation metric [1].

The abovementioned points, written as unit quaternions are:

$$Q = 0 + 0\mathbf{i} + 1\mathbf{j} + 0\mathbf{k},$$
$$P = 0 + 0.5\mathbf{i} + + 0.866025\mathbf{j} + 0\mathbf{k}, \text{ and}$$
$$R = -0.25 + 0.433013\mathbf{i} + 0.866025\mathbf{j} + 0\mathbf{k}.$$

Since a curve in S^3 is hard to visualize we have used Hopf fibration for describing it as a curve in S^2. Hopf fibration is defined as a mapping $h: S^3 \to S^2$:

$$h(x, y, z, u) = (x^2 + y^2 - z^2 - u^2, 2(xu + yz), 2(yu - xz)),$$

or in Hopf coordinates

$$h(\sin\theta\cos\phi, \sin\theta\sin\phi, \cos\theta\cos\eta, \cos\theta\sin\eta) =$$
$$= (-\cos 2\theta, \sin 2\theta\sin(\phi + \eta), -\sin 2\theta\cos(\phi + \eta)).$$

This mapping has applications in the field of rigid body mechanics, quantum information theory or in magnetic monopoles [17]. Figure 2 present an interpolation curve between the points Q, P and R.

Using the (7) and setting the tolerance $\epsilon = 10^{-5}$, as well as the step $\delta = 0.01$, interpolation curve reaches a final position at a moment $T = 10.89 + 10.17 = 21.06$.

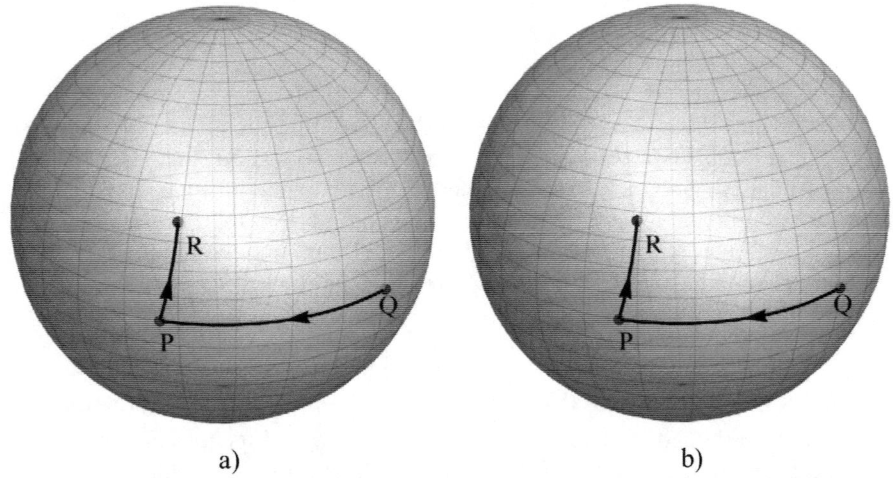

Figure 2. Multiple interpolation between three points: a) the algorithm and b) SLERP.

From closer inspection of Figure 2, we can see that our algorithm gives approximately the same results as the SLERP algorithm.

This method can also be applied for object's interpolation between two positions. We can display a rotating object in 3-dimensional space using the polar form of a unit quaternion, defined as $q = \cos\left(\frac{\varphi}{2}\right) + V\, sin\left(\frac{\varphi}{2}\right)$, where $V = \frac{q}{||q||}$ and $tg\left(\frac{\varphi}{2}\right) = \frac{||q||}{q_0}$. Using the polar form, the object's rotation can be defined by an axis of rotation V and the rotation angle φ around that axis [24]. Figure 3 represents an object's movement from a starting point $q_1 = 0.5 + 0.5\mathbf{i} + 0.5\mathbf{j} + 0.5\mathbf{k}$ to the final point $q_2 = 0.707107 + 0.707107\mathbf{i}$.

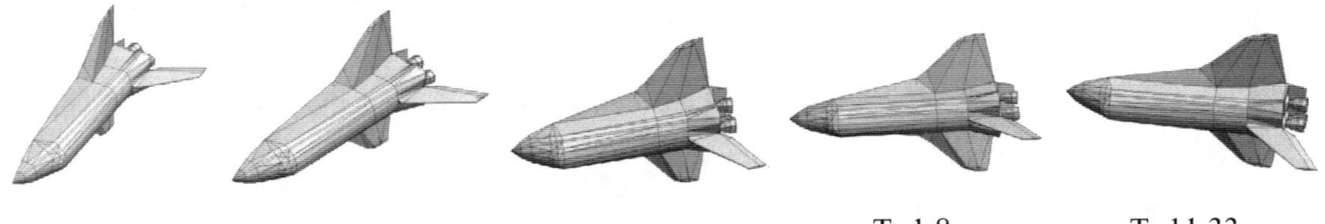

Figure 3. Interpolation of the object between two positions.

Object is shown in some specific positions displayed in Table 1.

Table 1. Quaternion values of objects positions in specific moments from Figure 3.

Moments T	Positions Q
$\boldsymbol{T_1 = 0.00}$	$q = 0.5 + 0.5\mathbf{i} + 0.5\mathbf{j} + 0.5\mathbf{k}$
$\boldsymbol{T_2 = 0.20}$	$q = 0.561 + 0.561\mathbf{i} + 0.430\mathbf{j} + 0.430\mathbf{k}$
$\boldsymbol{T_4 = 0.55}$	$q = 0.631 + 0.631\mathbf{i} + 0.320\mathbf{j} + 0.320\mathbf{k}$
$\boldsymbol{T_3 = 1.80}$	$q = 0.700 + 0.700\mathbf{i} + 0.096\mathbf{j} + 0.096\mathbf{k}$
$\boldsymbol{T_5 = 11.33}$	$q = 0.707107 + 0.707107\mathbf{i}$

IOP Conf. Series: Materials Science and Engineering 1298 (2023) 012022 doi:10.1088/1757-899X/1298/1/012022

From the results in Table 1 we see that our method produces a perfect spherical interpolation curve, because at each moment, model (7) preserves a unit sphere S^3. For better visualization, we have also provided a video simulation of the object's movement.[1]

4. Conclusion

In this paper, we have explored an alternative approach for interpolation on the $(n-1)$-dimensional unit sphere S^{n-1}, using the n-dimensional generalized Kuramoto model. We have shown through simulations that our algorithm can efficiently and accurately find the shortest and most direct interpolation curve that connects given two points on the unit sphere of any dimension. Our approach provides a promising direction for future research on this problem and its potential application in various fields such as virtual reality simulations, robotic arm path planning, and 3D model animation. Furthermore, our research has demonstrated the versatility of the Kuramoto model in solving complex problems beyond the study of synchronization phenomena.

References

[1] Dam E, Koch M and Lillholm M 1998 Quaternions, interpolation and animation CiteSeerX

[2] Gao J, Yang H and Xiao Y 2012 The quaternion theory and its application on the virtual reality Advances in Intelligent and Soft Computing 635–40 doi: https://doi.org/10.1007/978-3-642-29390-0_101

[3] Zhao X, Liu Y and Xie Z 2021 The application of acceleration-level quaternion interpolation in the visual-servoing process 2021 13th International Conference on Computer and Automation Engineering (ICCAE) doi: 10.1109/iccae51876.2021.9426103

[4] Haarbach A, Birdal T and Ilic S 2018 Survey of higher order rigid body motion interpolation methods for keyframe animation and continuous-time trajectory estimation 2018 International Conference on 3D Vision (3DV) doi: 10.1109/3DV.2018.00051

[5] Shoemake K 1985 Animating rotation with quaternion curves Proceedings of the 12th annual conference on Computer graphics and interactive techniques - SIGGRAPH '85 doi: 10.1145/325334.325242.

[6] Kuramoto Y 1975 Self-entrainment of a population of coupled non-linear oscillators International Symposium on Mathematical Problems in Theoretical Physics 420–2 doi: 10.1007/bfb0013365

[7] Cumin D and Unsworth C P 2007 Generalising the Kuramoto model for the study of neuronal synchronisation in the brain Physica D: Nonlinear Phenomena **226** 181–96 doi: 10.1016/j.physd.2006.12.004

[8] Scholz C H 2010 Large earthquake triggering, clustering, and the synchronization of faults Bulletin of the Seismological Society of America **100** 901–9 doi: 10.1785/0120090309

[9] Novikov A V and Benderskaya E N 2014 Oscillatory neural networks based on the Kuramoto model for Cluster Analysis Pattern Recognition and Image Analysis **24** 365–71 doi: 10.1134/S1054661814030146

[10] Arenas A, Díaz-Guilera A, Kurths J, Moreno Y and Zhou C 2008 Synchronization in complex networks Physics Reports **469** 93–153 doi: https://doi.org/10.1016/j.physrep.2008.09.002

[11] Guo Y, Zhang D, Li Z, Wang Q and Yu D 2021 Overviews on the applications of the Kuramoto model in modern power system analysis International Journal of Electrical Power & Energy Systems **129** 106804 doi: https://doi.org/10.1016/j.ijepes.2021.106804

[12] Crnkić A and Jaćimović V 2018 Data clustering based on quantum synchronization Natural Computing **18** 907–11 doi: 10.1007/s11047-018-9720-z

[13] Crnkić A, Ivanović I, Jaćimović V and Mijajlović N 2020 Swarms on the 3-sphere for online clustering of multivariate time series and data streams Future Generation Computer Systems **112** 11–7 doi: 10.1016/j.future.2020.05.018

[1] Supplementary video can be found at https://youtu.be/eUwIzKV3xqw

[14] Kapic Z, Crnkic A, Jacimovic V and Mijajlovic N 2021 A new dynamical model for solving rotation averaging problem 2021 20th International Symposium INFOTEH-JAHORINA (INFOTEH) doi: 10.1109/INFOTEH51037.2021.9400663

[15] Crnkić A and Kapić Z 2021 Interpolation on the special orthogonal group with high-dimensional Kuramoto model IOP Conference Series: Materials Science and Engineering **1208** 012037 doi: 10.1088/1757-899X/1208/1/012037

[16] Kapić Z and Crnkić A 2021 Interpolating rotations with non-abelian Kuramoto model on the 3-sphere Lecture Notes in Networks and Systems 610–6 doi: 10.1007/978-3-030-90055-7_48

[17] Lyons D W 2003 An elementary introduction to the Hopf fibration Mathematics Magazine **76** 87–98 doi: 10.1080/0025570X.2003.11953158

[18] Lohe M A 2009 Non-Abelian Kuramoto models and synchronization Journal of Physics A: Mathematical and Theoretical **42** 395101 doi: 10.1088/1751- 8113/42/39/395101

[19] Lohe M A 2019 Systems of matrix riccati equations, linear fractional transformations, partial integrability and synchronization Journal of Mathematical Physics **60** doi: 10.1063/1.5085248

[20] Jaćimović V and Crnkić A 2018 Low-dimensional dynamics in non-Abelian kuramoto model on the 3-sphere Chaos: An Interdisciplinary Journal of Nonlinear Science **28** 083105

[21] Sarlette A and Sepulchre R 2009 Consensus optimization on Manifolds SIAM Journal on Control and Optimization **48** 56–76 doi: 10.1137/060673400

[22] Markdahl J, Thunberg J Gonçalves J 2018 Almost Global Consensus on the n -Sphere IEEE Transactions on Automatic Control **63** 1664-75 doi: 10.1109/tac.2017.2752799

[23] Ha S-Y and Park H 2020 From the lohe tensor model to the lohe hermitian sphere model and emergent dynamics SIAM Journal on Applied Dynamical Systems **19** 1312–42 doi: 10.1137/19M1288553

[24] Sangwine S J and Bihan N L 2008 Quaternion polar representation with a complex modulus and complex argument inspired by the Cayley-Dickson form Advances in Applied Clifford Algebras **20** 111–20

IOP Conf. Series: Materials Science and Engineering 1298 (2023) 012023 doi:10.1088/1757-899X/1298/1/012023

Performance comparison of bilateral teleoperation control of direct control, PID control, and fuzzy logic control for water tank system

Una Drakulić

The University of Bihać, Technical Faculty of Bihać, B&H

E-mail: una.drakulic@unbi.ba

Abstract. Bilateral teleoperation has attracted significant research and application interests in a wide range of areas. The first and main use for bilateral teleoperation was to handle the dangerous and remote distance tasks such as space exploration and nuclear materials manipulation. In the past few years, bilateral teleoperation has found its way into other applications as a result of the development of control technologies and the latest breakthrough in artificial intelligence and machine learning. In this paper, the goal is to model bilateral teleoperation for direct control, PID control, and Fuzzy Logic control for the water tank system. The goal is to create a simulation for master-slave communication where the time delay is minimized to the optimal and accepted values. The experimental results obtained from the simulation show a fairly high accuracy in terms of all three ways of control modes, which highlights the effectiveness of the proposed system in the paper.

1. Introduction

Many fields such as nuclear, mining, space exploration, and medicine are using for years unilateral teleoperation. Unilateral teleoperation is controlling without force feedback, but if force feedback were included in those systems, it could improve the overall teleoperated task completion time and quality [1]. Incorporating force feedback into a control system is very challenging because there occurs a time delay, which may lead to inefficiency for the operator and instability for the system. When the operator can perceive the corresponding force feedback from distance located system (i.e. robot) this unilateral teleoperation becomes a bilateral teleoperation [2]. The robot directly manipulated by the human operator is called the master and the one located at the remote location is called the slave [3] [4]. A block diagram of the bilateral teleoperation system is shown in figure 1.

The main challenges in the bilateral teleoperation system are transmission delays and uncertainties (i.e. position tracking and force feedback) [6]. The transmission delays between master and slave are one of the most important parameters on which depends the control systems' performance [7]. Knowing the disadvantage of bilateral teleoperator systems, the aim of this research is to design and implement master-slave communication where the time delay for data transmission is minimized to the optimal and accepted values. This paper describes the design of a bilateral teleoperator system for controlling the level in the water tank. The water tank system is chosen for testing and implementing the bilateral teleoperation system because almost all industrial facilities have this type of system. The water level control is realized in three different

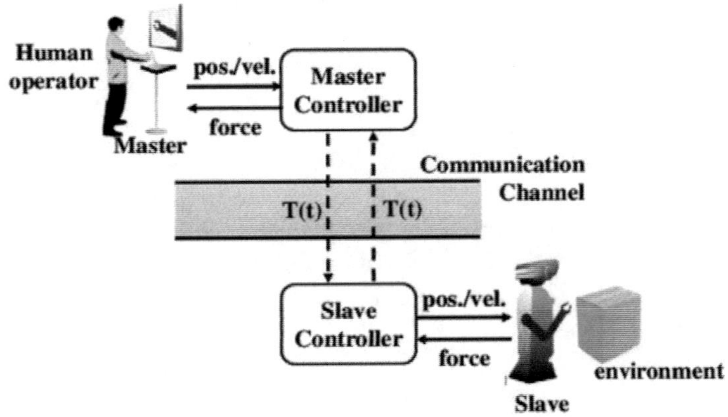

s1)

Figure 1. Block diagram of the bilateral teleoperation system [4].

ways (direct, using PID controller, and using Fuzzy Logic controller) in order to compare the time delay for each.

2. Modeling the bilateral teleoperation system for controlling level in water tank

In this section of the paper, the system for monitoring the level in the water tank is modeled.

Table 1. Given data for the system.

Water tank surface	$A = 2\,[m^2]$
Inlet and outlet valve	$mu = mi = 40\,[\frac{mm^2}{s}]$
Pump	$H = 1\,[m]$

In the figure 2 is shown the water tank system, where given data are in table 1.

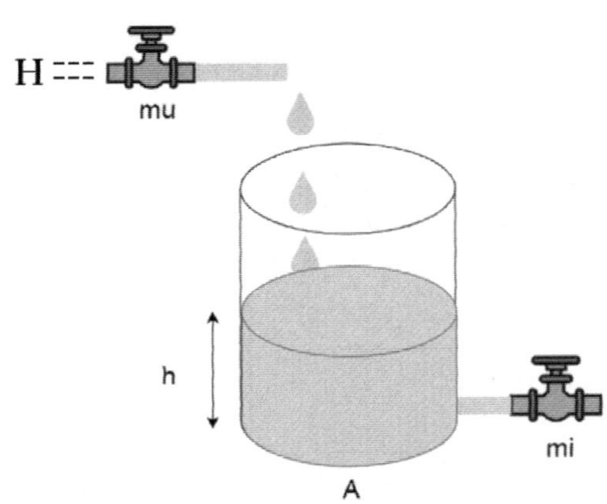

Figure 2. Water tank system.

IOP Conf. Series: Materials Science and Engineering 1298 (2023) 012023 doi:10.1088/1757-899X/1298/1/012023

Control of the level in the water tank is realized in three different ways:

- using the direct control method;
- designing the PID controller;
- designing a fuzzy logic controller.

Different control methods were used for the reason that a comparative analysis can be performed. This system is overall controlled using a bilateral teleoperation system, i.e. master-slave communication, where the goal is to, based on obtained results, determine the most efficient control mode. The most efficient control mode is the one where the time delay is minimized to the optimal and accepted values.

2.1. Direct control method for the water tank system

The water tank level control equation can be written as (1):

$$A \cdot \frac{dh(t)}{dt} = [\frac{\Delta H - h(t)}{mu} - \frac{h(t)}{mi}] \tag{1}$$

The measurement parameter is the water level h, (2):

$$\frac{dh(t)}{dt} = \frac{\Delta H}{A \cdot mu} - \frac{h(t)}{A \cdot mu} - \frac{h(t)}{A \cdot mi} \tag{2}$$

The expression (2) is modeled in Simulink/MATLAB (figure 3), and the obtained response is given in figure 4.

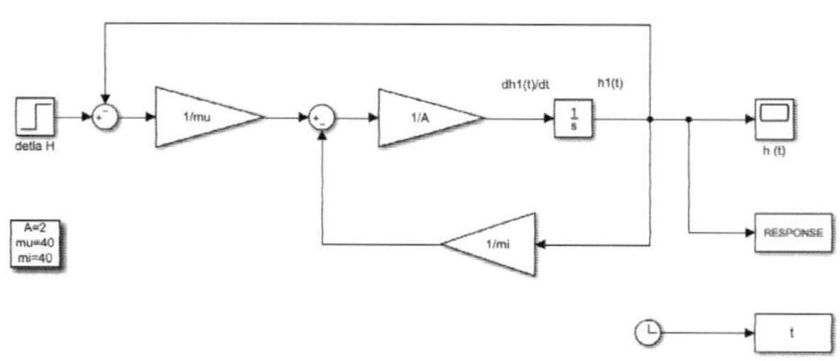

Figure 3. Simulink model for direct control method for level in the water tank system.

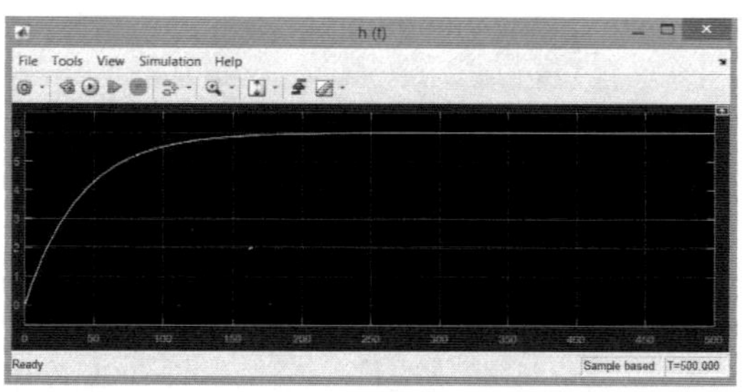

Figure 4. The obtained response for the model in figure 3.

2.2. Designing the PID controller for the water tank system

In order to determine the values of the parameters of the PID controller (Kp, Ki, and Kd), it is necessary to find the transfer function of the water tank system. As the transfer function represents the relationship between the output and input of the system, for the water tank system it is (3):

$$G(s) = \frac{h(s)}{\Delta H} \tag{3}$$

Need to apply the Laplace transformation to the expression (1) and than we get (4):

$$h(s) = \frac{1}{80s + 2} \tag{4}$$

If we present expression (4) in Simulink/MATLAB (figure 5), the obtained response is given in figure 6.

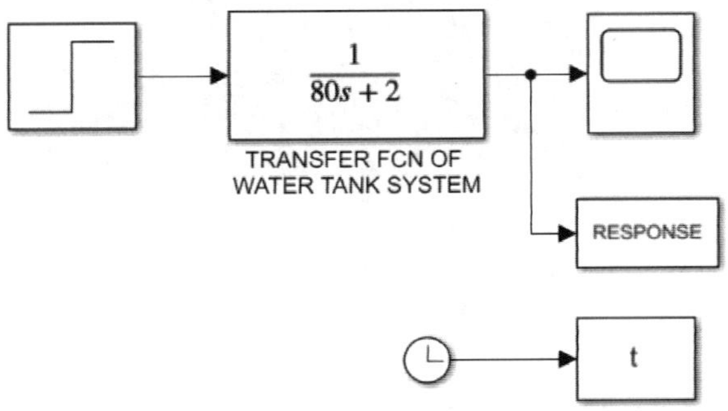

Figure 5. Simulink model for transfer function of the water tank system.

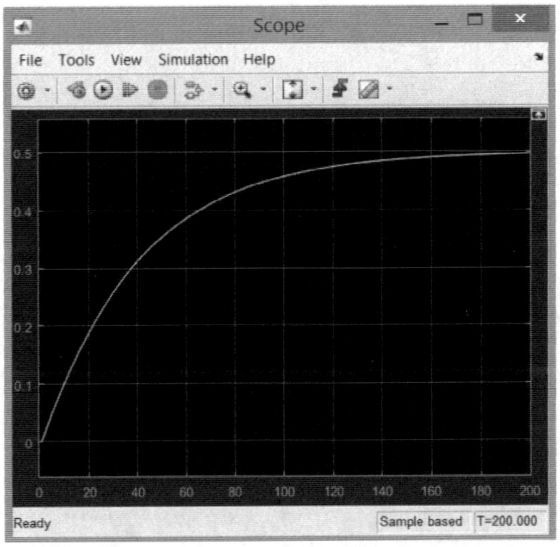

Figure 6. The obtained response for the model in figure 5.

This response (figure 6) is used to calculate the values of the PID controller parameters (Kp, Ki, and Kd) using the Ziegler-Nichols method (figure 7).

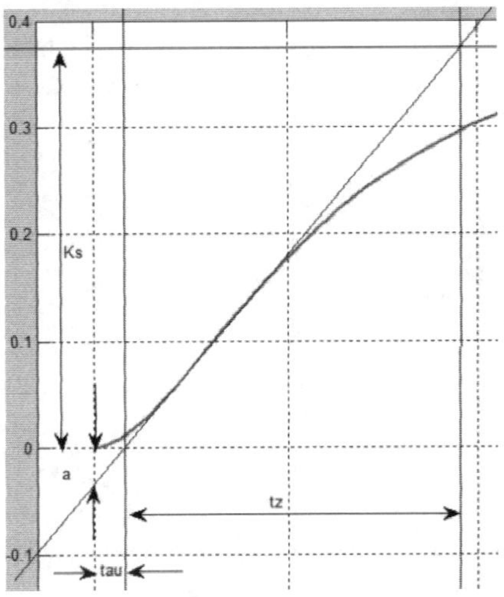

Figure 7. Response of expression (4) for calculating PID parameters using Ziegler-Nichols method.

$$a = \frac{Ks\tau}{tz} = 0.029 \quad Kp = \frac{1.2}{a} = 41.096 \quad Ki = \frac{0.6}{a\tau} = 1.447 \quad Kd = \frac{0.6\tau}{a} = 291.78 \qquad (5)$$

The values of the PID controller parameters (Kp, Ki, and Kd) using the Ziegler-Nichols method are given in (5).

Simulink model of designed PID controller for the water tank system is shown in figure 8.

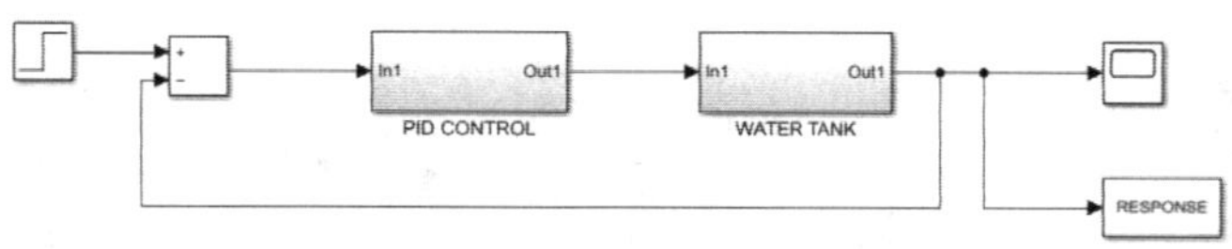

Figure 8. Simulink model of designed PID controller for the water tank system.

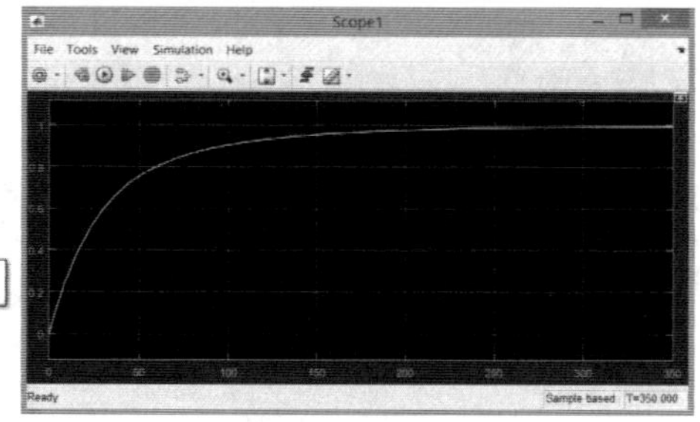

Figure 9. The obtained response for the model in figure 8.

The obtained response for the model in figure 8 is given in figure 9.

2.3. Designing a fuzzy logic controller for the water tank system

Mamdani fuzzy inference systems were used to create the Fuzzy controller. Since Mamdani systems have more intuitive and easier to understand rule bases than Sugeno, they are well-suited to expert system applications where the rules are created from human expert knowledge [8] [9]. To create a Fuzzy Logic controller, it is necessary to define the inputs to the system, the output from the system, and the rule base. The input to the system is the liquid level ('nivo') and the change in the liquid level ('promjena') in the water tank, and the output is the valve ('ventil'). The first input consists of five different membership activation functions: low, slightly low, good, slightly high, high ('nizak', 'malo nizak', 'dobar', 'malo visok', 'visok'). The interval is [-1, 1]. The second input consists of three different membership activation functions: negative, insignificant and positive ('negativna', 'neznatna', 'pozitivna'). The interval is [-0.27, 0.27]. The membership functions for inputs are shown in figures 10 and 11.

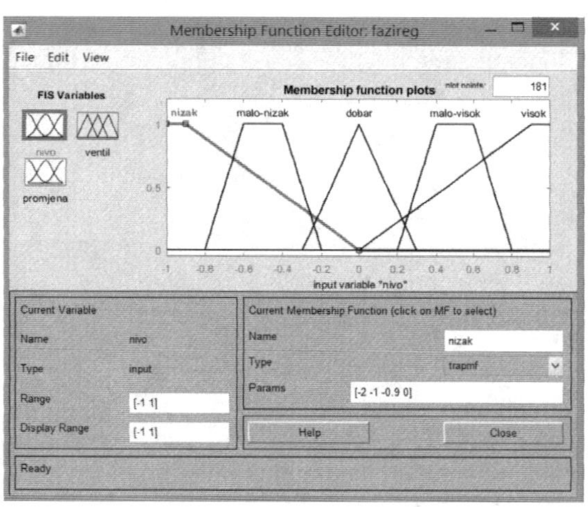

Figure 10. The membership functions for input 'nivo' (engl. level).

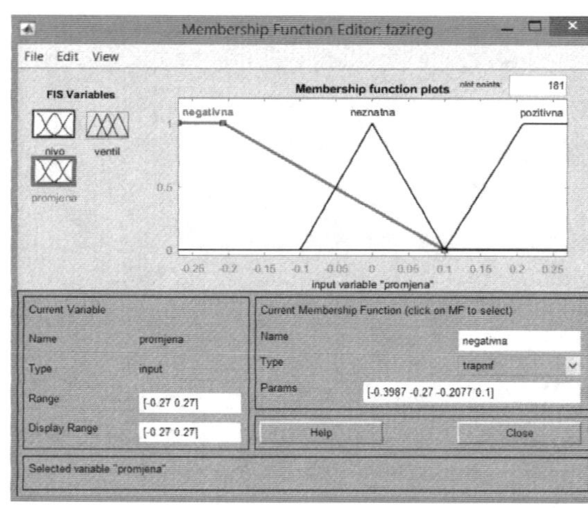

Figure 11. The membership functions for input 'promjena' (engl. change).

The output of the system is the change in the open/closed valve, which control the level of inflow/outflow of liquid from the water tank in order to maintain a constant level of liquid. The output consists of seven membership activation functions: close fast, close slowly, slightly closed, no change, slightly open, open slowly, open fast ('zatvori brzo', 'zatvori polako', 'malo zatvori', 'bez promjene', 'malo otvori', 'otvori polako', 'otvori brzo'). The interval is [-0.55 0.55]. The rule base consists of seven rules, namely:

1) If (level is high) then (valve is open fast)
2) If (level is low) then (valve is close fast)
3) If (the level is good) and (the change is negative) then (the valve is opened slowly)
4) If (the level is good) and (the change is positive) then (the valve is closed slowly)
5) If (the level is good) and (the change is insignificant) then (the valve is closed slowly)
6) If (the level is a slightly low) then (the valve is a slightly closed)
7) If (the level is a slightly high) then (the valve is a slightly open)

The output and the rule base are shown in figures 12 and 13.

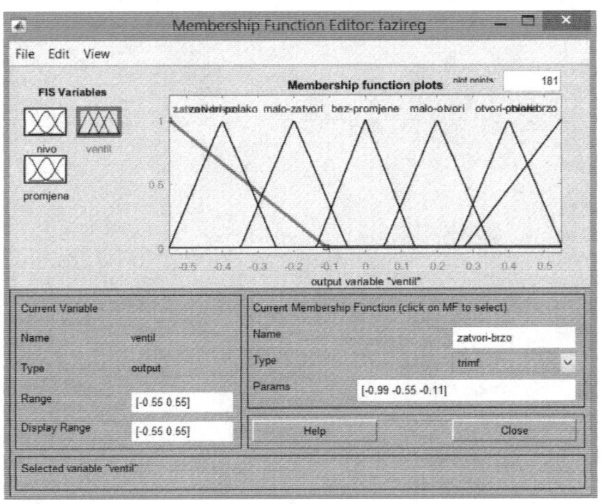

Figure 12. The membership functions for output 'ventil' (engl. valve).

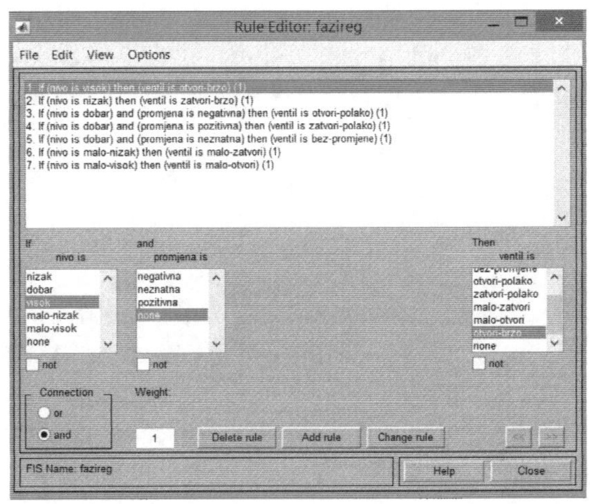

Figure 13. The Mamdani rule base.

Simulink model of designed fuzzy logic controller for the water tank system is shown in figure 14.

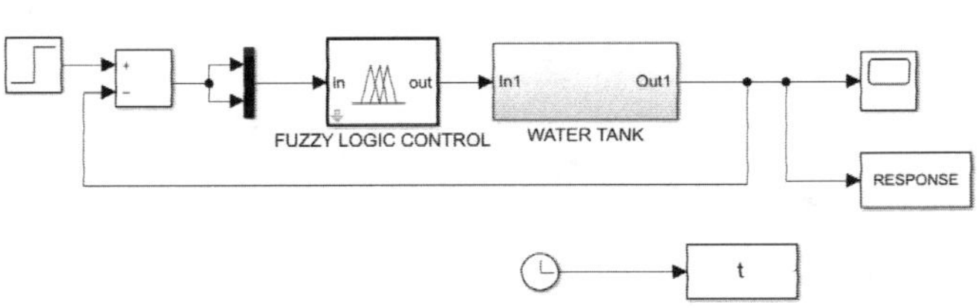

Figure 14. Simulink model of designed fuzzy logic controller for the water tank system.

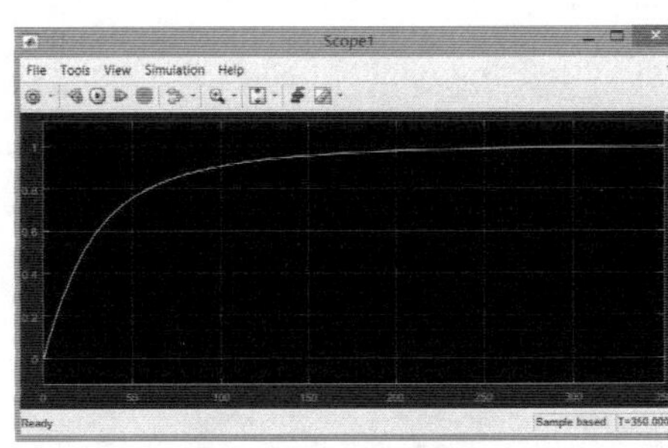

Figure 15. The obtained response for the model in figure 14.

The obtained response for the model in figure 14 is given in figure 15.

3. Application of wave variables for the bilateral teleoperator system

In this section of the paper, the application of wave variables for the bilateral teleoperator system is modeled. Figure 16 shows the Simulink model for the bilateral teleoperator system control of the water tank system without the application of wave variables.

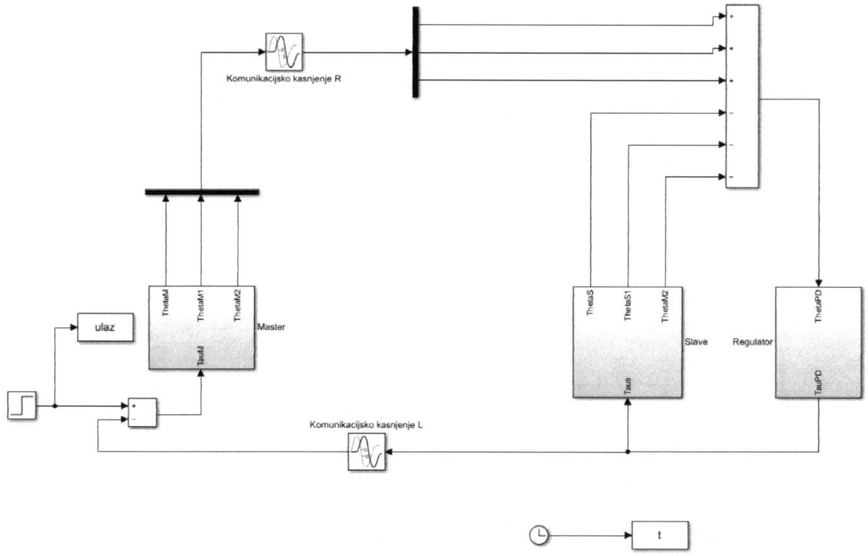

Figure 16. Simulink model for the bilateral system control of the level monitoring system in the water tank without the application of wave variables.

The figure 17 shown the master/slave subsystems.

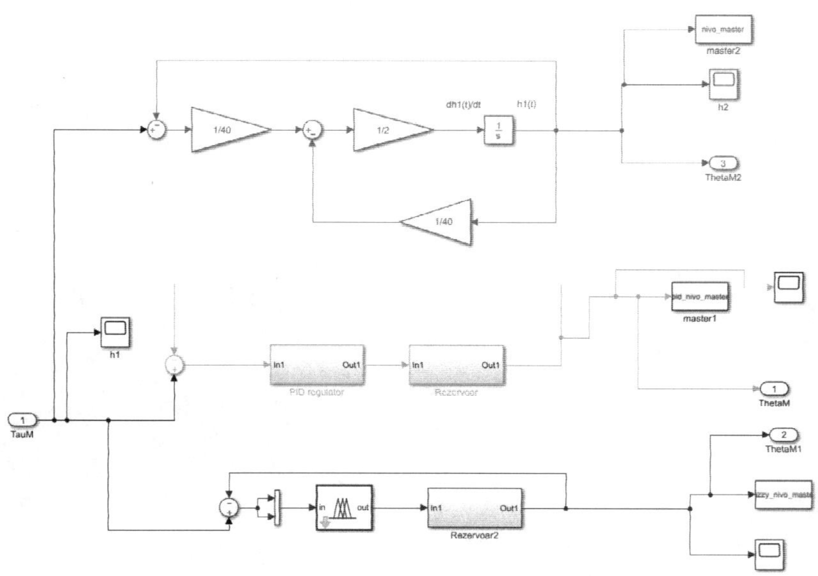

Figure 17. The master/slave subsystems.

Figures 18, 19 and 20 show the responses obtained for all three ways of controlling the liquid level in the water tank. Each figure consists of the master side response, the slave side response, and a comparison of the master and slave sides to compare the time delay.

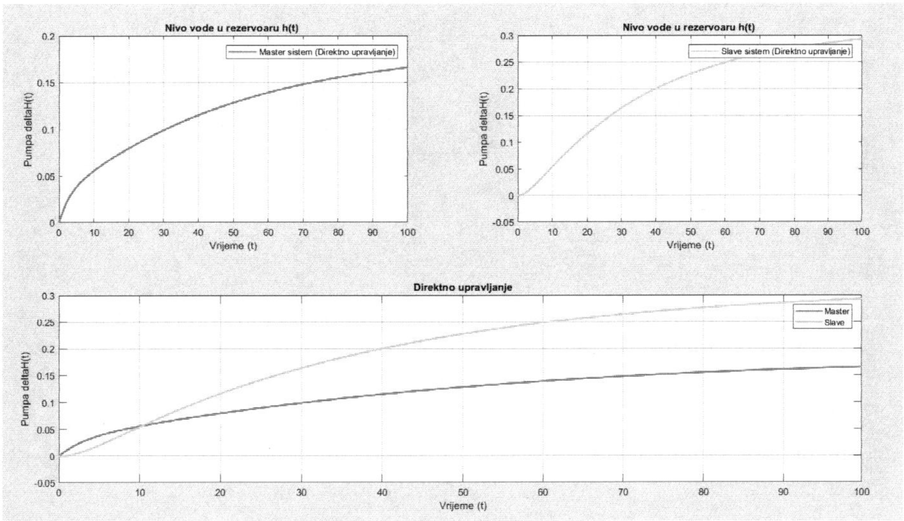

Figure 18. The response obtained for bilateral teleoperator system for direct control method with energy variables.

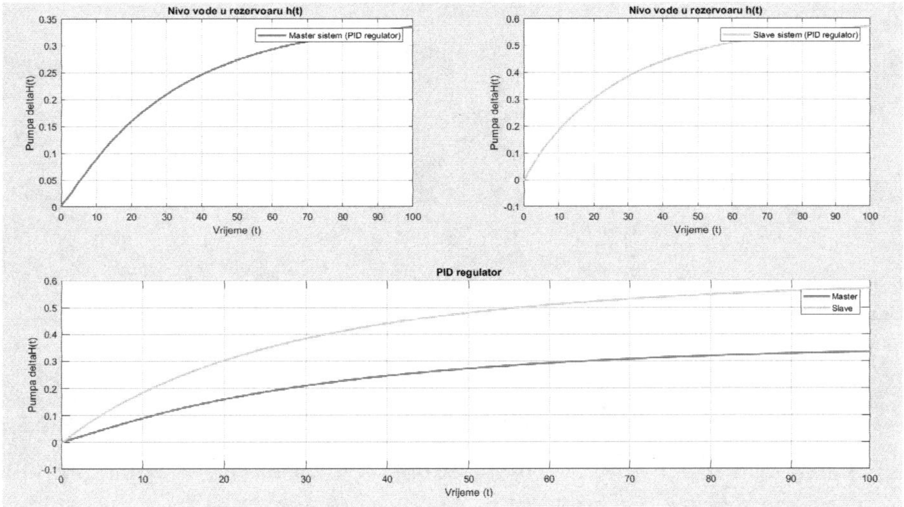

Figure 19. The response obtained for bilateral teleoperator system for PID control method with energy variables.

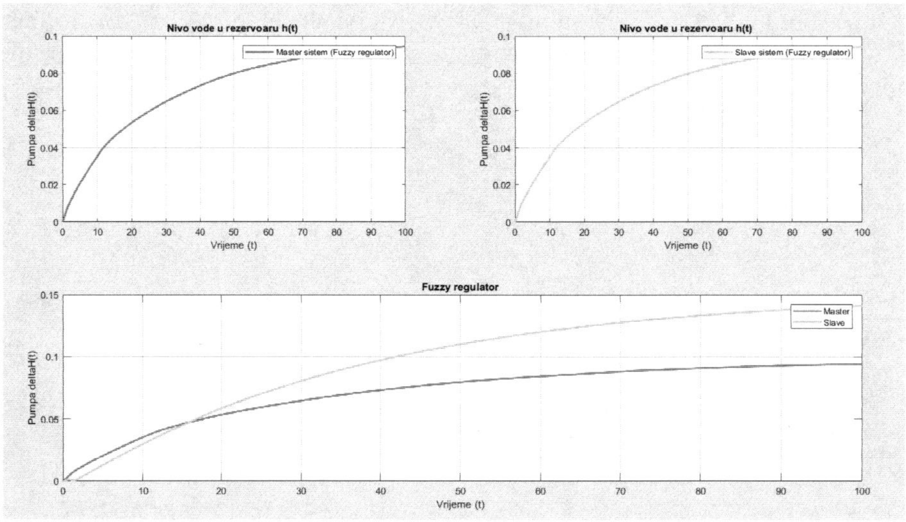

Figure 20. The response obtained for bilateral teleoperator system for fuzzy Logic control method with energy variables.

Figure 21 shows the Simulink model for the bilateral system control of the level monitoring system in the water tank with the application of wave variables.

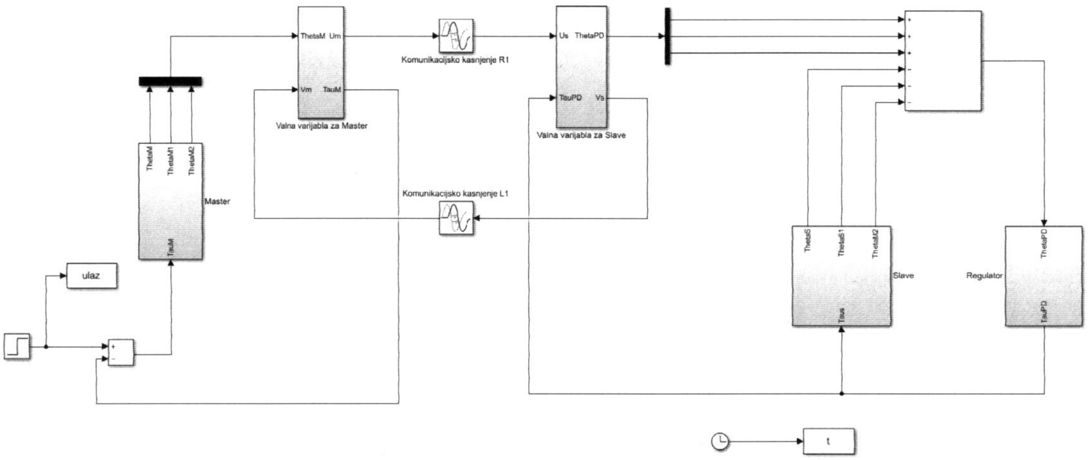

Figure 21. Simulink model for the bilateral system control of the level monitoring system in the water tank with application of wave variables.

The figure 22 shows the simulink model of transforming energy variables to wave variables, and the figure 23 shows the simulink model of transforming wave variables to energy variables.

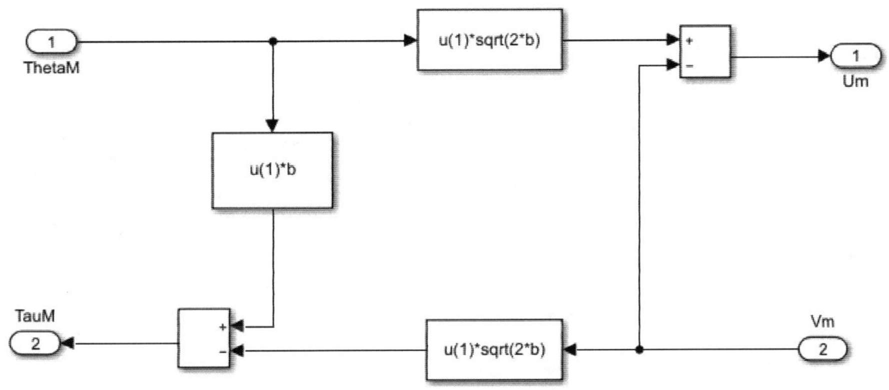

Figure 22. Simulink model of transforming energy variables to wave variables, on master side.

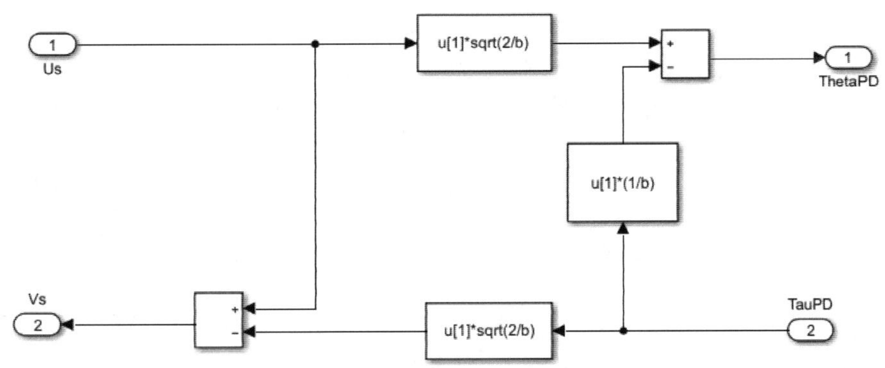

Figure 23. Simulink model of transforming wave variables to energy variables, on slave side.

Figures 24, 25 and 26 show the responses obtained for all three ways of controlling the liquid level in the water tank. Each figure consists of the master-side response, the slave-side response, and a comparison of the master and slave sides to compare the time delay.

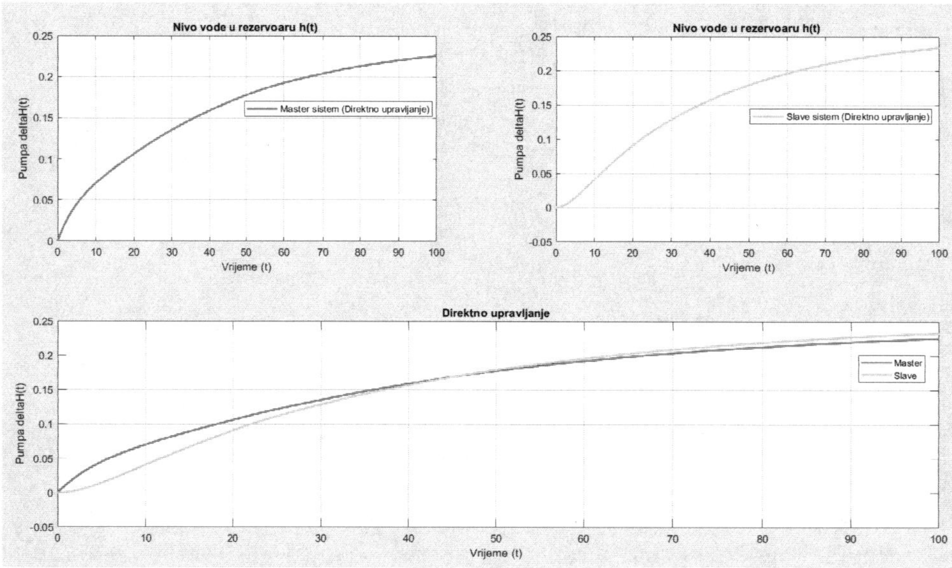

Figure 24. The response obtained for bilateral teleoperator system for direct control method with wave variables.

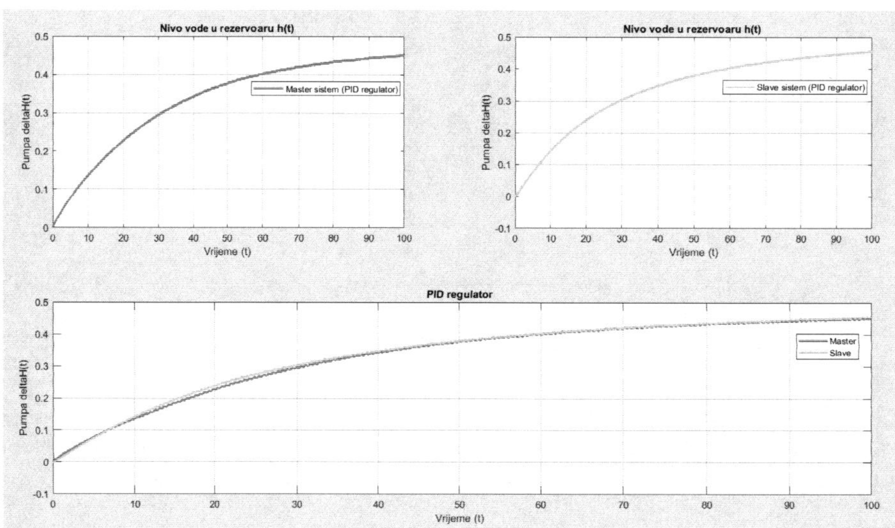

Figure 25. The response obtained for bilateral teleoperator system for PID control method with wave variables.

Based on the obtained results in figures 18, 19 and 20 and the results in figures 24, 25 and 26 it can be clearly seen that the time delay is significantly less when wave variables are applied to control the bilateral teleoperator system.

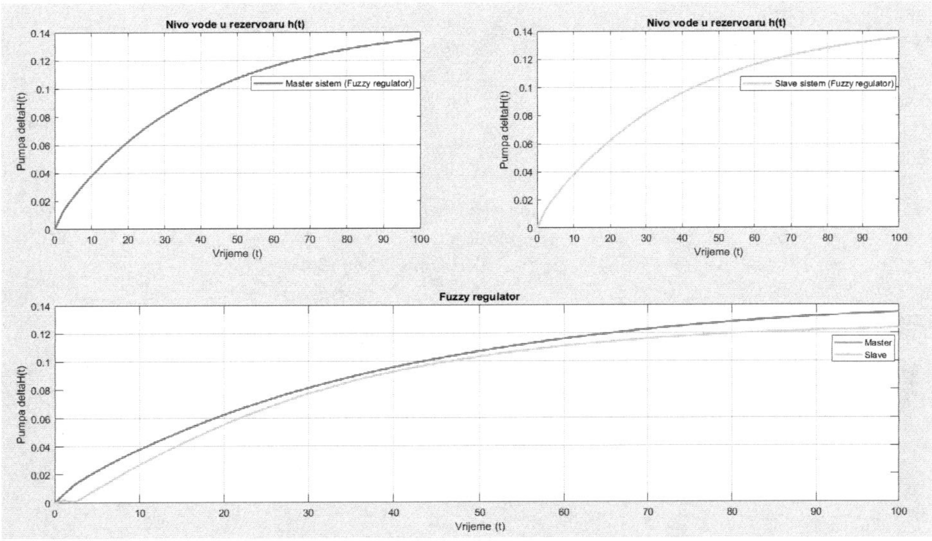

Figure 26. The response obtained for bilateral teleoperator system for fuzzy Logic control method with wave variables.

Also, from the obtained results it can be seen that the time delay is the least when the system is using the PID controller (figure 25).

4. Conclusion

Time delay during data transmission from/to master to/from the slave is one of the difficulties preventing bilateral teleoperation applications to be implemented. Time delays can be caused by the potentially large physical distance between the master and slave side of the system, the processing time of the controllers, the network congestion, the communication bandwidth, etc. will generally degrade the system's performance and cause instability. One of the ways to ensure the stability of the system in the presence of communication delays in the circular loop is the use of wave variables. The energy variables are transformed into wave variables and communication is done through the waves. In this paper, the goal was to find out which of the control methods (direct control, PID control, and Fuzzy Logic control) have the minimal and accepting value time delay using a bilateral teleoperation system to control the level in the water tank. The experimental results obtained from the simulations show that the minimal time delay for the projected bilateral teleoperation system is using PID control.

5. References

[1] Y. Michel, R. Rahal, C. Pacchierotti, P. R. Giordano and D. Lee, Bilateral Teleoperation With Adaptive Impedance Control for Contact Tasks, in IEEE Robotics and Automation Letters, vol. 6, no. 3, pp. 5429-5436, July 2021, doi: 10.1109/LRA.2021.3066974.

[2] F. Huang, Z. Chen and J. Gu, An enhanced wave-variable based four-channel bilateral teleoperation control design, 2018 IEEE 14th International Conference on Control and Automation (ICCA), Anchorage, AK, USA, 2018, pp. 180-185, doi: 10.1109/ICCA.2018.8444286.

[3] Edin Mujčić, Aljo Mujčić, Sandra Pajazetović, Internet-based teleoperation using wave variables and correction of position error, 2016 Osijek Cratia, International Conference on Smart Systems and Technologies, DOI: 11.1109/SST.2016.7765663 Edin Mujčić, Alma Oračević, Correction of Position Error in Internet-Based Teleoperation, Belgrade Serbia, 2012, Telecommunications Forum TELFOR, DOI: 10.1109/TELFOR.2012.6419161

[4] Huang, Fanghao & Zhang, Wei & Chen, Zheng & Tang, Jianzhong& Song, Wei & Zhu, Shiqiang. (2019). RBFNN-Based Adaptive Sliding Mode Control Design for Nonlinear Bilateral Teleoperation System Under Time-Varying Delays. IEEE Access. PP. 1-1. 10.1109/ACCESS.2019.2891887.

[5] A. Hodzic, Edin Mujčić, Teleoperation System Control Based on the Method for Supervisory Control With Variable Time Delay, Belgrade Serbia, 2016, Telecommunications Forum TELFOR, DOI: 10.1109/TELFOR.2015.7377480

[6] F. B. Chioson et al., Implementation of a Bilateral Teleoperation Haptic Feedback Controls to Robotic Minimally Invasive Surgery, 2020 IEEE 12th International Conference on Humanoid, Nanotechnology, Information Technology, Communication and Control, Environment, and Management (HNICEM), Manila, Philippines, 2020, pp. 1-5, doi: 10.1109/HNICEM51456.2020.9400009.

[7] R. Wang, C. Xia, W. Gu and K. Li, Fuzzy singularly perturbed model and stability analysis of bilateral teleoperation system, Proceedings of the 30th Chinese Control Conference, Yantai, China, 2011, pp. 3664-3668.

[8] F. Huang, Z. Chen, J. Gu and S. Zhu, A Novel Control Design of Bilateral Teleoperation System with Fuzzy Logic based General Environments, 2019 Chinese Control And Decision Conference (CCDC), Nanchang, China, 2019, pp. 5047-5052, doi: 10.1109/CCDC.2019.8833157.

IOP Conf. Series: Materials Science and Engineering 1298 (2023) 012024 doi:10.1088/1757-899X/1298/1/012024

Using a fuzzy logic controller in photovoltaic power systems for maximizing PV array efficiency

Una Drakulić[1], Branko Blanuša[2]

[1]The University of Bihać, Technical Faculty of Bihać, B&H
[2]Faculty of Electrical Engineering, Banja Luka, B&H

E-mail: una.drakulic@unbi.ba

Abstract. The goal of this work is to design a high-efficiency photovoltaic system (PV) with a fuzzy logic controller. This PV system consists of a PV panel, DC/DC Buck-Boost converter with a 24V DC load, and a fuzzy logic controller. Because the efficiency of normal solar PV modules is very low, we implemented the fuzzy logic controller. In this paper, we have proposed a system that is designed and simulated in MATLAB Simulink. In the fourth section of the paper are shown obtained results which show that using the fuzzy logic controller for photovoltaic systems has better performance. This controller can facilitate the PV array to reach the MPPT faster and provide more stable output power.

1. Introduction

Developed countries have based their development on fossil fuels such as oil, coal, and natural gas. This is the most common example of fossil fuels used for the production of heat energy. Fossil fuels are non-renewable resources, and as such cannot be used again and again [1]. If the fossil fuels continues to be used at the current level it won't be enough for the whole world, and in the end, it won't be any. A solution to this issue is the use of renewable resources as imposed energy, and in recent years more countries are trying to implement that [2]. The origin of most renewable energy originates directly or indirectly from the Sun. The energy of the Sun's rays, for example, which is known as solar energy, can be used directly for heating and lighting homes and other objects with solar collectors. Energy from the Sun is also stored in the biomass, which is converted into thermal energy by combustion [3].

Energy transition, green transition, and even green revolution are terms that are increasingly being used to denote the changes that have already been largely outlined in the new models of functioning of the energy system. One of the more environmentally friendly energy sources than traditional (fossil fuels, coal, and nuclear energy) is solar energy. Using Solar energy has become less expensive and more efficient over the years. Because of that, the use of photovoltaic systems among other renewable energy sources has increased. The PV system can be used directly as a stand-alone system to the load or indirectly as a grid-connected PV system via the electric grid [4]. The PV system's output power depends on irradiation and temperature, so the PV system requires a maximum power point tracking (MPPT) controller to maintain the service at the global maximum [5]. Different algorithms have been explained in detail in various journals and papers. Most of them are used to overcome the partial shading problem of PV systems. These techniques are mostly very complex with a high computational burden. They also have a bad impact on grid stability due to power interruption [6].

In order to solve the problem of partial shading, an MPPT algorithm based on a Fuzzy logic controller that can boost the output power when partial shading occurs, is proposed in this paper. Artificial intelligence methods such as fuzzy logic controllers are widely used in industrial settings. Fuzzy logic controllers are more efficient, reliable, and simple to design than pure MPPT controllers [7]. The second part of the paper describes the design of a fuzzy logic-controlled buck-boost DC-DC converter. A third part of the paper describes the implementation and simulation of the proposed system, as well as the obtained results. Finally, the paper concludes with a conclusion about the proposed system.

2. Design of Fuzzy Logic Controlled Buck-Boost DC-DC Converter

Compared to other control methods, fuzzy logic controllers have faster transient responses. Membership functions describe fuzzy sets, which are the main tools used by fuzzy logic controllers. The implementation of linguistic fuzzy rules by human operators does not require the estimation of mathematical model parameters, even for complex, nonlinear systems. Figure 1 shows the block diagram for photovoltaic power systems with the use of a fuzzy logic controller.

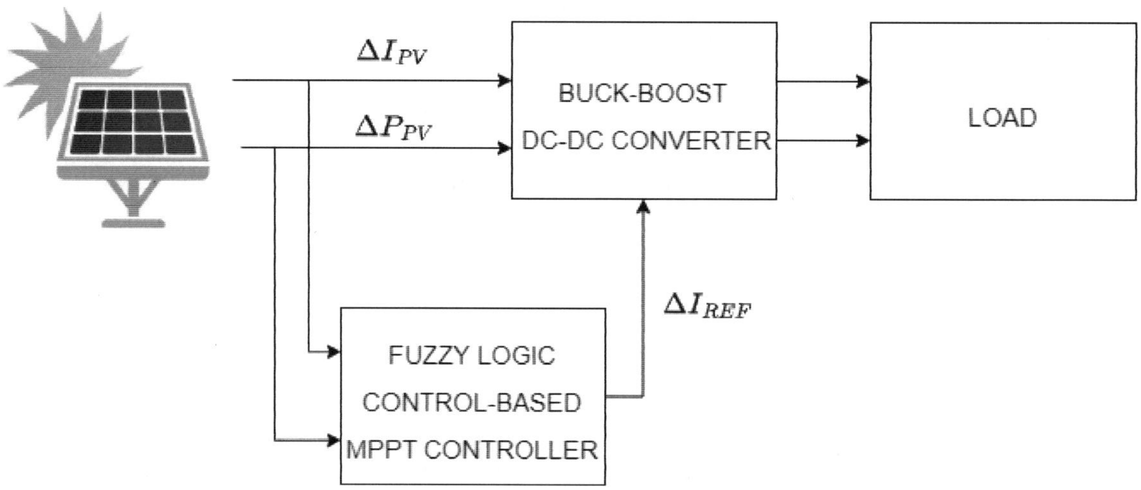

Figure 1. Block diagram of the PV systems with the use of a fuzzy logic controller.

The block named '*fuzzy logic control-based MPPT controller*' consists of fuzzy logic control based MPPT controller. The work algorithm for designing fuzzy logic control-based MPPT controller is shown in figure 2.

DC/DC converters using buck-boost technology are controlled using fuzzy logic control-based MPPT controllers. The control signal depends on its duty factor and frequency. As a result, the MPPT algorithm is implemented for controlling PV output voltage. In order to achieve maximum output power, the PV panel output voltage is changed in response to the cell temperature and solar radiation intensity. It is realized using duty factor control. A second function of the controller is to improve converter efficiency by controlling the switching frequency.

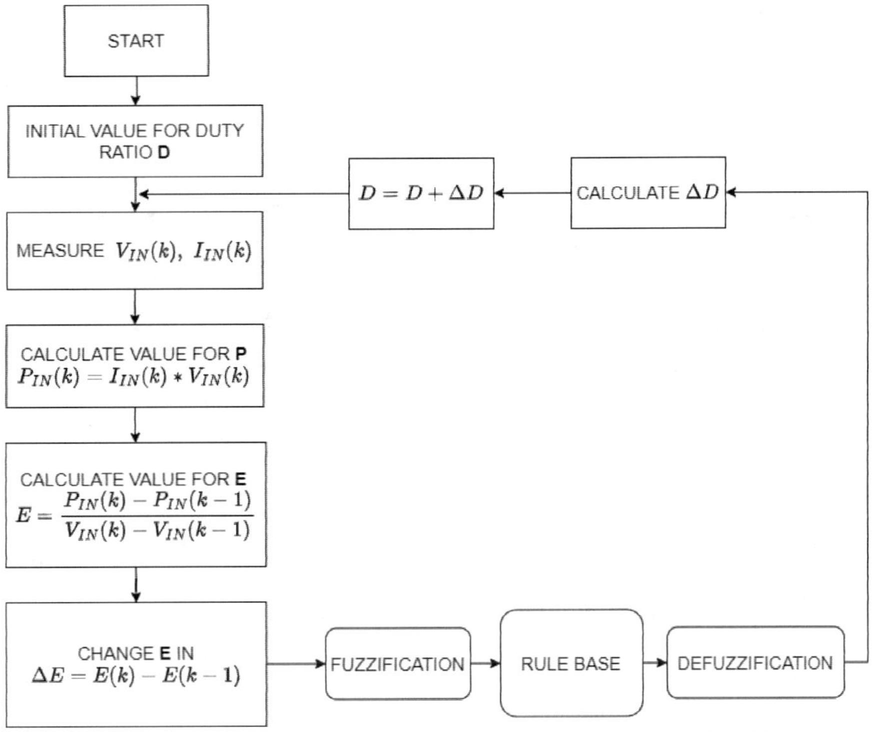

Figure 2. The work algorithm for designing fuzzy logic control based MPPT controller [7].

For the realization of the fuzzy logic-based MPPT controller is designed fuzzy logic system shown in figure 3.

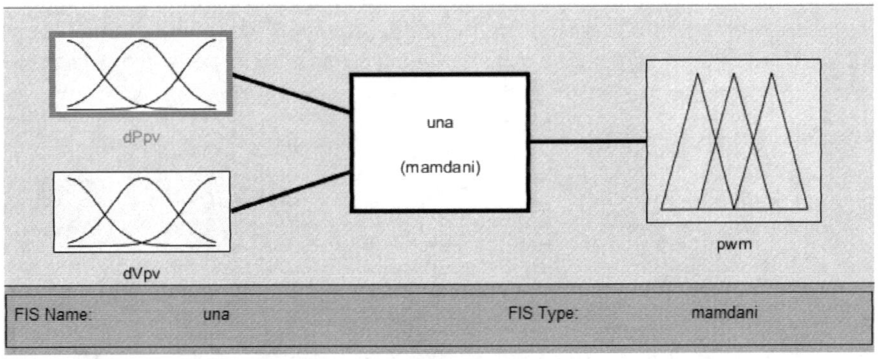

Figure 3. Designing the fuzzy logic system.

The fuzzy logic system was modeled using Matlab Logic Toolbox, ie FIS editor. The fuzzy logic system consists of fuzzification, rule evaluation, and defuzzification processes. In the process of fuzzification, the precise quantity is transformed into a fuzzy quantity, whereas in the process of defuzzification, the fuzzy quantity is transformed into a crisp quantity. Fuzzy logic is used to develop a mapping from a given input to an output in fuzzy inference. This mapping provides the base for making the decision and discovering patterns. In the figures 4, 5 and 6 the input and the output with membership functions are shown.

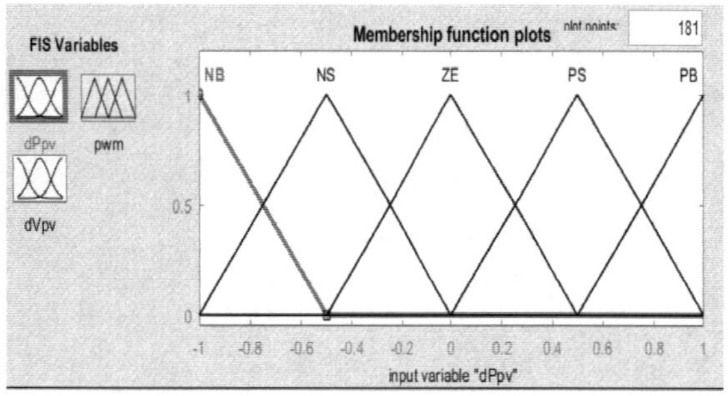

Figure 4. The membership functions for the first input value '*dPpv*'.

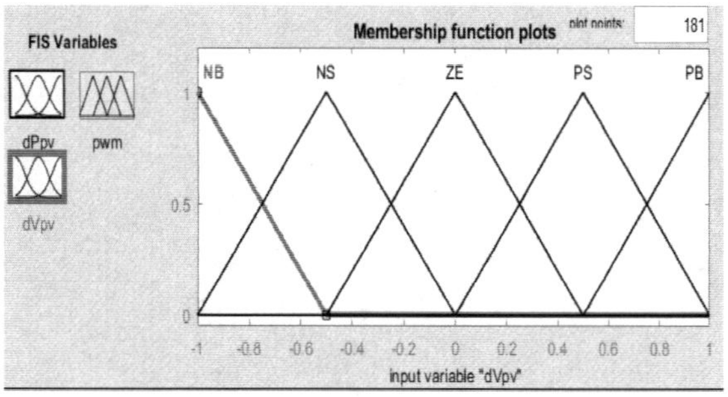

Figure 5. The membership functions for the second input value '*dVpv*'.

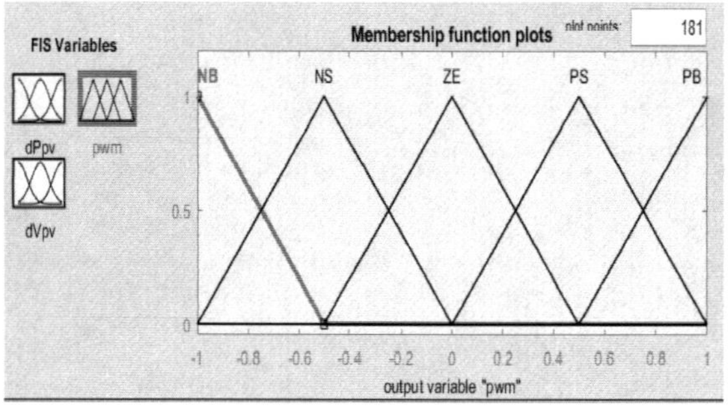

Figure 6. The membership functions for the output value '*pwm*'.

The membership functions have a vital role in the overall performance of fuzzy representation. The membership functions are presented as the building blocks of fuzzy logic. There are different types of membership functions such as trapezoidal, triangular, Gaussian, and others. These are the most used ones. Using the right shape and values is specific and important for a particular problem. The only condition a membership function must satisfy is it must vary between 0 and 1. For designing this system are chosen triangular membership functions in the range [-1, 1]. A Mamdani controller with a centroid defuzzification method is used. The rule base for the Mamdani interface is given in table 1:

Table 1. The rule base.

1.	IF	dPpv is NB	AND	dVpv is NB	THEN	pwm is PS	(1)
2.	IF	dPpv is NB	AND	dVpv is NS	THEN	pwm is PB	(1)
3.	IF	dPpv is NB	AND	dVpv is ZE	THEN	pwm is NB	(1)
4.	IF	dPpv is NB	AND	dVpv is PS	THEN	pwm is NB	(1)
5.	IF	dPpv is NB	AND	dVpv is PB	THEN	pwm is NS	(1)
6.	IF	dPpv is NS	AND	dVpv is NB	THEN	pwm is PS	(1)
7.	IF	dPpv is NS	AND	dVpv is NS	THEN	pwm is PS	(1)
8.	IF	dPpv is NS	AND	dVpv is ZE	THEN	pwm is NS	(1)
9.	IF	dPpv is NS	AND	dVpv is PS	THEN	pwm is NS	(1)
10.	IF	dPpv is NS	AND	dVpv is PB	THEN	pwm is NS	(1)
11.	IF	dPpv is ZE	AND	dVpv is NB	THEN	pwm is ZE	(1)
12.	IF	dPpv is ZE	AND	dVpv is NS	THEN	pwm is ZE	(1)
13.	IF	dPpv is ZE	AND	dVpv is ZE	THEN	pwm is ZE	(1)
14.	IF	dPpv is ZE	AND	dVpv is PS	THEN	pwm is ZE	(1)
15.	IF	dPpv is ZE	AND	dVpv is PB	THEN	pwm is ZE	(1)
16.	IF	dPpv is PS	AND	dVpv is NB	THEN	pwm is NS	(1)
17.	IF	dPpv is PS	AND	dVpv is NS	THEN	pwm is NS	(1)
18.	IF	dPpv is PS	AND	dVpv is ZE	THEN	pwm is PS	(1)
19.	IF	dPpv is PS	AND	dVpv is PS	THEN	pwm is PS	(1)
20.	IF	dPpv is PS	AND	dVpv is PB	THEN	pwm is PS	(1)
21.	IF	dPpv is PB	AND	dVpv is NB	THEN	pwm is NS	(1)
22.	IF	dPpv is PB	AND	dVpv is NS	THEN	pwm is NB	(1)
23.	IF	dPpv is PB	AND	dVpv is ZE	THEN	pwm is PB	(1)
24.	IF	dPpv is PB	AND	dVpv is PS	THEN	pwm is PB	(1)
25.	IF	dPpv is PB	AND	dVpv is PB	THEN	pwm is PS	(1)

Where, NB is Negative Big, NS is Negative Small, ZE is Zero, PS is Positive Small and PB is Positive Big.

3. Implementation and simulation of the proposed system

After designing fuzzy logic system, Simulink model for the proposed system is created (figure 7). In Simulink model for the PV Array is selected 1Soltech 1STH-215-P module with maximum power Pmax=213.15 W, open circuit voltage Voc=36.3 V, voltage at maximum power point Vmp=29 V, temperature coefficient of Voc=-0.36099 %/deg.C, short-circuit current Isc=7.4 A, current at maximum power point Imp=7.35 and temperature coefficient od Isc=0.102 %/deg.C. This PV Array module consists of 60 Ncell per module. The values used to project the photovoltaic power systems with fuzzy logic control are:

for D1 $Ron = 0.001\ \Omega, Vf = 0.8\ V, Rs = 500\ \Omega, Cs = 250\ e^{-9}\ F$
for D2 $Ron = 0.002\ \Omega, Vf = 0.5\ V, Rs = 500\ \Omega, Cs = 250\ e^{-9}\ F$
$C1 = C2 = 1000\ e^{-6}\ F$
$L = 10e^{-3}\ H$
for MOSFET $Ron = 0.02\ \Omega, Rd = 0.01\ \Omega, Rs = 1e^{5}\ \Omega, Cs = \infty$

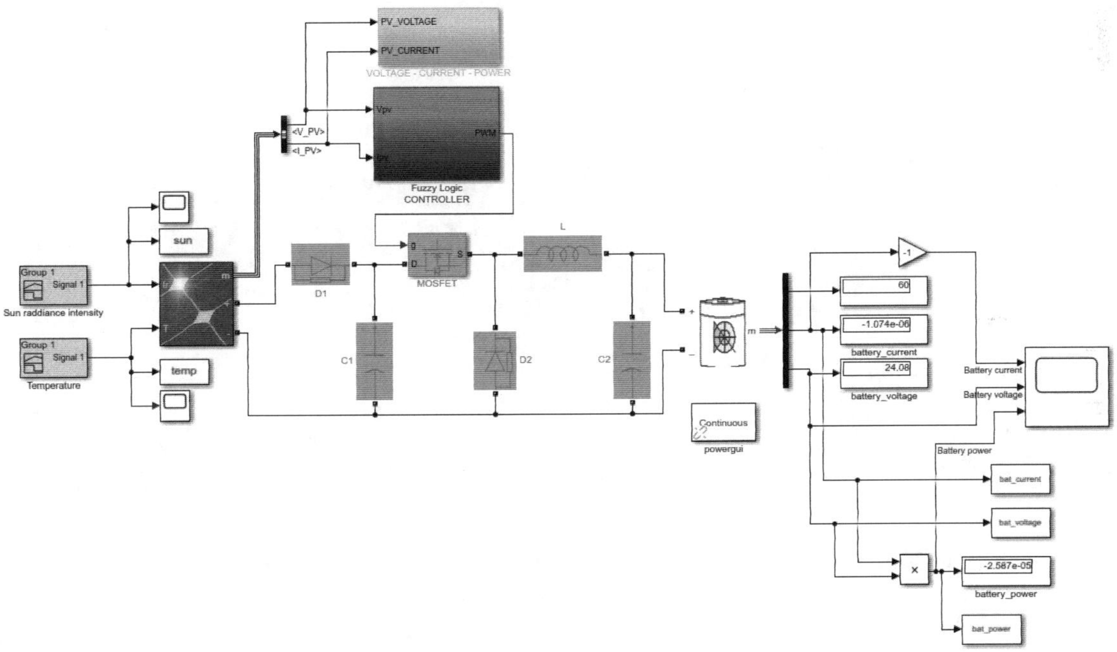

Figure 7. The Simulink model for the photovoltaic power systems with fuzzy logic control

Content of the subsystems are shown in figures 8 and 9.

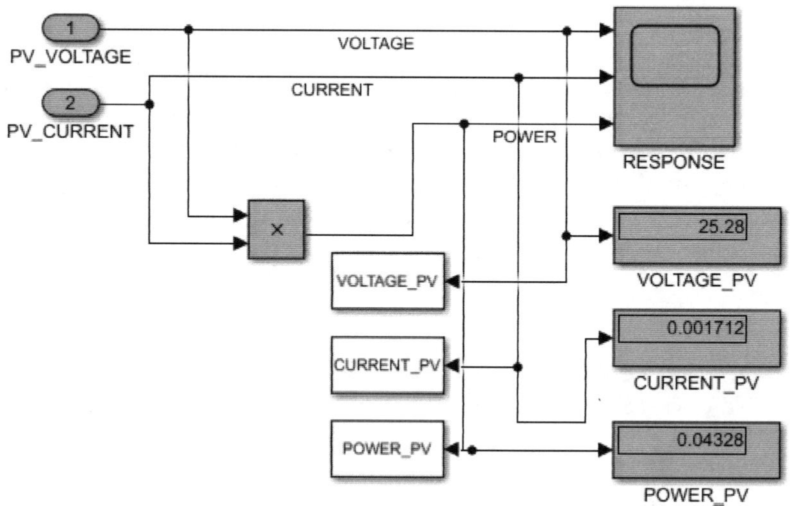

Figure 8. The Simulink model which measures voltage, current, and power from PV array.

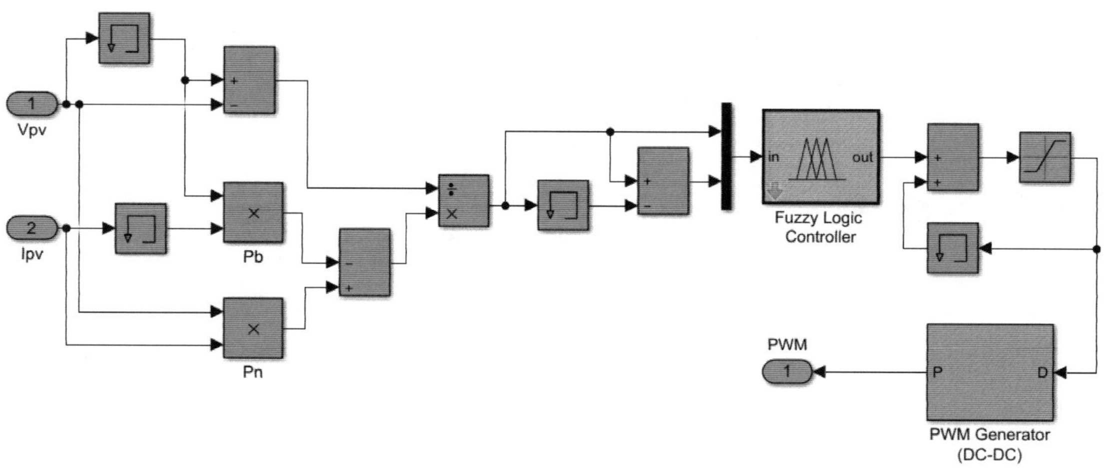

Figure 9. The Simulink model of subsystem block *'Fuzzy Logic CONTROLLER'*.

Using MPPT controllers in PV systems we get more power depending on weather and temperature. Also, more PV systems can be connected in the grid of PV modules in series which help to increase the voltage of the system.

Simulation time is T=1 sec, and after running the Simulink model, the obtained results are shown in figures 10, 11 and 12.

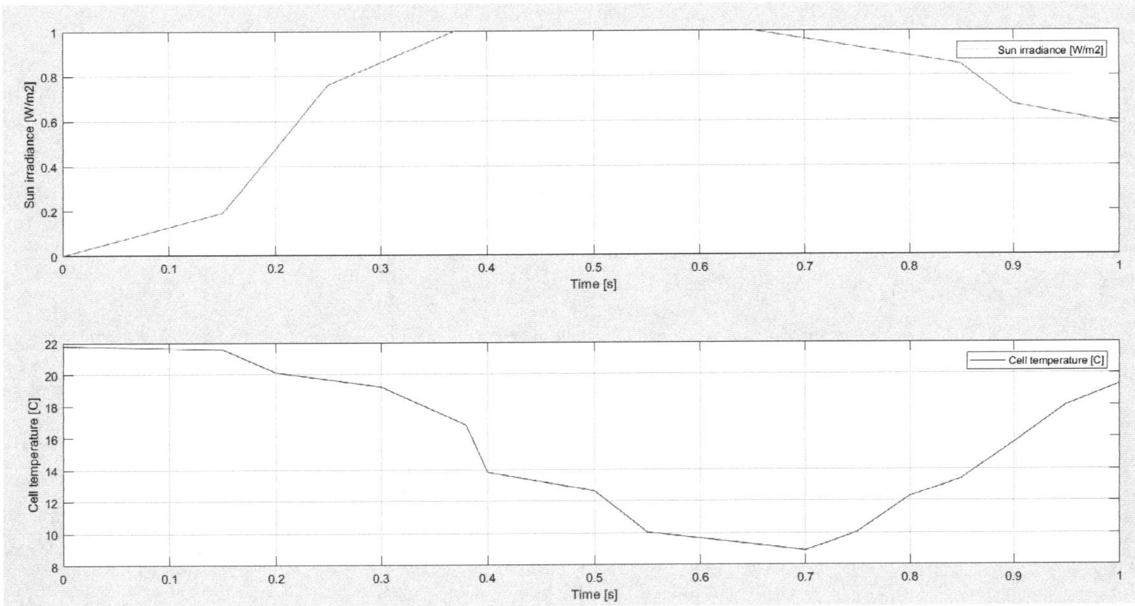

Figure 10. The Simulink model response for input parameters, i.e. for sun radiance intensity and cell temperature

The shape of the sun irradiance is chosen to be in the range between 0 and 1 by the y-axis which represents the full sunlight (1), the sunlight turns into shade (between 1-0) and the shade (0). Time is chosen to be T=1 sec. The shape of the temperature is chosen to be in the range between 9 and 22 C.deg. arbitrarily, and time is T=1 sec.

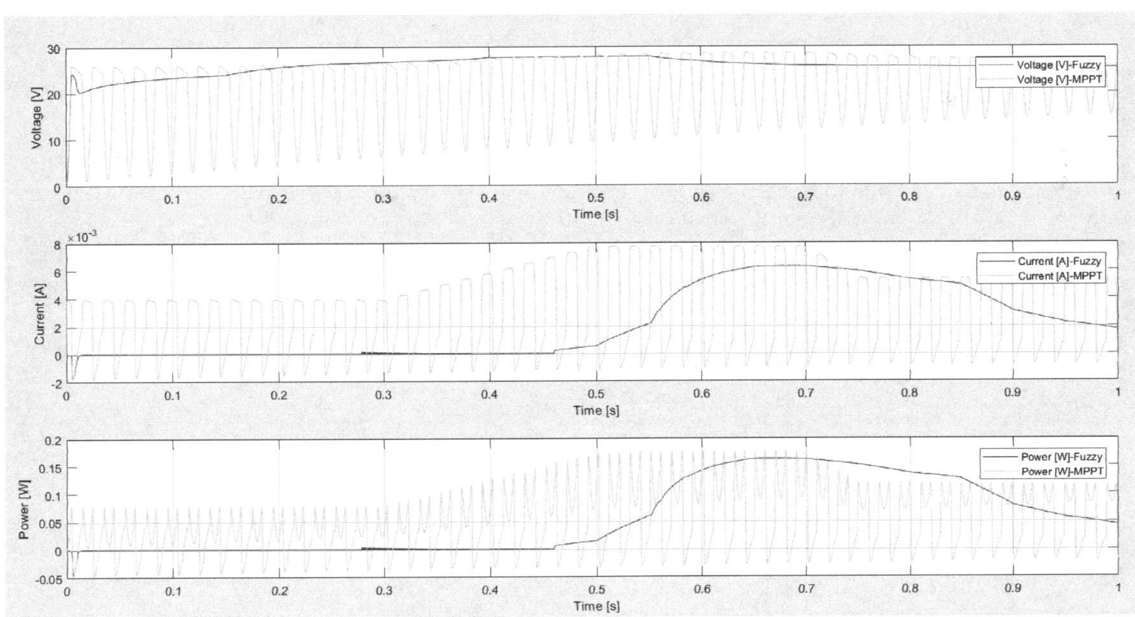

Figure 11. The Simulink model response for voltage (V), current (A) and for power (W) from the PV Array

The duration of the simulation is shorter than the real ambient changes of temperature and solar radiation. Since the switching frequency of the converter is 1000 Hz, the duration of the simulation that would take into account real changes of the ambient parameters would take a very long time. The

IOP Conf. Series: Materials Science and Engineering 1298 (2023) 012024 doi:10.1088/1757-899X/1298/1/012024

approach in the work does not affect on the generality of the analysis and even shows the fast operation of the MPPT algorithm with a fuzzy controller. The analysis takes into account different variants of temperature and solar radiation change, which are not directly correlated.

In the figure 11 are shown obtained response for voltage (V), current (A) and for power (W) from the PV Array. We see that the current almost fully follows the values from the input sun radiance intensity values, and that the voltage slowly increases values when the sun's radiance increases. Power is product of these two values, as can be seen in figure 11.

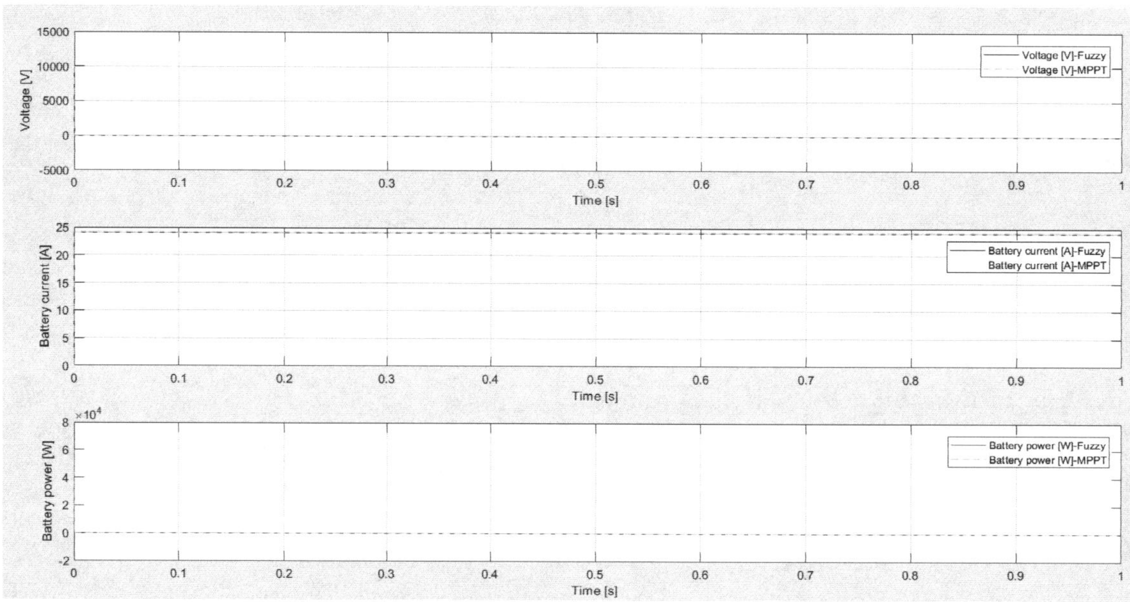

Figure 12. The Simulink model response for battery voltage (V), battery current (A) and for battery power (W) from the battery

For the load is used lead-acid battery with nominal voltage 24 V, rated capacity 100 Ah, for initial state-of-charge is set 60% and battery response time is set 60 s (see Figure 12).

For comparison reasons, the same Simulink model like in Figure 7, but instead of using the fuzzy logic controller is used MPPT charge controller block (see Figure 13) [7].

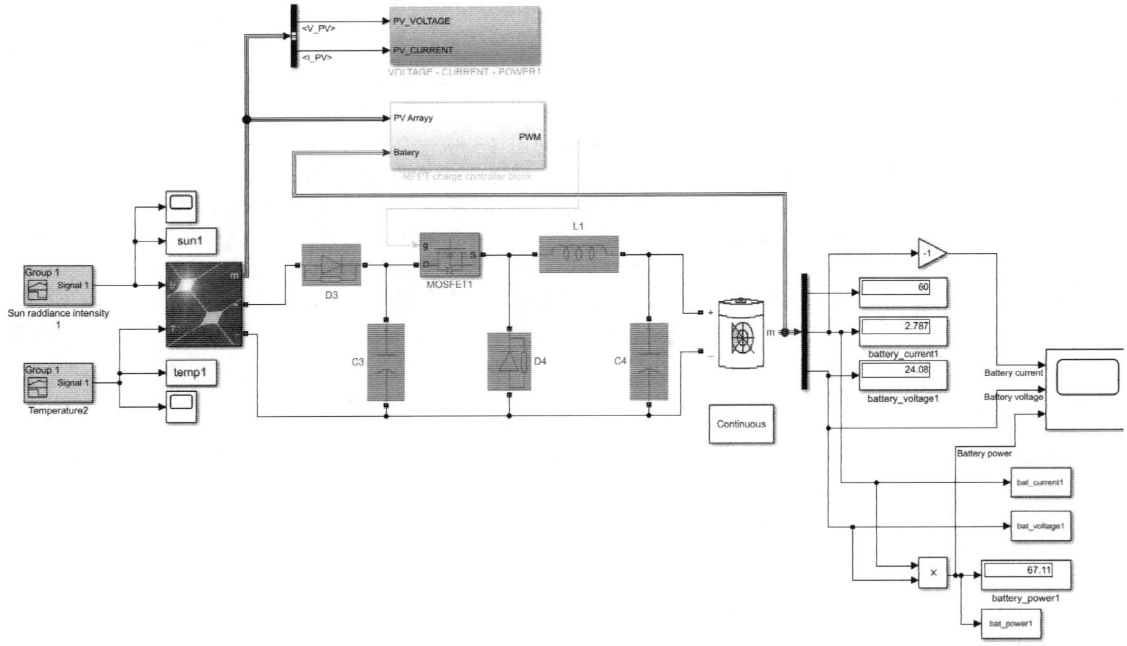

Figure 13. The Simulink model response with MPPT charge controller block

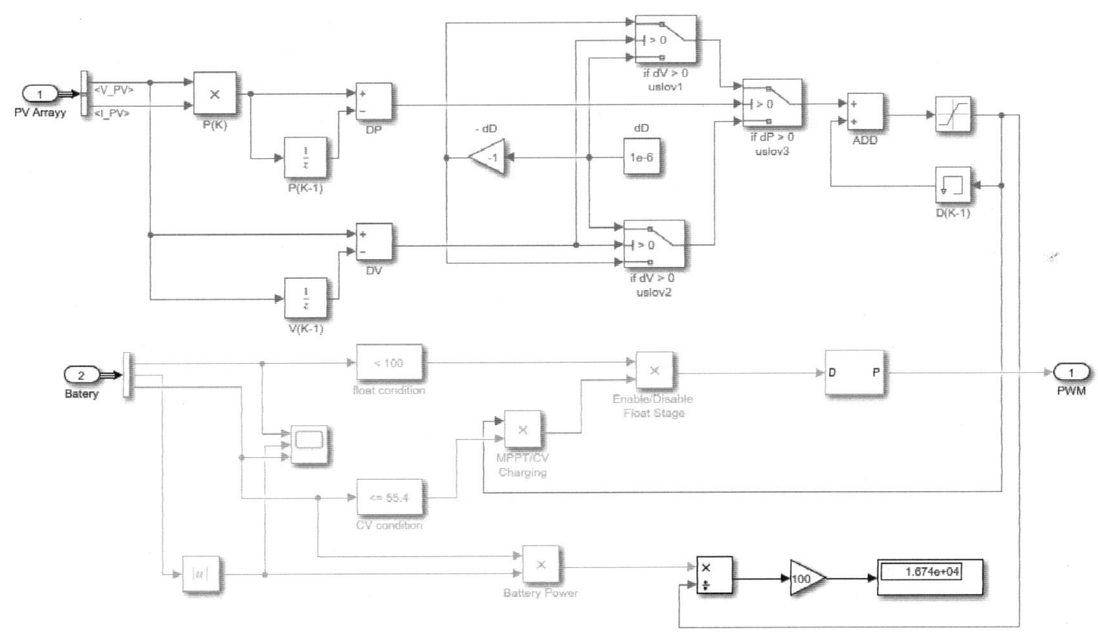

Figure 14. The Simulink model for subsystem named 'MPPT charge controller block'

Figure 14 shows the Simulink model for subsystem named 'MPPT charge controller block' in Figure 13. Disturb and observe, i.e., the P&O algorithm is based on the constant monitoring of the working power of the solar panel point, with the aim of increasing the points of maximum power. The result of the P&O algorithm is the value of the charge factor, which is constantly fed into the circuit by PWM signals, by means of which the operation of the DC/DC converter is controlled.

As can be seen in obtained results in Figure 11 (green response), the disadvantage of the P&O algorithm is that the operating point constantly oscillates around the reached maximum power of the solar panel. The biggest disadvantage of this algorithm is that, due to rapid changes in irradiance, it cannot correctly assess whether the power change is due to voltage changes or irradiance changes, so it may happen that the algorithm makes several consecutive changes in the opposite direction than expected and completely loses synchronization with the maximum value.

4. Conclusion

In this paper is described the design and simulation in MATLAB/Simulink software to model a fuzzy logic controller with Maximum Power Point Tracker (MPPT). This controller has been implemented to utilize the maximum available sunrays present in the environment. The system is more efficient when the MPPT is added to the Fuzzy logic controller. Obtained results from simulation prove that. In implemented Simulink model MPPT follows the precise MPPT point, which further increases the overall efficiency of the solar system. The MPPT based on fuzzy logic enables superior tracking and faster convergence speed.

5. References

[1] Ali O M Maka, Jamal M Alabid, *Solar energy technology and its roles in sustainable development*, (*Clean Energy*), Volume 6, Issue 3, June 2022, Pages 476–483

[2] P. P. Bezrukikh, S. M. Karabanov and P. P. Bezrukikh, *Renewable energy efficiency* (IEEE 16th International Conference on Environment and Electrical Engineering (EEEIC)) (Florence 2016) pp. 1-5

[3] Carlos Robels Algarin, John Taborada Giraldo, Omar Rodriguez Alvarez (Fuzzy Logic Based MPPT Controller for a PV System) (Energies 2017)

[4] F. Chekired, C. Larbes, D. Rekioua, F. Haddad. (Implementation of a MPPT fuzzy controller for photovoltaic systems on FPGA circuit) (*Energy Procedia*) Volume 6, 2011, Pg. 541-549

[5] M. Z. Abdullah, I. Sudiharto and R. P. Eviningsih *Photovoltaic System MPPT using Fuzzy Logic Controller* (International Seminar on Application for Technology of Information and Communication (iSemantic)) (Semarang, Indonesia, 2020) pp. 378-383

[6] Mostafa B., Ahmed A., Mostafa A., Michael G *PV System Control Using Fuzzy Logic Controller Employing Dynamic Saftety Margin under Normal and Partial Shading Conditions* (Article in Energies) (2021)

[7] Nasir Hussein Selman, Jawad Radhi Mahmood (Comparison Between Perturb&Observe, Incremental Conductance and Fuzzy Logic MPPT Techniques at Different Weather Conditions) (International Journal of Innovative Research in Science, Engineering and Technology) (July 2016) Vol. 5, Issue 7

Control of a Smart Security Object using an Arduino Platform and a GSM Controller

Azra Kapić and Amel Toroman

University of Bihać, Faculty of Technical Engineering, Bosnia and Herzegovina

E-mail: azra.kapic@unbi.ba

Abstract. A smart home is a term often used to define a residence that employs electronic control systems or simply "controllers" to integrate and activate various spatial detectors and functional subsystems embedded within the smart home, particularly as integrated parts of home automation. In this paper, the focus is on studying security and enhancing the technical protection system, rather than on the functions and capabilities of controllers for smart functionalities within a specific facility. The paper describes intelligent control of a security object, enabled using specific components. The advantage of the designed system lies in its practical application, offering increased security, facility monitoring, streamlined communication, and user-friendliness through the utilization of ICT technology and smart components.

1. Introduction

Nowadays, smart systems are an indispensable part of everyday use, and their security is one of the most important factors of such systems [1]. This paper shows one such system through practical implementation. The practical realization of the paper was realized as a modular part of three segments of paper, namely: *smart elevator control system*, *garage door control sensor system*, and *smart security system for controlling the residential building*, and around the entire building there is lighting that is controlled via mobile phone. The components required for the implementation of the system are the GSM (Global System for Mobile Communication) module (SIM900L), RFID (Radio Frequency Identification System), keyboard, temperature sensor, fan, stepper and servo motors, and other used components. The aim of the paper is to apply the theoretical part practically. Accordingly, the goal is to create a system that will serve to perform the specified task (controlling a residential building, controlling a garage door, and controlling an elevator inside the building).

The idea for the implementation of the paper is based on previous research, as follows.

In the research [2], the authors describe a security system that employs GSM technology and an Arduino control board for the continuous monitoring of sensors detecting intrusions, fire, and hazardous gases. When a threat is detected, the system activates visual, auditory, and phone-based alarms to alert the owner.

In the research [1], the authors describe a smart security box utilizing GSM technology for remote monitoring. This system employs sensors to detect unauthorized access attempts, triggering alarms and sending SMS notifications to the owner, while being controlled by an Arduino platform.

Within this research, a modular system consisting of three key components is presented. The first segment encompasses a smart security system for residential access control, utilizing RFID tags and passwords for authentication. The second segment focuses on a smart elevator control system, enabling users to call the elevator and access specific floors through GSM technology. The third segment implements a garage door control system, utilizing an ultrasonic sensor to detect vehicles and enabling remote outdoor lighting control via SMS messages. All three components are interlinked and managed using the GSM SIM 900L module.

The contribution of this research lies in presenting a modular system that encompasses the three key components. These components are interconnected and operated through the GSM SIM 900L module. This research introduces an integrated approach that combines diverse functionalities within a cohesive framework, offering an innovative solution for enhanced security, access control, and automation in residential and business environments.

2. Security of smart objects

Most systems in a smart home operate continuously, maintaining communication with the network [5]. Ensuring the security of these systems is crucial to protect them from potential attacks (Figure 1).

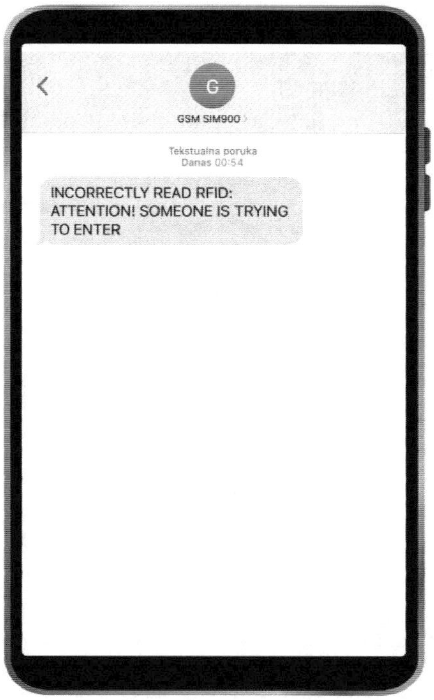

Figure 1. Display of notification about unauthorized entry

Access control is a system that enables the owner or a certain authority to control access to premises and resources in a physical object. In everyday life, access control is a common occurrence. A car door lock and ATM pin codes are also a form of access control. Implementing access control becomes of paramount importance when securing important, confidential, or sensitive information and equipment. Access control systems are utilized for monitoring and physical protection of buildings or specific areas within them. The primary task of the system is to permit access (entrance or passage) exclusively to authorized persons. Access control plays a significant role in preventing unwanted criminal acts and their potential consequences on the safety of people and property. Its purpose includes deterring perpetrators, early detection of attempted criminal activities, alerting the relevant authorities and intervention teams, and slowing down the perpetrators. Access control points can be doors, ramps, parking gates, elevators, or other physical barriers where access can be electronically controlled. Usually, the door serves as the primary access point. Electronic door access control may consist of several elements. In the simplest form, it involves an independent electric lock that the operator locks and unlocks with a switch. To automate this process, the role of the operator is replaced with a reader. The reader can be a keyboard, card reader or biometric reader. Typically, readers do not make access decisions, but instead transmit a code to the appropriate panels that compare the code with the entries on the access list. In most cases, only entry is controlled but some institutions also control exit (business buildings and facilities). In such cases, another reader is needed on the opposite side.

In situations where exit control is not required (free exit), devices known as " or request-to-exit (RTE) are used. RTE devices can be buttons or motion detectors. When the button is pressed or when the detector senses movement at the door, the alarm is instantly disarmed, and the door opens [6].

2.1. *Elements of the Access Control System*

Each access control system consists of several basic components: controller, reader and/or keyboard and electromechanical door elements.

Controllers are the main control unit to which card readers, exit buttons, magnetic contacts for door monitoring and executive elements, such as electrical receivers, are connected [7]. The controller stores the operating program of the monitored door, users with their authorizations and records all door events (the number of events depends on the size of the controller's memory). As a result, the controller can also work offline, i.e., without the need for a constant connection to a computer. Depending on the number of entrances for readers, it is also defined how many doors the controller controls. The most common are controllers for two readers, which can manage one door with double-sided control or two doors with one-sided control and free exit. Controllers for multiple readers are also possible, but when choosing a controller, one should consider the distance between the controlled doors, as the length of the cable for the readers and other elements on the door is limited. In addition to the reader, each door requires a minimum of two inputs (for a magnetic contact and an exit button) and at least one output for a lock. More advanced controllers provide more alarm inputs for additional functions and more outputs for connecting to other systems and signalling.

Card readers and keyboards are also elements of the Access Control System. The reader works by emitting low-intensity high-frequency energy (e.g., 125 kHz), which induces in the card the energy needed to read the unique code written in the microchip and send that code to the reader. Card readers can be divided into classic (non-intelligent), semi-intelligent and intelligent readers. Classic readers read the card number or PIN code and forward it to the controller. The Wiegand protocol is typically used to transfer data to the controller, but it is not uncommon to use the RS-232 or RS-485 protocol. Semi-intelligent readers have all inputs and outputs for door control (i.e., lock, door contact, exit button), but do not make any access decisions. When the user presents himself with a card or enters a PIN, the reader sends information to the controller and waits for a response. If there is a communication break between the controller and the reader, such a reader cannot perform its function or can be in a degraded mode. Most often, these readers are connected to the controller via the RS-232 protocol. Intelligent readers have all the inputs and outputs for controlling the door, but they also have memory and microprocessors, so they can independently make access decisions. In this case, the main controller sends any data updates to the reader and receives events from the reader. In addition to card readers for access control, it is also possible to use keyboards in such a way that the user can be identified by entering his code. Using only the keyboard for identification is only suitable for less risky applications, as the code can be given to anyone intentionally or unintentionally. More often, keyboards are used as additional protection in addition to the reader, i.e., it is also necessary to have a card and enter the appropriate code for the entrance permit. For high-risk applications, biometric identification is increasingly being used instead of card readers, which means that some of the person's unique biological characteristics are used. Passive (facial geometry, eye retina, fingerprint, hand geometry, hand veins) or active characteristics (signature dynamics, speech, facial expressions, manner of movement) or their combinations are measured. Identification is carried out in two stages: first, a biometric characteristic is scanned, and its digital representation is created, and then the coincidence of the individual's characteristic with the remembered characteristic is examined. The most common biometric identification today is fingerprint identification, which, due to its practicality and speed, is used in applications where speed is important, and a certain risk of false acceptance is acceptable.

2.2. The components used to realize the security of the created system

The security of the system is achieved by using of GSM SIM 900L module, RFID, and keyboard.

In the residential building module, security is ensured right at the building's entrance. Users cannot access the building unless they use the correct card and enter the correct password via the keyboard. Every action within the building triggers the GSM module to send a notification in the form of an SMS message to the specified number. The GSM module sends a notification when a user enters or leaves the facility, enables tracking of users within permissions.

In the elevator module, security measures include sending messages for requesting help from the elevator, where an SMS message is sent via the GSM module. If the elevator stops between floors, an automatic SMS message is sent to report the malfunction. Additionally, user can send SMS messages to the GSM module that are displayed on the LCD display inside the elevator so that the elevator users are informed about the occurrence and how to solve the malfunction itself.

GSM was developed with a moderate level of security service and was designed for user authentication with a pre-shared key. Communication between the system sender and the base station can be encrypted. The security model provides confidentiality and authenticity but has a limited authority that must not be rejected [8].

General Packet Radio Service (GPRS) serves as a packet-oriented mobile data service. It is used in 2G and 3G mobile communication system for GSM mobile devices whose connection is charged per minute. GPRS enables a data transfer speed of 56 to 114 Kbit/s. This system allows sending SMS messages and broadcasting messages, as well as supports the service of multimedia messages [8].

The GSM network is composed of several functional parts, which can be classified into three main parts [8]:

1. *Mobile station* - carried by the subscriber.
2. *Base station* - controls the radio connection with the mobile station.
3. *Network subsystem* - contains the MSC that establishes connections between users.

An RFID consists of two main components: a transponder/tag attached to the object to be identified, and a transceiver also known as an interrogator/reader. The reader consists of a radio frequency module and an antenna that generates a high frequency electromagnetic field. On the other hand, a tag is usually a passive device, meaning it does not contain a battery. Instead, it contains a microchip that stores and processes information, and an antenna for receiving and transmitting signals [8].

To read the information encoded on the tag, it is placed in the immediate vicinity of the reader (it does not have to be in the immediate line of sight of the reader). The reader generates an electromagnetic field that causes electrons to move through the tag's antenna and then "power" the chip. The powered chip inside the tag then responds by sending its stored information back to the reader in the form of another radio signal. This is called backscatter. The backscatter, or change in the electromagnetic/RF wave, is detected and interpreted by a reader that then sends the data to a computer or microcontroller [8].

The RC522 RFID reader module is designed to generate an electromagnetic field of 13.56 MHz, which it uses to communicate with RFID tags (ISO 14443A standard tags). The reader can communicate with the microcontroller via a 4-pin Serial Peripheral Interface (SPI) with a maximum data rate of 10 Mbps. It also supports communication via I2C and UART protocols [8].

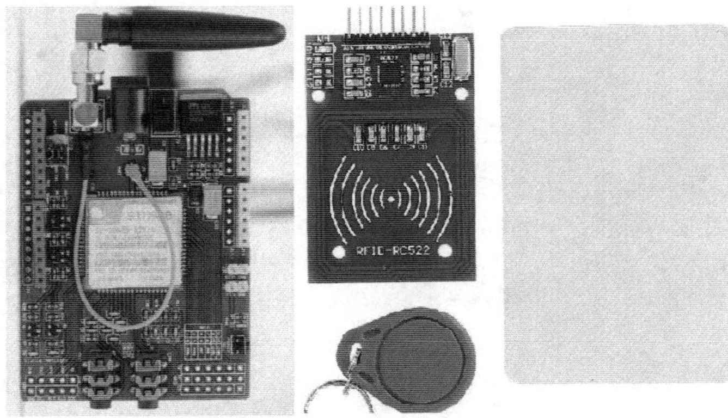

Figure 2. Display of GSM SIM900L module and RFID [8, 9]

Figure 2 shows the two most important components related to security and access control of the created system, namely GSM SIM900L and RFID.

3. System description

This research represents an advancement compared to the previous [1] and [2] studies. While [1] focuses on a smart security box with GSM technology for remote monitoring and detection of unauthorized access attempts, and [2] describes a security system with GSM technology for monitoring sensors detecting intrusions, fire, and hazardous gases, this research goes a step further. This paper introduces a modular system with three key components: a smart security system, a smart elevator control system, and a garage door control system. The integration of these segments enables a comprehensive approach to security, access control, and automation in residential and business environments, simultaneously utilizing the GSM SIM 900L module for management. This advanced approach makes this research more innovative and comprehensive compared to the previous ones, opening new possibilities for improving security and control systems.

Furthermore, this research highlights the potential for energy efficiency. The integration of different segments within this modular system enables the optimization of energy usage. For instance, the ability to control garage doors through an ultrasonic sensor means that the doors open only when a vehicle is detected, preventing unnecessary energy consumption.

The created system consists of three modular parts, which parts are: *a smart security system for controlling the residential building*, *smart elevator control system* and *sensor system for garage door control*. Each of these parts will be described below.

3.1. Smart security system for controlling a residential building

The method of opening the building's door was implemented using RFID, a keyboard and a button on the other side of the building. When one of the cards is touched to the reader, the corresponding LED and buzzer are activated. If an incorrect card is read, and an SMS message is sent to a specified number using the GSM SIM00L module, and a specific text is displayed on the screen. After a certain time, the access indicators are disabled.

If the authorized card is read, the green LED turns on, and the user is prompted to enter the password via the keyboard to access the facility. If the user enters the wrong password, an SMS message is sent via the GSM SIM900L module as a warning about unauthorized access. After entering the correct password, a servomotor is activated to open the facility's door certain time defined by the *delay()* function and then it automatically closes. When leaving the building, the user needs to press the button to activate the servo motor, and the door of the building opens. The action is accompanied by sending an SMS message via GSM module to a specified number.

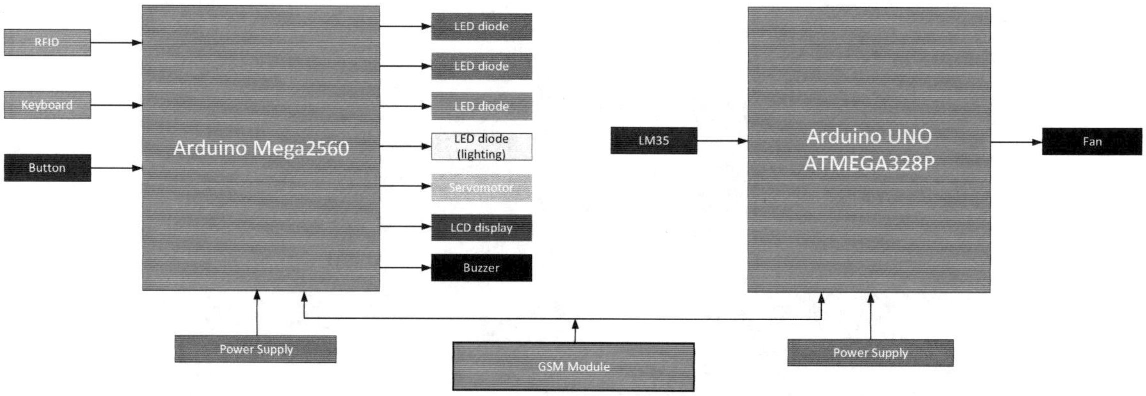

Figure 3. Block diagram of the residential building

The block diagram shown in Figure 3 is used to implement the previously described scenario.

3.2. Smart elevator control system

The method of opening the elevator door and the door on the first floor, as well as raising and lowering the elevator cabin is realized using two stepper motors and one servo motor. Each floor has a button to call the elevator, accompanied by LEDs indicating the direction in which the elevator is moving, and one LED signalling the current floor where the elevator has stopped. The buzzer is activated when the door is opened and the entire operation of the system is displayed as text on the LCD display.

Two additional buttons were installed to simulate two scenarios: stopping the elevator at the mezzanine and allowing users in the elevator to request help. Stopping at the mezzanine is achieved using the GSM SIM900L module. When the elevator stops at the mezzanine level, an automatic SMS message is sent via the GSM SIM900L module to inform about the malfunction, and a message is displayed on the LCD display. During this time, all other buttons are inactive to show the actual fault. It is necessary to first send an empty message, after which the current date and time are displayed on the LCD display, and after that it is necessary to send another message with the text "*TEAM SENT*". After a certain time defined with the *delay()* function, the elevator returns to the previous the floor he was on.

The simulation of installing an elevator on a mezzanine floor can be done from any floor, and the elevator will return to the previous floor it was on after the completed actions. If the user presses the button that represents the simulation of users requesting help inside the elevator, a specific text is displayed on the LCD display, but all functions remain active.

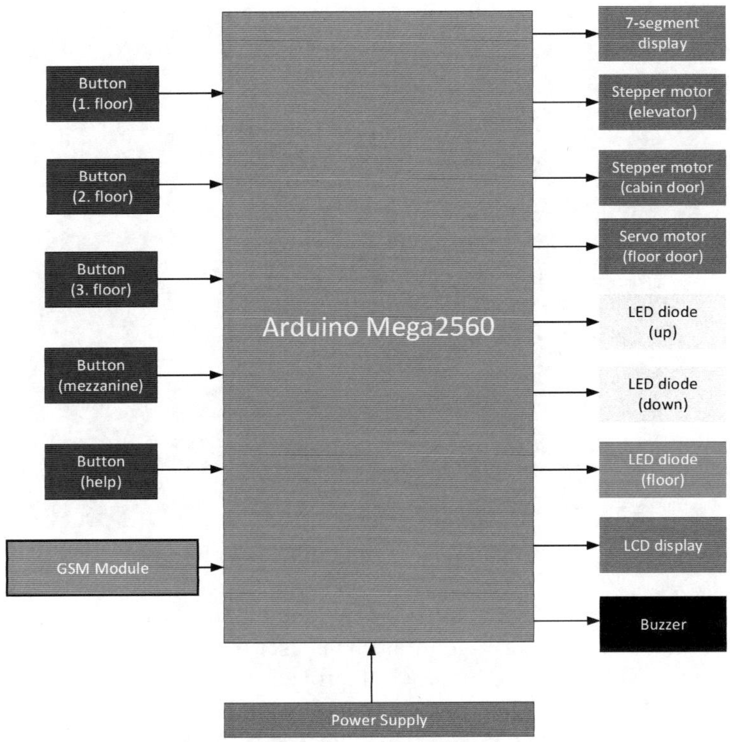

Figure 4. Block diagram of the elevator

Figure 4 shows the block diagram of the elevator with the indicated Input and Output elements, as well as the Power Supply.

3.3. Garage door sensor control system

In addition to the previously mentioned facilities, a garage was also included in the project, equipped with servomotors and ultrasonic sensors. The garage door opens when the ultrasonic sensor detects a car in front at a certain distance, and remains open for a certain time, defined by the *delay()* function.

The project also includes outdoor lighting, which is controlled by 6 LED diodes. They can be turned ON or OFF by sending an appropriate SMS message to the number provided in the GSM SIM900L module. After turning on or off the lighting, a confirmation message is sent to the GSM number associated with the SIM900L module.

4. Analysis of the functioning of the realizes System

When connecting the created project to the power supply, the status indicators are activated, and the text is displayed on the LCD screen while waiting for the user's action. The analysis of the system functioning can be viewed from two perspectives: the analysis of the system functioning when an authorized user accesses it and the analysis of the system functioning when an unauthorized user access it. Below are some of the steps for accessing one of the created modules.

4.1. Analysis of the functioning of the implemented system when the authorized user accesses it

If the user reads the correct card on the residential building and enters the correct password, the door of the residential building opens, the green LED is activated and the corresponding text for the action is displayed on the LCD screen. The blue LED is used to indicate whether the system is active, i.e., switched ON or OFF.

IOP Conf. Series: Materials Science and Engineering 1298 (2023) 012025 doi:10.1088/1757-899X/1298/1/012025

Figure 5. Display of the system after reading the correct data

The previously defined action is accompanied by sending an SMS message to specified number via the GSM SIM900L module. This message informs the user that someone has entered the facility using the correct data and that entry into the facility is approved.

Figure 6. Display of system status messages

The second message from Figure 6 is a notification about leaving the facility. Regarding the elevator module, if the user calls the elevator from the first floor to the second floor, the LEDs indicating upward movement (placed on the structure and the elevator cabin) are activated. Additionally, the floor number

IOP Conf. Series: Materials Science and Engineering 1298 (2023) 012025 doi:10.1088/1757-899X/1298/1/012025

to which the elevator is moving is displayed on the 7-segment display, and a message about the elevator's upward movement is shown on the LCD display.

Figure 7. View of the movement of the elevator cabin towards the second floor

After the elevator reaches the selected floor, the green LED of the floor is activated, and both the elevator cabin door and the floor's structure door are opened. The buzzer is also activated, and after a certain time the door are closed, awaiting the next user action.

The mentioned mezzanine button is placed on the elevator cabin. When this button is selected, 'E' (error) appears on the 7-segment display and the elevator moves to the mezzanine level. Upon reaching the mezzanine, an SMS message is automatically sent from the GSM module. The message must be answered with two messages: the first message is sent empty, and the second is the desired text defined by the code (both messages are printed on the LCD display to users in the elevator).

Figure 8. Display of messages with the mezzanine option

The first sent message is printed on the LCD display as the current date and time, and by sending the second message, the desired text is printed.

In addition to the described specific actions of the two modules, the third module created is a garage door sensor system, shown in Figure 9.

Figure 9. View of the garage door after detection

Lighting is placed around the created modules, which can be turned ON or OFF by sending an SMS message to the GSM module, shown in Figure 10.

Figure 10. Display of lighting on/off via SMS messages

If the message LIGHTON is sent, the lighting around the created system is turned ON, shown in Figure 11.

Figure 11. Display of lighting after switching on

The lighting is turned off by sending the SMS message LIGHTOFF to the number located in the GSM SIM900L module.

4.2. Analysis of the functioning of the realized system when an unauthorized user accesses it
If the user reads the wrong card on the residential building, the red LEDs are turned on and the buzzer is turned on as an alarm. A message is also sent from the GSM SIM900L module to specify number about an attempt to enter the facility with incorrect data.

Figure 12. View of the system after reading the wrong card

After reading the correct card, the keyboard is accessed so that the user can enter the password to access the facility. If the user enters the wrong password via the keyboard, an SMS message is sent to the user's number via the GSM SIM900L module "*WRONG PASSWORD ENTERED:*

ATTENTION! SOMEONE IS TRYING TO ENTER". The messages sent to the number represent current events at the facility and inform the user if someone is trying to enter the facility by using the wrong RFID card or entering the wrong password.

Figure 13. Display of messages from the GSM SIM900L module

The password consists of four numbers and each entry is indicated by an asterisk displayed on the LCD, showing the number of digits entered by the user.

Any attempt to enter the facility, using the wrong RFID card and the wrong password, is accompanied by the operation of the GSM module, where messages are automatically sent to the user's number, shown in Figure 13.

5. Conclusion

Today, a very important function of every facility is security. To implement the building's security system, it is necessary to use ICT technology and smart components. In this paper, we have created such system. Initially, the system was developed programmatically, and later a schematic representation was created in the Fritzing program. Once the program code was confirmed to be operating correctly, a physical model was built, with each component in its designated place.

The implementation of the system showed that the realized system is stable and enables the safety of the object on which it is implemented, and it can be implemented in the real world with minor changes in terms of components.

Based on the realized implementation, it is concluded that the definite advantage of the designed system lies in its actual application, offering improved security, object monitoring, easier communication, and simplicity of use using ICT technology and smart components. Additionally, this system is cost-effective and enables resource optimization, further contributing to its functionality and long-term sustainability. The only potential concern with such a system could be network congestion, but as technology advances, this issue is gradually diminishing over time, making it a lesser concern.

References

[1] Ayon, S.I. and Bin Shahadat, A.S. (2019) 'Smart security box using Arduino and GSM module', *2019 IEEE International Conference on Robotics, Automation, Artificial-intelligence and Internet-of-Things (RAAICON)* [Preprint]. doi:10.1109/raaicon48939.2019.70.

[2] Alumona, T.L., Oranugo, C.O., and Onyeyili, T.I., (2019) 'GSM based Smart Security System using Arduino', *IJARCCE*, **8**(10), pp. 32–42. doi:10.17148/ijarcce.2019.81005.

[3] *What is a smart home? - smart home energy* (2021) *Smart Home Energy*. Available at: https://smarthomeenergy.co.uk/what-smart-home/ (Accessed: July 2023).

[4] *History and technological development of elevators* (2021) Scribd. Available at: https://www.scribd.com/document/187003930/Istorija-i-Tehnoloski-Razvoj-Liftova (Accessed: June 2023).

[5] Sniatala, P., Iyengar, S.S. and Ramani, S.K. (2021) 'Smart objects in cyber-physical systems', *Evolution of Smart Sensing Ecosystems with Tamper Evident Security*, pp. 9–16. doi:10.1007/978-3-030-77764-7_2.

[6] Mandić, N., (2022) *Access control and cost management using RFID technology*, University of Split, Department of Electrical Engineering, Split.

[7] Norman, T. (2012) 'Access control system installation and commissioning', *Electronic Access Control*, pp. 393–399. doi:10.1016/b978-0-12-382028-0.00026-0.

[8] Subash, L., (2013) *Design and implementation of a smart Home system.*

[9] Want, R. (2006) 'An introduction to RFID technology', *IEEE Pervasive Computing*, **5**(1), pp. 25–33. doi:10.1109/mprv.2006.2.

Adaptive Fuzzy PID Car Control using a Bilateral Teleoperator

Amel Toroman[1] and Aljo Mujčić[2]

[1]University of Bihać, Faculty of Technical Engineering, Bosnia and Herzegovina
[2]University of Tuzla, Faculty of Electrical Engineering, Bosnia and Herzegovina

E-mail: amel.toroman@unbi.ba

Abstract. This paper investigates the application of Adaptive Fuzzy PID Control in the context of car control using a bilateral teleoperator. A bilateral teleoperator allows the operator to operate a remote car using his own controls and at the same time receive feedback on the car's condition. The goal of the research is the analysis and comparison of different control methods, including PID controller, Adaptive Fuzzy PID controller, and the use of energy and wave variables of the bilateral teleoperator. The paper presents a car control model by means of a bilateral teleoperator, which was implemented in a simulation environment. Then performance comparisons of different control methods were made. The PID controller was used as the basic method, while the Adaptive Fuzzy PID controller was additionally included to achieve system adaptivity. Also, the use of energy and wave variables of the bilateral teleoperator was additionally investigated. The results of the comparison show that the use of bilateral teleoperator wave variables gave the best response of the control system. The analysis of signal waveforms and frequencies enabled more precise monitoring of the car's condition and the detection of possible problems or instability in the system. This knowledge can be of great importance for improving the performance of car control by means of a bilateral teleoperator.

1. Introduction

In today's landscape, remote control is experiencing rapid and continuous development. Its application is expanding across various domains. The driving force behind this trend is the remarkable benefits offered by the Internet in comparison to other control methods. These advantages manifest in several ways: Internet-based control is cost-effective; it allows control from any corner of the world with an Internet connection; facilitates effortless exchange of information between diverse computer systems; boasts a user-friendly interface; supports a multitude of remote-controllable systems; and offers various other advantages.

Despite the advantages mentioned earlier in favor of remote control, it also brings several disadvantages. These drawbacks primarily stem from Internet limitations and the real-time nature of control systems. Real-time systems demand prompt and complete information. However, data transmission over the Internet introduces random delays and data loss due to traffic congestion. These drawbacks compromise the control system's capabilities, leading to instability. To address these challenges, ongoing efforts focus on developing methods to mitigate the impact of delays on control system operation.

Building upon previous research conducted by the author [11], this work extends the approach by incorporating a bilateral teleoperator [1-2] that utilizes wave variables. This method is demonstrated through an illustrative example involving a one-degree-of-freedom control system, specifically the car control system.

Utilizing wave transformations to transmit remote control signals, as discussed in this paper, effectively resolves the issue of transport delays.

The system's operation was assessed through simulation and experimental verification. Two approaches were compared: the direct control mode (employing an Adaptive Fuzzy PID controller without a bilateral teleoperator) and the utilization of a bilateral teleoperator featuring energy and wave variables. This comparison encompassed both the methods employed and the corresponding results obtained.

2. Remote Control Systems

Remote control is a technique of remote control of systems in which immediate (close) control is not feasible, because it is dangerous and/or unavailable [3-4].

Today it is represented by the so-called standard manipulation, i.e., wireless control and visual feedback via camera and monitor systems.

Teleoperation is a method for controlling a vehicle or system remotely. This technique has found extensive practical applications and has given rise to fresh perspectives on interacting with distant environments [5]. Teleoperation involves controlling a robot through human intelligence and requires a suitable fusion of human and machine components, known as the Human-Machine Interface [5].

Typically, a *teleoperation system* consists of two robot manipulators interconnected in a way that allows the operator to control one of the manipulators, referred to as the Master. The Master generates commands that are then transmitted to the other manipulator, known as the Slave, for execution [5].

The application of new computer technologies in teleoperation allows the operator to get a sense of presence in a remote area.

2.1. Bilateral teleoperators

The bilateral teleoperator consists of a Master and Slave system that are interconnected by a communication channel (Figure 1). The user acts on the parent manipulator, and the child manipulator that follows the parent manipulator acts on the environment. It is desirable that the user feels the return force because then he has a better feeling for the condition on the side of the Slave part of the bilateral teleoperator. With this type of control, the problem arises when the manipulators are located in different locations. Then the manipulators must be connected by a communication channel that creates a communication delay depending on the distance between them, the type of media for data transmission, etc. A special case is the use of the Internet for data transmission when there is a variable communication delay, the size of which depends on the distance between them.

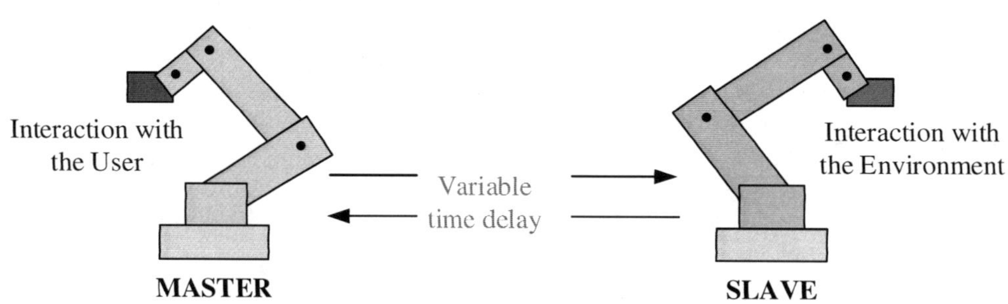

Figure 1. Structure of a bilateral teleoperator [3]

There is a Master and Slave part of the bilateral teleoperator, and a communication channel. Based on the difference between the speeds of the Master and Slave teleoperators, the regulator provides a force (moment) that acts on the Master and Slave teleoperators. The problem occurs when there is a communication delay in the communication channel.

2.1.1. Bilateral teleoperators with one degree of freedom. A system representing a mechanism with one degree of freedom will be represented by the transfer function of the car control system in the practical part of this work, Equation (10). A bilateral teleoperator will be created from the mentioned system, where the previously mentioned system (Master) will be bilaterally coupled to a similar (Slave) system by means of a communication link.

Figure 2 shows the Block diagram of a bilateral teleoperator with one degree of freedom.

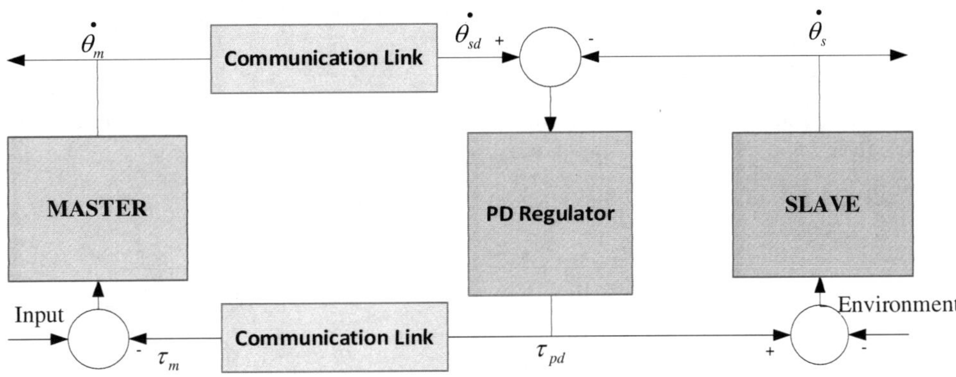

Figure 2. Block diagram of a bilateral teleoperator with one degree of freedom [5]

In order to regulate such a system, a PD regulator will be used whose transfer function is given as:

$$\tau_{pd} = \frac{B_{pd} \cdot s + K_{pd}}{s} \tag{1}$$

In order for one side of the bilateral teleoperator to monitor the other, the PD controller must be designed so that it is valid:

$$\tau_m = -\tau_{pd} \qquad \tau_s = \tau_{pd} \tag{2}$$

where is: τ_{pd} - moment of the PD regulator, K_{pd} - gain of the P regulator, B_{pd} - gain of the D regulator

The negative sign in Equation (2) is necessary to realize the interaction between the Master and Slave systems.

3. Wave variables

The Bilateral teleoperator may face instability in the presence of delays if energy variables (force and velocity) are employed for information transfer between the Master and Slave components. This finding is supported by research by Anderson and Spong (1989) [6] and Niemeyer and Slotine (1991) [7]. In light of these findings, they advocated for the utilization of wave variables for communication across the channel.

The bilateral guidance system can attain passivity (stability) even with arbitrary communication delay T by employing communication between the leading and following manipulators within the framework of wave variables (as depicted in the equation).

Wave variables constitute an expansion of passivity theory, introduced to address the influence on the system's passivity condition. A system is deemed passive if the energy input equals or exceeds the energy it outputs [3-4], implying that the system is (semi)dissipative, hence stable. Employing wave variables assures passivity by eliminating energy generation and establishing a sufficient criterion for system stability [8-10].

3.1. Transformations between energy and wave variables

Transformation of energy variables ($\dot{\boldsymbol{x}}\boldsymbol{F}$, $\boldsymbol{F}$) into wave variables ($\boldsymbol{u}$, $\boldsymbol{v}$) is given below.

$$b\dot{x}F = \frac{b}{2}(u - v)(u + v) \tag{3}$$

The parameter b defines the characteristic impedance that is associated with the wave variables and that directly affects the behavior of the system, and represents a positive number, i.e. $b>0$.

The transformation of wave variables into energy variables [3] is given by expressions (4) and (5):

$$\dot{x}_l = \sqrt{\frac{1}{2b}}\,(u_l - v_l), \qquad \dot{x}_r = -\sqrt{\frac{1}{2b}}\,(u_r - v_r) \tag{4}$$

$$F_l = \sqrt{\frac{b}{2}}\,(u_l + v_l), \qquad F_r = \sqrt{\frac{b}{2}}\,(u_r + v_r) \tag{5}$$

If the port speeds are given, then the wave transforms can be written as:

$$u_l = \sqrt{2b}\dot{x}_l + v_l, \qquad u_r = -\sqrt{2b}\dot{x}_r + v_r \tag{6}$$

$$F_l = b\dot{x}_l + \sqrt{2b}v_l, \qquad F_r = -b\dot{x}_r + \sqrt{2b}v_r \tag{7}$$

Where $\dot{x}_l$ is the energy output (variable) of the left (*master*) operator, and $\dot{x}_r$ is the energy output (variable) of the right (*slave*) operator. u_l represents the wave variable of the left (*master*) operator, while u_r represents the wave variable of the right (*slave*) operator. v_l stands for the velocity variable of the left (*master*) operator, and v_r represents the velocity variable of the right (*slave*) operator.

Applying the above Equations, a Block diagram of wave variables is obtained (Figure 3).

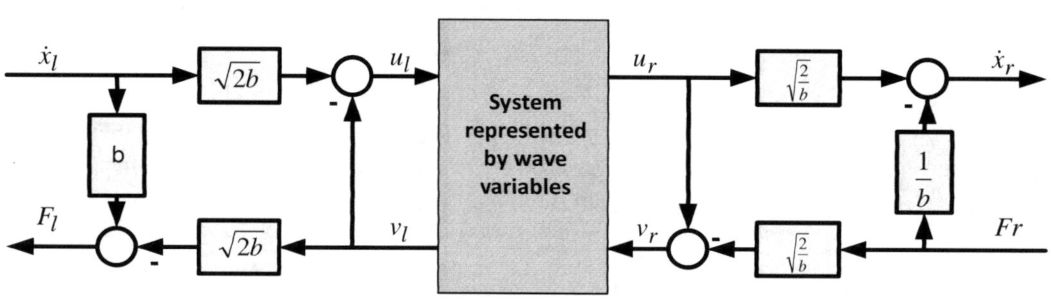

Figure 3. Block diagram of wave variables [5]

From energy variables to wave variables using wave transformations, left side in Figure 3, and then from wave variables again to energy variables using wave transformations, right side in Figure 3.

According to the Block diagram shown in Figure 3, simulation schemes of wave variables for the Master and Slave car control system were created.

3.2. Application of wave variables to a bilateral teleoperator with one degree of freedom

Increasing the communication delay between the Master and Slave components of the bilateral teleoperator triggers instability, rendering it unsuitable for scenarios where substantial communication delays prevail.

To mitigate this issue, the communication system needs to be passivated [5]. Prior research has demonstrated that wave variables guarantee the passivity of the communication system. When wave variables are utilized, the communication system becomes not only passive but also devoid of losses, irrespective of the communication delay [5].

Figure 4 shows a bilateral teleoperator with wave variables.

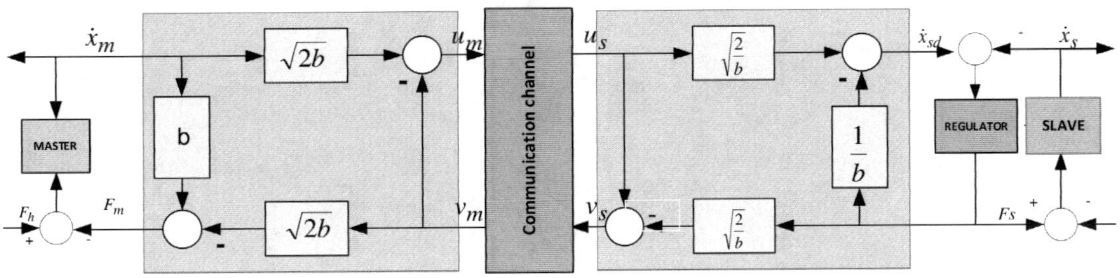

Figure 4. Bilateral teleoperator with wave transformations [5]

Where is :

- F_h force given by the operator,
- F_e force with which the environment acts on the Slave system,
- F_m force reflected back to the Master from the Slave system,
- F_s information force sent from Slave to Master system,
- $\dot{x}_{sd}$ speed at the output from the wave transformations on the side of the Slave system,
- u i v wave variables,
- $\dot{x}_m$ i $\dot{x}_s$ speeds of the Master and Slave systems respectively,
- b characteristic impedance.

By inserting wave transformations on the side of the Master and Slave systems, the passivity of the communication channel is achieved [5].

When both the Slave system and the communication channel exhibit passivity, the bilateral teleoperator attains stability. This is exemplified in the applied section of the paper, featuring a one-degree-of-freedom car control system employing a bilateral teleoperator with wave variables, as depicted in Figure 10.

4. Adaptive Fuzzy PID Car Control using a Bilateral Teleoperator

Automatic cruise control is an excellent example of a feedback control system found in many modern cars. The purpose of the cruise control system is to maintain a constant car speed despite external disturbances, such as wind changes or road gradients. This is achieved by measuring the speed of the car, i.e., the control of the car by comparing it with the desired or reference speed [11].

Figure 5. Mechanical car control system [11]

IOP Conf. Series: Materials Science and Engineering 1298 (2023) 012026 doi:10.1088/1757-899X/1298/1/012026

Figure 5 shows the mechanical system of car dynamics. A steering force, **f**, acts on a car of mass **m**. The force **f** represents the force generated at the road/tire interface. It is assumed that the resistance forces, **bv**, due to rolling resistance and wind resistance, vary linearly with the speed of the car, **v** act in the direction opposite to the movement of the vehicle.

On the basis of the above, a system of first-order mass dampers is obtained. Adding forces in the *x* direction by applying Newton's 2nd law leads to the system Equation (8):

$$m\frac{dv}{dt} + bv(t) = f(t) \tag{8}$$

Taking the Laplace transform of the differential Equation (8) and assuming zero initial conditions, we arrive at the transfer function of the cruise control system, Equation (9):

$$ms\dot{V}(s) + bV(s) = F(s) \tag{9}$$

$$G(s) = \frac{1}{ms + b} = \frac{1}{1000s + 50} \tag{10}$$

So, in this paper, the problem of driving a car, i.e., its movement in relation to a reference value, will be realized. The mass of the car is 1000 [kg], while the damping is 50 [Ns/m]. As a reference value, a sinusoid of amplitude 1.6 is supplied, and at the output it is desired to achieve the movement of the car in relation to this reference value.

5. Model Testing, Comparison, and Analysis of Results

A comparison of car movement was created using two types of control, i.e., using a PID controller and an Adaptive Fuzzy PID controller [12]. Based on the two previously mentioned control methods, a comparison of car movement without and with the use of a bilateral teleoperator was made. With the bilateral teleoperator, control was implemented using energy and wave variables.

As shown in Figure 2, the bilateral teleoperator uses a PD controller to control energy and wave variables. In addition to the PD controller, the PID controller was also employed in the study as a control system, along with its application in Adaptive Fuzzy control.

5.1. Car Control using PID and Adaptive Fuzzy PID Controller

Figure 6 shows the Car Control System using a PID Controller, as well as using an Adaptive Fuzzy PID controller. Two values were supplied as a reference signal, a step signal of amplitude 1, as well as a sinusoid of amplitude 1.6 and frequency 0.1 rad/s.

As can be seen from Figure 6, a sinusoid was chosen as the priority reference signal due to the dynamics of the car's movement.

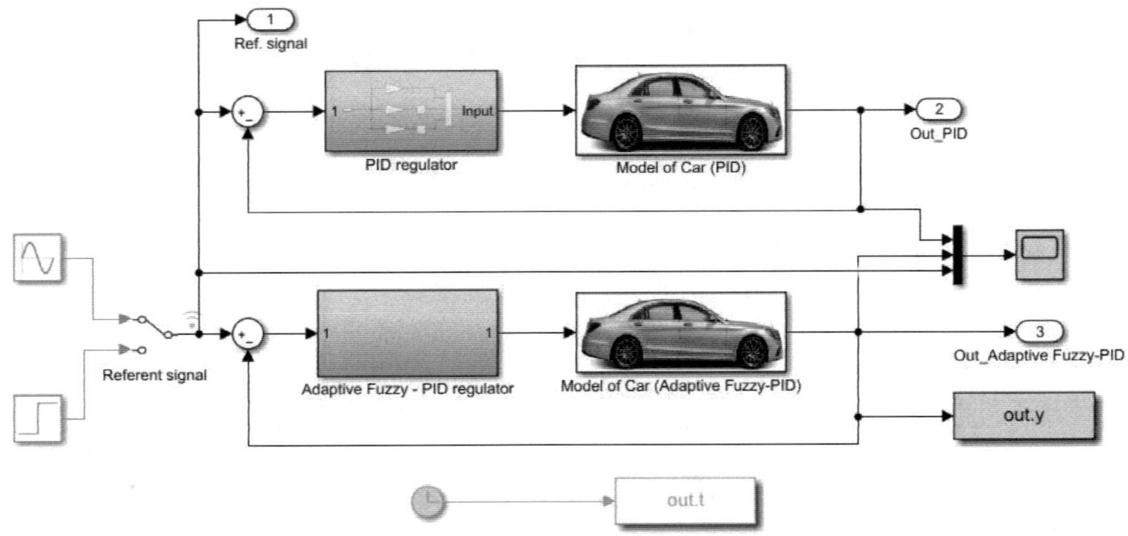

Figure 6. Car Control Model (PID and Adaptive Fuzzy PID Controller)

The parameters of the PID controller (K_p=750, K_i=396, K_d=631.05) were obtained using the Ziegler-Nichols method for open-loop control [13], which was applied to the car model, Equation (10).

To improve the created system, and to improve the results obtained using the PID controller, an analysis of the created system was performed using the Adaptive Fuzzy PID controller, whose model is shown in Figure 7.

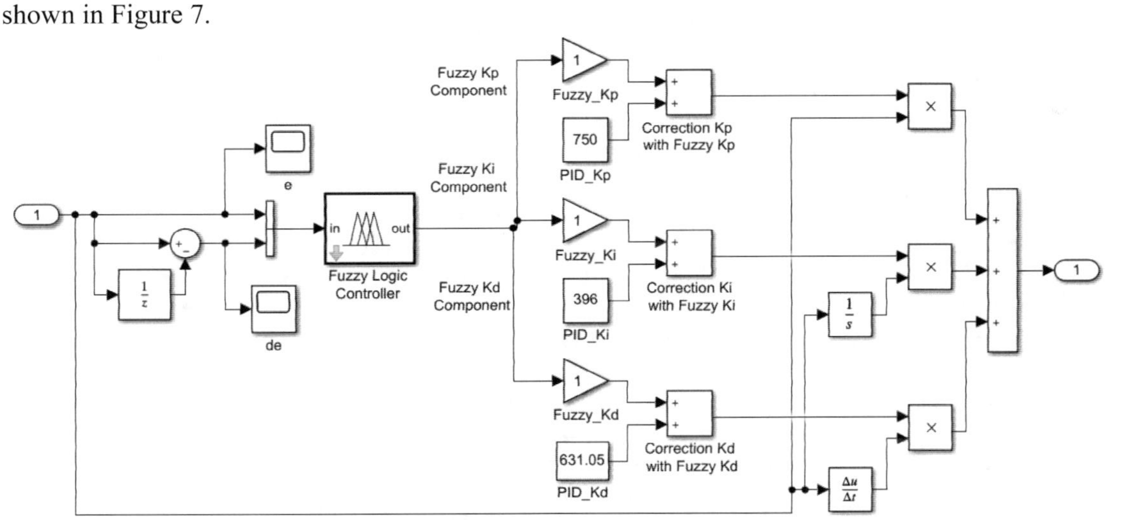

Figure 7. Adaptive Fuzzy PID controller [11]

The Fuzzy Controller was created with two inputs (*error* and *error change*) and one output (*control*), and 49 fuzzy rules were created for system control.

Based on the previously presented models and the entered parameters, a simulation of the car's control was obtained, lasting 100 seconds, as shown in Figure 8.

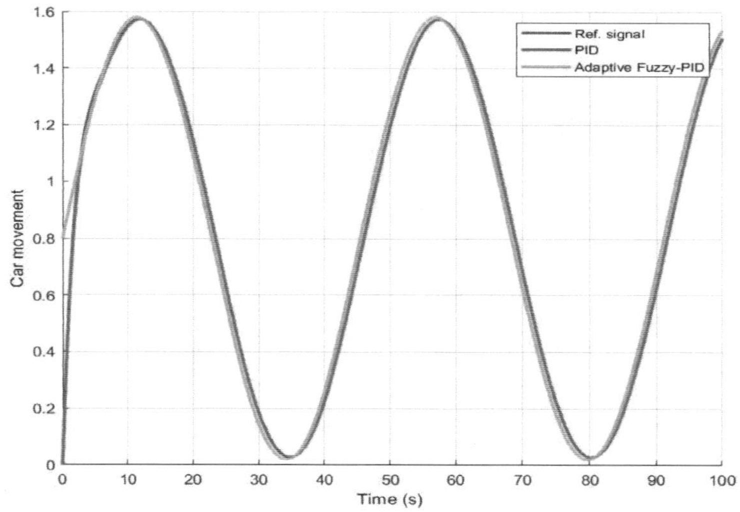

Figure 8. Displaying the results of the Car Control Model (PID and Adaptive Fuzzy PID controller)

As can be seen from the simulation results shown in Figure 8, the model is well designed, and the system delays are minimal compared to the reference signal.

As previously mentioned, the reference signal is a sinusoid of amplitude 1.6 and frequency 0.1 rad/s, and the created car control system follows the path of the reference signal to a good extent, which is shown in Figure 8 by the minimal delay of the received signals in relation to the reference signal, as using PID controller, as well as using an Adaptive Fuzzy PID controller.

5.2. Car Control using a Bilateral Teleoperator (Energy Variables)

Figure 9 illustrates a car control system implemented with a bilateral teleoperator employing energy variables. A sinusoidal signal with an amplitude of 1.6 and a frequency of 0.1 rad/s was employed as the reference signal, mirroring the approach adopted for the car control model utilizing both PID and Adaptive Fuzzy PID controllers. Delays of 2 seconds have been added at the entrance, as well as at the exits from the Master system.

In the bilateral teleoperators, the delays T_l (delay on the *Master* side) and T_r (delay on the *Slave* side) are often added or considered to compensate for the real communication delays that may occur in the communication between the Master and Slave parts of the system. There are several reasons for adding these delays:

- **Compensation for Communication Delay**: In the real world, the signal transmission speed across the communication channel between the Master and Slave parts of the bilateral teleoperator is not infinitely fast. There are certain delays due to the time required to transmit signals through the communication infrastructure such as networks or wireless connections. By adding the Tl and Tr delays to the model of the bilateral teleoperator, an attempt can be made to simulate these real delays to better understand how the system will react under real communication conditions.
- **System Stability**: Adding delays can help maintain system stability. If there is a delay in communication, it can affect the system dynamics and lead to oscillations or unpredictable reactions. By incorporating the Tl and Tr delays into the model, it is possible to test the system's stability under communication delay conditions and adjust controller parameters or algorithms to maintain stability.
- **More Realistic Modeling:** Adding delays to the model of the bilateral teleoperator enables a more realistic representation of the actual system's behavior under real communication conditions. This is important for assessing performance and system capabilities in the real world.

In conclusion, adding the Tl and Tr delays to the model of the bilateral teleoperator allows for a better understanding of the system's behaviour under real communication conditions and facilitates the optimization of parameters and algorithms to achieve the desired performance and system stability.

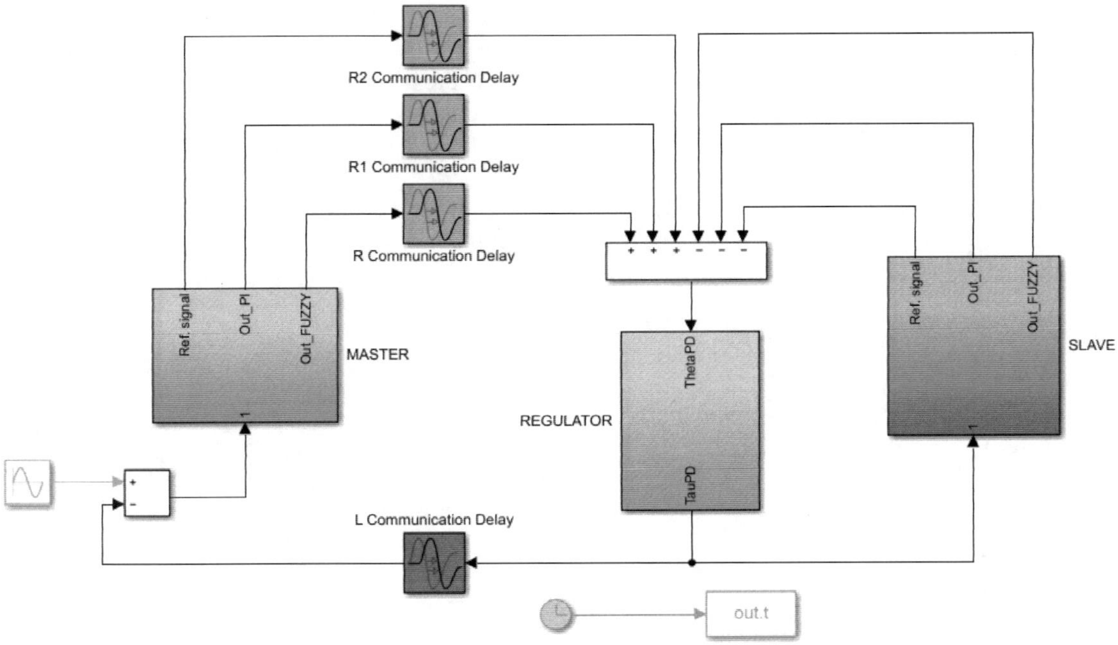

Figure 9. Car Control Model using a Bilateral Teleoperator (Energy Variables)

The Master and Slave system model is shown in Figure 6.

Based on the previously presented models and entered parameters, a simulation of the car's control was obtained using a bilateral teleoperator with energy variables, for a time duration of 100 seconds, as shown in Figure 10.

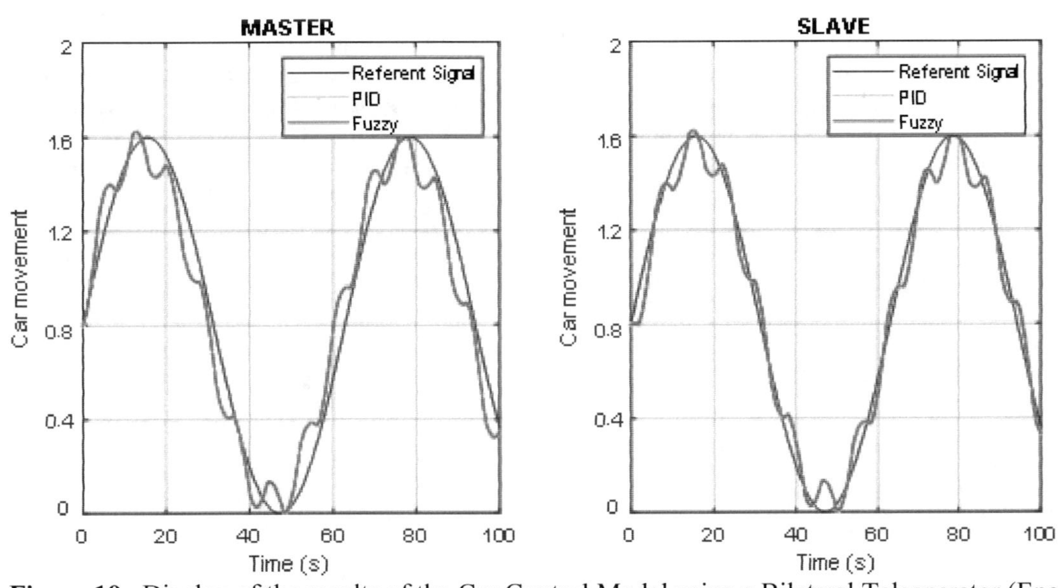

Figure 10. Display of the results of the Car Control Model using a Bilateral Teleoperator (Energy Values)

IOP Conf. Series: Materials Science and Engineering 1298 (2023) 012026 doi:10.1088/1757-899X/1298/1/012026

The simulation results presented in Figure 10 indicate that when employing a bilateral teleoperator with energy variables, minor delays are observed in relation to the reference signal. These delays are evident on both the Master and Slave sides of the system.

5.3. Car Control using a Bilateral Teleoperator (Wave Variables)

Figure 11 depicts the car control system implemented with a bilateral teleoperator employing wave variables. Similar to the car control models utilizing PID and Adaptive Fuzzy PID regulators, as well as the bilateral teleoperator with energy variables, a sinusoidal signal with an amplitude of 1.6 and a frequency of 0.1 rad/s was utilized as the reference signal. Delays of 2 seconds have been added at the entrance, as well as at the exits from the Master system.

According to the scheme of the bilateral teleoperator with wave transformations (Figure 4), the model shown in Figure 11 was created, adapted for the created car control system.

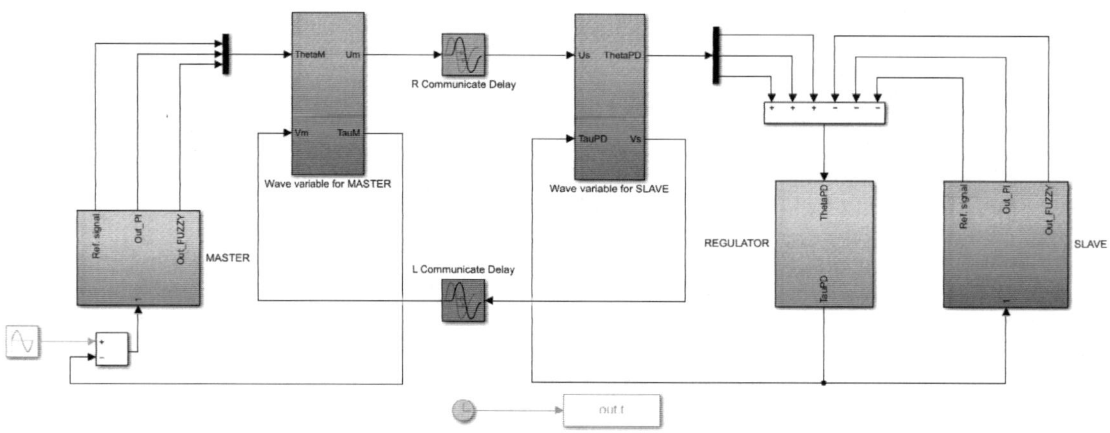

Figure 11. Car Control Model using Bilateral Teleoperator (Wave Variables)

As can be seen from Figure 11, the waveform variables are set on both the Master and Slave side of the system, to reduce the system delay.

Based on the block system of wave variables shown in Figure 4, models for wave variables were created both on the side of the Master and on the side of the Slave system.

The car control system implementing a bilateral teleoperator with wave variables employed a Master and Slave model (depicted in Figure 6), akin to the approach involving energy variables. Additionally, control was executed utilizing a PD regulator.

Utilizing the models and input parameters previously discussed, a simulation of the car's motion was conducted. This simulation employed a bilateral teleoperator featuring wave variables, spanning a time duration of 100 seconds, and the results are visualized in Figure 12.

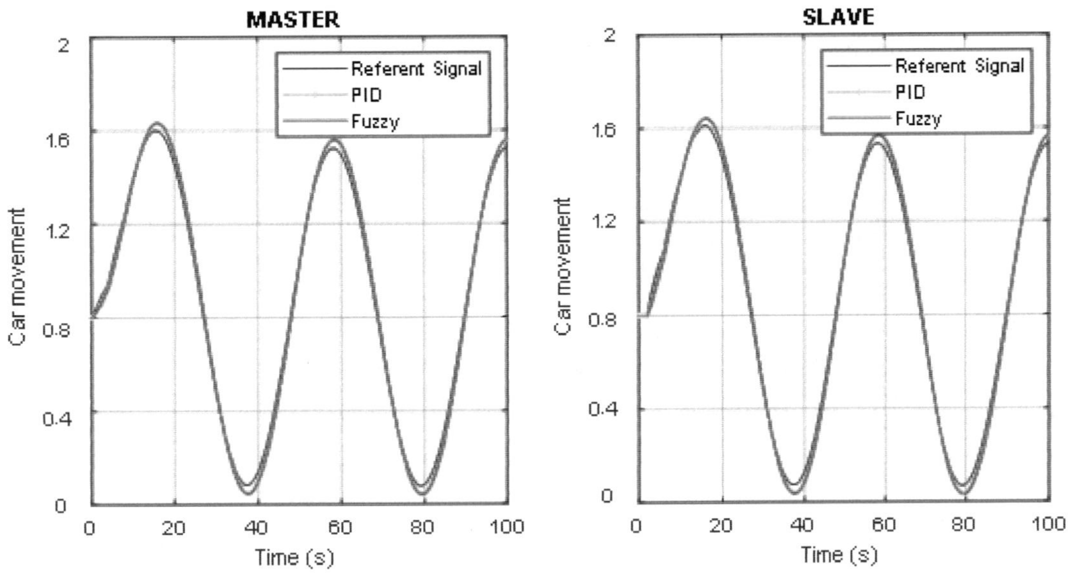

Figure 12. Display of the results of the Car Control Model using a Bilateral Teleoperator (Wave Variables)

Drawing from the simulation outcomes illustrated in Figure 12, the effects of employing wave variables within the bilateral teleoperator are evident. The car control system's response closely aligns with the reference signal, resulting in minimized delays. The achieved values successfully fulfil the targeted objective – eliminating signal delays in the car control system actualized using wave variables in the bilateral teleoperator.

6. Conclusion

This paper delves into the investigation of Adaptive Fuzzy PID control for a car using a bilateral teleoperator. The aim was to explore how this fusion of techniques could enhance car handling and contribute to a more comfortable ride.

Adaptive Fuzzy PID control is employed to fine-tune the PID controller, enhancing the precision and effectiveness of the control system. Fuzzy Logic facilitates the handling of uncertain and non-standard scenarios, while the PID controller ensures stability and rapid adjustments to altering driving conditions. Merging adaptability with Fuzzy Logic empowers the PID controller to be tailored and optimized for diverse driving conditions.

The role of the bilateral teleoperator in this paper is to facilitate remote control of the car using a teleoperation device. When coupled with Adaptive Fuzzy PID control, the bilateral teleoperator empowers the driver to effortlessly oversee the vehicle from a distant location. Meanwhile, the PID controller effectively adjusts to driver inputs and varying driving conditions.

Analyzing the outcomes reveals that the incorporation of wave variables within the bilateral teleoperator notably diminishes the delay between the Master and Slave systems. This achievement in delay reduction enables the driver to promptly receive real-time information about the car's status and react swiftly to alterations in driving conditions. Consequently, this enhances steering accuracy and responsiveness, ultimately contributing to safer and more pleasurable driving experiences.

Ultimately, research involving Adaptive Fuzzy PID control for cars, coupled with a bilateral teleoperator, indicates that integrating wave variables considerably curtails the delay between the Master and Slave systems, effectively minimizing it. Furthermore, energy variables assume a pivotal role in upholding system stability and dependability. Delving deeper into the exploration of these variables could pave the way for further advancements in the realm of car control.

In the feature, upcoming research will concentrate on refining the fusion of energy and wave variables to attain optimal performance within the bilateral teleoperator. Incorporating advanced communication and energy technologies has the potential to yield a more efficient and dependable car control system.

References

[1] Zheng, D., Pu, Y. and Bai, J. (2022) 'Analysis and control for bilateral teleoperation systems with part active power', *Communications in Computer and Information Science*, pp. 69–76. doi:10.1007/978-981-19-7946-0_7.

[2] Wang, H. (2021) 'Bilateral control of teleoperator systems with time-varying delay', *Automatica*, 134, p. 109707. doi:10.1016/j.automatica.2021.109707.

[3] Velagić, J. (2021) Remote Control System, *Faculty of Electrical Engineering Sarajevo*. Available at: https://people.etf.unsa.ba/~jvelagic/laras/dok/Predavanje%2012.pdf (Accessed:July 2023).

[4] Petrović, I. (2016) Control of Robotic System, *Faculty of Electrical Engineering and Computing Zagreb*. Available at: https://www.fer.hr/_download/repository/URS_Kovacic_1ciklus.pdf (Accessed: July 2023).

[5] Mujčić, E. (2013) 'Upravljanje teleoperacijskim sustavom zasnovano na valnim varijablama uz promjenjivo komunikacijsko kašnjenje', *Faculty of Electrical Engineering and Computing Zagreb*, Dissertation.

[6] Anderson, R.J. and Spong, M.W. (1989) 'Bilateral control of teleoperators with Time Delay', *IEEE Transactions on Automatic Control*, **34**(5), pp. 494–501. doi:10.1109/9.24201.

[7] Niemeyer, G. and Slotine, J.-J.E. (1991) 'Stable adaptive teleoperation', *IEEE Journal of Oceanic Engineering*, **16**(1), pp. 152–162. doi:10.1109/48.64895.

[8] Alis, M. T. (2007) Expansion and Implementation of the Wave Variable Method in Multiple Degre-OfFreedom System, *Dissertation*.

[9] Niemeyer, G. (1996) 'Using Wave Variables in Time Delayed Force Reflecting Teleoperation', *Dept of Aeronautics and Astronautics, Massachusetts Institute of Technology*. Ph. D Thesis.

[10] Petrović, I. (2009) Upravljanje robotskim sustavima, *Faculty of Electrical Engineering and Computing* Zagreb.

[11] Toroman, A. and Vojić, S. (2021) 'Adaptive car control system based on a predictive model', *IOP Conference Series: Materials Science and Engineering,* **1208**(1), p. 012040. doi:10.1088/1757-899x/1208/1/012040.

[12] Xin Feng (2004) 'Design of adaptive fuzzy PID tuner using optimization method', *Fifth World Congress on Intelligent Control and Automation (IEEE Cat. No.04EX788)* [Preprint]. doi:10.1109/wcica.2004.1342035.

[13] Stroski, P.N. (2019) 'PID controller tuning: Ziegler-Nichols Methods' *Electrical E, Library.com*. Available at: https://www.electricalelibrary.com/en/2019/11/06/pid-controller-tuning-ziegler-nichols-methods/ (Accessed: July 2023).

[14] Petrović, I. (2016) 'Control of Robotic System' *Faculty of Electrical Engineering and Computing Zagreb*. Available at: https://www.fer.unizg.hr/_download/repository/URS_Petrovi c_2015-16_Predavanje_02.pdf (Accessed: July 2023).

Raspberry Pi based Web Monitoring Smart Security System

Amel Toroman, Una Drakulić, Amel Džanić and Azra Kapić
University of Bihać, Faculty of Technical Engineering, Bosnia and Herzegovina

E-mail: amel.toroman@unbi.ba

Abstract. The surveillance system has been utilized over the years on public property in order to provide security and prevent unauthorized entrance. Lately, more private homeowners are choosing to implement security systems. There are many problems with the video surveillance system. These disadvantages are the indistinctiveness of the pictures/video and the need for a lot of storage space to save surveillance information. This paper describes the design and implementation of a low-cost Web Monitoring System based on Raspberry Pi. Also, a web application is designed for the purposes of controlling the camera and for live streaming. The live stream from a web camera can be viewed from any web browser, even mobile, in real-time. Controlling the camera is possible with the web application using four commands (up/down, left/right). Also, the camera moves depending on the detection of an object located in the visible area of the PIR Motion sensor, thereby issuing a warning message about object detection.

1. Introduction

Nowadays, system monitoring is an important factor from the aspect of security. There are many problems with the video surveillance system. These disadvantages are the indistinctiveness of the pictures/video and the need for a lot of storage space to save surveillance information [1].

Modern surveillance security systems use motion detection [2-3]. The motion detection algorithm for live camera streaming allows the system to analyze incoming images from cameras and recognize when movement occurs. When movement occurs the surveillance security system starts to record video and store the surveillance information. In research [3] authors describe the monitor and control based on the Raspberry Pi for designing smart homes through IoT [6]. In research [4] authors write about IoT-based smart human traffic monitoring systems [7] using Raspberry Pi. In research [5] authors describe the use of Raspberry Pi and IoT for traffic lights in the smart city. These papers are describing the use of the Raspberry Pi for the surveillance system for different purposes.

This paper describes the design and implementation of the Web Monitoring Security System based on Raspberry Pi. The proposed system is a low-cost and uses a designed web application for controlling the camera and for live streaming. Using the website, it is possible to perform remote monitoring and control of the camera, and in addition to have an overview of the live streaming.

On the web application there are commands for controlling the camera in the desired directions (up/down, left/right). Also, the camera will move in case the PIR (Passive Infrared) Motion sensor detects any of the objects in its visible area. In case of detection of an object using the PIR Motion sensor, a message that the object has been detected will be displayed on the web application, otherwise a message that the object has not been detected will be displayed.

2. Design and implementation

The block diagram for the proposed system is shown in Figure 1. Figure 1 shows that the inputs for the Raspberry Pi microprocessor are the state of four buttons, which are defined on a designed web application. These buttons are defined and programmed to enable the user to choose the camera direction (up/down, left/right). Also, two ULN2003A modules which are connected to step motors are used because one is needed for tilt the camera and other for pan the camera. Power supply for system is +5V.

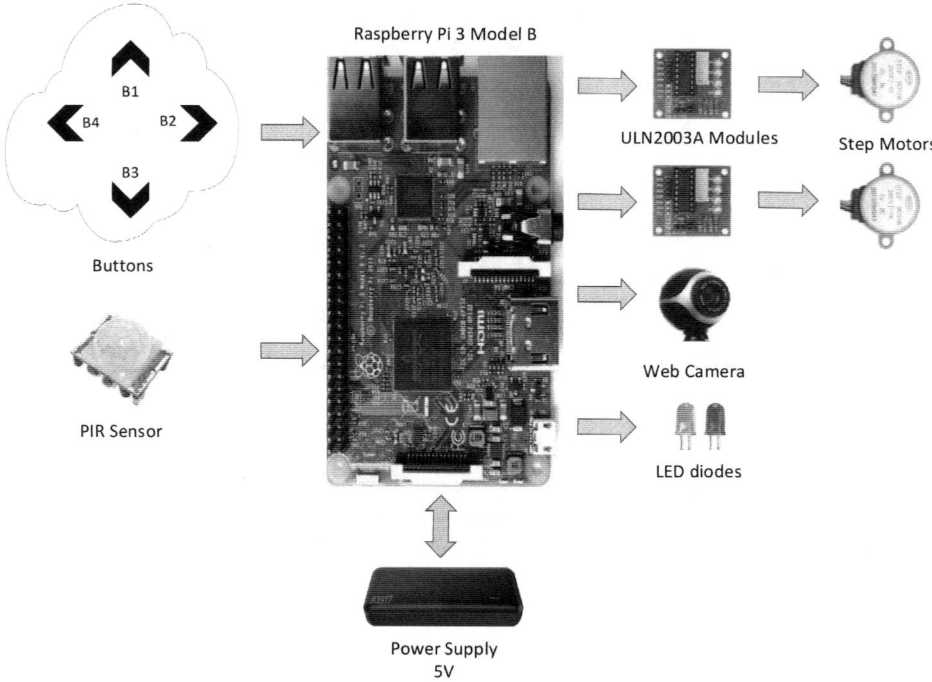

Figure 1. Block diagram for a Raspberry Pi based Web Monitoring System

The proposed system uses a wireless communication for receiving/transmitting the data from/to designed web application, and serial communication for transmitting the data to UNL2003A modules and camera. The working principle of the proposed system is shown using algorithm in Figure 2.

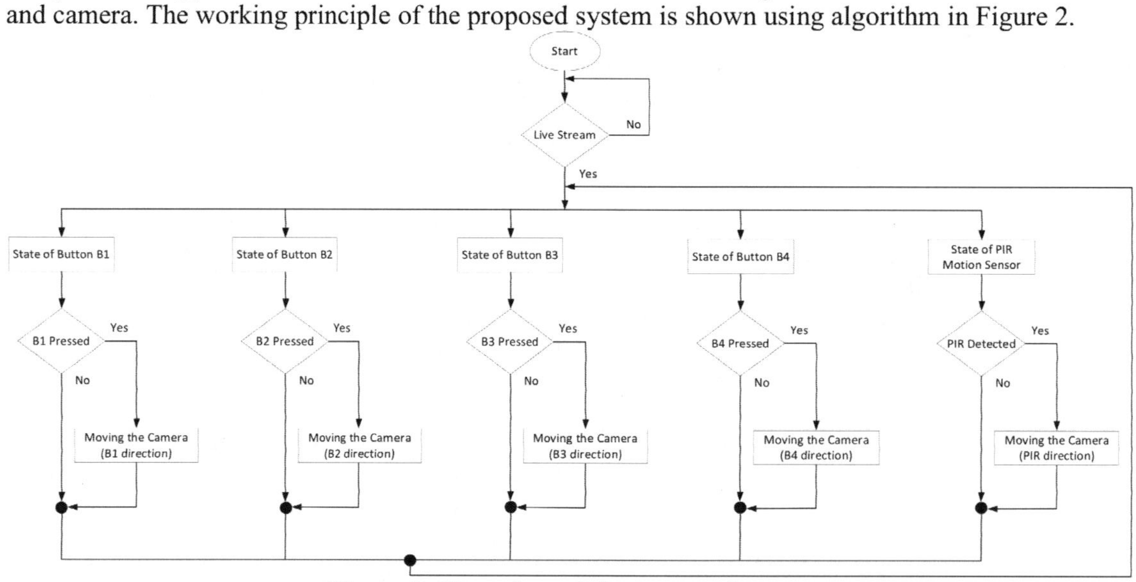

Figure 2. Flow chart of the proposed system

IOP Conf. Series: Materials Science and Engineering 1298 (2023) 012027 doi:10.1088/1757-899X/1298/1/012027

Based on the flow diagram in Figure 2, it can be seen that the live camera feed is continuous and in real time. If the user wants to move the camera up, down, left or right, he selects button B1, B3, B4 or B2, or activates the PIR Motion sensor. After the user presses the desired button, the Raspberry Pi Microprocessor receives the command for that direction, processes it, and passes it to the UNL2003A modules to rotate the stepper motors. Stepper motors will move and tilt the camera in the selected direction. The camera will also move if the PIR Motion sensor is activated. If the object is in the visible area of the PIR Motion sensor, the camera will be directed towards the specified object, and it will move as the object moves. If the PIR Motion sensor has detected an object, it will be signalled by turning on the green LED, otherwise, the red LED will turn on.

For practical implementation, the Raspberry Pi 3 Model B Microprocessor was used. Two UNL2003A modules were connected to the Raspberry Pi development breadboard, one on pins 26, 24, 22, and 18, and the other on pins 23, 21, 19, and 15. These modules are used for the correct operation of the stepper motor. A Web camera is connected to the Raspberry Pi via a USB port. The designed system with control part is shown in Figure 3. The PIR Motion sensor is connected to pin 17 of the Raspberry PI Microprocessor, while the LEDs (green, red) are connected to pins 36 and 38.

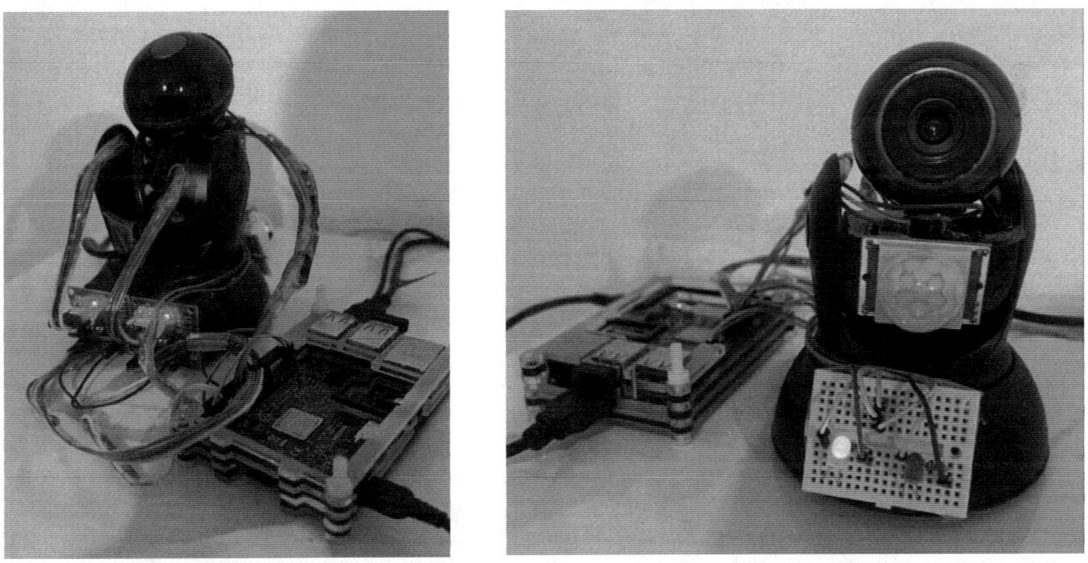

a) b)

Figure 3. The designed system with control part: a) back side b) front side

The housing of project is small, lightweight, and compact. It is designed using a 3D printer. In housing are placed two UNL2003A modules, two step motors, and on top is placed a camera. An additional power supply is not installed on the case, since the power supply system is created from the Raspberry Pi controller.

The placement of step motors inside the housing is shown in Figure 4. One step motor is used to pan the camera, and the second for tilt the camera, and the stepper motors are controlled via ULN2003A modules. The angles of possible rotations are shown in Figure 4 b).

IOP Conf. Series: Materials Science and Engineering 1298 (2023) 012027 doi:10.1088/1757-899X/1298/1/012027

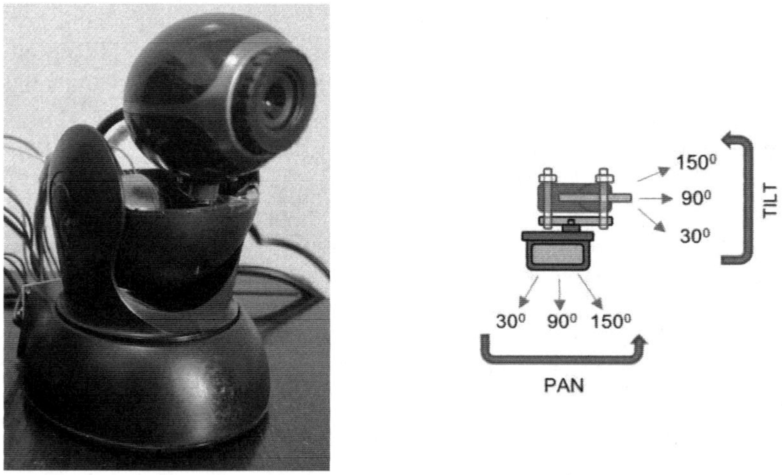

a) b)

Figure 4. The placement of step motors to pan and tilt the camera: a) final look b) rotation angles

The different sensors can be used for motion detection, and the most commonly used are PIR Motion sensors, Microwave (radar) sensors and Ultrasonic sensor. Here is an overview of their characteristics, advantages and disadvantages:

1. **PIR (Passive Infrared) sensors**: *Characteristics*: These sensors detect changes in thermal radiation, which means that they register movement based on changes in the temperature of objects in the environment. *Advantages*: They are effective for detecting people and animals, respond well to rapid temperature changes and are usually cheaper. *Disadvantages*: PIR sensors are not always accurate in identifying the type of movement or the number of objects, and are not able to provide information about the shape or size of an object.
2. **Microwave (radar) sensors**: *Characteristics*: Microwave sensors emit microwaves and analyse reflections from objects in the environment. They register changes in the speed of objects, which makes them effective for motion detection. *Advantages*: Microwave sensors can work through obstacles and weather, are not sensitive to light or darkness, and can detect a large number of objects. *Disadvantages*: These sensors are usually more expensive than PIR sensors, and there is a possibility of false positives due to reflections from walls, doors or windows.
3. **Ultrasonic sensors**: *Characteristics*: They emit ultrasonic waves and measure the time it takes for them to bounce off objects and return. Based on this time, the presence and distance of objects can be determined. *Advantages*: They accurately detect movement and can measure the distance from the object. They do not react to temperature changes. *Disadvantages*: There is a possibility of reflections that can cause false results. Also, exposure to loud sounds can cause interference with the sensors. The choice of sensor depends on the specific requirements of the project.

In this project, the integration of a PIR Motion sensor with a camera and a Raspberry Pi platform enables an innovative approach to object detection and tracking. Due to its best characteristics, the PIR Motion sensor was chosen for use in the practical implementation of this work.

a) b)

Figure 5. The placement of PIR Motion sensor on the camera: a) final look b) angles of detectable area

The PIR Motion sensor is located on the camera housing, which is used for motion detection. In Figure 5 a) you can see the position of the PIR Motion sensor on the camera housing, while in Figure 5 b) you can see the position of the visibility angles that cover the operation of the PIR sensor. Based on the above, it can be seen that the visible, that is the working range of the PIR Motion sensor is up to 110°, and adjustable range is from 3 meters to 7 meters. That is, the minimum area covered by the PIR Motion sensor from a distance is 3 meters, and the maximum is 7 meters.

The PIR Motion sensor plays a key role in fast and efficient motion detection in the camera's surroundings. This sensor stands out for its features and advantages over other detection sensors, providing a high level of reliability and energy efficiency. The advantage of PIR Motion sensors:

Fast motion detection: PIR Motion sensors react to changes in thermal radiation, enabling immediate detection of the presence of objects. This ability is crucial for quickly pointing the camera towards detected objects.

Energy efficiency: PIR Motion sensors are passive, which means they do not require energy to emit a signal. This reduces power consumption and makes them ideal for power-constrained applications such as Raspberry Pi devices.

Accuracy of detection: PIR Motion sensors are aimed at detecting thermal changes. This reduces the likelihood of false positives that other sensors, such as ultrasonic sensors, can cause.

Simple integration: The implementation of the PIR Motion sensor in the project is relatively simple, it does not require complex calibrations and additional components. This makes development easier and speeds up project creation time.

Also, the web application is designed for controlling the direction of camera and for Live Stream. The design is very simple and user friendly (see Figure 16 and Figure 17).

This web application was designed in Python using the *Flask framework*. Template *index.html* is in charge of the entire interaction through the implementation of the form with the *POST method*, which is used to control camera direction. This *POST method* is given below in Figure 6.

```
<form method="post" action="{{ url_for('tasks') }}">
  <table>
    <tr>
      <td></td>
```

```
      <td><button class="btn btn-outline-dark" value="up" name="u"><span style="font-
size: x-large;">B1</span></button></td>
      <td></td>
    </tr>
    <tr>
      <td><button class="btn btn-outline-dark" value="left" name="l"><span style="font-
size: x-large;">B4</span></button></td>
      <td></td>
      <td><button class="btn btn-outline-dark" value="right" name="r"><span
style="font-size: x-large;">B2</span></button></td>
    </tr>
    <tr>
      <td></td>
      <td><button class="btn btn-outline-dark" value="down" name="d"><span style="font-
size: x-large;">B3</span></button></td>
      <td></td>
    </tr>
  </table>
</form>
```

Figure 6. POST method for control camera direction

From the given code, it can be seen that by pressing the desired button, the user calls the function tasks in Python, which then, based on the value of the name attribute of the button, recognizes in which direction to move the camera. The entire code of the function is in the attachment shown in Figure 7.

```
@app.route('/requests',methods=['POST','GET'])
def tasks():
    if request.method == 'POST':
        if request.form.get('u')=='up':
            print('Command for UP')
            for i in range (1,15,1):
                 GPIO.output(motor_channel, (GPIO.HIGH,GPIO.LOW,GPIO.LOW,GPIO.HIGH))
                        sleep(0.02)
              GPIO.output(motor_channel, (GPIO.LOW,GPIO.LOW,GPIO.HIGH,GPIO.HIGH))
              sleep(0.02)
              GPIO.output(motor_channel, (GPIO.LOW,GPIO.HIGH,GPIO.HIGH,GPIO.LOW))
              sleep(0.02)
              GPIO.output(motor_channel, (GPIO.HIGH,GPIO.HIGH,GPIO.LOW,GPIO.LOW))
              sleep(0.02)
        elif request.form.get('l') =='left':
            print('Command for LEFT')
            for i in range (1,15,1):
              GPIO.output(motor_channel2, (GPIO.HIGH,GPIO.LOW,GPIO.LOW,GPIO.HIGH))
              sleep(0.02)
              GPIO.output(motor_channel2, (GPIO.LOW,GPIO.LOW,GPIO.HIGH,GPIO.HIGH))
              sleep(0.02)
              GPIO.output(motor_channel2, (GPIO.LOW,GPIO.HIGH,GPIO.HIGH,GPIO.LOW))
              sleep(0.02)
              GPIO.output(motor_channel2, (GPIO.HIGH,GPIO.HIGH,GPIO.LOW,GPIO.LOW))
              sleep(0.02)
        elif request.form.get('r') == 'right':
            print('Command for RIGHT')
            for i in range (1,15,1):
              GPIO.output(motor_channel2, (GPIO.HIGH,GPIO.LOW,GPIO.LOW,GPIO.HIGH))
              sleep(0.02)
              GPIO.output(motor_channel2, (GPIO.HIGH,GPIO.HIGH,GPIO.LOW,GPIO.LOW))
              sleep(0.02)
              GPIO.output(motor_channel2, (GPIO.LOW,GPIO.HIGH,GPIO.HIGH,GPIO.LOW))
              sleep(0.02)
              GPIO.output(motor_channel2, (GPIO.LOW,GPIO.LOW,GPIO.HIGH,GPIO.HIGH))
              sleep(0.02)
        elif request.form.get('d') == 'down':
            print('Command for DOWN')
            for i in range (1,15,1):
              GPIO.output(motor_channel, (GPIO.HIGH,GPIO.LOW,GPIO.LOW,GPIO.HIGH))
              sleep(0.02)
              GPIO.output(motor_channel, (GPIO.HIGH,GPIO.HIGH,GPIO.LOW,GPIO.LOW))
              sleep(0.02)
```

```
              GPIO.output(motor_channel,(GPIO.LOW,GPIO.HIGH,GPIO.HIGH,GPIO.LOW))
              sleep(0.02)
              GPIO.output(motor_channel,(GPIO.LOW,GPIO.LOW,GPIO.HIGH,GPIO.HIGH))
              sleep(0.02)
    elif request.method=='GET':
        return render_template('index.html')
    return render_template('index.html')
```

Figure 7. Function code to control the camera pan motors

As can be seen from the code given in Figure 7 the motors move the camera relative to the selected position (tilt motor moves the camera up/down and pan motor moves the camera left/right).

```
import RPi.GPIO as GPIO
GPIO.setwarnings(False)
GPIO.setmode(GPIO.BOARD)
GPIO.setup(13, GPIO.IN)          #Read output from PIR motion sensor
GPIO.setup(36, GPIO.OUT)         #LED output pin (green)
GPIO.setup(38, GPIO.OUT)         #LED output pin (red)
while True:
i=GPIO.input(13)
if i==0:                         #When output from motion sensor is LOW
print "The Object is not detected!",i
GPIO.output(38)                  #Turn ON red LED
elif i==1:                       #When output from motion sensor is HIGH
print "The Object is detected!",i
GPIO.output(36)                  #Turn ON green LED
```

Figure 8. Function code to control the PIR Motion sensor

As can be seen from Figure 8, the PIR Motion sensor is connected to pin 13. Activating the PIR Motion sensor turns on the green LED (pin 36) and the message "The Object is detected!" is displayed, while in the case when the PIR sensor is not active the red LED (pin 38) is on and the message "The Object is not detected" is displayed.

Pictures from 9 to 13 shows the functional code for object control and tracking using "dlib". **Tracker** in this context represents an object created from the **dlib.correlation_tracker()** class. This class enables object tracking in a series of images using the correlation technique.

```
import RPi.GPIO as GPIO
import time
import cv2
import dlib
from imutils.video import VideoStream

# Configuration of GPIO and PIR pins for DC motors
PIR_PIN = 13
MOTOR_UP_DOWN_PIN = 15
MOTOR_LEFT_RIGHT_PIN = 16
GPIO.setup(MOTOR_UP_DOWN_PIN, GPIO.OUT)
GPIO.setup(MOTOR_LEFT_RIGHT_PIN, GPIO.OUT)
GPIO.setup(PIR_PIN, GPIO.IN)
GPIO.setmode(GPIO.BCM)

# Motor control function
def control_of_motors(up_down, left_right):
    GPIO.output(MOTOR_UP_DOWN_PIN, up_down)
    GPIO.output(MOTOR_LEFT_RIGHT_PIN, left_right)

# Camera initialization
camera = VideoStream(src=0).start()
time.sleep(2.0)  # Pause for camera initialization

# Initialization of object tracking
tracker = None
track_started = False
```

Figure 9. Function code to tracking and control the object using dlib (initialization of components)

Figure 9 shows the necessary libraries are imported and GPIO settings are configured for the PIR sensor and motors. The camera is initialized using the **VideoStream** class from the **imutils** library. **tracker** and **track_started** variables are also defined.

```
try:
    while True:
        if GPIO.input(PIR_PIN):
            print("The Object is detected!")

            if not track_started:
                track_started = True
                tracker = dlib.correlation_tracker()
```

Figure 10. Function code to tracking and control the object using dlib (Main loop)

An infinite loop is entered, responding to motion detected by the PIR Motion sensor. If motion is detected by the PIR Motion sensor and the variable **track_started** is not true, it is indicated that tracking has not yet been initialized. The variable **track_started** is set to true, and a **tracker** object is created from the class **dlib.correlation_tracker()** (Figure 10).

```
            frame = camera.read()
            frame = cv2.flip(frame, 1)  # Image mirroring horizontally

            gray_frame = cv2.cvtColor(frame, cv2.COLOR_BGR2GRAY)
```

Figure 11. Loading a picture from the camera and conversion to grayscale for object tracking

The current frame is read from the camera. The **cv2.flip()** function is also employed to horizontally flip the frame, which can aid in determining the tracking direction. The current frame is transformed into a grayscale image **(gray_frame)** to enable object tracking (Figure 11).

```
            if tracker:
                rect = dlib.rectangle(100, 100, 200, 200)  # Initial coordinates
                tracker.start_track(gray_frame, rect)
```

Figure 12. Creating a rectangle that represents the initial position for object tracking

If the **tracker** object is initialized (meaning **track_started** is true), a rectangle (**rect**) is created that represents the initial position for object tracking (Figure 12).

```
                for i in range(10):  # Number of iterations
                    tracker.update(gray_frame)
                    pos = tracker.get_position()
                    x = int(pos.left())
                    y = int(pos.top())
                    w = int(pos.width())
                    h = int(pos.height())
                    cv2.rectangle(frame, (x, y), (x + w, y + h), (0, 255, 0), 2)
                    control_of_motors(y + h < frame.shape[0]/2, x + w < frame.shape[1]/2)
                    time.sleep(0.1)  # Pause

            cv2.imshow("Object tracking", frame)
            cv2.waitKey(1)
except KeyboardInterrupt:
    camera.stop()
    GPIO.cleanup()
    cv2.destroyAllWindows()
```

Figure 13. Object tracking and camera pointing and image display

Within this loop (**for i in range(10):**), the **tracker** object is updated. The position of the tracked object is obtained and utilized to draw a rectangle on the frame. Additionally, based on the object's position, control over the motors is exerted to modify the camera's orientation. The loop operates for a specific count of iterations, which can be adapted as per your needs (Figure 13).

Also, template *index.html* has a Live Stream implemented through the *img* tag that calls the *video_feed* function. The code of this function and img tag is below (Figure 14).

```
<img src="{{url_for('video_feed')}}" width="90%" alt="Picture from camera">
```

Figure 14. Function code for Live Stream from camera

The *video_feed* function generates the response using the *gen_frames* function which generates a new image every two milliseconds.

```
@app.route('/video_feed')

def video_feed():
    return Response(gen_frames(), mimetype='multipart/x-mixed-replace;
boundary=frame')
```

Figure 15. Function code for generate image (video_feed function)

Based on the above code, a video stream is generated on the web page by generating an image every 2 milliseconds using the *video_feed* function.

3. Testing the proposed system

This project is a cost-effective way to run a security camera that you can see over the network and have it too, so it's managed from a web application.

Remember that the Raspberry Pi is not powerful, so if you try to do too much by adding too many high-definition cameras, performance will decrease. That said, it's still a great way to build your own affordable camera network, as shown in this paper.

Testing the Web Monitoring System based on Raspberry Pi was performed for 48 hours. The designed web application (Figure 16 and Figure 17) is used to rotate the camera in the desired direction. During the testing time proposed system was working correctly as expected.

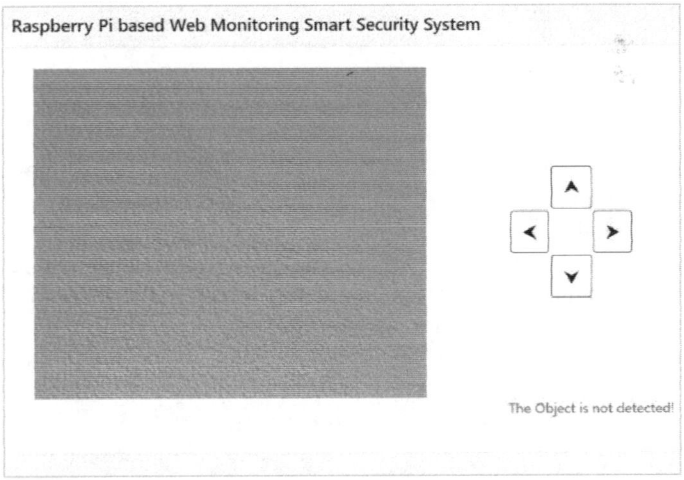

Figure 16. The designed web application for proposed system (The Object is not detected)

Figure 17. The designed web application for proposed system (The Object is detected)

The Web Monitoring System based on Raspberry Pi can be proposed to use in wide areas in society, for private or public needs. The advantage of the created system is that the user can see the display from the video camera in real-time at any time, and use the buttons B1, B2, B3 and B4 to control the camera in the desired direction.

Combined with the camera and the Raspberry Pi platform, the PIR Motion sensor adds value to the project by enabling interaction between motion detection and camera movement. Based on the detection, the system can automatically react and perform the desired actions, such as directing the camera towards the detected objects. Besides, LED indication and on-screen messages add interactivity and a better user experience.

The combination of PIR Motion sensor, camera and Raspberry Pi provides an efficient and smart way to detect objects and track them in various applications, such as surveillance, security or automation. This project highlights the importance of PIR Motion sensors as a key element for fast, accurate and energy-efficient motion detection and interaction with other system components.

4. Conclusion

Surveillance is the main thing that everyone wants and requires to protect their belongings or monitor their places. In this paper, we proposed a mobile surveillance camera using a Raspberry Pi and a web camera. The movement of the proposed surveillance camera can be controlled using the designed web browser and Python scripting language. It can cover a wide area. The proposed architecture is portable and mobile. It is widely used in areas with high human risk and is also used in industry for continuous monitoring of flawless working machines. Moving the camera can be done using the web application where using the navigation commands it is possible to move the camera in the desired direction.

The web monitoring system based on Raspberry Pi is cost-efficiently, easy to implement, user-friendly, and long-lasting. It's vital to have 24/7 access to monitor private or public areas and properties because it gives us a detailed look at every aspect of presence and can give us an early warning about potential problems.

The created Web Cam Control System using Raspberry Pi proved to be a stable system, it does not take up a lot of network and memory resources.

The disadvantage would be if, due to the limitations of the Raspberry Pi controller, more devices were added, such as more cameras, which would further choke the system and the network in which it works.

PIR Motion sensors enable the safe side of the implemented system, considering that if any of the objects in the PIR Motion sensor's field of view is detected, the camera is directed towards the object and messages about object detection is displayed on the web application and the green LED is turned on. If the object is outside the visible range of the PIR Motion sensor, a message is displayed that no object has been detected, and the camera can be moved with some buttons (up/down, left/right).

Acknowledgments
This research was part of the subject of doctoral studies, under the late professor Edin Mujčić at the University of Bihać, Technical faculty. Professor Edin Mujčić died on January 6, 2023. We want to dedicate this research to him as a Thank-You for everything he taught us.

References
[1] Wilkinson, M., Bell, M.C. and Morison, J.I. (2021) 'A raspberry pi-based camera system and image processing procedure for low cost and long-term monitoring of Forest Canopy Dynamics', *Methods in Ecology and Evolution* [Preprint]. doi:10.1111/2041-210x.13610.
[2] Kumar, K.N., Natraj, H. and Jacob, T.P. (2017) 'Motion activated security camera using Raspberry Pi', *2017 International Conference on Communication and Signal Processing (ICCSP)* [Preprint]. doi:10.1109/iccsp.2017.8286658.
[3] Wang, Y., Chen, C. and Zhang, W. (2023) 'Design of smart home control system based on internet of things', *2023 IEEE 6th Information Technology,Networking,Electronic and Automation Control Conference (ITNEC)* [Preprint]. doi:10.1109/itnec56291.2023.10082440.
[4] Musharaf Hussain, M.M. et al. (2022) 'IOT based smart human traffic monitoring system using Raspberry Pi', *Intelligent Computing & Optimization*, pp. 361–372. doi:10.1007/978-3-031-19958-5_34.
[5] D.S. Sharma, V. Giradkar, A. Sanap and S. Sarolkar. (2018) 'IoT Based Traffic Light Controller in Smart City', *International Journal of Advanced Research in Electrical, Electronics and Instrumentation Engineering, Vol. 7, Issue 1,* ISSN (Online): 2278 – 8875, doi: 10.15662/IJAREEIE.2018.0701014.
[6] Karthikamani, R. and Rajaguru, H. (2021) 'IOT based smart irrigation system using Raspberry Pi', *2021 Smart Technologies, Communication and Robotics (STCR)* [Preprint]. doi:10.1109/stcr51658.2021.9588877.
[7] Mujčić, E., Drakulić, U. and Toroman, A. (2022) 'Intelligent Traffic Light Control System using fuzzy logic', *Advanced Technologies, Systems, and Applications VII*, pp. 751–763. doi:10.1007/978-3-031-17697-5_58.

IOP Conf. Series: Materials Science and Engineering 1298 (2023) 012028 doi:10.1088/1757-899X/1298/1/012028

How fragile are buildings in Bosnia and Herzegovina once exposed to earthquake motion?

Naida Ademović

University of Sarajevo-Faculty of Civil Engineering, Patriotske lige 30, 71 000 Sarajevo, Bosnia and Herzegovina

naida.ademovic@gf.unsa.ba; naidadem@yahoo.com

Abstract. The 2020 earthquakes that hit the Croatian capital city Zagreb and Petrinja revealed the (un)preparedness of the country once exposed to strong earthquake motion. Three years after, numerous buildings out of 25,000 that have been heavily damaged and destroyed still have not been reconstructed. The effect of the earthquake was felt as well in the bordering cities and towns in Bosnia and Herzegovina (BIH). Seismic assessment of the existing buildings in BIH is done by individual researchers and there is no defined methodology not policy for such activities. This paper aims to present the work that has been done in this field up to now and to give guidelines for the further work that needs to be conducted. The starting point was the calculation of the earthquake risk for the entire Bosnia and Herzegovina based on the rapid assessment taking into account the vulnerability of buildings, seismic hazard, and population exposure. A more detailed assessment was conducted for the three largest cities in BIH, specifically Sarajevo, Banja Luka, and Tuzla. Typical residential masonry structures were identified, and detailed calculations were conducted, leading to the need for their strengthening. To have a more detailed picture much work has to be conducted.

1. Introduction

From US$ 14 billion in 1985 to more than US$ 140 billion in 2014, annual damages from earthquake action have escalated due to population growth around the world during the previous 30 years. Metropolitan areas susceptible to seismic danger are growing increasingly as the population in seismically-prone areas grows. In contrast to countries with low and high GDP, the analysis revealed that middle-income countries in the last two decades were at a higher risk (Fig. 1). Bosnia and Herzegovina is classified as having a lower-middle income based on the data that is currently available [1]. In recent decades, when seismic events have become more frequent, evaluating the seismic safety of existing buildings has become one of the most crucial considerations. Because numerous elements affect historical structures' overall seismic performance, characterizing them is an extremely difficult undertaking. Particularly, the subpar seismic behavior of such structures is a significant contributor to tragedies. The vulnerability is now one of the key elements of input data models that authorities use to create guidelines, code requirements, and catastrophe mitigation and prevention actions. The process of choosing an evaluation approach for seismic vulnerability will be influenced by several factors. It depends on a variety of factors, including the study's nature and purpose, the information at hand, the unique features and characteristics of individual structures or groups of buildings, the best method of assessment (qualitative or quantitative) [2], the organization of data collecting, and the involvement of decision-makers. In order to obtain the overall picture it is recommended first to conduct a rapid assessment of a bigger region, once this is done, cities would be identified where a more detailed assessment of buildings should be conducted. In that sense, the next step would be the identification of certain block and building typologies, leading to a detailed assessment of typical buildings on one hand and the other hand, the assessment of most endangered buildings that can be identified through the, for example, Rapid Visual Screening (RVS) developed in Canada by Rainer et al. [3].

Figure 1. As a percentage of GDP, average annual direct losses from natural disasters [1].

2. State-of-art regarding seismic assessment in Bosnia and Herzegovina

There are essentially three different evaluation approaches that can be used to determine a building's seismic vulnerability: analytical, heuristic, and empirical. Bosnia and Herzegovina is within six countries that lack building inventory data [4]. This to a large extent limits the applicability of various existing methods developed for the vulnerability assessment of existing buildings. The Vulnerability Index Method was developed in Italy, the foundation of which is Vulnerability functions generated from a building typology's vulnerability index and documented damage relationship, as well as two other methodologies recently developed by the RISK-UE WP4 Project (used for Bosnia and Herzegovina by [5][6][7], the LM1 (empirical) and the LM2 (mechanical) [8] methodologies. They are based on, respectively, Damage Probability Matrices related to Building Typology Matrices (from observed damage), Capacity Spectra and Fragility Models (from numerical models), and Building Typology Matrices.

An improved rapid risk (i-RAPID) assessment for Bosnia and Herzegovina was conducted by Ademović et al. [9], where the most endangered areas were identified, as cities of Banja Luka, the capital of Bosnia and Herzegovina-Sarajevo, and Zenica, as well cities like Tuzla and its surroundings in the northeast; Mostar, Ljubuški, Čapljina, and Trebinje in Hercegovina, as indicated in Figure 2.

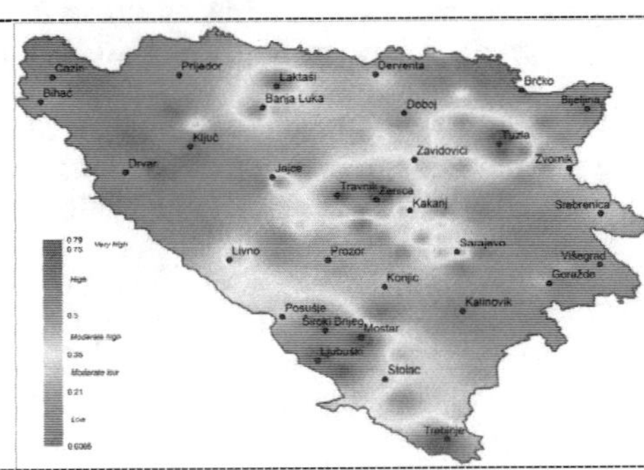

Figure 2. Relative seismic risk of Bosnia and Herzegovina [9].

The three most populated cities (Sarajevo, Banja Luka, and Tuzla) were identified as having a high seismic risk to earthquake ground motion. This has led to further work by the authors [5],[7]. Based on the assessment conducted in [9] the highest vulnerability among these three cities, regarding population and buildings is Tuzla. For Banja Luka, the vulnerability for buildings is very close to 1 (0 to 1 level of vulnerability; 0-minimum and 1-maximum level of vulnerability), whereas the vulnerability for the population is a bit higher than 0.8. For the capital city of Bosnia and Herzegovina, Sarajevo, the situation is vice versa with respect to Banja Luka.

The architectural and structural properties collected between 2014 and 2016 were included in the typology database, along with the energy efficiency of the buildings surveyed across the entire territory of BIH in accordance with the methodology and predefined criteria1 to assess the national building stock's energy efficiency [17]. Due to the lack of an official and systematic register on the residential building stock in the country, this effort began the first-of-its-kind study of a sample of 13,044 residential buildings, ranging from single-family homes to apartment buildings and residential high-rises. Banja Luka, Tuzla, and Sarajevo all used this database.

Three categories of structures are selected from the database for the assessment of structures in Sarajevo and Banja Luka: confined masonry, unreinforced masonry (URM) with rigid floors, and unreinforced masonry with a rigid diaphragm [5]. In this case, site effects were neglected. The database was narrowed down to a sample of 2933 buildings for this study since it only included unreinforced (with different types of floors) and confined masonry structures constructed in Sarajevo and Banja Luka constructed between 1918 and 2014. Based on calculations, it was projected that URM with wooden floors would sustain minor damage at intensity level VII according to Mercalli–Cancani–Sieberg (MCS) scale (the most likely level of damage), moderate damage at intensity level VIII, and significant to severe damage at intensity level IX. On the other hand, URM with stiff floors would only experience minor damage for intensity levels VII and VIII, and substantial damage may be anticipated for intensity levels IX. Confined masonry showed the best behavior, and would experience minor damage when the intensity is between VIII and IX. The number of floors was the behavior modification that had the biggest impact on how the vulnerability class changed.

A sample of two hundred and three buildings, comprising reinforced dual concrete buildings, URM with flexible and stiff floors, and confined masonry buildings, was created by meticulously analyzing the available database collection of structures in Tuzla. The selected sample's building era spans from 1961 to 2014. Residential buildings, either for private or communal habitation, were analyzed [7]. The city of Tuzla is very specific as during its history and valid construction regulations the seismic intensity changed so this had to be taken into account during calculations and analysis of the structures. Structures were built in nearly equal amounts over the two time periods, from 1983 to 1991 (26.1%) and from 1992 to 2014 (26.6%), while the greatest number of structures were built between 1964 and 1982 (42.9%). In this specific case, the goal was to research the impact of site effects on structures' behavior once exposed to earthquake ground motion. Incorporation of the site effects directly influences the vulnerability class. For example, if site effects were not taken into account 50% of masonry structures with flexible floors were classified as class A, as per EMS-98 [18] and the same percentage as class B. However, once site effects were taken into consideration all structures became class A (more vulnerable). It is interesting to note that once the site effects are taken into consideration that no building was found in class E and only six buildings could be identified as class D, while the largest number of buildings now became B. It is clear, that with the inclusion of site effects the buildings become more vulnerable as has been as well confirmed by [18][19]. It is possible to draw the conclusion from this study's findings that the site impacts are a major factor in determining how vulnerable existing structures in Tuzla City's urban area are to moderate to high-intensity ground motion. Inaccurate results would result in underestimating damages and overestimating the capacity of existing buildings if site influences were disregarded.

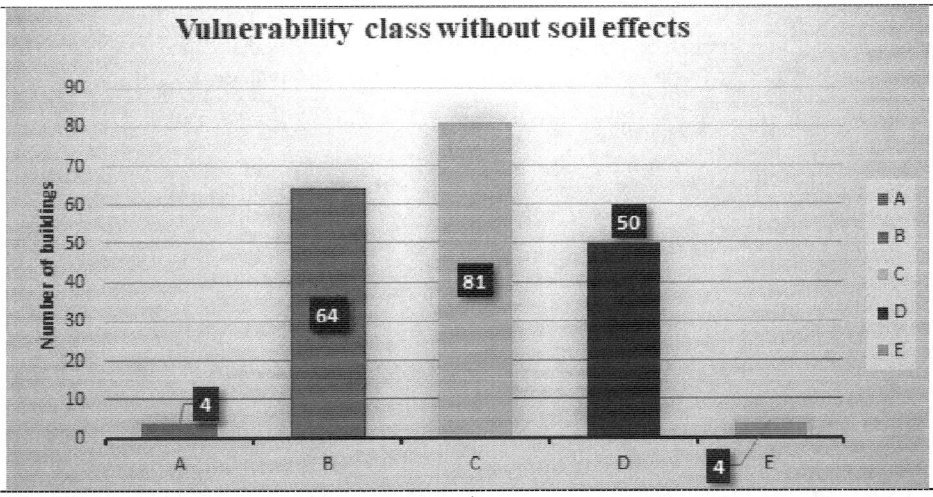

a)

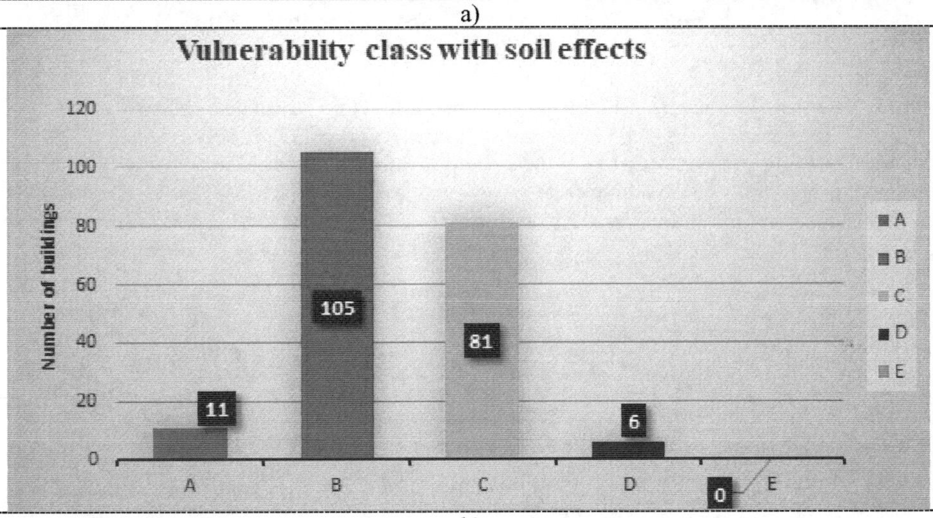

b)

Figure 3. Vulnerability class: a) structures on rock-no soil effect, b) soil type taken into consideration

And finally in order to make a detailed calculation of a typical building a structure in Sarajevo was selected, which represents a typical building constructed in the 50s and 60s of the XIX century in ex-Yugoslavia. One of the reasons why this specific building was selected was the fact that this type of building was heavily damaged during the 1963 Skoplje earthquake. The building is 38.0 m long by 13.0 m wide, has 7 storeys, and is a total of 19.6 m high. As shown in Figure 4, the structure is made up mostly of load-bearing walls running in the transverse direction (Y direction). The slabs have been constructed from semi-prefabricated components. The numerous apertures significantly reduce the load capacity of the longitudinal walls (X direction). Walls in red color represent the loadbearing walls, while the partition walls are marked in green color. The transverse walls are constructed up of a façade (0.125m thick hollow bricks) and a load-bearing masonry wall made of solid brick 0.25 m thick. The 0.25 m thick inner bearing walls are constructed entirely of solid brick. Bricks with uniform size are joined together with cement mortar. The slabs are built of concrete semi-prefabricated elements. Reinforced concrete was used to construct the basement walls. Both outer longitudinal and inner perpendicular walls are 0.30 m thick, while the thickness of the two inner walls is thicker by 8 cm [20].

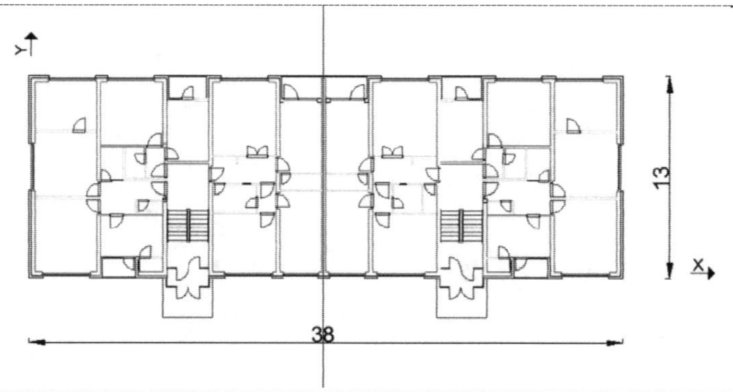

Figure 4. Floor plan of the elaborated building [20]

After conducting a pushover analysis it was noticed that the capacity of the structure in the X direction is very low. Due to the structure's very low X-direction resistance, the first crack formed at 4.5% of the force, whereas the highest horizontal-to-vertical coefficient only reached 9%. Identified failures were bending and compression at the ground level (where the masonry construction starts, as the basement is made of reinforced concrete walls), and this caused the façade walls to fail. A similar situation was observed in the building that was heavily damaged during the 1963 Skopje earthquake (see Figure 5).

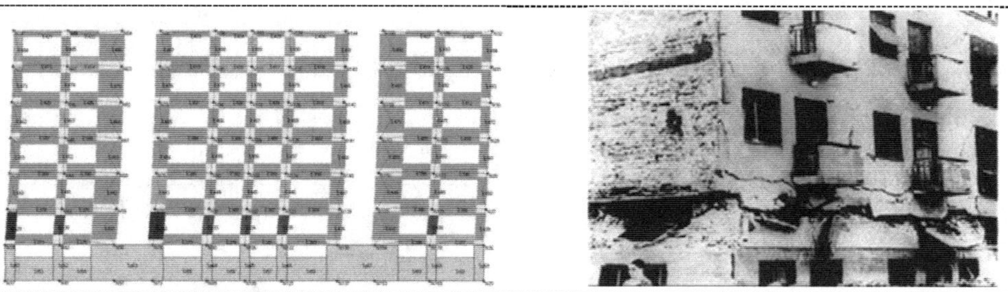

a) b)

Figure 5. a) Damage state at failure for facade walls; b) Damage after Skopje earthquake (WHE, 2002) [21]

Several strengthening procedures were suggested in the work [22] and they have been examined. Generally, new walls had to be constructed with the aim of increasing the stiffness in the X direction. However, the isolated damage at the ground floor level was not solved by this strengthening measure alone. So additional measures had to be incorporated. By adding ties at the level of the ground floor, this was partially resolved. Shear failure and local crushing, however, continued to be an issue. Local strengthening using fiber-reinforced polymers (FRP) was applied on the ground level with the aim to solve this issue. After this strengthening measure was applied, the only damage associated with this kind of strengthening is bending damage. A detailed strengthening was done in [23]. Unfortunately, the strengthening procedure used could not solve the issue of this type of structure's seismic fragility for larger earthquake motions, for Peak Ground Acceleration (PGA) equal to 0.2g, and for the soil type C where the structure is located. The strengthened structure has an acceptable behavior if exposed to PGA=0.1g; but, as the PGA rises, the bending damage is concentrated within the strengthened building's lower levels, seriously damaging the masonry spandrels.

3. Conclusion

The fragility of buildings in Bosnia and Herzegovina to earthquakes varies depending on factors such as construction practices, building codes, and seismic hazards. Traditional construction techniques and the presence of older buildings pose significant vulnerabilities. It was shown that masonry structures with flexible floors are more fragile compared to the other structural typologies investigated in various research. However, by implementing retrofitting measures, improving construction practices, raising public awareness, and conducting regular building inspections, the resilience of buildings can be enhanced. It is crucial for Bosnia and Herzegovina to prioritize seismic-resistant design and construction practices to mitigate potential damage, protect lives, and ensure the long-term safety of its building stock in the face of seismic events.

References

[1] Cummis JD and Mahul O 2009 Catastrophe risk financing in developing countries: principles for public intervention The World Bank Washington

[2] Preciado A, Ramirez-Gaytan A, Salido-Ruiz RA, Caro-Becerra JL and Lujan-Godinez R 2015 Earthquake risk assessment methods of unreinforced masonry structures: hazard and vulnerability *Earthq. Struct* **9**(4) pp 719–733 https://doi.org/10.12989/eas.2015.9.4.719

[3] Rainer JH, Allen DE and Jablonski AM 1993 Manual for screening of buildings for seismic investigation *Institute for Research in Construction National Research Council* Ottawa Canada

[4] NERA 2011 Report D7.2 State of the knowledge of building inventory data in Europe EUCENTRE

[5] Ademović N, Hadzima-Nyarko M and Zagora N 2020 Seismic vulnerability assessment of masonry buildings in Banja Luka and Sarajevo (Bosnia and Herzegovina) using the macroseismic model *Bull. Earthq. Eng.* **18**(8) pp 3897–3933 https://doi.org/10.1007/s10518-020-00846-8

[6] Ademovic N, Zagora N and Hadzima-Nyarko M 2021 Seismic Vulnerability Analysis in Urban and Rural Regions of Visoko, BIH, *12th Days of BHAAAS in Bosnia and Herzegovina Advanced Technologies, Systems, and Applications VI, Proc. Int., Symp on Innovative and Interdisciplinary Applications of Advanced Technologies* https://doi.org/10.1007/978-3-030-90055-7

[7] Ademovic N, Hadzima-Nyarko M and Zagora N 2022 Influence of site effects on the seismic vulnerability of masonry and reinforced concrete buildings in Tuzla (Bosnia and Herzegovina). *Bull. Earthq. Eng.* **20** pp 2643–2681 https://doi.org/10.1007/s10518-022-01321-2

[8] Khemis A, Athmani A and Ademović N 2023 Rapid Application of the RISK-UE LM2 Method for the Seismic Vulnerability Analysis of the Algerian Masonry Buildings *Int. J. Archit.* DOI: 10.1080/15583058.2023.2195379

[9] Ademović N, Kalman Šipoš T and Hadzima-Nyarko M 2020 Rapid assessment of earthquake risk for Bosnia and Herzegovina" *Bull. Earthq. Eng.* **18**(5) pp 1835–1863 https://doi.org/10.1007/s10518-019-00775-1

[10] Giovinazzi S and Lagomarsino S 2004 A macroseismic method for the vulnerability assessment of buildings. *Proc. 13th World Conference on Earthquake Engineering* Vancouver BC Canada paper no 896 pp 1–16

[11] Grünthal G 1998 European macroseismic scale 1998 *Cahiers du Centre Européen de Géodynamique et de Séismologie* **15** pp 1–99

[12] Guéguen P, Michel C and Le Corre L 2007 A simplified approach for vulnerability assessment in moderate-tolow seismic hazard regions: application to Grenoble (France) *Bull. Earthq. Eng.* **5**(3) pp 467–490 https://doi.org/10.1007/s10518-007-9036-3

[13] Hadzima-Nyarko M, Pavić G and Lešić M 2016 Seismic vulnerability of old confined masonry buildings in Osijek, Croatia *Earthq. Struct.* **11**(4) pp 629–648 https://doi.org/10.12989/eas.2016.11.4.000

[14] Podestà S and Romano C 2014 A Macroseismic Method for Vulnerability Assessment of Rationalist Architectural Heritage *Procedia Econ. Financ.* **18** pp 173–180 doi:10.1016/S2212-5671(14)00928-9

[15] Lestuzzi P, Podestà S, Luchini C, Garofano A, Kazantzidou-Firtinidou D, Bozzano C, Bischof P, Haffter A and Rouiller JD 2016 Seismic vulnerability assessment at urban scale for two typical Swiss cities using Risk-UE methodology *Nat. Hazards* **84** doi:10.1007/s11069-016-2420-z

[16] Ródenas JL, García-Ayllón S and Tomás A 2018 Estimation of the Buildings Seismic Vulnerability: A Methodological Proposal for Planning Ante-Earthquake Scenarios in Urban Areas *Appl. Sci.* **8** 1208. https://doi.org/10.3390/app8071208

[17] Arnautović-Aksić D et al 2016 Typology of residential buildings in Bosnia and Herzegovina Faculty of Architecture, University of Sarajevo, Sarajevo German Agency for International Cooperation ed D Arnautović-Aksić and N Zagora

[18] Stewart JP, Fenves GL and Seed RB 1999 Seismic soil-structure interaction in buildings. I: analytical methods *J. Geotech. Geoenviron. Eng.* **125**(1) pp 26–37 https://doi.org/10.1061/(ASCE)1090-0241(1999)125:1(26)

[19] Chieffo N and Formisano A 2019 The influence of geo-hazard effects on the physical vulnerability assessment of the built heritage: an application in a district of Naples *Buildings* **9**(1)26 https://doi.org/10. 3390/buildings9 010026

[20] Ademovic N, Hrasnica M and Oliveira DV 2013 Pushover analysis and failure pattern of a typical masonry residential building in Bosnia and Herzegovina *Eng. Struct.* **50** pp 13–29 doi:10.1016/j.engstruct.2012.11.031

[21] WHE 2002 World Housing Encyclopedia Unreinforced Brick Masonry Apartment Building, Report No. 73 Ljubljana Housing Report

[22] Ademović N and Hrasnica M 2014 Seismic Assessment and Strengthening of an Existing Multi-Storey Masonry Building in Sarajevo, Bosnia and Herzegovina, *Proc. of 2nd European Conference on Earthquake Engineering and Seismology Joint Event of the 15th European Conference on Earthquake Engineering and the 34th General Assembly of the European Seismological Commission*, ISBN: 9781510810211, pp. 6966–6976, Publication: European Association for Earthquake Engineering (EAEE)

[23] Ademović N, Oliveira DV and Lourenco PB 2019 Seismic Evaluation and Strengthening of an Existing Masonry Building in Sarajevo BiH *Buildings* **9**(2) 30 https://doi.org/10.3390/buildings9020030

IOP Conf. Series: Materials Science and Engineering 1298 (2023) 012029 doi:10.1088/1757-899X/1298/1/012029

Evaluation of reality and standards for fire evacuation routes in buildings – *case study*

Lamija Balić[1], Ahmed El Sayed[1], Sanin Džidić[2] and Adnan Novalić[1]

[1]Department of Architecture, International Burch University, Francuske revolucije bb, 71210 Ilidža, Bosnia and Herzegovina
[2] Department of Civil Engineering, University of Bihać, Dr Irfana Ljubijankića b.b., Bihać, Bosnia and Herzegovina
E-mail:lamija.balic43@gmail.com

Abstract. Any building, at any time could experience a fire. Because of that, occupants need a fire evacuation path that is designed according to the design standards as they are unable to be entirely prepared for fire emergency scenarios. However, the majority of these standards were developed using empirical judgments. The aim of the paper is to determine whether the fire escape exit route design takes into account the demands and behaviour of building occupants or if it only considers the recommendations and regulations already in place. One university building has been chosen for qualitative research, and as a result, students and teachers are the primary target population. The following factors are taken into consideration throughout the research: the number of students and professors, the movement speed of building occupants and persons with special needs, the plan, the number of floors, the width and length of corridors, the location and type of stairways, as well as type, position and geometry of doors. The findings of this case study indicated that younger occupants can safely exit the building, however older occupants and students with disabilities, as well as teaching personnel need more time for evacuation across the same evacuation paths in fire situation. In this regard, the findings of this research paper offer conclusions that might be useful in design and organization of such and similar buildings and environments.

1. Introduction

In the early stages of a fire, the people in a building typically have to either rely on themselves, or be rescued by others in their immediate vicinity [1]. As that, it can be concluded that the means of escape provision is a crucial aspect of fire safety requirements for safeguarding life not only in a fire occurrence but also from other threats [2]. The term, fire safety, refers to fire prevention, limiting the spread of fire and smoke, extinguishing a fire and the chance of a quick and safe exit [1]. Human behaviour during this initial phase is, therefore, a crucial factor in terms of survival [3, 4]. It can be defined as the actions that people take based upon their perception of the situation, their intention to act, and the considerations involved before these actions are carried out [1]. In lieu of data and theory, evacuation models (and users) make assumptions and simplifications about occupant behaviour, which can inappropriately characterize the time it actually takes to evacuate a building [5]. Also, case-studies show that occupants often are found to be incapable to escape in time [6] because there is a lack of available data and theory on occupant behaviour for use by evacuation models to estimate evacuation time results and their uncertainty [5]. In cases where assumptions lead to evacuation estimates that are either too optimistic or too conservative, buildings and procedures can be designed with either insufficient or unnecessary (and costly) egress routes and fire protection/notification systems [5]. From that, it can be concluded that a majority of standards that were established from empirical judgements, are not always scientifically proven [7, 8].

That highlights the question of whether the design of fire escape exit routes is there to suit the needs and behaviour of the building inhabitants or is there only to satisfy the recommendations and regulations

IOP Conf. Series: Materials Science and Engineering 1298 (2023) 012029 doi:10.1088/1757-899X/1298/1/012029

that are already in place. The functionality of the design in emergency scenarios is in doubt if it is purely the product of existing regulations and recommendations, and it poses a risk to the lives of inhabitants who are using fire escape exit routes in the event of a fire. Therefore, it is better to let the fire safety of buildings be based upon actual human behaviour in fire [6]. However, the aims and objectives of this study would be to examine whether human behaviour during a fire evacuation was considered while designing following existing recommendations and regulations, to analyse whether evacuation routes are only intended for regular people or whether people with disabilities could benefit from them, and to examine the function of fire staircases in the case of evacuation. The literature review contributed to the definition of potential problems that need to be considered in further research. The problems can be outlined as the following: (1) measures currently required by law do not always provide the support that people in burning buildings need [1]; (2) regulations focus on how people enter a building but not on how to evacuate safely [9]; (3) occupants often not behave accordingly the assumptions in the regulations [6]; (4) the effectiveness of the exit components is strongly affected by the building use and the population involved [10]; (5) designers and fire safety engineers only think of “average”, healthy and agile people in evacuations [9]; (6) in many buildings and infrastructures today, self-rescue is difficult for persons with special needs, e.g. wheelchair users [9].

2. Research Methodology

Qualitative research will be performed to determine whether the current recommendations and regulations followed in designing fire escape exit routes satisfy the needs of the university's occupants or if they are merely there to satisfy the recommendations and regulations. The Medical Academy Faculty of Public Health of the Lithuanian University of Health Sciences has been chosen for the case study. The focus target group of students and professors of various genders and ages was chosen because for them there was more information available than for other age groups. The following factors are taken into consideration during the research: national building codes, faculty type, building shape, number of floors, location and type of staircases, corridor width and length, door area, number of students and professors, and people with special needs' movement rates. However, it is crucial to point out that actually the faculty type plays a vital role in the research because each faculty type has distinct functions, content, and functional organization in terms of the placement of various spaces within the faculty's structure. Having mentioned all of that, the research questions that be formulated in this manner:

1. Did the university building's design take into account how people would behave during an evacuation in the event of a fire?
2. Are evacuation routes intended to be used by both regular people and individuals with disabilities?
3. In a fire emergency situation, what function do fire evacuation stairs serve?

3. Case Study of the Faculty of Public Health at the Lithuanian University of Health Sciences

As was already noted, the 5,217 m^2 Medical Academy Faculty of Public Health at the Lithuanian University of Health Sciences was chosen for the case study. A single volume measuring 49 m by 44 m and 11.88 m tall, the building housing education and scientific research has three floors above ground and one below. A staff car garage with 35 parking spaces, service areas, and storage rooms can be found on the underground floor. Auditoriums with adjustable sizes for lectures and group study are located on the ground floor. On the ground level, there is also a 60 m^2 open seating area. The seven departments of the faculty are served by administrative offices, research labs, and lecture halls on the two higher floors [10].

Figure 1. Lithuanian University of Health Sciences Medical Academy Faculty of Public Health [11].

The structure was designed to accommodate up to 550 students daily and 60 permanent staff members [11]. The following factors need to be taken into account when analysing the time required for the occupants to successfully exit the building from the furthest point of the building to the outside: the occupant's age, the average movement speed of a regular person and a person with special needs under stressful conditions, the area and width of the fire exit doors, the width of the corridors, the occupants' shoulder widths, and the building's height. A key element in evacuation models is walking speed that changes depending on a number of variables, including the type of walking [12], the walking environment [13], the type of occupant [14], and the location [15] (various types of outdoor and interior locations). Table 1 lists the walking speeds according to the relevant factors.

Table 1. Walking speed according to influencing factors [12, 13, 14, 15].

Influencing factors		**Speed (m/s)**	**Range (m/s)**
Walking type	Free move		1.2 ~ 1.8
	Exit move		0.8 ~ 1.5
Walking conditions for corridors, doorway on ship	Low	1.40	
	Optimum	0.70	
	Moderate	0.30	
	crush	0.10	
Place type	Public place		0.51 ~ 1.27
	High-rise	1.05	0.57 ~ 1.20
	apartment	0.95	0.56 ~ 1.12
Occupant type	Children	1.08	
	Female elderly	1.04	
	Male elderly	1.05	
	Elderly	1.04	
	Female adult	1.24	
	Male adult	1.30	

	Adult	1.27

The wheelchair type [17], the type of subject with disability [18], and the gender [16] all affect the wheelchair user's walking pace. Table 2 lists the walking speeds for people with disabilities.

Table 2. Walking speed for occupants with disabilities [16, 17, 18].

Factors		Speed (m/s)	Range (m/s)
Wheelchair	Man	1.06	
	Woman	1.06	
High rise building	Electric wheelchair	0.89	
	Manual wheelchair	0.69	
	Crutches	0.94	
	Walking stick	0.81	
	No disability	1.24	
Subjects with disability	All subjects with disability	1.0	0.10 ~ 1.77
	With locomotion	0.80	0.10 ~ 1.68
	No aid	0.95	0.24 ~ 1.68
	Crutches	0.94	0.63 ~ 1.35
	Cane	0.81	0.26 ~ 1.60
	Walker/Roller	0.57	0.10 ~ 1.02
	Without locomotion disability	1.25	0.82 ~ 1.77
	Unassisted wheelchair	0.89	0.85 ~ 0.83
	Assisted ambulant	0.78	0.21 ~ 1.40
	Assisted wheelchair	1.30	0.84 ~ 1.98

The width of the occupants' shoulders is a key factor in determining how they will exit the burning space, namely whether they will leave the area one at a time, two at once, etc. The shoulder measurements of various age groups are shown in Table 3.

Table 3. Shoulder widths of different age groups [19, 20].

Age Group	Shoulder Width (cm)
Children	35.0
Young men	41.0
Young Women	38.0
Middle-aged men	41.9
Middle-aged women	39.5
Elderly men	40.5
Elderly women	39.0

In addition to the information already mentioned, it is crucial to emphasize that the 1.40 m-wide corridors inside the Lithuanian University of Health Sciences Medical Academy Faculty of Public Health lead to fire escape doors that are 90 cm x 200 cm in size and are situated at the top of the staircases. Two 12-m^2 closed U-shaped fire escape stairs are situated on the east side of the building's corners on the last floor. Two elevators and open linear stairs are situated in a 45 m^2 area on the west side. The unenclosed stairs and elevators exit to the main entrance hall at ground floor level whereas the enclosed staircases have a direct access to the outside. The staircases have a 1.20 m width. The use of open staircases and elevators will not be considered in the event of a fire because of the possibility that smoke and gases will fill them first. In Figure 2, the rooms are identified by numbers, and the distances between each room and a secure outdoor space have been calculated.

Figure 2. Second floor plan with numbered paths and movements path towards fire escape staircases.

The distance from the fire exit door to the outdoor area is: from the second to first floor (11.30 m) and from the first floor to the ground floor where the exit of the fire escape staircases is located (14.30 m). The total number and gender of students and teaching staff (Table 4) and the total number and gender of students and teaching staff with disabilities (Table 5) will be assumed in order to determine the average amount of time in worst and best case scenario required to safely exit the building from the second floor to the outside.

Table 4. Total number and gender of students and teaching staff on second floor used for calculations.

Number and gender of students	Number and gender of teaching staff	Number and gender of teaching staff
21-Young men	13-Middle-aged men	9-Elderly men
21-Young women	10-Middle-aged women	10-Elderly women

Table 5. Total number and gender of students and teaching staff with disabilities on second floor used for calculations.

Number and gender of students	Number and gender of teaching staff
3-Young men	2-Middle-aged men
3- Young women	2-Elderly women

After determining the number, gender, and age of the occupants, Tables 1, 2, and 3 will be used to calculate movement speed and shoulder widths:

Table 6. Movement speed and shoulder widths of students and teaching staff.

Age Group	Movement Speed (m/s)	Shoulder Width (cm)
Young men	1.32	41.00
Young women	1.27	38.00
Middle-aged men	1.30	41.90
Middle-aged women	1.24	39.50
Elderly man	1.05	40.50
Elderly women	1.04	39.00

However, the movement speed will differ for students and teaching staff with disabilities depending on whether they require assistance from another occupant or not when using the fire escape exit routes. A wheelchair's width is also a key factor in evaluating if a fire escape exit door is sufficient for it so that the wheelchair may pass through it and safely leave. As a result, the calculations will utilize a wheelchair with an average width of 63.50 cm [21].

Table 7. Movement speed of students and teaching staff with disabilities, with (out) assistance of another occupant and shoulder widths.

Age Group	Movement Speed without assistance (m/s)	Movement Speed with assistance (m/s)	Shoulder Width (cm)
Young men			41.00
Young women			38.00
Middle-aged men			41.90
Middle-aged women	1.06	1.30	39.50
Elderly men			40.50
Elderly women			39.00

The evacuation time will be calculated by dividing the length of the path that the occupant needs to pass from the particular room to the outdoor area and the movement speed of the occupant depending on the gender and age as follows:

$$e = l/v \tag{1}$$

where:

e – evacuation time (s)
l – length of the path that the occupant needs to pass from the particular room to the outdoor area (m)
v - movement speed of the occupant depending on the gender and age (m/s).

The teaching staff moves at varying speeds depending on the age group and gender, as shown in Table 6. However, for calculations, the age group with the slowest movement speed of 1.04 m/s (Table 6) and fastest movement speed 1.30 m/s (Table 6) will be taken into consideration as, in the case of emergency, that age group could enter any of mentioned university rooms. As that, the evacuation time for teaching staff in worst and best case scenario will be calculated.

Table 8. Calculations of evacuation time for teaching staff in worst and best case scenario.

Room number	Route length from the specified room to outdoor area (m)	Movement speed in worst case scenario (m/s)	Movement speed in best case scenario (m/s)	Evacuation time needed in worst case scenario (s)	Evacuation time needed in best case scenario (s)
1	50.30			48.36	38.69
2	40.30			38.75	31.00
3	48.10			46.25	37.00
4	53.60			51.53	41.23
5	56.60			54.42	43.53
6	60.60			58.26	46.61
7	72.60			69.80	55.84
8	71.60			68.84	55.07
9	59.90			57.59	46.07
10	50.20			48.26	38.61
11	45.90			44.13	35.30
12	45.30			43.55	34.84
13	39.10			36.73	30.07
14	38.20			36.73	29.38
15	34.40			33.07	26.46
16	53.00			50.96	40.76
17	46.40			44.61	35.69
18	45.30			43.55	34.84
19	39.30			37.78	30.23
20	38.20			36.73	29.38
21	33.60			32.30	25.84
22	39.60	1.04	1.30	38.07	30.46
23	46.30			44.51	35.61
24	54.60			52.50	42.00
25	57.70			55.48	44.38
26	58.80			56.53	45.23
27	64.80			62.30	49.84
28	78.50			75.48	60.38
29	80.40			77.30	61.84
30	69.00			66.34	53.07
31	61.20			58.84	47.07
32	65.10			62.59	50.07
33	58.00			55.76	44.61
34	58.00			55.76	44.61
35	51.60			49.61	39.69
36	50.60			48.65	38.92
37	50.60			48.65	38.92
38	44.60			42.88	34.30
39	43.40			41.73	33.38
40	43.40			41.73	33.38
41	38.80			37.30	29.84
42	43.90			42.21	33.76

According to the findings, it will take the teaching staff between 32.30 and more than 77.30 seconds to safely reach the outdoor area in worst case scenario. In the best case scenario, the outdoor area will be safely reached between 25.84 and 61.84 seconds. Additionally, the corridor's 1.40 m width allowed for parallel passage between occupants despite their shoulder width (Table 3). The 1.20 m-wide staircases and 90 cm-wide fire escape doors each permitted two people to pass through at once. However, Table 6 demonstrates that pupils' movement rates vary depending on their age group and gender. However, for calculations, the age group with the slowest movement speed of 1.27 m/s (Table 6) and fastest movement speed 1.32 m/s (Table 6) will be taken into consideration as, in the case of emergency, that age group could enter the mentioned university rooms. As a result, the worst-case and best-case scenario evacuation time for students will be determined.

Table 9. Calculations of evacuation time for students in worst and best case scenario.

Room number	Route length from the specified room to outdoor area (m)	Movement speed in worst case scenario (m/s)	Movement speed in best case scenario (m/s)	Evacuation time needed in worst case scenario (s)	Evacuation time needed in best case scenario (s)
1	50.30			39.60	38.10
2	40.30			31.73	30.53
3	48.10			37.87	36.43
4	53.60			42.20	40.60
5	56.60			44.56	42.87
6	60.60			47.71	45.90
7	72.60			57.16	55.00
8	71.60			56.37	54.24
9	59.90			47.16	45.37
10	50.20			39.52	38.03
11	45.90			36.14	34.77
12	45.30	1.27	1.32	35.66	34.31
13	39.10			30.78	29.62
14	38.20			30.07	28.93
15	34.40			27.08	26.06
16	53.00			41.73	40.15
17	46.40			36.53	35.15
18	45.30			35.66	34.31
19	39.30			30.94	29.77
20	38.20			30.07	28.93
21	33.60			26.45	25.45
22	39.60			31.18	30.00
23	46.30			36.45	35.07
24	54.60			42.99	41.36
25	57.70			45.43	43.71

26	58.80			46.29	44.54
27	64.80			51.02	49.09
28	78.50			61.81	59.46
29	80.40			63.30	60.90
30	69.00			54.33	52.27
31	61.20			48.18	46.36
32	65.10			51.25	49.31
33	58.00	1.27	1.32	45.66	43.93
34	58.00			45.66	43.93
35	51.60			40.62	39.09
36	50.60			39.84	38.33
37	50.60			39.84	38.33
38	44.60			35.11	33.78
39	43.40			34.17	32.87
40	43.40			34.17	32.87
41	38.80			30.55	29.39
42	43.90			34.56	33.25

In this instance, the findings showed that it takes students between 26.45 and 63.30 seconds to arrive safely at the outdoor area in the worst case scenario. However, in the best case scenario the time needed to safely reach the outdoor area is between 25.45 and 60.90 seconds. Additionally, just like with teaching staff, the corridor's 1.40 m width allowed for parallel movement through the space based on occupant shoulder width (Table 3). The 1.20 m-wide staircases and 90 cm-wide fire escape doors each permitted two people to pass through at once. With 1.30 m/s or 1.06 m/s without the help of another occupant, the movement speeds of students and teaching staff with disabilities are the same for all age groups and genders, according to Table 7. However, it is clear from Table 7 that the mobility speeds of students and teaching staff with disabilities are the same for each age group and gender, either with aid from another occupant at 1.30 m/s or without it at 1.06 m/s. Given that all age groups can enter any of the aforementioned university rooms in an emergency and that movement speeds are the same for all age groups, any age group can be considered for calculations.

Table 10. Calculations of evacuation time for students and teaching staff with disability.

Room number	Route length from the specified room to outdoor area (m)	Movement speed without assistance (m/s)	Movement speed with assistance (m/s)	Evacuation time needed without assistance (s)	Evacuation time needed with assistance (s)
1	50.30			23.30	19.00
2	40.30			13.86	11.30
3	48.10			21.22	17.30
4	53.60			26.41	21.53
5	56.60			29.24	23.84
6	60.60			33.01	26.92
7	72.60	1.06	1.30	44.33	36.15
8	71.60			43.39	35.38
9	59.90			32.35	26.38
10	50.20			23.20	18.92
11	45.90			19.15	15.61
12	45.30			18.58	15.15

13	39.10			12.73	10.38
14	38.20			11.88	9.69
15	34.40			8.30	6.76
16	53.00			25.84	21.07
17	46.40			19.62	16.00
18	45.30			18.58	15.15
19	39.30			12.92	10.53
20	38.20			11.88	9.69
21	33.60			7.54	6.15
22	39.60			13.20	10.76
23	46.30			19.52	15.92
24	54.60			27.35	22.30
25	57.70			30.28	24.69
26	58.80			31.32	25.53
27	64.80	1.06	1.30	36.98	30.15
28	78.50			49.90	40.69
29	80.40			51.69	42.15
30	69.00			40.94	33.38
31	61.20			33.58	27.38
32	65.10			37.26	30.38
33	58.00			30.56	24.92
34	58.00			30.56	24.92
35	51.60			24.52	20.00
36	50.60			23.58	19.23
37	50.60			23.58	19.23
38	44.60			17.92	14.61
39	43.40			16.79	13.69
40	43.40			16.79	13.69
41	38.80			12.45	10.15
42	43.90			17.26	14.07

According to the findings, it takes students and teaching staff with disabilities between 7.54 and 49.90 seconds to reach the fire escape stairs on their own, and between 6.15 and 42.15 seconds with another occupant's help. Additionally, due to the corridor's 1.40 m width, a wheelchair may pass through it parallel without any help. However, due to the width of the wheelchair and the width of the person providing assistance, parallel movement is not possible if the wheelchair user needs help from another occupant. Given that wheelchairs had a width of 63.50 cm, the 90 cm fire escape doors barely allowed wheelchair users to pass through them one at a time. Up until the moment where the teaching staff and students with disabilities arrive at the fire evacuation staircases, all calculations have been completed. Since they could potentially be first filled with smoke and gases, the unclosed stairways and elevators will not be taken into consideration for usage, as it was previously mentioned. The requirement to use the fire evacuation staircases stems from this. Evacuation chairs, sheets, and sleds are common evacuation devices [22]. One or two people could also employ carrying techniques. It is significant to note that each approach necessitates moving at a different pace in order to get to the outdoor area. The manner of evacuating students and teaching staff with disabilities, as well as the amount of time required to do so until reaching a safer outdoor area, are not the focus of current research, but they unquestionably represent an area that requires further investigation.

IOP Conf. Series: Materials Science and Engineering 1298 (2023) 012029 doi:10.1088/1757-899X/1298/1/012029

4. Discussion and Results

A fire can quickly spiral out of control after only 30 seconds of ignition. The temperature of the building will have significantly increased after the first full minute of a fire, and smoke will be swiftly spreading [23]. The statistics showing the students and teaching staff velocities show that it takes, on average, between 26.45 and 77.30 seconds to leave the building safely in worst case scenario, and between 25.45 and 60.90 seconds in best case scenario. The movement of occupants through the hallway, fire escape doors, and staircases was influenced by the shoulder width and the width of the corridor.

However, older members of the teaching staff take longer than a minute to exit the building safely from the farthest point on the second floor, although younger occupants can do so in a minute over the same distance. However, due to their slow pace and health conditions, older members of the teaching staff run the risk of breathing in fire smoke. On the other hand, younger occupants move more quickly, so if they can get to a safe outdoor place faster, they can escape breathing in fire smoke. By demonstrating that the previous recommendations and regulations did not take into account human behaviour and movement speed while functionally structuring the areas within the faculty building, the first research question is addressed.

However, it takes students and teaching staff with disabilities between 7.54 and 49.90 seconds to reach the fire escape staircases on their own and between 6.15 and 42.15 seconds with the help of another occupants. Given that wheelchairs had a width of 63.50 cm, the 90-cm fire escape doors barely allowed wheelchair users to pass through them one at a time. Since the time required to evacuate a student and teaching staff with disabilities to a safe outdoor area was not the focus of the study, if the maximum time of 49.90 seconds is used, it means that it will take more than a minute to reach the fire escape stairs and almost a minute to reach the outdoor area. As a result, it is crucial in architectural design to consider evacuation models for students and teaching staff with disabilities from the second-floor room's exit to the outside because, according to calculations, if adequate evacuation methods are not anticipated, people's lives may be in danger. One approach would be to create safe haven rooms where students and teaching staff with disabilities would be protected from the fire for a predetermined amount of time before assistance was provided to safely exit the building. In order to prevent smoke and fire from entering the not large safe haven rooms, concrete should be utilized for all room components, including the walls, ceiling, and floor. Additionally, the doors should be fire resistant for at least 120 minutes. This provides an answer to the second study question since it shows that the evacuation routes are not made to accommodate the needs and movement speed of people with disabilities in an emergency fire situation.

However, as was already indicated, the open stairways and elevators could not be used in an emergency fire evacuation because they would be the ones affected by fire first. Additionally, rather than having to go outside, the main hall on the ground floor can be reached by using them. Additionally, open stairs can allow gases and flames to move quickly and easily vertically, even from floors below and above. This form of staircases might be used by occupants and senior teaching staff with disabilities alike by being enclosed and having a direct link to the outdoors, while also offering evacuation from the farthest areas on the second floor till the safe outdoor area. As a result of the discussion about the function of fire evacuation staircases in this particular case study, the third research issue is resolved.

5. Conclusion

When assumptions lead to unduly optimistic or underestimated evacuation forecasts, buildings and procedures may be built with insufficient or superfluous (and expensive) fire protection/notification systems. This begs the question of whether the design of the fire escape exit routes is there to meet the demands and behaviour of the building inhabitants or is there only to meet the existing recommendations and rules. The results demonstrated that because of their slower movement pace, the older teaching staff at the institution required more than one minute to safely escape. Younger occupants (students), on the other hand, required less time to travel the same distance.

However, in accordance with the amount of time required to safely exit the building and enter the outside, students and faculty staff with disabilities are included in the previously mentioned group of senior faculty staff. The unclosed staircases that were not taken into account during the evacuation would aid in a better and faster evacuation, especially for students and faculty staff with disabilities as well as other faculty members who are placed in rooms close to this staircases. However, incorporating safe haven rooms into the building's initial architectural design would boost the likelihood of a successful evacuation of students and faculty staff with disabilities and potentially save many lives. It is advisable to consider improved architectural solutions that consider the security challenges of the building, as well as the safety of users in challenging security challenges and building utilization, when developing the architectural design of such, similar, and other public buildings. As all of the rooms could be accessed and utilized by all of the aforementioned age groups and genders, the research paper's conclusions could aid in the functional organization and design of spaces inside the institution (as well as other types of buildings with similar floor plans). The number of persons entering the halls and exiting the rooms was not taken into account during the research because calculations were made as the subjects left the rooms singly and in pairs. These criteria should also be considered in future research. Additionally, there needs to be more research done on the best way to evacuate students and faculty members with disabilities as well as the amount of time needed to go to a safer outdoor place.

References

[1] Kobes M, Helsloot I, De Vries B and Post J G 2010 Building safety and human behaviour in fire: A literature review *Fire Safety Journal.* **45(1)** 1-11
[2] Samad M H A, Taib N and Ying C S 2017 *Means of escape provisions and evacuation simulation of public building in Malaysia and Singapore* AIP Conference Proceedings. (Vol.1892, No. 1, p. 180002). AIP Publishing LLC
[3] Purser D A and Bensilum M 2001 *Quantification of behaviour for engineering design standards and escape time calculations* Safety Science 38 (2001) pp 157–182
[4] Pires T T 2005 An approach for modeling human cognitive behavior in evacuation models *Fire Safety Journal* **40** pp 177–189
[5] Kuligowski E D and Kuligowski E D 2009 *The process of human behavior in fires* (Gaithersburg: US Department of Commerce, National Institute of Standards and Technology) pp. 627-632
[6] Kobes M, Helsloot I, De Vries B, Oberijé N and Rosmuller N 2007 *Fire response performance in a hotel. Behavioural research. Proceedings of the 11th International fire science and engineering conference, Interflam* pp 1429-34
[7] Frantzich H 1994 *A Model for Performance-based Design of Escape Routes*, (Lund:Department of Fire Safety Engineering, Lund Institute of Technology, Lund University, Sweden)
[8] Fahy R F and Proulxm G 2001 *Toward creating a database on delay times to start evacuation and walking speeds for use in evacuation modelling* (Boston:Second International Symposium on Human Behaviour in Fire) pp 175–183
[9] Künzer L, Hofinger G and Zinke R 2020 Safe Evacuation for All A top 10 List of Requirements *Collective Dynamics*, **5** pp 197-204
[10] Ronchi E and Nilsson D 2013 Fire evacuation in high-rise buildings: a review of human behaviour and modelling research *Fire science reviews* **2(1)** pp 1-21
[11] Luco Andreas, "Lithuanian University of Health Sciences Medical Academy Faculty of Public Health / Architectural Bureau G.Natkevicius and Partners," ArchDaily, 18 12 2022. [Online]. Available: https://www.archdaily.com/993678/lithuanian-university-of-health-sciences-medical-academy-faculty-of-public-health-architectural-bureau-gnatkevicius-and-partners. [Accessed 20 01 2023]
[12] Fang Z, Wang P, Chen D H, Chen D H, Duan J X and Hu Z R 2004 The development of evaluation software of safety evacuation for high buildings *Fire science and technology* **23(5)** pp 439-442
[13] Lee D, Kim H, Park J H and Park B J 2003 The current status and future issues in human

evacuation from ships *Safety Science* **41(10)** pp 861-876
[14] Yeo S K and He Y 2009 Commuter characteristics in mass rapid transit stations in Singapore *Fire safety journal* **44(2)** pp 183-191
[15] Chen T, Song W, Fan W C, Lu S and Yao B 2003 Pedestrian evacuation flow from hallway to stairs *CIB REPORT* pp 79-86
[16] Tsuchiya S, Hasemi Y and Furukawa Y 2007 Evacuation characteristics of group with wheelchair users *Proceedings of the Asia-Oceania Symposium on Fire Science and Technology*
[17] Pelechano N and Malkawi A 2008 Evacuation simulation models: Challenges in modeling high rise building evacuation with cellular automata approaches *Automation in construction* **17(4)** pp 377-385
[18] Proulx G 2002 *Evacuation planning for occupants with disability* (Ottawa, Canada: Fire Risk Management Program, Institute for Research in Construction, National Research Council Canada)
[19] Shields T J and Boyce K E 2000 A study of evacuation from large retail stores *Fire safety journal* **35(1)** pp 25-49
[20] Liang Y L, Zhang X and He S H 2004 Measurements and analyses on the body date of 6~ 12 school-age children *JOURNAL-XIAN UNIVERSITY OF ENGINEERING SCIENCE AND TECHNOLOGY* **18** pp 115-120
[21] Bryan "Wheelchairs Dimensions and Drawings," 08 04 2020 [Online]. Available: https://www.dimensions.com/element/wheelchairs. [Accessed 20 01 2023]
[22] Vanderbilt University Office of Emergency Management "Evacuation for Persons With Disabilities," [Online]. Available: https://emergency.vanderbilt.edu/vu/quick-ref-guides/evacuation-disabled.php/. [Accessed 20 01 2023]
[23] Scutum, London "How Fast Does a Fire Spread and How to Prevent It," 12 12 2022 [Online]. Availble: https://www.scutumlondon.co.uk/help-advice/how-fast-does-a-fire-spread/. [Accessed 20 01 2023]

IOP Conf. Series: Materials Science and Engineering 1298 (2023) 012030 doi:10.1088/1757-899X/1298/1/012030

Spalling of concrete caused by ASR – a review

Sanin Džidić[1] and Irfan Bidževič[2]

[1]Department of Civil Engineering, Faculty of Technical Sciences, University of Bihać, Dr. Irfana Ljubijankića bb, 77000 Bihać, Bosnia and Herzegovina
[2]Department of Civil Engineering, Faculty of Engineering, Natural and Medical Sciences, Study Program: Civil Engineering, International Burch University, 71000 Sarajevo, Bosnia and Herzegovina

ninsa_d@hotmail.com

Abstract. A very well-known fact is that cracks in concrete members can allow water to enter the member, thus jeopardizing the concrete member integrity by changing the chemical composition of concrete and causing the aggregates to expand. This initiates the consequent chemical reactions that may evoke spalling of concrete thus creating a cycle of undesirable occurrences in concrete. As a final result, the concrete member can be deteriorated affecting the serviceability, but sometimes even leading the reduction of its load bearing capacity. An overlooked cause of concrete spalling of a chemical nature is the spalling caused by the Alkali-aggregate reaction. This is colloquially known as "concrete cancer". It is a specific type of a reaction between alkali hydroxides and specific types of aggregate that occurs in concrete. Two main types of alkali-aggregate reactions are the alkali-silica reaction (ASR) and the alkali-carbonate reaction (ACR). The main goal of this research was to review the alkali-silica reaction in concrete as well as the mechanism of aggregate expansion it causes, which ultimately results in concrete spalling. Finally, the ways of prevention the alkali-silica reaction were reviewed and discussed.

1. Introduction

A specific type of reaction between alkali hydroxides and certain types of aggregate occurs in concrete. Portland cement is usually the main source of hydroxides. This type of reaction is called the alkali-aggregate reaction, often times colloquially referred to as the „concrete cancer". The alkali-aggregate reactions (AAR) can be classified as:

- Alkali-silica reaction (ASR)
- Alkali-carbonate reaction (AAR)

The alkali-aggregate reaction occurs between the hydroxyl ions (OH^-) which originate from the alkalis in the hydraulic cement mostly (sodium or potassium) and certain constituents of some aggregates. The AAR is classified depending on those constituents of aggregates that participate in the reaction. If the aggregate is of a carbonaceous nature such as calcitic dolostone or dolomitic limestones, the main components that participate in the AAR are the carbonates such as $CaMg(CO_3)_2$, or Na_2CO_3, and that is the alkali-carbonate reaction. On the other hand, if the reaction occurs between the (OH^-) and silica that originates from certain siliceous rocks and minerals such as opal, chert, quartz or acidic volcanic glass, the reaction is classified as alkali-silica reaction [1].

An example of an Alkali-carbonate reaction:

IOP Conf. Series: Materials Science and Engineering 1298 (2023) 012030 doi:10.1088/1757-899X/1298/1/012030

$$CaMg(CO_3)_2 + 2NaOH \rightarrow Mg(OH)_2 + Na2CO_3 + CaCO_3 \quad (1)$$

where $CaMg(CO_3)_2$ is the dolomite found in the aggregate, sodium hydroxide is the alkaline found in hydraulic cement in most cases, but can originate from other sources, and the products are brucite ($Mg(OH)_2$) and calcium carbonate ($CaCO_3$), along with some sodium carbonate (Na_2CO_3).

An example of an Alkali silica reaction:

$$Ca(OH)_2 + H_4SiO_4 \rightarrow CaH_2SiO_4 \text{ x } 2H_2O \quad (2)$$

The H_4SiO_4 is, in fact, the amorphous silica found in the aggregate, or the equivalent to SiO_2 x $2H_2O$. The final product of the ASR (CaH_2SiO_4 x $2H_2O$) is the calcium silicate hydrate, or in cement chemists' notation C-S-H.

It is worth mentioning that ASR is far more likely to occur than ACR. A product of ASR is in gel state (ASR gel) and is fairly hygroscopic, which means that it absorbs water and swells. This swelling is the main cause of cracking of concrete [2].

The initial result of the breakdown of the structure is the gel composed of the alkali- silicate solution, which depends on the moisture content. The concentration of hydroxy ions decreases, along with the pH, as silica dissolves. The pH value plays a big role since the silica continues dissolving until it reaches the pH equilibrium. The silica concentration in the solution gets decreased due to the calcium abundance, which provides additional source of hydroxyl ions (Figure 1) [1].

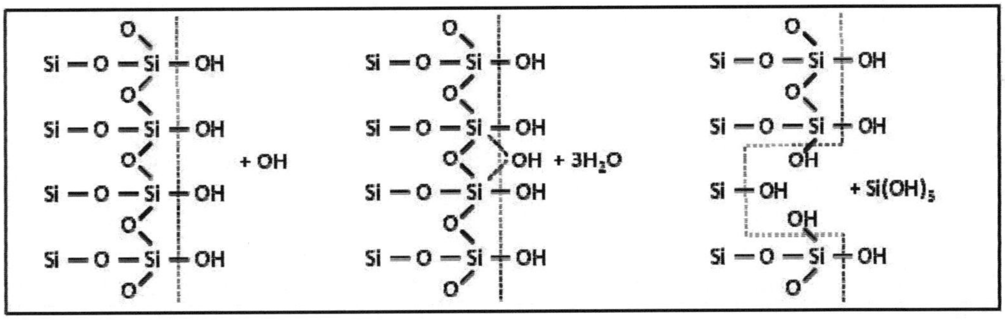

Figure 1. Mechanism of dissolution of silica due to attack by hydroxyl ions. Dotted line represents the interface between silica and water [3].

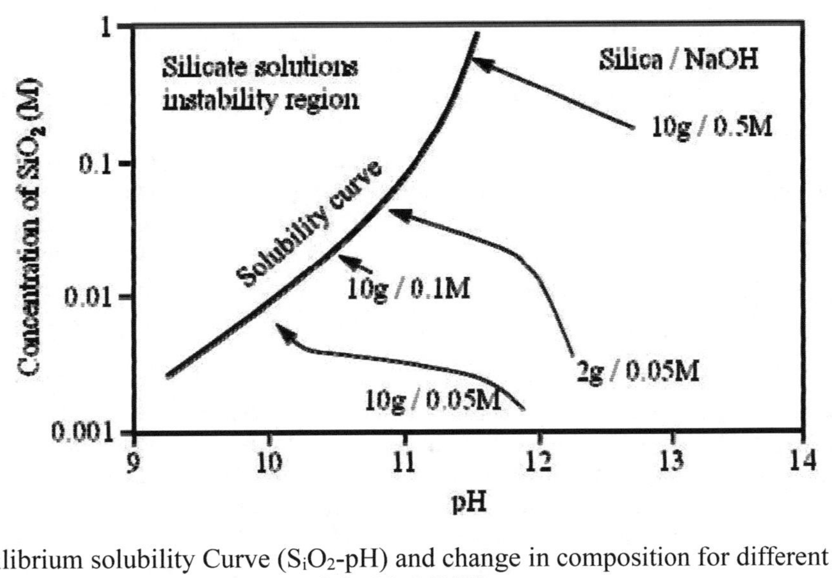

Figure 2. Equilibrium solubility Curve (S_iO_2-pH) and change in composition for different solutions of S_iO_2-NaOH [3].

2. Mechanism of aggregate expansion

Figure 3 depicts a petrographic thin slice of a concrete sample that is affected by ASR, seen under a transmitted light microscope. Cement paste contains a piece of flint sand incorporated in it (see left). The fissure in the surrounding cement mixture has grown larger and wider as a result of the flint particle's reaction, enlargement, and splitting. Alkali-silica gel reaction product partially fills the crack inside the aggregate and paste [1].

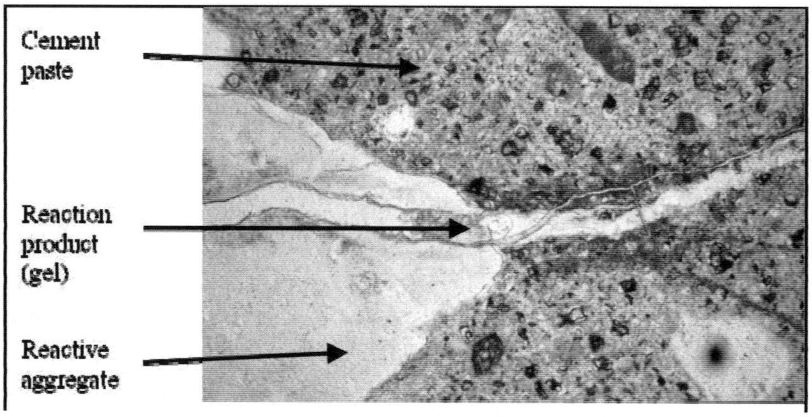

Figure 3. Thin concrete section affected by ASR (field of view approximately 5mm) [2].

Even though the chemical reactions that contribute to this phenomenon are accepted as a mechanism of expansion, other mechanisms are thought to contribute as well, proposed by Thomas, et all [1]:

According to Hansen the surrounding cement paste around reactive grains works as a semi-permeable barrier, not allowing complex silicate ions to flow, but allowing water or pore solution. The reacting grain pulls water inside, where it has the lowest chemical potential. This forms an osmotic pressure cell, and increases the hydrostatic pressure already applied to the paste, causing fissures of the mortar surrounding the grain [4].

McGowan proposed that hydraulic pressure would be reduced, and further expansion would be prevented by the surrounding cement paste "membrane" breaking as a result of ASR. They proposed an alternative mechanism that considers the alkali silica-gel physically absorbing water and expanding as a result [5].

Powers & Steinour issued a compromise by stating that both osmotic and imbibition pressures may be produced, with respect to the alkali-silicate combination being a fluid or solid phase. According to their theory, depending on its makeup, the reaction product itself may function as a semi-permeable barrier [6].

Regardless of the approach, the main cause of swelling is the addition of water to a region where the presence of a solute or adsorption reduces the region's free energy. Numerous experiments have found that considerable expansion only happens when a sufficient amount of calcium hydroxide [$Ca(OH)_2$] is present. Silica dissolves and stays in solution in systems with a lot of alkali hydroxides and reactive silica but no calcium hydroxide. Although the specific function calcium plays in gel expansion is still unknown, several processes have been postulated in the Alkali Alkali-aggregate reactivity facts book [1].

Although the exact function(s) of calcium hydroxide are unclear, it is obvious that Ca is required for harmful reactions to take place. Therefore, decreasing the availability of calcium should reduce the expansion caused by ASR, such as by consuming $Ca(OH)_2$ through pozzolanic processes [1].

According to Conrow, the amount of $Ca(OH)_2$ created by the cement may be connected to the expansion of concretes including siliceous sand-gravel, and the favorable effects of pozzolan are related to the latter material's propensity to interact with $Ca(OH)_2$. The positive benefits of pozzolans in avoiding unacceptable expansion may be explained by the simple consumption of calcium hydroxide in a reaction with a pozzolan. [7, 1].

3. Interrelation between the aggregate size and its expansion

Reactive aggregate particle size can have a significant impact on the amount of mortar expansion, as Stanton demonstrated in his seminal work on ASR. Figure 4 (a) depicts the growth of mortar specimen made of reactive siliceous magnesian limestone that were divided into different sizes. As expected, mortars constructed with the 180-600 micron fraction expanded more than mortars made with the bigger fractions (0.6-2 mm and 2-6.7 mm), even at younger stages. This occurred because the reactive silica has a bigger surface area and is therefore more accessible. Surprisingly, the mortar bars made with reactive aggregate passing a 180 micron pore size showed no adverse reactions. The following two theories were proposed by Stanton, as to why there is no expansion when the aggregate is in a finely-divided form: (1) the response is distributed evenly throughout the bulk without producing strong expanding pressures; or (2) Before the mortar hardens, the reaction is virtually done. Even though the exact size of reactive aggregate size that mitigates the expansion has not been clearly defined, other investigations have later provided comparable findings [8, 1].

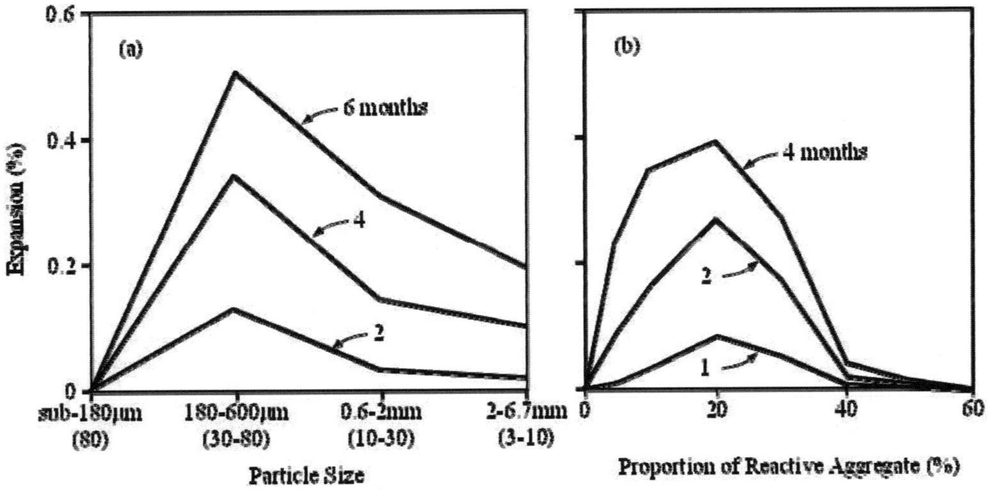

Figure 4. Effect of aggregate size (a) and proportion (b) on expansion due to ASR [2].

Recent research by Fanijo et. al. [9] show that, not only does different aggregate size govern the ASR, but the different size of an aggregate of a different origin influences the probability of the ASR occurrence. The data collected in the research by Fanijo et. al. is presented in the following table:

Table 1. Reported aggregates and their size fraction that are highly prone to ASR, adapted from [9].

Aggregates evaluated for ASR	Aggregate size prone to ASR
Siliceous magnesium limestone containing opal & chalcedony	0.17-0.6mm
opal aggregates	0.02-0.125mm
Opaline aggregate particles	0.07-0.85mm
Various siliceous aggregates	0.15mm
Reactive siliceous aggregates	0.15-0.30mm
Siliceous limestone	1.25-3.25mm
Alpine chlorotic schist aggregate	2 to 4mm
Soda lime glass aggregates	1.18-2.36mm
Spherical glass particles using the mortar bar specimens	2-3mm

4. The Pessimum Effect

The increase of expansion is not always proportionate with the reactive aggregate content. This is another phenomenon observed by Stanton. The expansion of mortar bars is shown in Figure 10. Mortar bars which contained siliceous magnesian limestone were stored over water. In this experiment water played the role of reactive aggregate proportion [8].

The specimen comprised of the reactive sand aggregate had an expansion of 20%. The percentage of expansion lessens by further increasing the amount of reactive aggregate. In specimen that contained more than 60% of reactive sand no notable expansion was observed. The described phenomenon is called the "Pessimum effect" and is characteristic to aggregates of a highly reactive nature. The theory proposed by Hobbs is founded on two assumptions. First off, it is presumable that as the system's ratio of reactive silica and alkali content (SiO_2/Na_2O) rises above the maximum, the gel formed (reaction product) is of a lower alkali-silica ratio (Na_2Oe/SiO_2), which lowers its affinity for absorbing water and, as a result, lowers its capacity to swell. Second, if the content of reactive silica is very high, it is thought that gel forms quickly before the concrete becomes rigid enough to sustain damage. Alkalis are quickly consumed by the rapid gel formation, and once the concrete has dried and become hard, there is insufficient reaction to cause harm [10].

Dent Glasser [3] demonstrated that the amount of dissolved silica is dependent on the silica to alkali ratio (SiO_2/Na_2Oe). As it was previously discussed, the concentration of silica in solution rises in a system of reactive silica and alkali hydroxides up until the point where the equilibrium solubility curve (SiO_2-pH) is achieved, as presented in Figure 2. The silica concentration rises when the silica content is low, dissolving all of the silica before the solubility curve is reached. When the solubility curve is reached in a mixture with a very high silica content, the concentration rises up to the inflection point value on the graph. On the other hand, excess silica from the solution reacts with hydroxyl ions (OH^-) and lowers the pH value. As a result, the amount of silica in solution decreases. Since the entire quantity of SiO_2 dissolves when the solubility curve is achieved, an ideal silica to alkali ratio exists (SiO_2/Na_2O_e) in which silica concentration is at maximum. This represents the lowest reactive silica content in terms of mortar or concrete expansion [11, 1].

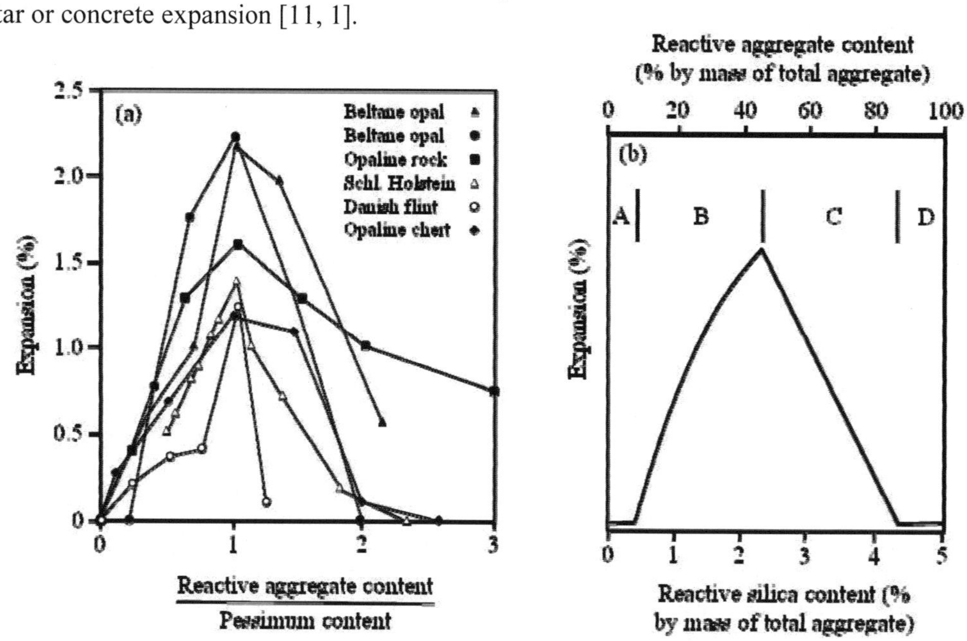

Figure 5. (1) The effect of pessimum for five different opaline material containing aggregates and (2) Four portions of the pessimum [2].

Table 2. Regions of pessimum behavior [10].

Region	Effect	Explanation
A	Reaction but no cracking	Insufficient gel forms to cause damage.
B	Reaction, cracking – excess of alkalis	Expansion continues until all the reactive silica is depleted. Composition of gel is probably independent of the alkali/ reactive silica ratio.
C	Reaction, cracking – excess of reactive silica	Expansion continues until the alkali level is insufficient to sustain ASR. Alkali/silica ratio and swelling capacity of gel decreases with increasing silica content.
D	Reaction but no cracking	Reactive silica content very high and reaction so rapid that the gel forms before concrete has hardened sufficiently and there is no damage. Copious quantities of gel can be observed.

The bottom line is, mentioned in the Alkali-aggregate Reactivity Facts Book [1], that without the following three conditions AAR will not be initiated nor sustained:

1. Presence of reactive mineral forms in aggregate materials
2. Alkali hydroxide ([Na+,K+-OH-]) concentration must be high in the concrete pore fluid
3. Enough moisture must be engaged

Different responses can be expected within different concrete elements affected by AAR. The versatility depends of the conditions stated above.

Common visual effects of the alkali-silica rection are recognizable by [1]:

- Cracking
- Surface pop-outs
- Surface discoloration and gel exudations
- Localized crushing of concrete
- Deformation, relative shifts, and dislocation brought on by expansion
- Extrusion of joint (sealant) material

A maximum concrete alkali content has been used in several specifications as a control for expansion reactive aggregate containing concrete. According to Nixon and Sims, different nations and agencies have set maximum allowable alkali contents from 2.5 to 4.5 kg/m^3 Na_2O_e, including the allowable alkali content, which may vary in dependence of reactivity of the aggregate [12].

According to a recent research paper by Fernandes et al. [13] 'Sodium equivalent' is now a common way to describe alkali content. Numerous concrete structures with ASR degradation were built with high-alkali cement that included more than 1.00 wt% Na_2O_e. To reduce the degradation of concrete when reactive aggregates are used, Portland cement standards frequently have an alkali concentration limit of 0.60 wt%. However, some buildings constructed with an alkalis content below this threshold nonetheless sustain considerable harm.

The value 0.658 denotes the atomic mass ratio of sodium chloride (62.0 g/mol) to potassium chloride (94.20 g/mol), which implicitly assumes that sodium and potassium behave in the same ways under the circumstances present in ASR concrete.

5. Preventing the Alkali-Silica reaction

Numerous laboratory studies have demonstrated that, if the recycled glass is crushed to a specific fineness, it is possible to fully avoid the expansion of mortar or concrete that includes using highly reactive aggregate made of recycled glass. In fact, using larger pieces of the same glass as (reactive) aggregates can help prevent the damming alkali-silica reaction when glass is crushed to a high fineness. However, if the glass has a high alkali content, this method might not work as well. According to Jin et al. as the reactivity of the glass rises, the pessimum size, or the size portion that produces the largest expansion when tested in mortar, decreases. [14, 15, 16].

As a preventive approach against ASR, Iceland has employed the strategy of employing ground reactive aggregate, a mixed cement that contains up to 9% pulverized rhyolite and Portland cement clinker for limiting expansion when used with reactive rhyolitic materials [17, 2].

Some evidence has shown that reactive silica may still cause adverse reactions at aggregate sizes finer than 30 μm concentrations. An observation about the mortar bars expansion made of mortar which is comprised of reactive opal and high-alkali cement, where size ranges from 20 to 30 μm is made in the Alkali-aggregate reactivity facts book (the fraction was separated by sedimentation). It is known to the author that this is the isolated instance in which ASR has been harmed by this particle size range of reactive silica (e.g., 30 μm), and subsequent studies have not supported this behavior.

A few strategies that can help with ASR mitigation in concrete include avoiding reactive aggregates, regulating the alkali level of concrete, employing additional cementing ingredients, and using lithium-based compounds. [1]. The following chapters debates each of these individually. Fundamental factors for harmful ASR occurrence in concrete are:

- Adequate reactive silica content in aggregates
- Sufficient alkali concentration in Portland cement primarily
- Adequate moisture content

By eliminating at least one of the factors the chances of ASR occurring are reduced. Water exclusion from concrete is not a good option for concrete since it disrupts the already designed or required w/c ratio. Hence, the remaining options are:

- Avoiding reactive aggregate use
- Minimization of the amount of alkalis in PC
- Implementation of supplementary cementing materials (SCM)
- Lithium-based compounds implementation

The main difference between the mechanisms of lithium and SCMs is that the SCMs consume alkalis from PC and in that way obstruct their ability to react with the aggregate and lithium compounds change the nature of the reaction product [1].

Another way of preventing the alkali-silica reaction is the use of silica fumes, which are already used up to some extent in the high-performance concrete. The mechanism is described in the Advanced Concrete Technology book [18]. Calcium silicate hydrates are created when silica fume and calcium hydroxide combine, decreasing the pH and reducing the likelihood of reactions brought on by high alkalis. The K and Na ions involved in the creation of the calcium silicate hydrates are locked in the matrix and are unable to interact with any possible siliceous aggregates.

5.1. Use of non-reactive aggregates

Another way to address the non-reactive aggregates is non-deleteriously-reactive aggregates since all aggregates are reactive, but with different products in different conditions and circumstances. Although competent and careful testing is necessary to make sure aggregate sources are not reactive, this is definitely a viable method when it comes to preventing the damage induced by ASR. Petrographic analysis, mortar or concrete expansion testing and field performance can all be used to achieve this. To make sure that the composition (and reactivity) inside a pit or a quarry does not change, regular testing should be performed. There is no need to take any additional safety measures to prevent ASR if the aggregate is not of a reactive nature. The use of non-reactive aggregates is not always practical for a variety of reasons, including, as mentioned in the Alkali-aggregate reactivity facts book [19, 20, 1]:

- Non-reactive aggregates are unavailable in the area, and shipping non-reactive materials from elsewhere would be prohibitively expensive.
- Compared to non-reactive materials, reactive aggregates are easily accessible, abundant, less expensive, and have a lower environmental impact. They are also perfectly suitable for concrete.
- Test results and laboratories are not always viable, as well as unclear results
- Every single locally accessible material fails the extremely accelerated test (regardless of true reactivity)

Despite the use of non-reactive aggregates, there are some situations that call for extra caution. This includes critical structures (prominent constructions, for example, or those with a lengthy design life) and members exposed to violent environment (any environment that provides source of alkalis such as seawater and de-icing salts) [1].

5.2. Limiting the alkali content of concrete

According to Stanton's pioneering research on ASR from 1940, expansive reaction is unexpected when the cement's alkali content is less than 0.60% of Na_2Oe. On the other hand, further studies revealed that the use of low-alkali cements could cause harmful ASR as in the laboratory so in the field [21, 22, 23]. Despite this, the USA accepted the maximum limit of reactive aggregates in cement, and it is listed as an optional limit in ASTM 150 Standard Specification for Portland Cement when concrete contains detrimentally reactive aggregate [23, 22, 21].

Limiting the cement's alkali content ignores the concrete's cement content, which in combination with the alkali content of cement determines the concrete's overall alkali content and is thought to be a better indicator of expansion risk in cases of reactive aggregate implementation. The effect of different alkali contents on expansion on concrete mixtures of different cement contents is depicted in Figure 7. The relationship demonstrates that, as opposed to the cement alkali level alone, the governing alkali content depends on both cement content and the alkali content (alkali level) in the concrete.

Figure 7 shows the concrete expansion in prisms, for three different reactive aggregates and their dependence of alkali content of tested concrete. It is observable that for harmful expansion in the concrete prism test the needed threshold of alkali content is variable, with values for the aggregates shown ranging from roughly 3.0 kg/m^3 (5 lb/yd^3) Na_2O_e to more than 5.0 kg/m^3 Na_2O_e [24, 1].

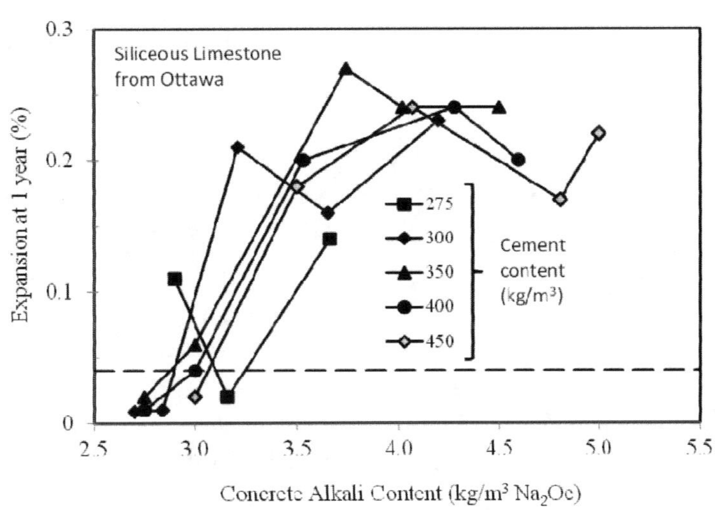

Figure 6. Expansion affected by alkali content contained in concrete [2].

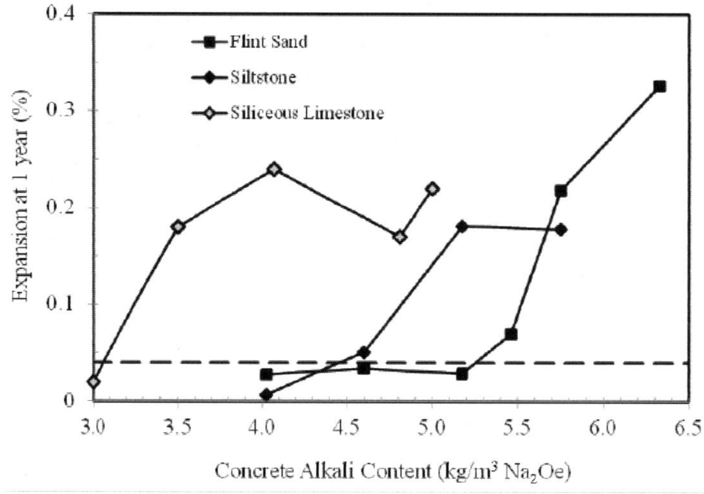

Figure 7. How alkali content affects the prism expansion (stored over water at 38 °C, 100 °F) with different aggregates [25].

Maximum alkali content in concrete is used in multiple specifications as a limit on how much concrete with reactive aggregates can expand. Various nations and organizations have set maximum permissible alkali contents between 2.5 and 4.5 kg/m³ (4.2 to 7.5 lb/yd³) Na_2O_e, according to Nixon and Sims, although aggregate reactivity dictates the allowable alkali content. The alkali content threshold for aggregates in concrete cannot be guaranteed by traditional concrete prism tests, due to alkali content in concrete during exposure. Figure 8 shows the expansion of concrete block specimen (0.38 x 0.38 x 0.71 m) comprised of reactive sand and various alkali levels. The aforementioned blocks are stored on an outdoor exposure site. The results are compared with concrete expansion of concrete prisms, stored over water, made in the same mix (ASTM C 1923). The tests show a greater alkali content threshold than what can be seen in elements of larger dimensions kept in the field, it is to be concluded [12, 1, 25].

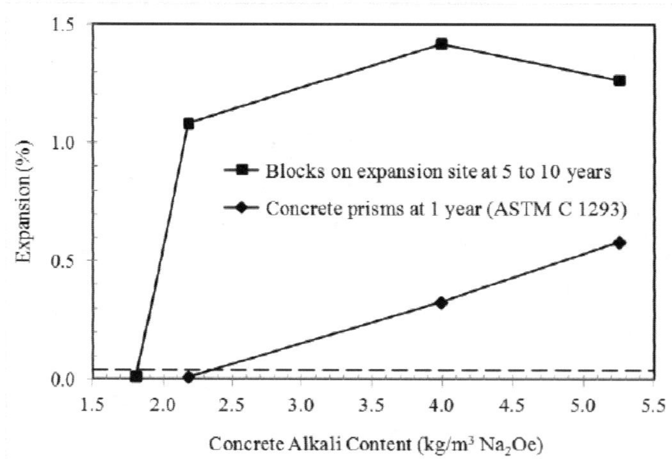

Figure 8. Blocks placed outside and prisms stored above water expand differently depending on the alkali level, according to research (unpublished data from the University of Texas at Austin) [2].

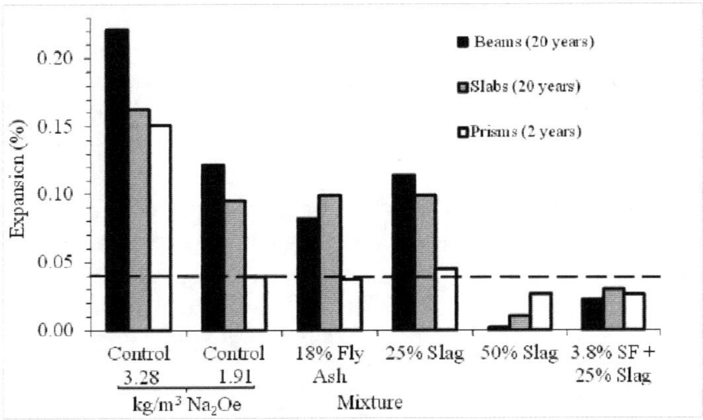

Figure 9. Concrete blocks and slabs expansion at the Kingston exposure site as well as the laboratory's expanding concrete prisms [27].

Alkali migration has been linked to popouts at slab surfaces and can happen very early in the life of concrete. Alkalis may be transported towards the surface in freshly laid concrete together with the bleed water, and this migration may be aggravated both while setting is in progress and when setting process is finished, by the evaporation of water from the slab's surface, particularly in hot, dry weather. A tiny piece of mortar with a conical form on top of the responding aggregate, known as a popout, may break due to the development of gel and the expansion of aggregates in the immediate area of the concrete as a result of the concentration of alkalis at the surface. Alkali migration has been linked to popouts on slab surfaces in the early stages of concrete's life and can happen at any time [7, 28, 2].

5.3. Use of supplementary cementing materials (SCMs)

Almost all supplementary cementing materials (SCMs) are usable in this capacity as long as they are used at a level of replacement that is high enough to effectively control expansion brought on by the alkali-silica reaction. The following factors typically require larger content of SCM:

- The amount of alkali that can be obtained from Portland cement rises (The amount of SCM needed will probably rise if large alkalis are present in the aggregates, such as feldspars or greywackes, or from outside sources.)
- The SCMs alkali content rises.
- The SCMs CaO/SiO_2 ratio rises.
- The aggregate becomes more reactive.

The following ranges (modified from Thomas and Folliard) represent the typical ranges of the SCM content necessary to mitigate harmful expansion caused by the alkali-silica reaction [2]:

Table 3. Required levels of SCM [2].

Type of SCM	Level required (%)
Low-calcium fly ash (< 8% CaO)	20 to 30
Moderate-calcium fly ash (8-20% CaO)	25 to 35
High-calcium fly ash (> 20% CaO)	40 to 60
Silica fume	8 to 15
Slag	35 to 65
Metakaolin	10 to 20

However, under unusual circumstances, the required level of SCM might be higher than these values (e.g., aggregate with a high alkali availability, including that which comes from the aggregate, and particularly reactive aggregate), concrete exposed to high alkali concentrations in service, and critical structure with prolonged service life) [1].

Recent research [29, 30, 31] proves that incorporating silica fume into the concrete mixture is efficient in mitigating ASR since it reduces concrete porosity because of the smaller particle size and hinders the ingress of moisture needed for the reaction. Replacing Portland cement by 0.15 wt% silica flume or even 0.30 wt% fly ash significantly helps mitigate the occurrence of ASR.

Another research [32] shows that if cement content is partially replaced by 30 percent of silica fume the final bar expansion can be reduced by 55%. The lithology, the environment, and the reactivity of the stone aggregates can all have a major impact on the alkali-silica reaction within the examined concrete structures.

Over a period of 50 years, there is a complex interaction between the reactive alkalis, aggregates, and humidity that essentially determines how far the deterioration progresses [33].

5.4. Use of Lithium

McCoy and Caldwell were the first to show that control of the destructive expansion caused by the alkali-silica reaction (ASR) in mortar and concrete could be controlled by the use of lithium. Out of the more than 100 chemical compounds tested, they demonstrated that a number of lithium salts (such as LiCl, Li_2CO_3, LiF, Li_2SiO_3, $LiNO_3$, and Li_2SO_4) were the most promising and could almost completely prevent the expansion of mortar containing Pyrex glass if they were used at levels of replacement that were high enough. Numerous studies conducted since then have supported this earlier discovery [28, 34].

Lithium belongs to the first group of elements in the PSE which are alkali metals (like Na and K), which may sound contradictory that Li-compounds act as ASR suppressant. Although many hypotheses have been put forth, the exact mechanism by which lithium regulates ASR is unknown. Reactive silica reacts with lithium salts similarly as it reacts with Na and K, but the resulting product is an insoluble lithium-silicate barely predetermined to absorb water and swell. This is the most straightforward and widely accepted explanation. Lithium silicate surrounds the reactive aggregate particles and shields the reactive silica beneath from alkali hydroxides' "attack." [34, 1]

McCoy and Caldwell's initial research demonstrated that the lithium amount needed for expansion control was a function of the presence of alkalis different from sodium and potassium in the system, and they came to the conclusion that Pyrex glass contributes to suppression of mortar bars expansion effectively, as long as the lithium-to-sodium-plus-potassium molar ratio was greater than 0.74, or [Li]/[Na+K] > 0.74. Later on, many researchers have shown a connection between available alkali content and required lithium content, but it has been discovered that the minimum value of [Li]/[Na+K] depends on numerous factors, including the form of lithium, the characteristics of the reactive aggregate, and possibly the test method used [34, 35]

Lithium compounds generally have positive effects, but lithium nitrate ($LiNO_3$) is thought to most effectively reduce ASR. The product is currently offered as a 30% solution of $LiNO_3$ and is commercially available from various companies in North America under the name "ASR-suppressing admixture." It takes 4.6 gallons of $LiNO_3$ solution per 1 kg of Na_2O_e to achieve a lithium-to-sodium-plus-potassium molar ratio of [Li]/[Na+K] = 0.74 using a 30% solution of $LiNO_3$ (2.082 L of solution per 0.45 kg Na_2O_e). The "standard dose" of lithium nitrate solution has been described as this.

Aggregate type and its effect on the quantity of lithium needed for expansion suppression caused by ASR has been highlighted in research by Tremblay. Figure 10 depicts the expansion of concrete prisms over two years with various lithium doses (the standard dose is [Li]/[Na+K] = 0.74), 12 different reactive aggregates, and 1 non-reactive aggregate (NF). For 6 of the 12 aggregates, 75% to 100% of the usual dosage was sufficient to suppress growth (0.040% at 2 years). It was necessary to use 125% to 150% of the standard dose for 3 of the aggregates, but even at 150% of the standard dose, expansion could not be stopped for the other 3 aggregates [1, 36].

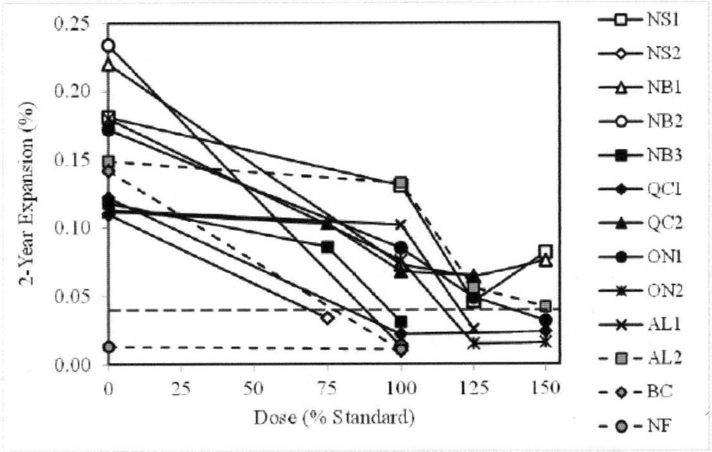

Figure 10. Lithium dose's impact on the expansion of concrete containing various reactive aggregates [2].

Alkali-silica reaction cannot be controlled by a universally prescribed single dose because the effect of lithium highly depends on the aggregate. The minimum lithium dose is established by testing lithium particularly with the chosen aggregate. It is advised to use the concrete prism test for this purpose because there is currently no agreement on whether accelerated tests are appropriate for determining the proper lithium dose.

6. Conclusion

One of the major causes of concrete spalling that this paper focuses on is the alkali-silica reaction (ASR), colloquially called "concrete cancer". ASR is mainly caused by concrete that has a high pH value. The reaction produces a gelous matter which is composed of Na, K, Ca, and Si, and forms around and within the aggregate. Water from the surrounding cement paste and eventually external water (atmospheric) that enters the concrete makes the gel expand. The swelling causes cracking and spalling of the concrete. This can be mitigated mostly by introducing pH-lowering constituents into the cement mixture such as SCMs (supplementary cementing materials) such as fly ash, silica fume, and slag. SCMs can also prevent sulfate attacks. Aggregate size plays another role in ASR. It has been observed that the lesser the particle size, the smaller the expansion, especially at younger ages [2]. Sufficient moisture, reactive mineral aggregates, and high concentrations of alkali hydroxides contribute to the ASR the most [2]. Another way to avoid ASR is the use of lithium.

The Alkali-Silica reaction can be prevented by limiting the alkali content of concrete, by the use of supplementary cementing materials (SCMs) and by the use of non-reactive aggregates (such as carbonate aggregates). By eliminating fundamental factors for harmful ASR content such as sufficient quantity of reactive silica in aggregates, sufficient alkali concentration in Portland cement and sufficient moisture, the chances of ASR occurrence are significantly reduced. Water exclusion from concrete is not advised.

Some studies point out that, in order to mitigate the ASR, the cement alkali content must be kept less than 0.60% of Na_2O_e, but other studies instigate that the low alkali cements cause harmful ASR. That is where the competing alkali constituents are most recognizable. When it comes to HSC, silica fume is a key constituent, which increases the chances for ASR, but not if kept inside the favorable threshold. In other words, mitigating the ASR by limiting the alkali content must be executed very carefully and inside the favorable or allowed thresholds. Further experimental research is recommended.

One of the ways of controlling or mitigating the ASR is the use of supplementary cementing materials and lithium, as mentioned. In order to control the expansion the lithium-to-sodium-plus-potassium molar ratio is the key and it must be greater than 0,74 ($[Li]/[Na+K] > 0.74$). This, again, relates to the “competing constituents” mechanism. The most effective lithium compound, in reducing the ASR, is lithium nitrate ($LiNO_3$). It is commercially available and is offered as a 30% solution. To create a lithium-to-sodium-plus-potassium molar ratio of [Li]/[Na+K] = 0.74 using a 30% solution of $LiNO_3$ (0.55 gallons of solution per 1 lb Na_2Oe), it needs 4.6 gallons of $LiNO_3$ solution per 1 kg of Na_2O_e. Because the impact of lithium is greatly dependent on the aggregate, the alkali-silica reaction cannot be regulated by a single dose that is generally provided. Testing lithium specifically with the selected aggregate determines the minimal lithium dose. There is presently no consensus on whether accelerated tests are acceptable for figuring out the right lithium dose, hence it is suggested to utilize the concrete prism test instead.

References

[1] Thomas M D A at all 2006 Test methods for evaluating preventive measures for controlling expansion due to alkali-silica reaction in concrete *Cement and Concrete Research*, 36(10): p1842-1856

[2] Thomas M A at all, 2013 Alkali-Aggregate Reactivity (AAR) *Facts Book*. Springfield VA, USA

[3] Dent Glasser L and N K 1984 The chemistry of alkali-aggregate reacions *Cement and Concrete Research 1-9*

[4] Hansen W 1944 Studies relating to the mechanism by which the alkali-aggregate reaction proceeds in concrete *Journal of the American Concrete Institute* 213-227

[5] McGowan J K and Vivian H E 1952 Studies in cement-aggregate reaction: Correlation between crack development and expansion of mortars *Australian Journal of applied science* p228-232

[6] Powers T and Steinour H 1955 An investigation of some published researches on alkali-aggregate reaction part 1: The chemical reactions and mechanisms of expansion *Journal of the American Concrete Institute* p497-516

[7] Conrow A 1952 Studios of abnormal expansion of portland cement concrete *Proceedings of the American Society for Testing and Materials* pp 1205-1227

[8] Stanton T 1940 Expansion of concrete through reaction between cement and aggregate *Proceedings of the American Society of Civil Engineers* p1781-1811

[9] Fanijo E O at all 2021 Alkali-silica reaction (ASR) in concrete structures: Mechanisms, effects and evaluation test methods adopted in the United States *Case Studies in Construction Materials Volume 15,* e00563, ISSN 2214-5095, https://doi.org/10.1016/j.cscm.2021.e00563.

[10] Hobbs D W 1988 Alkali-Silica Reaction in Concrete London

[11] Džidić S 2015 Otpornost betonskih konstrukcija na požar *International Burch University Sarajevo ISBN 978-9958-834-47-9 COBISS.BH-ID 22444550* Sarajevo Bosnia and Herzegovina

[12] Nixon P a 1992 Alkali Aggregate Reaction - Accelerated Tests Interim Report and Summary of National Specifications *Proceedings of the 9th International Conference on Alkali-Aggregate Reaction in Concrete Vol. 2 p.731-738. Slough: The Concrete Society*

[13] Fernandes I and Broekmans, M A T M 2013 Alkali–Silica Reactions: An Overview. Part I. *Metallogr. Microstruct. Anal.* 2, p 257–267 https://doi.org/10.1007/s13632-013-0085-5

[14] Cong D L 2004 ASR-induced surface defects *Proceedings of the 12th International Conference on alkali-aggregate reaction in concrete* October 15-19 p 1142-1147

[15] Shayan A and Xu A 2004 Value-added utilisation of waste glass in concrete *Cement and Concrete Research 34* p81-89

[16] Jin W Meyer C and Baxter S 2000 Effects of fly ash on pore solution composition in calcium and sodium chloride-bearing mortars *Cement and Concrete Research 18* 2000 p763-773

[17] Asgeirsson H and Gudmundsson G 1979 Pozzolanica activity of silica dust *Cement and Concrete Research 9* p249-259

[18] Lewis R at all 2003 Cementitious additions, Advanced Concrete Technology *Butterworth-Heinemann* p3-66, ISBN 9780750656863, https://doi.org/10.1016/B978-075065686-3/50279-2

[19] C1293 A 2nd Standard Test Method for Determination of Length Change of Concrete Due to Alkali-Silica Reaction *West Conshohocken: ASTM International*

[20] Carpenter A J and Cramer S M 1999 Mitigation of ASR in pavement patch concrete that incorporates highly reactive fine aggregate *Transportation Research Record 1668* p60-67
[21] Wang H and Gillot J 1991 Mechanism of Alkali-Silica Reaction and the Significance of Calcium Hydroxide *Cement and Concrete Research* p647-654
[22] Stark D 1980 Alkali-Silica Reactivity: Some Reconsiderations *Cement, Concrete and Aggregates* p92-94
[23] Blaikie N Bowling A and Carse A 1996 The assessment & management of alkali-silica reaction in the Gordon River Power Development intake tower *Proceedings of the 10th International Conference on Alkali-Aggregate Reaction* pp 500-507 Melbourne
[24] Lindgård J A C 2012 Alkali-silica reactions (ASR): Literature review on parameters influencing laboratory performance testing *Cement and Concrete Research 40* p223-243
[25] Thomas M D A 1996b Review of the effect of fly ash and slag on alkali-aggregate reaction in concrete *Building Research Establishment Report, BR314, Construction Research Communications Ltd* Watford UK 1996
[26] Thomas M D Nixon P J and Pettifer K 1991 The effect of pfa on alkali-silica reaction *Second CANMET/ACI Conference on the Durability of Concrete* pp. 919-940 Detroit: American Concrete Institute
[27] McDonald C A Rogers C and Hooton R D 2012 "The relationship between laboratory and field expansion – observations at the Kingston outdoor exposure site for ASR after twenty years *Proceedings of the 14th International Conference on Alkali-Aggregate Reaction in Concrete* Austin TX US
[28] Landgren R a 2002 Surface Popouts Caused by Alkali-Aggregate Reactions *Research and Development Bulletin RD121 Portland Cement Association Skokie* 13
[29] Elia H Rock L and Skinner A P 2023 Impact of Supplementary Cementitious Materials on Alkali-Silica Reactivity of Concrete, *Research Gate*
[30] Akhnoukh A K 2020 The use of micro and nano-sized particles in increasing concrete durability *Particulate Science and Technology Journal, Vol. 38, No. 5, pp529-534* https://doi.org/10.1080/02726351.2018.1467988
[31] Akhnoukhv A K and Elia H 2019 Developing high performance concrete for precast/prestressed concrete industry *Elsevier Journal of Case Studies in Construction Materials, Vol. 11* https://doi.org/10.1016/j.cscm.2019.e00290
[32] Akhnoukh A K 2023 Improving Concrete Infrastructure Project Conditions by Mitigating Alkali–Silica Reactivity of Fine Aggregates *Construction Materials, 3(2), pp233-243*
[33] Paglia C S Antonietti S and Corredig G 2023 The environmental degradation of cement-based materials due to alkali-silica reaction: case studies *Journal of Materials and Polymer Science ISSN 2832-9384 Uniscience Publishers Volume 3 Issue 1*
[34] Feng X T 2005 Studies on lithium salts to mitigate ASR-induced expansion in new concrete: a critical review *Cement and Concrete Research* 1789-1796
[35] McCoy W a 1951 New approach to inhibiting alkali-aggregate expansion *Journal of the American Concrete Institute 22(9)* p693-706.
[36] Tremblay C B 2008 Use of the Accelerated Mortar Bar Test to Evaluate the Effectiveness of $LiNO_3$ Against Alkali-Silica Reaction - Part 2: Comparison with Results from the Concrete Prism Test *Journal of ASTM International (JAI) Online ISSN: 1546-962X*, 21

IOP Conf. Series: Materials Science and Engineering 1298 (2023) 012031 doi:10.1088/1757-899X/1298/1/012031

Estimation of channel and flood plain roughness using HEC-RAS model: A case study of the Veseočica River, Bosnia and Herzegovina

Hata Milišić and Emina Hadžić

University of Sarajevo - Faculty of Civil Engineering, Bosnia and Herzegovina

hata.milisic@gmail.com

Abstract. Modeling flow in open channels is essential for determining channel capacity and predicting flood events. The key segment for creating flood hazard and risk maps is a high-quality hydraulic model of the flow in the main bed with its tributaries. Several software packages have been developed for hydrodynamic simulations of flood events. In addition to the geometric characteristics of the watercourse, the equations of a one-dimensional flow model include a parameter that connects all influences that provide resistance to the flow of water. Determining the roughness of a natural channel is one of the most demanding procedures for the development of hydraulic models. Therefore, it is necessary to calibrate and validate the Manning coefficients of the channel roughness using simulation models. In the present study, the estimation of channel and floodplain roughness for the Veseočica River in Bosnia and Herzegovina was performed using the HEC-RAS hydrodynamic model. After calibration and verification of the model, flood hazard maps were created for flood waves with return periods of 20, 100, and 500 years. ArcGIS and HEC-GeoRAS software tools were used to create the flood risk maps.

1. Introduction

Floods are the most common and dangerous natural hazard that cannot totally be avoided. However, some human activities (such as expanding populations and financial investments on floodplains and decreasing natural water retention via land use) as well as climate change are responsible for an increase in the frequency of floods and their negative impacts [1]. Numerous recent studies have been carried out to explore impacts of climate change on flooding in many parts of the world based on observations and/or model simulations [2], such as China [3,4], the United States [5,6], Europe [7,8], as well as Australia [9,10]. However, studies may draw different and even contradictory conclusions about changes in flooding in the same region. Floods are causing increasing havoc in our rapidly urbanizing world, with disproportionally high impacts on the poorest and most vulnerable. Effective adaptation strategies are needed, which combine flood protection infrastructure, nature-based solutions, and risk financing schemes to manage floods and buffer their economic impacts [11]. Adaptive measures to increased flooding risk includes early warning methods, enhancing water retention systems through urban green infrastructure and preserving natural upland ecosystems to store water and slow water flow and good land-use planning including no-build zones [11]. A number of studies on a global scale have shown that the trend of increasing flood magnitude for the most extreme events will continue over the 21st century for many regions of the world [12].

The EU Directive on the Assessment and Management of Flood Risks (Flood Directive) was adopted by the European Parliament and Council of the European Union in 2007 [13] as a measure of protection against floods. Its goal is to create a framework for the assessment and management of flood risks in order to lessen the negative effects of floods on people's health, the environment, cultural heritage, and economic activities.Measures and activities that entirely renounce the flood "control" strategy in favor of the flood risk management strategy have been defined. Three crucial actions are involved in managing flood risk [14].

- doing an initial analysis of the flood risk in river basins and the surrounding coastal waters to find "focal points" that may have a large flood risk,
- creating risk and hazard maps for these "hot spots"; and
- the development of a flood risk management strategy that outlines actions to lower risks and possible outcomes based on preparation, protection, and prevention

Flood hazard and risk maps are used to spatially identify and depict locations with various levels of flood risk [15]. The requirements and expectations of the final user must be taken into account while making flood maps. For the aim of communicating complicated messages regarding floods and flood risk, this is crucial. For a region that has been determined to potentially be at danger of flooding, flood risk maps are created based on hydrological and hydraulic evaluations. In essence, a scenario with a designated region (surface) that would be inundated in some flood probability (for chosen scenarios) is required [15].

Hazard and flood risk maps can be created in several ways. The most basic approach is the application of field surveys, local knowledge, and/or stories in order to create a flood map for a specific event or series of events. Obviously, such maps contain a high degree of uncertainty and cover only the extent of flooding. A more sophisticated approach is the application of remote sensing results. In this case, the extent of flooding for a certain event is plotted on the map, not necessarily at the time of maximum flooding. This methodology is much more accurate and, in some cases, can also provide information on flood depth if combined with a digital terrain model (DTM).

Hydrodynamic models are widely used nowadays to produce maps of flood hazards (and risks). To understand flood risks and hazards more quickly and thoroughly, hydrodynamic model application can make a significant contribution. The floodplain's borders must be established using hydraulic calculations utilizing 1D and/or 2D flow models, regardless of the kind of flood.

In engineering practice, one-dimensional (1D) models are typically used to predict the effects of flood occurrences. A succession of cross-sections that illustrate the topography of the main watercourse bed and floodplains make up one-dimensional models in general. Each cross-section's water levels are estimated based on linear unsteady flow equations for the input data, where the water level in the cross-section is horizontal. It is essentially necessary to supply information on the topography of the terrain in preset cross-sections in order to create a linear flow model. When embankments or high topography limit the width of a floodplain, linear models perform well and may be relied upon to provide accurate estimates of hydraulic quantities when a flood wave comes [16].

Due of complicated topographical circumstances and varying flow direction, linear models have difficulty simulating plane flow in floodplains. The speed and ease with which they may be created make linear models the most advantageous, while their biggest drawback is their inability to effectively describe the hydrodynamics of the flow in situations when there are obvious changes in the flow directions [16].

The usage of plane models has a substantial benefit when it is impossible to employ linear models to simulate a flow that greatly deviates from a linear flow. Planar models, on the other hand, require continuous topographic data that spans the whole region for which the hydraulic model is constructed, as opposed to line models, which characterize the topography of the terrain using a series of cross-sections. In order to discretize the region under consideration spatially, plane models most frequently use a computational grid where each computational cell contains information about the topography of the terrain it covers. Since the accuracy of the findings is most strongly influenced by the quality of

IOP Conf. Series: Materials Science and Engineering 1298 (2023) 012031 doi:10.1088/1757-899X/1298/1/012031

topographical data, obtaining high-quality topographical data is also the main issue when developing plane flow models [16].

Therefore, the benefits of plane modeling may be seen in the accuracy of the outcomes, especially when there is natural flow direction fluctuation. When compared to linear models, the drawbacks may be seen in the greater pool of data needed to create the model, as well as in the time and computer resources needed to run the simulation.

The creation of a combined model, or the so-called "coupling" of 1D and 2D models, is a strategy that reflects a compromise in order to take use of the benefits, or to make up for the drawbacks of linear and planar flow models [16]. In this instance, a linear model is used to simulate the flow's almost one-dimensional main bed, while planar models are used to simulate the flow's naturally flat portions of the considered area (floods area) [16].

2. 1D HEC-RAS Simulation and Flood Inundation Mapping in HEC-RAS

2.1. Model Description

Hydrodynamic simulations frequently use the HEC-RAS model, one of the most used flood modeling programs. Both the program and its calibration accuracy are readily available. In addition to simulating sediment movement, water temperature, and quality, this model was created to simulate 1D steady flow and 2D unsteady river flow [17]. The model includes algorithms for geometric and hydraulic computations for a network of both naturally occurring and artificially created river channels. Research mainly focuses on the HEC-RAS model's capacity to run 1D river flow and 2D flood inundation, as well as the combined approach, particularly the examination of flood inundation caused by a dam or levee break. There are definitely some areas where 2D modeling can produce better results than 1D modeling, and there are also situations in which 1D modeling can produce just as good or better results than 2D models with less effort and computational requirements. If you do not have detailed overbank and channel bathymetry, or if you only have detailed cross sections at representative distances apart, many of the benefits of the 2D model will not be realized due to the poor accuracy of the terrain data. Unfortunately, there is a very large range of situations that fall into a gray area, and one could list the positive and negative aspects of both methodologies for specific applications [17]. In river flood analysis, both the two-dimensional model [18, 19] and the 1D model of HEC-RAS have been extensively employed [20–22]. The most recent version of the HEC-RAS 5.0.7 software package now includes the capability to execute combined 1D and 2D unsteady flow routing inside the unsteady flow model. This feature enables work on a wider river system and restricts the use of 2D modeling to places that need a greater degree of hydrodynamic precision [23, 24].

Figures 1 and 2 show the case of a complex bed in which a 1D model (for the basic bed) and a 2D hydraulic calculation model (for inundations) should be combined [25].

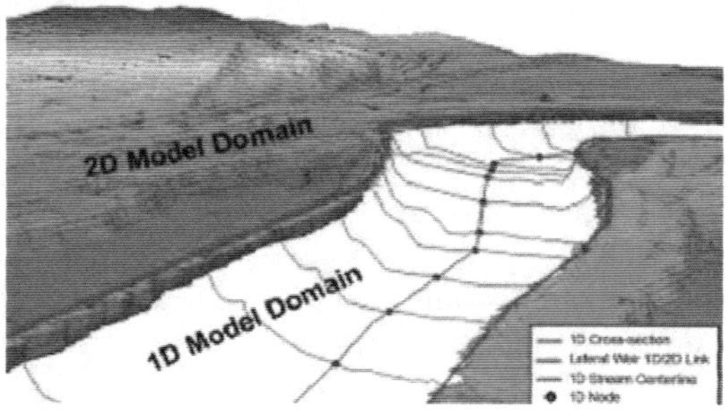

Figure 1. An example of a complex bed that should be analysed with 1D and 2D models [25].

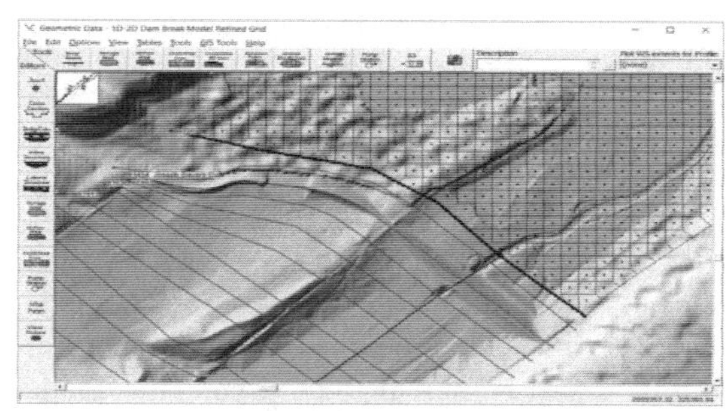

Figure 2. Utilizing Lateral Structure Models to Connect a 2D Flow Area to a 1D River Reach [25].

With the aid of this software, flood inundation and flood mapping of the low-relief area affected by flooding can be easily traced to provide a greater response to several return interval storm actions and has the ability to transform the predicted data result into a strategy opinion magnitude of submerging, according to research done on various literatures [26–29] related to floodplain mapping and the use of HEC-RAS and ArcGIS. The use of HEC-RAS software is regarded as a crucial instrument for urban and municipal expansion, emergency response plans, flood insurance, and natural research [30]. When used in conjunction with GIS technologies, the model is thought to be a valuable tool for calculating the depth of the water. GIS software makes it simple to compute flood-inundated regions and create velocity maps. By utilizing physical features with different modeling methodologies and the accessibility of multiple records, data, etc., the HEC-RAS program is effective for researching impending risky scenarios [30]. On the American Army Corps of Engineers website, the most recent HEC-RAS version is freely accessible under a public domain license [31].

2.2. The Saint-Venant equations

Natural watercourses are characterized by turbulent, spatially three-dimensional, and temporally variable-unsteady flow. For easier analysis, natural flows are reduced to planar ones by depth averaging (two-dimensional in the horizontal plane) flows, and they, by further normalization by the width of the bed, into linear (spatially one-dimensional) flows.

The basic equations of linear (1D) unsteady flow in open natural channels (The Saint-Venant) are the continuity equation and the dynamic equation. The Saint-Venant equation (Equations (1)–(2)), first formulated by De St. Venant (1871) is almost always used to model the flow [32].

The continuity equation for 1D flow is written as:

$$\frac{\partial Q}{\partial x} + \frac{\partial A}{\partial t} + q = 0 \tag{1}$$

Where Q is the flow rate, A is the cross-sectional area, and q is the lateral inflow.

The momentum equation for the 1D flow is given as

$$\frac{\partial Q}{\partial t} + \frac{\partial}{\partial x}\left(\frac{Q^2}{A}\right) + gA\frac{\partial H}{\partial x}\left(S_0 + S_f\right) = 0 \tag{2}$$

Where A = cross-sectional area normal to the flow; Q = discharge; g = acceleration due to gravity; H = elevation of the water surface above a specified datum, also called stage; S_0 = bed slope; S_f = energy slope; t = temporal coordinate; and x = longitudinal coordinate.

The most effective and widely used method for solving the one-dimensional unsteady flow equations is the four-point implicit scheme, often known as the box scheme. In semi-implicit form, which translates to a weighting factor (θ) of 0.6 for unsteady flow simulation, it has been demonstrated that this numerical scheme is totally non-dissipative but only marginally stable. Users can choose any value between 0.6 and 1 in HEC-RAS, while the default value is 1 [33].

3. Study

3.1. Study Area

Particularly in 2010 and 2014, Bosnia and Herzegovina repeatedly suffered major water damage. This was largely due to the inadequacy of the nation's infrastructure for flood protection as well as its flood warning and prevention systems. The Sava River's tributaries, the Drina, Una, Ukrina, Bosna, and Vrbas, frequently experience flooding [34].

3.1.1. The Vrbas River Basin. The basin of the Vrbas River. With a total area of 6,288.5 km^2 and an estimated 900 km2 of karst, the Vrbas River is a right tributary of the Sava. At 1,715 meters above sea level, the springhead of the Vrbas River is situated on the southern slope of the Vranica Mountain. At a height of 90 meters, the Vrbas empties into the Sava River close to Srbac. Between its source and its mouth, the Vrbas River reaches over a length of 235 km. The mainstream has an average slope of 6 m/km, which gives it an elevated velocity and makes it useful for energy production. The catchments of 11 of the Vrbas's tributaries are greater than 100 km^2.

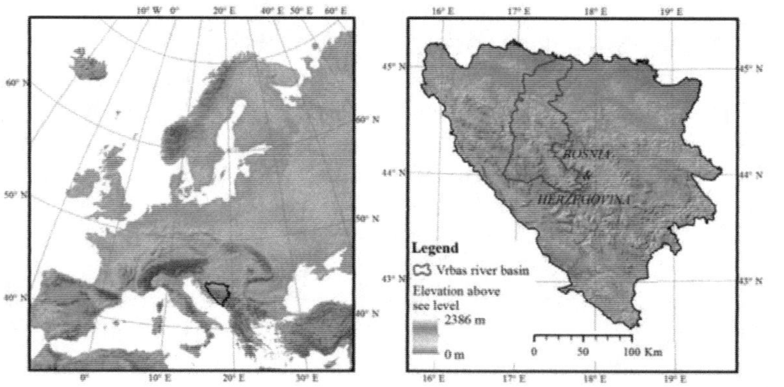

Figure 3. Location of study area - the Vrbas River Basin (Bosnia and Herzegovina) [34].

3.1.2. The Veseočica River. The river Veseoica is a tributary of the river Vrbas and is located in the municipality of Bugojno (Figure 4). The stream is about 7 km long, and the area of its basin is about 135 km2. Duboka and Prljanica are two rivers that form Veseoica. Veseoica damages the river bank during the transitional seasons when it overflows. These floods are essentially the result of the pluvial-marine regime of the Duboka River [35].

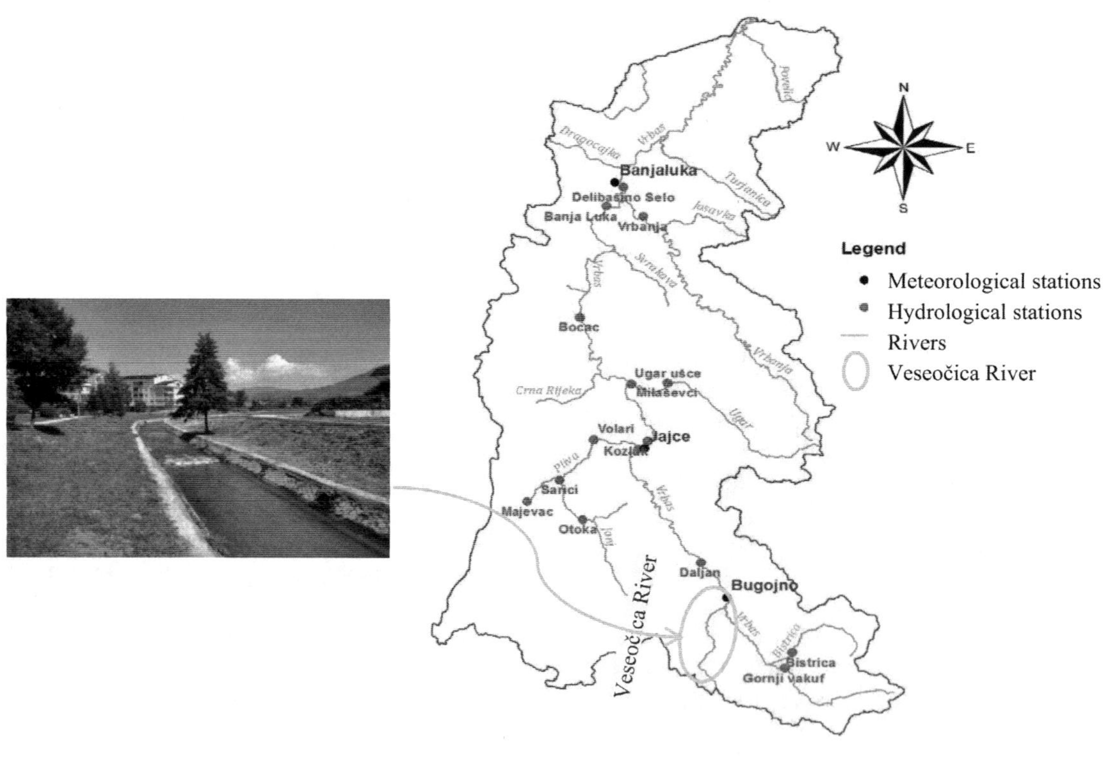

Figure 4. Location of study area: The Vrbas River Basin and the Veseočica River [35].

In the Bugojno municipality, many riverbeds are underdeveloped and unable to manage a higher flow, which is why water flows from riverbeds (Vrbas, Veseoica, and Porinica) to coastal areas and floods extremely important areas along watercourses. In the municipality of Bugojno, during times of high precipitation and melting snow, fresh springs appear that develop into streams and torrents that harm property and endanger lives, particularly in the areas flowing through the city, the suburbs, and rural communities. Floods with these characteristics happened in December 1906, November 1925, November 1969, November 1975, November 1981, and April 2001, March and April 2004, and January 2010 [35]. Overall, the flood protection works so far have been partial and have focused primarily on the protection of only the central parts of some settlements. Relatively small results have been achieved, so no settlement has a secure level of flood protection.

3.2. *Data collection and processing*

Reliable and detailed input data is crucial for the development of any model. In this study, the data used to develop the hydraulic model of the Veseočica River will be presented. In the following, the basic input data (modeling area, geodetic foundations, and boundary conditions – hydrological data) will be given for creating a hydraulic model of the Veseočica River.

3.2.1. Geometric Data. The geometry of the watercourse should be detailed enough to describe the hydraulic situation of the considered area. Cross sections should be placed at representative locations to capture important changes in geometry. The following data were used to define the geometry of the Veseočica river bed [36]:

- Cross-section profiles were used to record the Veseoica River's main bed, together with a LIDAR image of the valley area (with a greater inundation zone on both banks) and
- Aero photo background of the LIDAR recording area (with a basic trough and inundation).

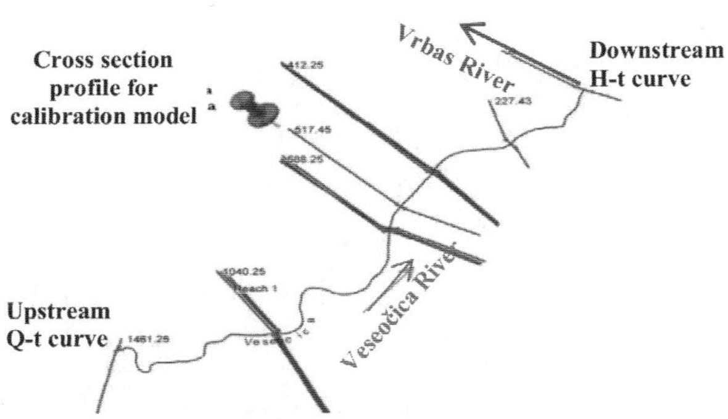

Figure 5. Geometry of the study river reach in HEC-RAS.

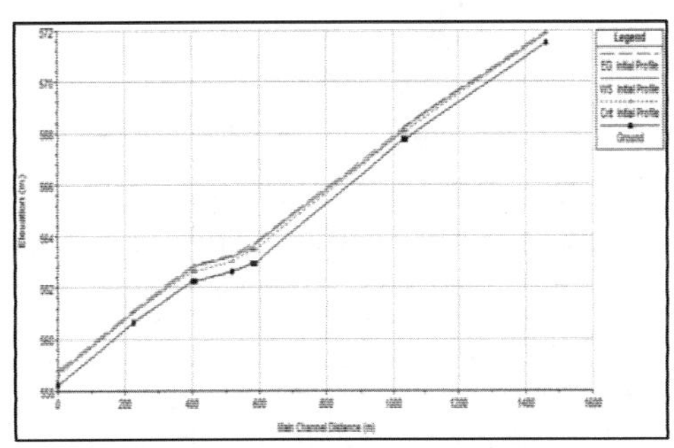

Figure 6. Longitudinal views of the study river reach in HEC-RAS.

When combined with the geometry of the LIDAR survey of the inundation terrain, the cross profiles that were acquired for hydraulic analysis permitted the building of a reliable geometrical terrain model for the Veseočica River [36]. In this way, the mathematical model's input parameter, which is necessary for running simulations of hydrodynamic flow, was collected. Calculating hydraulic parameters such as the depth, speed, and other features as well as the height of the water table and the degree of flooding is required. The ArcGIS software extension HEC-GeoRAS 5.0 was used for all HEC-RAS preprocessing (definition of transverse sections, channel and margins, and geometric data) [31].The relevant section was around 1.46 kilometers long in total. 13 cross profiles made up the computation section (see Figure 5), from which the longitudinal profile of the Veseočica River (see Figure 6) was created. In this part of the watercourse, there are three bridges. Because the objects on the watercourse are not included in the calculation, the roughness coefficients include the local resistances of the bridges.

3.2.2. Hydrological data. Data on the hydrological regime of watercourses, that is, on water levels and flows, are important bases for hydraulic modeling and the creation of flood hazard maps. The Veseočica River usually overflows its banks during transitional seasons and causes significant damage. The values of the maximum flows for the characteristic probabilities of occurrence are listed in Table 1 [36].

Table 1. Maximum discharge values for expected return periods (years) on the Veseočica River [36].

Return period (years)	Maximum discharge/ Q_{max} (m^3/s)
20	31.10
100	51.96
500	76.71

3.2.3. Flow boundary conditions /Upstream and downstream/. Floods for two years, 2014 and 2015, were selected for model calibration and verification as peak discharges under flood conditions. The flow hydrograph (**Q-t curve**) for 2015 was used as the upper boundary condition on the most upstream profile (Figure 7), and the stage hydrograph (**H-t curve** - *at the confluence Veseočica River with the Vrbas river*) was used as the downstream condition (Figure 8) for the calibration hydraulic model [36].

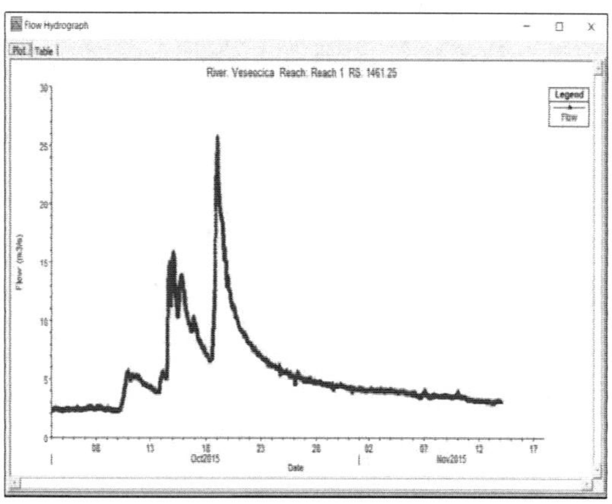

Figure 7. Flow Hydrograph – upstream boundary condition [36].

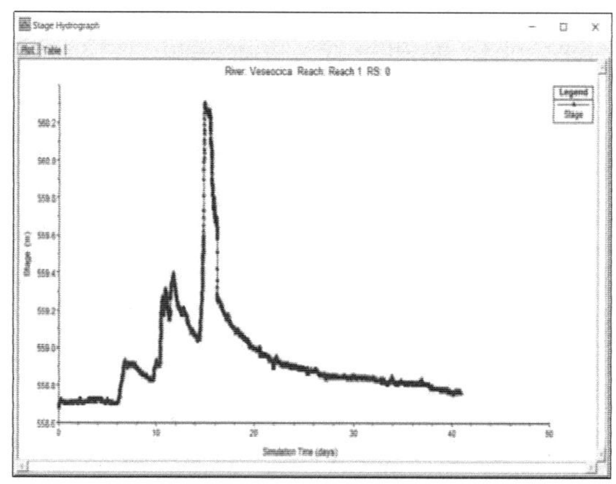

Figure 8. Stage Hydrograph – downstream boundary condition [36].

3.2.4. Manning's roughness coefficient. Manning's roughness coefficient is a very important calibration parameter [37-39]. Channel roughness is not a constant parameter, and it varies along the river depending on the variation in channel characteristics along the flow. Manning's coefficient "n" varies significantly and depends on a large number of factors, such as surface roughness, vegetation and overgrowth of the bed, channel irregularities, branching or merging of watercourses, obstacles, size and shape of the channel, channel slope, water height and flow, seasonal changes at different seasons, influence of temperature, erosion, and suspended material. The exact values of Manning's roughness coefficient are unknown for the Veseočica River, and the selection of appropriate values is of great importance for the accuracy of the calculation of flow in the open watercourse.

Table 2 shows the values of the calculated resistance coefficients based on which the hydrodynamic model of the Veseočica River was calibrated in the section under consideration. The table shows the roughness coefficient values along the watercourse, where "n_1" represents the roughness coefficient of the left bank, "n_2" the roughness coefficient of the main bed, and "n_3" is the roughness coefficient of the right bank. To calibrate the model over the entire height of the bed (i.e., over the entire amplitude of the water level (low, medium, and high water)), it was necessary to more precisely define the n_2* roughness coefficient of the main bed, that is, the resistance coefficients in the lower and middle zones of the bed.

Table 2. Values of Manning's coefficient of roughness by cross-section height for profile P6.

	Manning's coefficient of roughness		
Water level on profile P6	Left bank n_1	Main bad n_2^*	Right bank n_3
Low water: 563.00 mn.m		0.050	
Medium water: 564.00 mn.m		0.035	
High water: 565.00 mn.m	0.035	0.020	0.035

The model was calibrated to high-water values and input data. As the bed consists of large, gravelly materials, it is subject to changes, especially when it comes to large nonstationarity. Given that the morphology of the watercourse is subject to changes owing to cleaning, regulating the watercourse, etc., there is also a change in the Manning roughness coefficient. In this case, the geometry of the watercourse should be verified for each subsequent use of the model.

4. Results and discussion

4.1. Model calibration and estimation of Manning's n values

Hydraulic models must be calibrated and validated against floods of various magnitudes, with particular emphasis on the largest observed flood event. This should be combined with hydrological-hydraulic model calibration. It should also be rigorously tested using scenario testing to examine the robustness of calibrated parameters. In this study, the model was calibrated and verified based on recorded floods (2015 and 2014, respectively).

The timing of the event in respect to morphological and vegetative changes was taken into account during calibration. There is always some uncertainty in the assessment of "n" of a river's reach, which is why a variety of approaches are used [39]. One technique for calculating "n" is calibration using the inflow and outflow hydrographs that were actually observed in a river stretch. The chosen flow routing model's computed outflow hydrograph must match the observed outflow hydrograph for the calibration procedure to be successful [39]. This is done by modifying "n" to the proper value. The current calibration techniques typically take into account uniform roughness along a river's reach. This work introduces a new method for calibrating "n" that can distinguish between distinct river reaches with differing "n" even when there is only one recorded inflow and matching outflow hydrograph available.

Because there is no gauging station on the Veseočica River, the model calibration was performed in relation to the results of the numerical model MIKE11 [36] for the selected cross-section profile P6 at station 0+517,45 m for the period October 4, 2015, to November 14, 2015. The calibrated values of Manning's roughness coefficient for the considered watercourse segment range in the main bed from 0.05 for low, average, and 0.02 for high waters and 0.035 for the inundation area. The calibration results were in good agreement with the results of the numerical model MIKE11 [36]. The calibration results are shown in Figure 9.

Model verification was performed in relation to the results of the numerical model MIKE11 [36] for the period from April 7, 2014, to June 6, 2014. The results of the model verification for cross-section profile P6 at station 0+517.45 m are shown in Figure 10. Good agreement between the results of the numerical model MIKE11 and the HEC-RAS model simulation result was determined by the coefficient of determination (R^2). The 2014 extreme event calibration gave coefficient of determination (R^2) values of $R^2 = 0.9979$, and for the 2015 validation event, the estimated R^2 was 0.9965.

Considering that verification is debatable in the case of watercourses whose morphology is subject to changes over time, it is necessary to make possible changes and re-verify the model. As the model calibration was performed for 2015, it is possible that the morphology of the watercourse may change due to cleaning, regulation of the watercourse bed, etc., which may affect the results of the model.

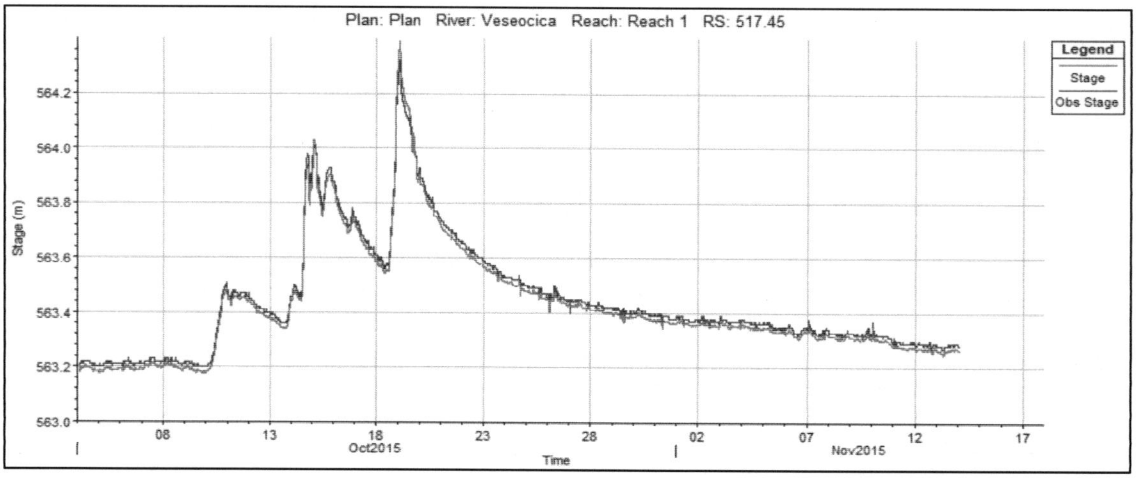

Figure 9. Model calibration: comparative view of the results of the numerical model MIKE11 and the simulated Water surface elevation at the cross-section profile P6 (4.10.–14.11.2015).

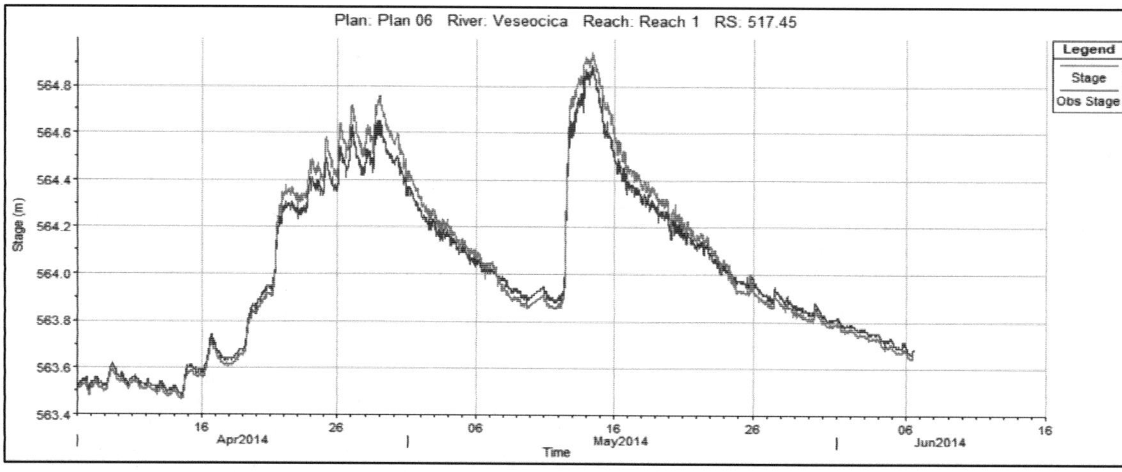

Figure 10. Model verification: a comparative view of the results of the numerical model MIKE11 and the simulated Water surface elevation at the cross-section profile P6 (7.04.–06.06.2015).

5. Sensitivity analysis of Manning's roughness coefficient

Sensitivity analysis of the roughness coefficient was performed for the 1D model to determine its influence on the model results. Given that Manning's roughness coefficient is an important calibration parameter and that the model is sensitive to its change, it is necessary to change the roughness coefficient according to the cross-section height. This was done only for profile P6 ("517.45 m); therefore, the roughness coefficient was selected for low, medium, and high waters. In addition, to obtain an even better agreement between the results, it is possible for each cross-section profile to change the roughness coefficient according to the height of the cross-section.

The results are presented in Figures 11, 12, 13, and 14, where it can be seen that the change in Manning's roughness coefficient by cross-section height provides the best agreement between the results of the HEC-RAS numerical model and the results of the MIKE11 numerical model.

Table 3. Presentation of the coefficient of determination for different "n" on the profile P6

Cross-section P6 – 0+517.45"	Coefficient of determination R^2
n = 0.050 - the same for the main bed	0.9953
n = 0.035 - the same for the main bed	0.9972
n = 0.020 - the same for the main bed	0.9945
Variable "n_2*" by cross-section height for the main bed	0.9979

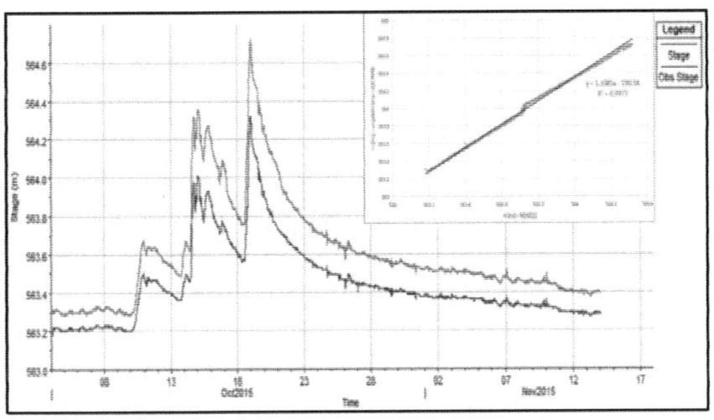

Figure 11. Flow Hydrograph for n=0.050.

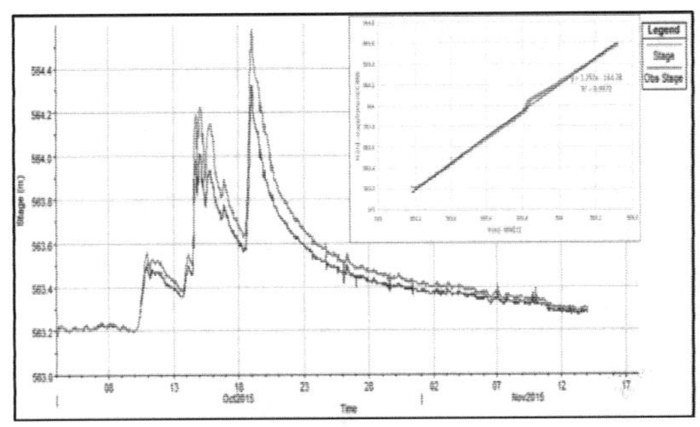

Figure 12. Flow Hydrograph for n = 0.035.

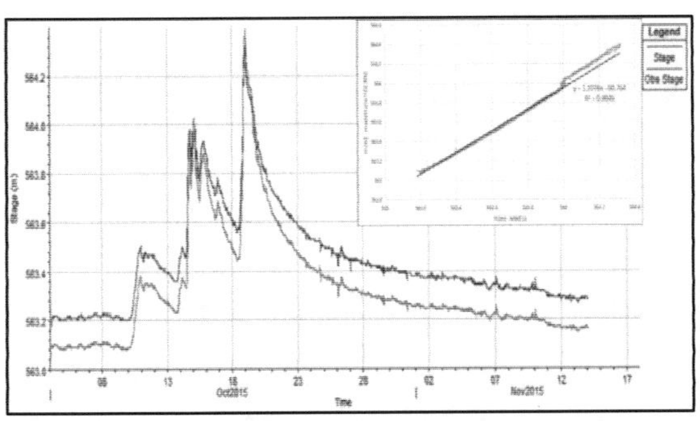

Figure 13. Flow Hydrograph for n=0.020.

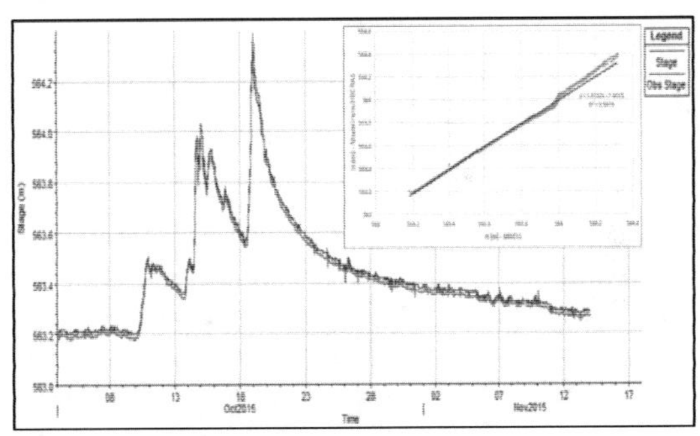

Figure 14. Flow Hydrograph for variable n^* = 0.05 - 0.035- 0.02.

6. Mapping of flood inundation areas of the Veseočica River

After the hydraulic model calibration, validation, and sensitivity testing, hydrographs Q20, Q100, and Q500 were run through the hydraulic model to generate appropriate flood maps. In this study, we developed the first flood vulnerability assessment for floods with 5% (20-year), 1% (100-year), and 0.2% (500-year) recurrence intervals within the Veseočica River floodplain based on a Light Detection and Ranging (LiDAR)-derived Digital Elevation Model (DEM) and 1D HEC-RAS hydraulic modeling.

These maps show the submergence areas in the plan maps, which are shown in Figure 15 for the floodplain and water depth. The range of the flooded area is 0.0898–0.1708 km^2.

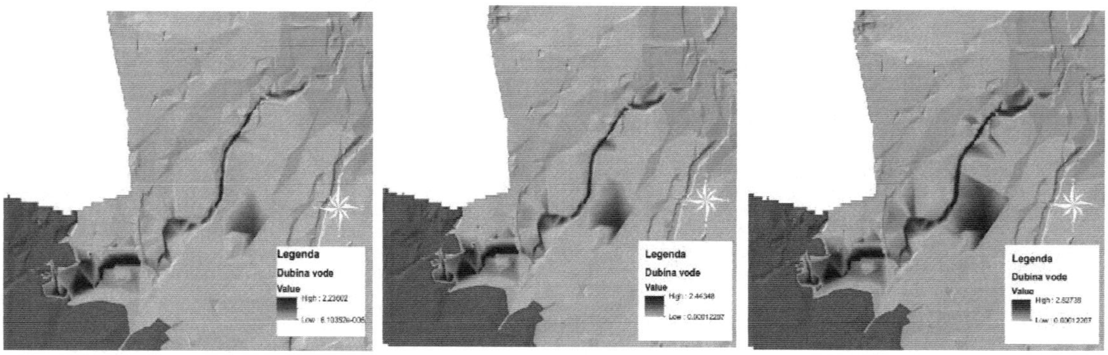

Figure 15. Flood maps of the Veseočica River basin computed by HEC-RAS for high waters with a return period of 20 (left), 100 (in the middle) and 500 years (right).

7. Conclusions

River flooding resulted in significant economic as well as human losses. As a result, river nature research is required. The complexity of river behavior makes computer models useful instruments for researching and simulating river behavior at the lowest possible cost. The effects of floods can be predicted using hydraulic models.

Today, in engineering practice, software packages with different capabilities, such as MIKE and HEC-RAS, are most often used for hydraulic calculations of stationary and non-stationary flows in networks of open river flows with irregular geometries. The most popular software, owing to its availability and applicability, is HEC-RAS. Its advantage is its compatibility with other today's standard software packages, such as CAD and ArcGIS, which prepare the input data for the hydraulic model.

In this study, the HEC-RAS 1D model was developed for the river reach Veseočica (the left tributary of the Vrbas River in Bosnia and Herzegovina), incorporating selected cross sections. Using the measured stage and discharge data in the selected cross-section, the HEC-RAS model was calibrated to determine the appropriate Manning roughness values for the selected cross-section. After the calibration and verification of the Veseočica River model, flood hazard maps were created for return periods of high water events of 1/20, 1/100, and 1/500 years. To analyze the flooded areas for different return periods, the stationary flow hydraulic model was imported into the ArcGIS/Hec-GeoRAS software, where the size of the flooded area was calculated for each return period. The size of the flooded area ranged from 0.0898 to 0.1708 km^2.

The key contribution of this research should be reflected in the planning and project activities in the watershed, because it should be the starting point for the creation of project and planning-spatial documentation in the watershed, in coordinating the prevention and assessment of flood damage, and in the establishment of security measures in areas with flood risk, as is practiced in the countries of the European Union.

8. References

[1] Jerome Glago F. (2021) Flood Disaster Hazards; Causes, Impacts and Management: A State-of-the-Art Review [Internet]. *Natural Hazards - Impacts, Adjustments and Resilience. IntechOpen*; Available from: http://dx.doi.org/10.5772/intechopen.95048

[2] Aryal, A.; Acharya, A.; Kalra, A. (2022) Assessing the Implication of Climate Change to Forecast Future Flood Using CMIP6 Climate Projections and HEC-RAS Modeling. *Forecasting* 4, 582-603. https://doi.org/10.3390/forecast4030032

[3] Li, J., Chen, Y. D., Zhang, L., Zhang, Q., Chiew, F. H. S., (2016) Future changes in 648 floods and water availability across China: linkage with changing climate and 649 uncertainties. *Journal of Hydrometeorology* **17**, 1295-1314. 650 https://doi.org/10.1175/JHM-D-15-0074.1

[4] Zhang, Q., Gu, X., Singh, V. P., Xiao, M., (2014) Flood frequency analysis with 779 consideration

of hydrological alterations: Changing properties, causes and 780 implications. *Journal of Hydrology* 519, 803-813. 781 http://dx.doi.org/10.1016/j.jhydrol.2014.08.011

[5] Archfield, S. A., Hirsch, R. M., Viglione, A., Blöschl, G., 2016. Fragmented patterns 535 of flood change across the United States. *Geophysical Research Letters* 43, 536 10232-10239. https://doi.org/10.1002/2016GL070590

[6] Mallakpour, I., Villarini, G., (2016) Investigating the relationship between the 673 frequency of flooding over the central United States and large-scale climate. 674 *Advances in Water Resources* 92, 159-171. 675 http://dx.doi.org/10.1016/j.advwatres.2016.04.008

[7] Bierkens, M. F. P., and L. P. H. van Beek, (2009) Seasonal predictability of European 547 discharge: NAO and hydrological response time. *Journal of Hydrometeorology* 548 10, 953–968. https://doi.org/10.1175/2009JHM1034.1 549

[8] Blöschl, G., Hall, J., Parajka, J., (2017) Changing climate shifts timing of European 26 550 floods. *Sciences* 357, 588-590. https://doi.org/10.1126/science.aan2506

[9] Liu, J., Zhang, Y. Q., (2017) Multi-temporal clustering of continental floods and 659 associated atmospheric circulations. *Journal of Hydrology* 555, 744-759. 31 660 https://doi.org/10.1016/j.jhydrol.2017.10.072

[10] Gu, Xihui, et al. (2020) The changing nature and projection of floods across Australia. *Journal of Hydrology* 584, 124703.

[11] Jongman, B., (2018) Aadaptation to rising flood risk. *Nat Commun* 9, 1986 https://doi.org/10.1038/s41467-018-04396-1

[12] Eccles, R., Zhang, H., Hamilton, D., Trancoso, R., and Syktus, J., (2021) Impacts of climate change on streamflow and floodplain inundation in a coastal subtropical catchment. *Advances in Water Resources* 147 103825 1-15. https://doi.org/10.1016/j.advwatres.2020.103825

[13] European Union. (2007) Directive 2007/60/EC of the European Parliament and of the Council of 23 October 2007 on the Assessment and the Management of Flood Risks (OJ L 288, 6.11.2007).

[14] https://www.icpdr.org/main/activities-projects/implementation-eu-floods-directive

[15] Van Alphen, J., Martini, F., Loat, R., Slomp, R. and Passchier, R., (2009) Flood risk mapping in Europe, experiences and best practices. *Journal of Flood Risk Management*, 2: 285-292. https://doi.org/10.1111/j.1753-318X.2009.01045.x

[16] Betsholtz N., Nordlof B., (2017) Potentials and limitations of 1D, 2D and coupled 1D-2D flood modeling in HECRAS – A case study on Hoje river, Lund - Sweeden.

[17] Brunner et al., (2015) Combined 1D and 2D Modeling with HEC-RAS. World Environmental and Water Resources Congress http://dx.doi.org/10.1061/9780784479162.141

[18] Quiroga, V.M.; Kure, S.; Udo, K.; Mano, A. (2016) Application of 2D numerical simulation for the analysis of the February 2014 Bolivian Amazonia flood: Application of the new HEC-RAS version 5. *RIBAGUA* 3, 25–33 https://doi.org/10.1016/j.riba.2015.12.001

[19] Kadir, M.A.A.; Abustan, I.; Razak, M.F.A. (2019) 2D Flood Inundation Simulation based on a Large-Scale Physical Model using Course Numerical Grid Method. *Int. J. GEOMATE* 17, 230–236. https://doi.org/10.21660/2019.59.icee17

[20] Zainalfikry, M.K., Ab Ghani, A., Zakaria, N.A., Chan, N.W. (2020). HEC-RAS One-Dimensional Hydrodynamic Modelling for Recent Major Flood Events in Pahang River. In: Mohamed Nazri, F. (eds) *Proceedings of AICCE'19*. AICCE 2019. *Lecture Notes in Civil Engineering*, vol 53. Springer, Cham. https://doi.org/10.1007/978-3-030-32816-0_83

[21] Hadžić E., Mulaomorević-Šeta A., Milišić H., Lazović N. (2018). Estimation of Peak Flood Discharge for an Ungauged River and Application of 1D Hec-Ras Model in Design of Water Levels. In: Hadžikadić M., Avdaković S. (eds) *Advanced Technologies, Systems, and Applications II. IAT 2017. Lecture Notes in Networks and Systems*, vol 28. Springer, Cham https://doi.org/10.1007/978-3-319-71321-2_52

[22] Hadžič, E., Aronica Tito, G., Milišić, H., Šuvalija, S., Ključanin, S., Šarić, A., Sulejmanović, S., and Mujić, F., (2022) Development of flood hazard and risk maps in Bosnia and Herzegovina,

key study of the Zujevina River. *Coupled Systems Mechanics*, Vol. 11, No. 6 (2022) 505-524. https://doi.org/10.12989/csm.2022.11.6.505

[23] Horritt M, Bates P (2002) Evaluation of 1D and 2D numerical models for predicting river flood inundation. *J Hydrol* 268(1–4):87–99 https://doi.org/10.1016/s0022-1694(02)00121-x

[24] Yadav, R. and Yadav, S. M.: (2021) Assessment of 1D/2D coupled model for prediction of flood inundation, *EGU General Assembly 2021*, online, 19–30 Apr 2021, EGU21-4881, https://doi.org/10.5194/egusphere-egu21-4881

[25] Tayefi, V.; Lane, S. N.; Hardy, R. J.; Yu, D. (2007) A comparison of one-and two-dimensional approaches to modelling flood inundation over complex upland floodplains. *Hydrological Processes: An International Journal*, 21(23), 3190-3202. https://doi.org/10.1002/hyp.6523.

[26] Yang J., Townsend R.D., Daneshfar B., (2006), Applying the HEC-RAS model and GIS techniques in floodplain delineation, *Canadian Journal of Civil Engineering*, **33**, 19-28,

[27] Milišić, H., Hadžić, E., Šuvalija, S., Jahić, E. (2021). Floodplain Mapping Using HEC-RAS and Lidar Data: A Case Study of Bistrica River (Vrbas River Basin in B&H). In: Karabegović, I. (eds) *New Technologies, Development and Application IV. NT 2021*. (pp 1093-1103) Lecture Notes in Networks and Systems, vol 233. Springer, Cham. https://doi.org/10.1007/978-3-030-75275-0_121

[28] Azouagh, A.; El Bardai, R.; Hilal, I.; El Messari, J.S. (2018) Integration of GIS and HEC-RAS in Flood modeling of Martil River (Northern Morocco). *Eur. Sci.* J. 2018, 14, 12.

[29] Eric C. Tate, David R. Maidment, Francisco Olivera and David J. Anderson (2002)"Creating a Terrain Model for Floodplain Mapping" *Journal of Hydrologic Engineering-ASCE*, volume 7(02), 2002, Pg.100-108. https://doi.org/10.1061/(ASCE)1084-0699(2002)7:2(100)

[30] F.A.Maniyar , J.P.Bhatt, (2015)Literature Study on Hydraulic modelling of floodplain mapping, IJRET: *International Journal of Research in Engineering and Technology* eISSN: 2319-1163 | pISSN: 2321-7308, Volume: 04 Issue: 11 | Nov-2015, Available @ http://www.ijret.org

[31] US Army Corps of Engineers, https://www.hec.usace.army.mil/software/hec-ras/features.aspx

[32] Akbari, G.; Firoozi, B. (2010) Implicit and explicit numerical solution of Saint-Venant equations for simulating flood wave in natural rivers. *In 5th National Congress on Civil Engineering*,7. Iran: Feridowsi University of Mashhad.

[33] US Army Corps of Engineers (2016), HEC-RAS-River Analysis System, Hydraulic Reference Manual, Version 5.0.

[34] Tošić, R., Lovrić, N., Dragićević,S., & Manojlović,S. (2018) Assessment of torrential flood susceptibility using gis matrix method: case study - Vrbas river basin (B&H), *Carpathian Journal of Earth and Environmental Sciences*, Vol. 13, No. 2, p. 369 - 382;

[35] Šanjta A. (2020) Numeričke simulacije tečenja u prirodnim vodotocima, *Master rad*, Građevinski fakultet Univerziteta u Sarajevu.

[36] UNDPBIH-16-026-VRBAS (2016) Zavod za vodoprivredu Bijeljina i Sarajevo __Hydraulic Report: Izrada mapa opasnosti i mapa rizika od poplava na slivu rijeke Vrbas u BiH – Razvoj hidrauličkog modela rijeke Vrbas.

[37] P. Parhi, (2013) "HEC-RAS Model for Mannnig's Roughness: A Case Study," *Open Journal of Modern Hydrology*, Vol. 3 No. 3, 2013, pp. 97-101. doi: 10.4236/ojmh.2013.33013.

[38] P. K. Parhi, R. N. Sankhua and G. P. Roy,(2012) "Calibration of Channel Roughness of Mahanadi River (India) Using HEC-RAS Model," *Journal of Water Resources and Protection*, Vol. 4, No. 10, 2012, pp. 847-850. doi:10.4236/jwarp.2012.410098

[39] P. V. Timbadiya, P. L. Patel and P. D. Porey, (2011) "Calibration of HEC-RAS Model on Prediction of Flood for Lower Tapi River, India," *Journal of Water Resources and Protection,* Vol. 3, No. 11, 2011, pp. 805-811. doi:10.4236/jwarp.2011.311090

IOP Conf. Series: Materials Science and Engineering 1298 (2023) 012032 doi:10.1088/1757-899X/1298/1/012032

One experimental research with cement substitution by fly ash

Sanin Džidić[1] **and Arnel Popržенović**[1]

[1]Department of Civil Engineering, Faculty of Technical Sciences, University of Bihać, Dr. Irfana Ljubijankića bb, 77000 Bihać, Bosnia and Herzegovina

ninsa_d@hotmail.com

Abstract. Compared to all other construction materials, concrete is by far the most utilized construction material worldwide today. Billions of tons of natural virgin materials have been mined and processed worldwide to produce concrete. Such industrial processes and facilities leave significant unfavourable ecological traces in nature. In addition, it is assumed that the concrete production industry together with the cement industry emits seven percent of the total CO_2 emissions in the world. For every ton of cement produced, 0.9 ton of CO_2 is released into the atmosphere. It is almost certain that there will be no reduction in concrete production in the world in the coming years, but the question arises as to how and in what way to reduce the use of cement in concrete production. One of the mineral additives to the binder is fly ash, which has pozzolanic properties and is produced as a by-product of thermal power plants and is available in Bosnia and Herzegovina. This research considers questions whether and to what extent it is possible to replace cement with fly ash and how this substitution affects the compressive strength of concrete using referenced concrete mix with local concrete ingredients.

1. Introduction

Modern trends in construction, with the entry into the 21st century, constantly require designing and building of advanced quality, more durable and more economical structures. In this regard, materials with improved physical and mechanical properties, their advanced rheological and exploitation possibilities including application of new technologies along with material science developments become trend [1]. Application of fly ash as mineral admixture to cement in the concrete industry is not a new technology but contributes to the new developments and quality of construction products. The benefits of the introducing of fly ash to the concrete mixture are related to the technical aspects, but also to economy and ecology [2]. Today, cement is main binder in concrete technology, while history of concrete remembers binders made of jaggery, lead, jute, rice husk etc. Adding fly ash to cement gained popularity for its sustainability, since it is possible to materialize a variety of concrete applications and at the same time to be efficient in resources management and durability [3].

Concrete is the most used material in construction today. Its price and properties are the main reasons for its application. However, its production, as well as the production of its ingredients, leaves a large negative impact on the environment [4]. Nowadays, the cement substitution with fly ash becomes more important from the ecological aspect wishing to preserve the planet Earth, since significant quantity of CO_2 is produced by cement industry. Production of every ton of cement yields approximately 0.9 tons of CO_2. Every cubic meter of concrete contains about ten percent by weight of cement [5].

Fly ash is produced as a by-product in modern coal-fired thermal power plants, while coal must first be ground. Therefore, fly ash is a fine powder that mainly consists of spherical particles, obtained from the combustion of coal, possibly with secondary combustion materials [6]. Considering that fly ash is an industrial by-product (waste), its use as a cement substitute for concrete production is an additional benefit in preserving the environment.

Research conducted so far has shown that the influence of fly ash on the cement-based composite material is extremely good and desirable, especially if discuss cements that have their application in aggressive environments [6].

The possibility to produce mortars and concrete substituting cement by industrial byproducts that are locally available was confirmed by much research, including [7].

Fly ash improves the properties of concrete in two ways: by the physical effect associated with the increase in the proportion of fine particles, since fly ash particles are smaller than cement particles and by pozzolanic and/or hydraulic reactions. Fly ash reduces the need for water and water extraction, improves workability, impermeability, and slows down the release of heat of hydration [8].

However, taking into account that concrete mixes differ in origin and type of aggregates, type, manufacturer and origin of cement, type, origin and technology of mineral additives such as fly ash, even chemical characteristics of water, quantitative ratios in the concrete mix — research on the effect of cement substitution with fly ash for each individual existing prescribed mix does not necessarily have the same results and effects to the properties of fresh or hardened concrete for each specified mix. It is enough just to change the origin of the fly ash (as a by-product from different thermal power plants) or other concrete ingredients (e.g. local aggregates), so that the results are not necessarily the same.

2. Research methodology

There is some research on the use of fly ash in the production of concrete as a mineral additive to the binder in Bosnia and Herzegovina. Otherwise, there is also an indisposition among the professionals about substituting part of the cement with fly ash. However, new trends in environmental protection, ecological construction, global trends in the reduction of release of greenhouse gases and ground pollution, as well as in the rationalization of business costs make this topic more current. The aim of this research is to point out the possibilities of substituting cement with fly ash and its effect on the compressive strength of the existing concrete mix that is already in use.

An experimental method was used, where one concrete mix design with only cement as a binder is taken as a reference, and then four other mix designs with cement substitution with fly ash with a share of 10-40 percent in relation to the total amount of binder are considered. After 28 days, the compressive strength of each mix designs with fly ash is determined and compared with the compressive strength of reference mix design. In this sense, the following research questions are considered: Does the substitution of cement with fly ash in the production of concrete have environmental effects? and "To what extent and how does the substitution of cement with fly ash affect the compressive strength of concrete made of existing referenced concrete mix?

3. Fly ash

Fly ash is a fine powder that is produced as a by-product of burning coal in thermal power plants (it is considered waste material of thermal power plants). Fly ash is a pozzolanic substance that contains aluminum and silicate material that together with water forms a binding mixture. When mixed with lime and water, it forms a com-pound similar to Portland cement, so it is considered a suitable addition to the cement manufacturing process and a useful material in the concrete manufacturing industry.

The use of fly ash as a pozzolanic ingredient was recognized as early as 1914, but it was first mentioned in literature in 1937 and is called a mineral additive. According to their origin, mineral additives can be divided into natural (pozzolana, diatom earth, volcanic ash) and industrial (fly ash, blast furnace slag, silica fume, rice husk ash). The chemical composition of fly ash depends on the type of coal used and the technological conditions of burning. In the specifications of fly ash, two types stand out and they differ according to the share of pozzolanic active silicon dioxide (S_iO_2) and aluminum oxide (Al_2O_3), that is, but the share of calcium oxide (CaO) as well. In relation to the proportion of CaO, low proportion of CaO in fly ash means better quality, while a high proportion of CaO leads to lower quality [8].

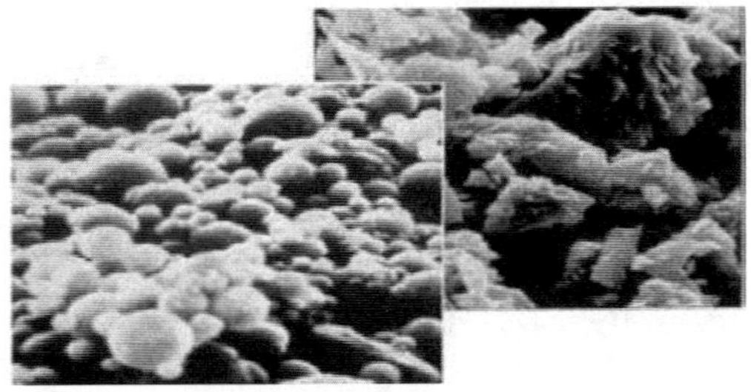

Figure 1. Microscopic comparison of fly ash particles (left) with cement particles (right) [8].

"American Society for Testing and Materials (ASTM C618 – 12a), classifies fly ash into two classes, Class C and Class F." [9]. The main difference between these classes is the amount of calcium, silicon, aluminum oxide and iron.

"Combustion of harder, older anthracite and bituminous coal usually produces class F fly ash. This fly ash is pozzolanic in nature and contains less than 7 percent of lime (CaO). Possessing pozzolanic properties, vitreous silica and alumina, Class F fly ash requires a basic binder, such as Portland cement, quicklime, or hydrated lime – which when mixed with water react to produce cementitious compounds." [8].

Fly ash produced by burning younger lignite or subbituminous coal, besides having pozzolanic properties, also has some self-binding properties. In the presence of water, Class C fly ash hardens and gains strength over time. Class C ash generally contains more than 20 percent lime (CaO). Unlike Class F, Class C self-binding fly ash does not require an activator. Typically, class F fly ash is used in doses of 15 to 25 percent of the mass of binder material, while class C fly ash is used in doses of 15 to 40 percent [8].

Fly ash is recognized as an environmentally friendly material in contrast to Port-land cement, whose production requires the release of a large amount of heat. Concrete with fly ash also requires less water than Portland cement and is easier to use in cold conditions. Other advantages include high workability, resistance to low temperatures, reduces issues with cracks in concrete, reduces the temperature of concrete during hydration, allows a lower ratio of water-cement ratio compared to mixtures without the use of fly ash, reduces CO_2 emissions, and is considered a non-shrinking material and produces dense concrete with a smooth surface.

One of the greatest advantages of concrete with the addition of fly ash is with massive concrete, where the problem of increased temperature during hydration is pronounced. In large blocks of massive concrete structures (e.g. dams), the temperature of the concrete inside the element can be higher than 60° C. Fly ash releases less heat during hydration compared to Portland cement, and because of this property, it is most often used in massive concrete [8].

Tests have also shown that fly ash improves the workability properties of concrete. It is also known that concrete with fly ash continues to gain compressive strength well beyond 28 days. [10, 11]. This long-term hardening improves the structure of the concrete and thus contributes to the impermeability of the concrete [11].

4. Materials used and experimental concrete mix designs

The concrete mix design is a process with goal to select suitable ingredients of concrete and achieve their optimal proportions in a mix to satisfy the required strength, durability, and workability, but taking in consideration the economic dimension as well. Properties of fresh and hardened concrete depend upon many factors: quality and quantity of ingredients, batching and mixing, transportation of concrete to the site, statical and dimensional stability of the formwork and scaffolding, mechanical installation and compaction and curing of concrete up to the age of 28 days, but weather conditions as well [4]. In this regard, this research explores the influence of partial substitution of cement with fly ash and its application with local materials [12] on compressive strength of such concrete.

4.1. Cement

The cement material used in this experimental research is Portland Cement CEM IV/B – W 32.5 N, produced by the Kakanj cement factory (Heidelberg Materials) in Bosnia and Herzegovina. The manufacturer states that CEM IV/B – W 32.5 N is intended for the preparation of concrete and mortar for all construction works that are carried out under prescribed construction conditions for reinforced and plain concrete structures of lower strength classes. It contains 45-64 percent of Portland cement clinker, 36-55 percent of calcium fly ash addition and up to 5 percent of natural gypsum as a binding regulator. This combination of clinker and admixtures gives this cement a slowed hydration reaction, lower heat of hydration, expected strength, good concrete workability, easier installation and increased chemical resistance compared to pure Portland cement. The characteristic values of chemical properties of cement CEM IV/B – W 32.5 N and EN 197-1 [13] requirements are given in Table 1.

Table 1. Chemical properties of cement CEM IV/B – W 32.5 N [14].

	CEM IV/B-W 32.5 N	EN 197-1 Requirements
SO_3 (%)	2.80	≤ 3.5
Cl (%)	0.01	≤ 0.10

The characteristic values of physical and mechanical properties of cement CEM IV/B – W 32.5 N and EN 197-1 requirements are given in Table 2.

Table 2. Physical and mechanical properties of cement CEM IV/B – W 32.5 N [14].

	CEM IV/B-W 32.5 N	EN 197-1 Requirements
Initial setting time (min)	210	≥ 75
Soundness (expansion) (mm)	1.0	≤ 10
Early compressive strength at 7 days (MPa)	25	≥ 16
Standard compressive strength at 28 days (MPa)	41	$\geq 32.5 \leq 52.5$

4.2. Aggregate

The aggregate used in this experiment is limestone from the "Pritoka" quarry near Bihac, Bosnia and Herzegovina. Chemical analysis of the stone shows the following structure of components S_iO_2 – 1.05 percent; Fe_2O_3 – 0.24 percent; Al_2O_3 – 3.60 percent; CaO – 54.26 percent; MgO – 1.35 percent; SO_3 – 0.26 percent and loss on annealing – 39.86 percent. Chemical and mineralogic-petrographic analysis confirm that it is a microcrystalline limestone. The following aggregate grain sizes were used in this experiment: 0–4, 4 –8, 8–16.

4.3. Water

For the purpose of the experiment, water from the city potable water supply system in Bihac from the Klokot source was used, which has a pH value of less than 7 and is safe for human consumption. This means that the water from the Klokot spring does not contain chemical or other impurities that could lead to unwanted, harmful chemical reactions for concrete production. In all test mixtures of this experiment, a fixed water/binder ratio of 0.45 was used.

4.4. Fly ash

The fly ash used for this research is fly ash from the Kakanj Thermal Power Plant (Bosnia and Herzegovina), which belongs to calcium ash according to EN 197 – 1:2011, since CaO content is greater than 5 percent.

"According to the ASTM C 618, fly ash from the Kakanj Power Plant belongs to class C ash - ashes resulting from the combustion of lower quality coal (brown and lignite). The specific surface area of ash is 2490 g/cm^2, and the specific density is 2.70 g/cm^3, which is within the limits of the value of class C ash, which ranges from 2.6 to 2.7 g/cm^3. Ash ignition loss is 0.95 percent, which is within the limits prescribed by EN 197 - 1:2011 (max. 5 percent) and ASTM C618 (max. 6.0 percent). The content of SO_3 in the ash is 1.57 percent and it is within the permissible limits below 5.0 percent (ASTM C618). It is advantageous that the ash has a low content of free CaO (0.96 percent), since it is harmful to cement, because it causes subsequent expansion and cracking of cement composites. Grain size analysis shows that fly ash consists of very small particles. The largest percentage is particles with a diameter of 16.6 to 19 μm (6.6 percent). "Also, 50 percent of the particles have a diameter smaller than 7.86 μm, and 90 percent of the particles have a diameter smaller than 22.3 μm" [15].

Figure 2. Fly ash from Thermal Power Plant Kakanj, Bosnia and Herzegovina.

4.5. Concrete mix designs

In this research, four concrete mixtures were designed (based upon the initial existing concrete mix without fly ash) for the determination of the influence of cement substitution with fly ash on the compressive strength of concrete. For each concrete mix, three cube test samples were made (15 cm x 15 cm x 15 cm) in accordance with EN 12390-2 [16]. The first concrete mix is referenced mix without cement substitution with fly ash and serves as a reference mix, while the other four mixes contain 10-40 percent fly ash.

4.5.1. Mix design 1- Reference mix.

Mix design 1 - Concrete mixture of aggregate, cement, and water - w/c 0.45

Table 3. Mix design 1.

Material	For 1 m^3 of concrete
Cement	385 kg
Water	173 l
w/c ratio	0.45
Aggregate (0-4 mm)	883 kg
Aggregate (4-8 mm)	471.75 kg
Aggregate (8-16 mm)	651.43 kg

4.5.2. Mix design 2.

Mix design 2 - Concrete mixture of aggregate, cement, water and fly ash (10% of the binder) - w/b 0.45

Table 4. Mix design 2.

Material	For 1 m^3 of concrete
Cement	346.5 kg
Water	173 l
w/b ratio	0.45
Fly ash (10% of binder)	38.5 kg
Aggregate (0-4 mm)	883 kg
Aggregate (4-8 mm)	471.75 kg
Aggregate (8-16 mm)	651.43 kg

4.5.3. Mix design 3.

Mix design 3 - Concrete mixture of aggregate, cement, water and fly ash (20% of the binder) - w/b 0.45

Table 5. Mix design 3.

Material	For 1 m^3 of concrete
Cement	308 kg
Water	173 l
w/b ratio	0.45
Fly ash (20% of binder)	77 kg
Aggregate (0-4 mm)	883 kg
Aggregate (4-8 mm)	471.75 kg
Aggregate (8-16 mm)	651.43 kg

4.5.4. Mix design 4.

Mix design 4 - Concrete mixture of aggregate, cement, water and fly ash (30% of the binder) - w/b 0.45

Table 6. Mix design 4.

Material	For 1 m^3 of concrete
Cement	269.5 kg
Water	173 l
w/b ratio	0.45
Fly ash (30% of binder)	115.5 kg
Aggregate (0-4 mm)	883 kg
Aggregate (4-8 mm)	471.75 kg
Aggregate (8-16 mm)	651.43 kg

4.5.5. Mix design 5.

Mix design 5- Concrete mixture of aggregate, cement, water and fly ash (40% of the binder) - w/b 0.45

Table 7. Mix design 5.

Material	For 1 m^3 of concrete
Cement	230 kg
Water	173 l
w/b ratio	0.45
Fly ash (40% of binder)	155 kg
Aggregate (0-4 mm)	883 kg
Aggregate (4-8 mm)	471.75 kg
Aggregate (8-16 mm)	651.43 kg

IOP Conf. Series: Materials Science and Engineering 1298 (2023) 012032 doi:10.1088/1757-899X/1298/1/012032

5. Conducting experiment

The concrete cub test samples were prepared and cured for this experiment in accordance with EN 12390-2:2000 [16].

Figure 3. Fresh concrete test samples.

Figure 4. Curing concrete test samples (left) and test samples ready for compression strength testing after 28 days (right).

The compressive strength tests of the samples were conducted in accordance with EN 12390-3:2019 [17].

Figure 5. Compressive strength test (left) and failure mode (right).

6. Experimental results

The following results on compressive strength were obtained from this experiment:

6.1. Mix design 1 – reference mix

Table 8. Test results of Mix Design 1.

Mix Design 1	Age od specimen	Mass of specimen	Area	Maximal force at failure	Compressive strength
	days	kg	mm^2	kN	MPa
Specimen 1	28	8.34	22500	1223	54.36
Specimen 2	28	8.14	22500	1173	52.13
Specimen 3	28	8.20	22500	895	39.78
Mean value		8.25		1097.00	48.76

According to the criteria stated in EN 206:2013+A2:2021, [18], Annex B, Table B1, Mix Design 1 is classified as concrete class C 30/37. As concrete specimens are produced in cubic shape, for determination of concrete class characteristic compressive strength (f_{ck}) value is 37 MPa.

Table 9. EN 206:2013+A2:2021, Annex B, Table B1, Identity Criteria for Compressive Strength [18].

Number *n* of test results for compressive strength from the defined volume of concrete	Criterion 1	Criterion 2
	Mean of n results (f_{cm}) N/mm²	*Any individual test result (f_{ci}) N/mm²*
1	Not applicable	$\geq f_{ck} - 4$
2 to 4	$\geq f_{ck} + 1$	$\geq f_{ck} - 4$
5 to 6	$\geq f_{ck} + 2$	$\geq f_{ck} - 4$

EN 206:2013+A2:2021, Annex B, Table B1, states that mean value (f_{cm}) must be equal or greater than f_{ck} + 1 (Criterion 1), thus:

- Mean value – 48.76 MPa > 38 MPa (C 30/37) and > 46 MPa (C 35/45)

and individual test result f_{ci} must be equal or greater than f_{ck} – 4, therefore:

- Specimen 1 – 54.36 MPa > 33 MPa (C 30/37) and > 41 MPa (C 35/45)
- Specimen 2 – 52.13 MPa > 33 MPa (C 30/37) and > 41 MPa (C 35/45)
- Specimen 3 – 39.78 MPa > 33 MPa (C 30/37) and < 41 MPa (C 35/45),

so, this concrete does not meet the requirements of Criterion 2 (Specimen 3) for concrete class C 35/45, but it meets both criteria for concrete class C 30/37 and is classified in concrete class C 30/37.

6.2. Mix design 2 – mix with fly ash (10 percent of the binding material)

Mix Design 2 is classified as concrete class C 30/37. Concrete class characteristic compressive strength (f_{ck}) value is 37 MPa.

Table 10. Test results of Mix Design 2.

Mix Design 2	Age od specimen	Mass of specimen	Area	Maximal force at failure	Compressive strength
	days	kg	mm^2	kN	MPa

Specimen 1	28	8.34	22500	890	39.60
Specimen 2	28	8.30	22500	1170	52.00
Specimen 3	28	8.22	22500	905	40.22
Mean value		8.30		988.33	44.00

- Mean value – 44.00 MPa > 38 MPa (C 30/37) and < 46 MPa (C 35/45)
- Specimen 1 – 39.60 MPa > 33 MPa (C 30/37) and < 41 MPa (C 35/45)
- Specimen 2 – 52.00 MPa > 33 MPa (C 30/37) and > 41 MPa (C 35/45)
- Specimen 3 – 40.22 MPa > 33 MPa (C 30/37) and < 41 MPa (C 35/45).

This concrete does not meet the requirements of Criterion 1 and Criterion 2 (Specimens 1 and 3) for concrete class C 35/45, but it meets both criteria for concrete class C 30/37 and is classified in concrete class C 30/37.

6.3. Mix design 3 – mix with fly ash (20 percent of the binding material)

Mix Design 3 is classified as concrete class C 30/37. Concrete class characteristic compressive strength (f_{ck}) value is 37 MPa.

Table 11. Test results of Mix Design 3.

Mix Design 3	Age od specimen	Mass of specimen	Area	Maximal force at failure	Compressive strength
	days	kg	mm^2	kN	MPa
Specimen 1	28	8.08	22500	995	44.30
Specimen 2	28	8.20	22500	1015	45.20
Specimen 3	28	8.14	22500	915	40.70
Mean value		8.14		975.00	43.40

- Mean value – 43.40 MPa > 38 MPa (C 30/37) and < 46 MPa (C 35/45)
- Specimen 1 – 44.30 MPa > 33 MPa (C 30/37) and > 41 MPa (C 35/45)
- Specimen 2 – 45.20 MPa > 33 MPa (C 30/37) and > 41 MPa (C 35/45)
- Specimen 3 – 40.70 MPa > 33 MPa (C 30/37) and < 41 MPa (C 35/45)

The concrete from Mix Design 3 does not meet the requirements of Criterion 1 and Criterion 2 (Specimen 3) for concrete class C 35/45, but it meets both criteria for concrete class C 30/37 and is classified in concrete class C 30/37.

6.4. Mix design 4 – mix with fly ash (30 percent of the binding material)

Mix Design 4 is classified as concrete class C 25/30. Concrete class characteristic compressive strength (f_{ck}) value is 30 MPa.

Table 12. Test results of Mix Design 4.

Mix Design 4	Age od specimen	Mass of specimen	Area	Maximal force at failure	Compressive strength
	days	kg	mm^2	kN	MPa
Specimen 1	28	8.18	22500	780	34.70
Specimen 2	28	7.94	22500	885	39.40
Specimen 3	28	8.10	22500	875	40.70
Mean value		8.08		846.70	37.70

IOP Conf. Series: Materials Science and Engineering 1298 (2023) 012032 doi:10.1088/1757-899X/1298/1/012032

- Mean value – 37.70 MPa > 31 MPa (C 25/30) and < 38 MPa (C 30/37)
- Specimen 1 – 34.70 MPa > 26 MPa (C 25/30) and > 33 MPa (C 30/37)
- Specimen 2 – 39.40 MPa > 26 MPa (C 25/30) and > 33 MPa (C 30/37)
- Specimen 3 – 40.70 MPa > 26 MPa (C 25/30) and > 33 MPa (C 30/37)

The concrete from Mix Design 4 does not meet the requirements of Criterion 1 for concrete class C 30/37, but it meets both criteria for concrete class C 25/30 and is classified in concrete class C 25/30.

6.5. Mix design 5 – mix with fly ash (40 percent of the binding material)

Mix Design 5 is classified as concrete class C 25/30. Concrete class characteristic compressive strength (f_{ck}) value is 30 MPa.

Table 13. Test results of Mix Design 5.

Mix Design 5	Age od specimen	Mass of specimen	Area	Maximal force at failure	Compressive strength
	days	kg	mm^2	kN	MPa
Specimen 1	28	8.28	22500	775	34.50
Specimen 2	28	8.16	22500	670	30.00
Specimen 3	28	8.12	22500	815	36.30
Mean value		8.20		753.33	33.60

- Mean value – 33.60 MPa > 31 MPa (C 25/30) and < 38 MPa (C 30/37)
- Specimen 1 – 34.50 MPa > 26 MPa (C 25/30) and > 33 MPa (C 30/37)
- Specimen 2 – 30.00 MPa > 26 MPa (C 25/30) and < 33 MPa (C 30/37)
- Specimen 3 – 36.30 MPa > 26 MPa and > 33 MPa (C 30/37)

The concrete from Mix Design 5 does not meet the requirements of Criterion 1 and Criterion 2 (Specimen 2) for concrete class C 30/37, but it meets both criteria for concrete class C 25/30 and is classified in concrete class C 25/30.

6.6. Graphic presentation of experimental results for all five mix designs

The test results showed that design mix 2 and 3 (mixes with 10 and 20 percent fly ash content) retain the same compressive strength class as the reference mix (with no fly ash). They have approximately 90 percent compressive strength compared to the reference mix (Design mix 1). Design mix 4 and 5 have 77 percent and 69 percent of compressive strength respectively, compared to the reference mix. Design mix 2 and 3 retain concrete class C30/37, while design mix 4 and 5 are classified as concrete class C25/30. These results are in line with research conclusion exposed in [10] for concrete compressive strength at 28 days.

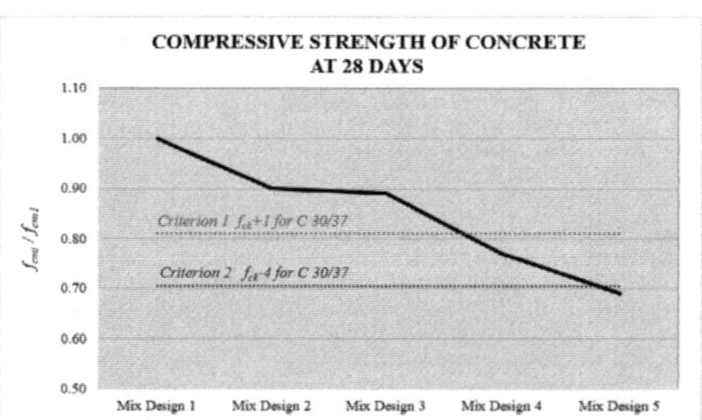

Figure 6. Graphic presentation of experimental results for all five mix designs.

7. Conclusion

This research showed that it is possible to substitute cement with fly ash referencing to concrete mix specified in Chapter 4 and that the most optimal substitution of cement by fly ash is in the range of 10 to 20 percent of total binder mass to keep same concrete class. When 30 percent and higher substitution of cement with fly ash was done, there was a significant drop in concrete compressive strength.

New research could be conducted to precise percentage of substitution of cement with fly ash above 20 percent of substitution in increments of 1 percent, but for other concrete classes as well. Taking in consideration that applied cement CEM IV/B – W 32.5 N in this research already contains the fly ash (cement in bags), the testing of this pure Portland cement (that is supplied directly to concrete factories) would be of more more benefit.

At the same time, the results of this research refer exclusively and are limited to concrete mixes with the applied type and manufacturer of cement, aggregates from the mentioned quarry, water from the mentioned water supply and previously de-fined fly ash.

However, this research contributes to the further research in application of fly ash and substitution of cement by fly ash in concrete and its influence to the concrete class, as well as other properties of fresh and hardened concrete, by varying different concrete ingredients. Significant research is needed in future for other type and origin of cements in Bosnia and Herzegovina, aggregates from different quarries throughout country, other origins of fly ash, other prescribed or design mix, but also to the rheological properties of such concrete as well as behavior of concrete structures made of concrete with fly ash.

This research also contributes to raising the awareness of the professional community at local level (when local aggregates and water are used in the production of concrete) about the potential of partial replacement of cement with fly ash in the production of concrete, benefits and shortcomings. Very often, local professional communities are not fully exposed to the scientific findings.

The substitution of cement with fly ash in the technology of concrete production is favorable from environmental aspects as well. During the production of cement, a large amount of CO_2 is released. On the other hand, fly ash as a by-product of industrial processes, which without being used in other technologies, would have to be disposed of in adequate landfills, which further contribute to the pollution of the natural environment. In this regard, any substitution of cement by fly ash (by-product) directly contributes to the decrease in consumption and production of cement, thus decreasing release of CO_2 in atmosphere. By substituting cement with fly ash, the price of concrete per unit volume is reduced, since fly ash is significantly less expensive than cement. The aforementioned reasons synergistically contribute to sustainability and stimulate the application of revitalized existing towards new concepts in construction.

References

[1] Sahinagic-Isovic M 2015 Posebne vrste betona – mikroarmirani betoni *Univerzitet, "Dzemal Bijedic", Gradjevinski fakultet Mostar Bosnia and Herzegovina* ISBN 978-9958-604-87-4

[2] Ghazali N at all 2019 Utilization of Fly Ash in Construction *IOP Conf. Series: Materials Science and Engineering 601 (2019) 012023 IOP Publishing* doi:10.1088/1757-899X/601/1/012023

[3] Naidu, P V and Pandey P K 2014 Replacement of Cement in Concrete *International Journal of Environmental Research and Development ISSN 2249-3131 Volume 4 Number 1 pp 91-98. © Research India Publications India*

[4] Cecez M and Sahinagic-Isovic M 2021 Primjena crvenog mulja u gradjevinarstvu *Univerzitet "Dzemal Bijedic" Gradjevinski fakultet Mostar Bosnia and Herzegovina* ISBN978-9926-434-54-0

[5] Dzidic S 2014 Can Concrete be Green in Bosnia and Herzegovina? *The 5th International Symposium on Sustainable Development ISSD 2014 Proceedings* International Burch University Sarajevo Bosnia and Herzegovina

[6] Djuzelic E 2019 Dodaci vezivu betonskih mješavina *12th International Scientific Conference on Production Engineering-Development and Modernization of Production RIM 2019 ISSN*

2566-3257 Year 22 No. 12 University of Bihac, Faculty of Technical Engineering Sarajevo, Bosnia and Herzegovina
[7] Cecez M and Sahinagic-Isovic M 2019 Mortars with addition of local industrial by-products. *Gradjevinar 71 (2019) 1 1-7 Zagreb Croatia* DOI: https://doi.org/10.14256/JCE.2358.2018
[8] Lal Gambhir M 2013 Concrete Technology: Theory and Practice *McGraw Hill Education (India) Private Limited* ISBN 1259062554 9781259062551
[9] ASTM C618-05: Standard Specification for Coal Fly Ash and Raw or Calcined Natural Pozzolan for Use in Concrete *2022 Edition ASTM Headquarters 100 Barr Harbor Drive P.O. Box C 700 West Conshohocken* PA 19428-2959 USA
[10] Mujkanovic A at all 2021 Unprocessed calcareous fly ash as a partial cement replacement in self-compacting concrete *Journal of Sustainable Construction Materials and Technologies Vol. 6 Issue 2 pp. 44–52 Published by Yıldız Technical University Press İstanbul Turkey* DOI: 10.29187/jscmt.2021.59
[11] Xiaohong W and Yueqing Y 2012 The Research on a New Cleaner Production Model for Flying Ash Aerated Concrete Production *Proceeding of the 24th Chinese Control and Decision Conference (CCDC)* pp. 429-434
[12] Sahinagic-Isovic M at all 2016 Osobine samougradjujuceg betona sa dodatkom mikrosilike i leteceg pepela *The 6th International Conference: Civil Engineering – Science & Practice GNP 2016 Proceedings* University of Montenegro Faculty of Civil Engineering ISBN 978-86-82707-30-1 COBISS.CG-ID 29599504 pp 765-772 Zabljak, Montenegro
[13] EN 197-1:2011: Cement - Part 1: Composition, specifications and conformity criteria for common cements *European Committee for Standardization (CEN)* Management Centre: Avenue Marnix 17 B-1000 Brussels Belgium
[14] Salihbegovic N CEM IV/B-W 32,5 N za individualnu gradnju *Kakanj cement-Heidelberg Cement group* https://www.heidelbergcement.ba/
[15] Busatlic I 2009 Testing quality of fly ash from TPP Kakanj in function of its use in cement industry *Naučno-stručni skup sa sa međunarodnim učešćem "KVALITET 2009" Neum*, Bosnia and Herzegovina pp 471-476
[16] EN 12390-2:2000: Testing hardened concrete - Part 2: Making and curing specimens for strength tests *European Committee for Standardization (CEN)* Management Centre: Rue de Stassart B-1000 Brussels Belgium

[17] EN 12390-3:2019: Testing hardened concrete - Part 3: Compressive strength of test specimens. *European Committee for Standardization (CEN)* CEN-CENELEC Management Centre: Rue de la Science B-1040 Brussels, Belgium
[18] EN 206:2013+A2:2021, Concrete. Specification, performance, production and conformity. *European Committee for Standardization (CEN)* CEN-CENELEC Management Centre: Rue de la Science B-1040 Brussels Belgium

Experimental and Analytical Deformation Analysis of Truss Members

Damir Hodžić and Elma Veladžić

University of Bihać, Faculty of Technical Engineering Bihać, Bosnia and Herzegovina

E-mail: emelija15@gmail.com

Abstract. Paper presents comparative analysis of experimental and analytical results of mechanical properties on truss members. Deformations are experimentally determined on the model of plexiglass truss, using six strain gauges. Strain gauges are measuring the deformation of truss members at the pre-defined measuring points. Knowing modulus of the elasticity and Hooke's law, it is possible to calculate the stress values. The experimental stress values will be compared with the values obtained by the software and the results will be analyzed.

1. Introduction

Theoretical methods of solving problems in the design and calculation of structures use numerous assumptions and simplified calculation models. Input data related to geometry and material characteristics can often deviate from actual values. The result of such a calculation is an aproximation of the behavior of a real construction, and the actual behavior can only be determined by testing [1]. In construction testing as we know it today, it is often set as a condition that the testing does not cause damage to the construction, which is why non-destructive testing methods are applied [2].

In this paper, the deformations will be experimentally determined on the model of a plexiglass truss, with the use of strain gauges. Six strain gauges will be measuring the deformation of truss members at the pre-defined measuring points. With the help of the Hooke's law and by knowing the modulus of elasticity, we will be able to calculate the stress values. The experimental stress values will be compared with the values obtained by the software and the results will be analyzed.

2. Strain gauges

In 1856, William Thomson noted that strain gauges work on the principle of electrical resistance of the current flowing through them, where under the influence of mechanical load there is a change in electrical resistance [3]. At the same time, Lord Kelvin proved that the electrical resistance of a copper-iron wire changes when it is streched, which represents the basic principle of the strain gauges.

The strain gauge consists of a wire bent several times and then glued with a special glue to its supporting element, which can be made of paper, synthetic mass and metal foil. The bent wire is glued over the supporting element to the construction which is exposed to the action of the load, as a result of which it is deformed [4].

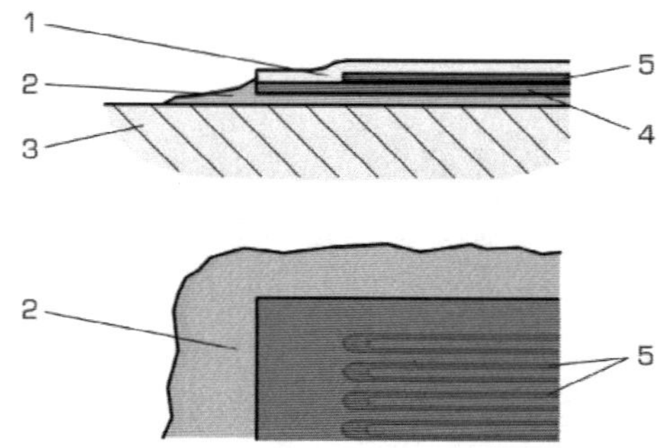

Figure 1. Parts of a strain gauge: 1- cover sheet; 2-glue; 3- construction to be tested; 4-bearing element of the strain gauge; 5-bend wire in the supporting element of the strain gauge [5].

The simple use of strain gauges and the possibility of their direct application to the structure being tested are responsible for not only their wide distribution and application, but also for their status as the most used technique in the experimental analysis of structures. Strain gauges are used for testing all kinds of objects and structures, from nuclear power plants, dams, rockets – to determining the stress and deformation state of human bones [6].

2.1. The relationship between elongation and change in electrical resistance

The principle of measuring deformations is based on the property of the wire to change its electrical resistance in proportion to the change in lenght. The longitudinal direction of the strain gauge is set so that it coincides with the direction for which the deformations will be measured. The deformation of the structure is transfered to the sensitive wire part of the strain gauge, whereby the operation of the strain gauge is based on a linear relationship between the change in electrical resistance and mechanical dilatation (elongation). The analytical proof of Lord Kelvin's experiment starts from the expression for electrical resistance [7]:

$$R = \rho \cdot \frac{L}{A} = \rho \cdot \frac{L}{D^2 \cdot \frac{\pi}{4}} = \rho \cdot \frac{L}{D^2 \cdot c} \tag{1}$$

where:
R – resistance in the wire,
L – wire length,
ρ – specific conductor resistance,
A – cross-sectional area,
D – diameter,
c – constant $\pi/4$.

By differentiating the expresion above, we get:

$$dR = \frac{1}{c \cdot D^2}\left[(Ld\rho + \rho dL) - 2 \cdot \rho \cdot L \cdot \frac{dD}{D}\right] \tag{2}$$

Mutual division of the expressions gives us:

$$\frac{dR}{R} = \frac{dL}{L} - 2 \cdot \frac{dD}{D} + \frac{d\rho}{\rho} \tag{3}$$

where:
$\frac{dL}{L} = \varepsilon$ – longitudinal deformation of the wire,

$\frac{dD}{D} = \varepsilon_p$ – transverse deformation of the wire.

Expression (3) can be written as:

$$\frac{dR/R}{dL/L} = 1 - 2\cdot\frac{dD/D}{dL/L} + \frac{d\rho/\rho}{dL/L} \tag{4}$$

$K = \frac{dR/R}{dL/L}$ – strain gauge coefficient,

$\upsilon = -\frac{dD/D}{dL/L}$ – Poisson's coefficient,

from which it follows:

$$K = \frac{dR/R}{dL/L} = \frac{dR/R}{\varepsilon} = 1 + 2\cdot\upsilon + \frac{d\rho/\rho}{dL/L} \tag{5}$$

Finally, the relationship between the change in resistance in the wire and the deformation can be expressed as:

$$\varepsilon = \frac{1}{K}\cdot\frac{\Delta R}{R} \tag{6}$$

ΔR is determined during the measurement, while R and K are the characteristics of the strain gauge. Expression (6) is analytically proven experiment of Lord Kelvin that the electrical resistance of the wire from which the strain gauge is made of changes with the elongation of the wire [7].

3. Experimental determination of deformations

For the experimental determination of the deformations on the model of plexiglass truss, six strain gauges were used, with the pre-defined measuring points and dimensions in millimeters shown in figure 2. The modulus of elasticity of plexiglass is E = 3300 MPa.

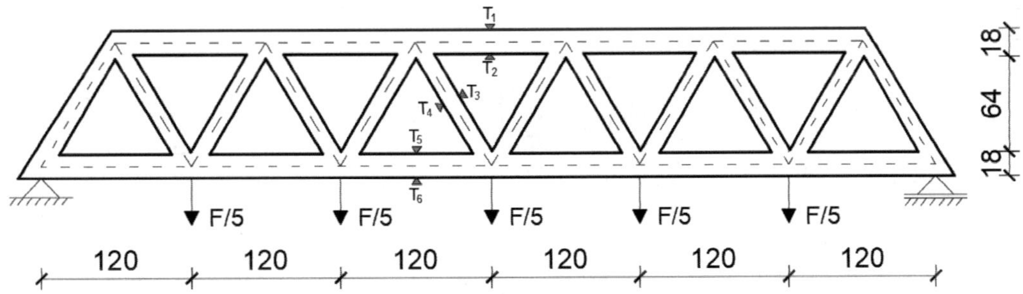

Figure 2. Truss [8].

The geometrical characteristics of truss members are shown in table 1.

Table 1. The geometrical characteristics of truss members.

Upper and lower edge:	Filling:
b/h = 10.5/18.0 mm	b/h = 10.5/15.0 mm
A = 189.0 mm^2	A = 157.5 mm^2
I = 5103.0 mm^4	I = 2953.125 mm^2
W = 567.0 mm^2	W = 393.75 mm^2

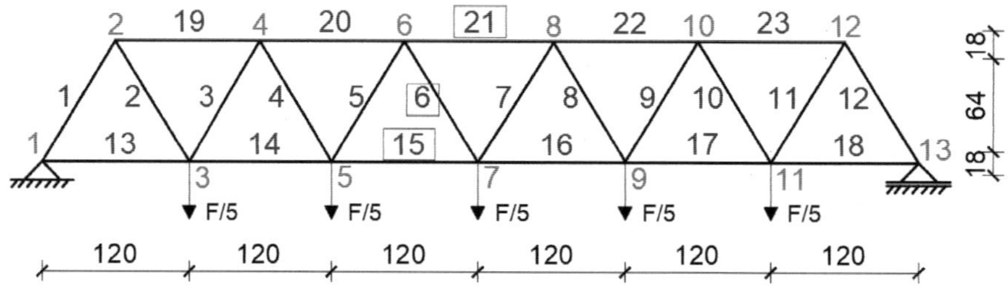

Figure 3. Truss model [8].

The truss model is loaded according to figure 3., in the nodes 3, 5, 7, 9 and 11, with fifths of the force value F = 100 N, which is F/5 = 100/5 = 20 N in each of the mentioned nodes. The load was applied in stages, and the deformation values were measured.

Table 2. Measurement results for truss model.

Measuring points	Load stage				
	F = 0 N	F = 50 N	F = 0 N	**F = 100 N**	F = 0 N
EOT T_1	10	-84	/	**-238**	/
EOT T_2	10	-50	/	**-165**	/
EOT T_3	10	37	/	**25**	/
EOT T_4	10	59	/	**54**	/
EOT T_5	10	111	/	**192**	/
EOT T_6	10	128	/	**230**	/

By using Hooke's law, since we know the modulus of elasticity of plexiglass, after obtaining the deformation values at the measuring points, we can determine the stress for the analysed rods.

$$\sigma = E \cdot \varepsilon \tag{7}$$

Table 3. Stress values.

rod	upper edge	lower edge
15	$\sigma_5 = 3300 \cdot 192 \cdot 10^{-6} = 0.634$ MPa	$\sigma_6 = 3300 \cdot 230 \cdot 10^{-6} = 0.759$ MPa
6	$\sigma_3 = 3300 \cdot 25 \cdot 10^{-6} = 0.083$ MPa	$\sigma_4 = 3300 \cdot 54 \cdot 10^{-6} = 0.178$ MPa
21	$\sigma_1 = 3300 \cdot -238 \cdot 10^{-6} = -0.785$ MPa	$\sigma_2 = 3300 \cdot -165 \cdot 10^{-6} = -0.545$ MPa

4. Analytical determination of deformations

In this chapter, the deformation values obtained in the SCIA ENGINEER 19.1 software package for static calculations of structural elements will be prestented. A statically determined truss made of plexiglass was analyzed. The truss model is loaded in software according to figure 3, in the nodes 3, 5, 7, 9 and 11, with fifths of the force value F = 100 N, which is F/5 = 100/5 = 20 N in each of the mentioned nodes. Nodes and rods of the truss will be marked according to the figure 4.

IOP Conf. Series: Materials Science and Engineering 1298 (2023) 012033 doi:10.1088/1757-899X/1298/1/012033

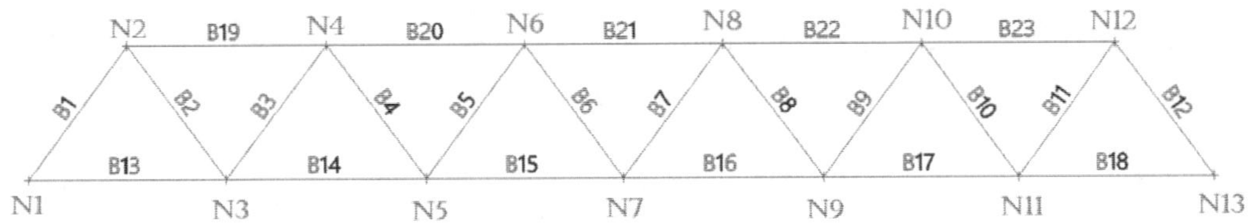

Figure 4. Truss in the local coordinate system [8].

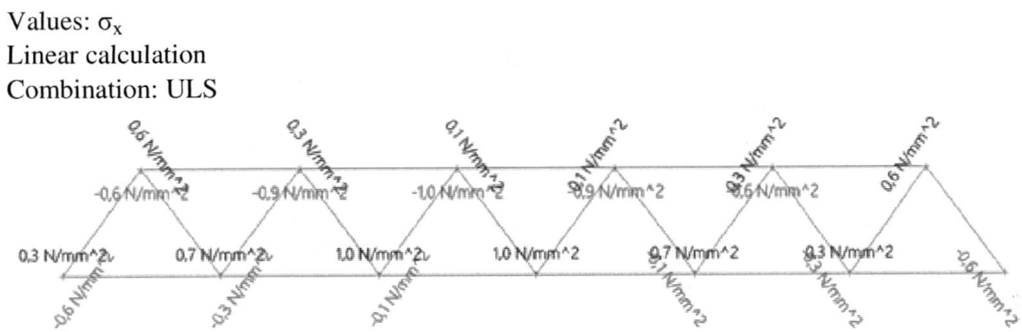

Figure 5. Stress values σ_x (SCIA 19.1) [8].

Table 4. Stress values determined by software.

Rod	Load case	$\sigma\left[\frac{N}{mm^2} = MPa\right]$
B1	ULS	-0.6
B2	ULS	0.6
B3	ULS	-0.3
B4	ULS	0.3
B5	ULS	-0.1
B6	ULS	0.1
B7	ULS	0.1
B8	ULS	-0.1
B9	ULS	0.3
B10	ULS	-0.3
B11	ULS	0.6
B12	ULS	-0.6
B13	ULS	0.3
B14	ULS	0.7
B15	ULS	1.0
B16	ULS	1.0
B17	ULS	0.7
B18	ULS	0.3
B19	ULS	-0.6
B20	ULS	-0.9
B21	ULS	-1.0
B22	ULS	-0.9
B23	ULS	-0.6

IOP Conf. Series: Materials Science and Engineering 1298 (2023) 012033 doi:10.1088/1757-899X/1298/1/012033

5. The results

Figure 6 shows a comparison of the truss stress values obtained by software and experimentally.

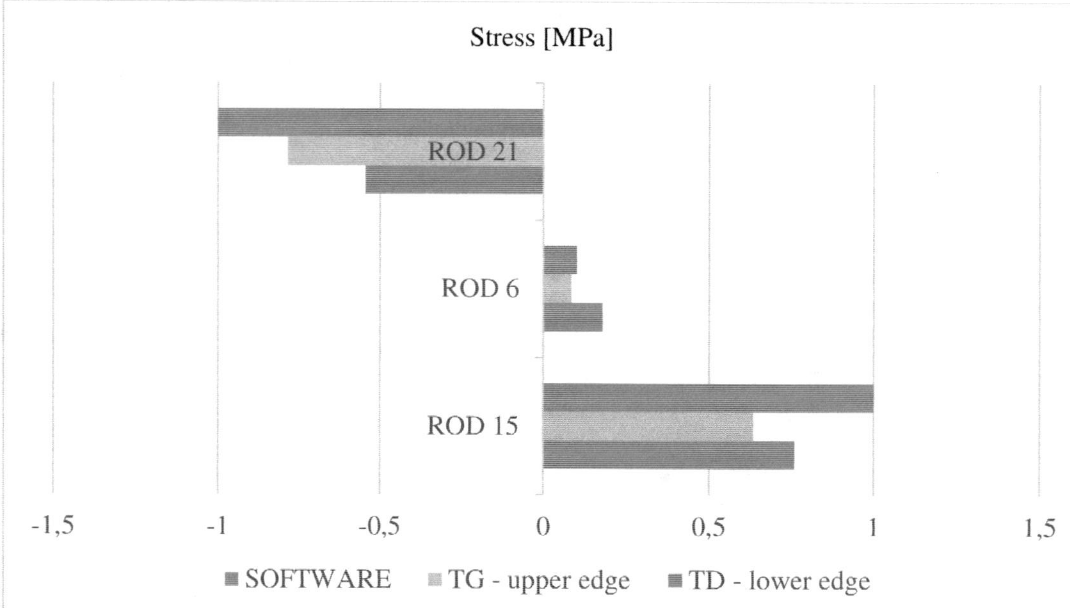

Figure 6. Comparison of truss stress values (software-experimental) [8].

Table 5. Extreme stress values.

ROD	21		6		15	
EXPERIMENTAL	T1	T2	T3	T4	T5	T6
	-0.785	-0.545	0.083	**0.178**	0.634	0.759
SOFTWARE	**-1.0**		0.1		**1.0**	

By comparing the results shown in the table 5 and analyzing them, we can conclude:

- the measured stress in rods 15 and 21 is lower than stress values obtained by the software,
- the measured stress in rod 6 exceeds the stress value obtained by the software.

6. Conclusion

As input data in SCIA software, the dead weight of the truss and the load from the concetrated force (F) are classified as constant load and are multiplied by a partial safety factor of 1.35. The values of the measured stresses in rods 15 and 21 are lower than the values obtained by the software. However, the value of the measured stress in rod 6 exceeds the value of the stress obtained by the software.

Using strain gauge T4, glued to the lower edge of the rod 6, a stress of 0.178 MPa was calculated, while the stress value obtained by the software is 0.1 MPa. We can conclude that the actual stress value of rod 6 determined by experiment does not correspond to the approximation of the stress value obtained through the software.

7. References

[1] Đuranović N 2009 *Uvod u ispitivanje konstrukcija sa primjerima* (Podgorica: Građevinski fakultet)

[2] Damjanović D 2018 *Ispitivanje konstrukcija*, lecture script (Zagreb: Građevinski fakultet)

[3] Radojković M 1976 *Ispitivanje konstrukcija* (Beograd, Građevinski fakultet, Univerzitet u Beogradu)

[4] Vukotić R 1998 *Ispitivanje konstrukcija* (Beograd: Građevinski fakultet, Univerzitet u Beogradu)

[5] https://www.gunt.de/en/products/engineering-mechanics-and-engineering-design/strength-of-materials/experimental-stress-and-strain-analysis/strain-gauge-training-system/021.10000/fl100/glct-1:pa-148:ca-13:pr-319 visited 02.09.2022.

[6] Đuranović N 2008 *Eksperimentalna analiza konstrukcija mjernim trakama* (Podgorica: Građevinski fakultet)

[7] Dolček V, Karabegović I, Martinović D, Jurković M, Blagojević D, Bogdan Š and Bjelonja I 2004 *Elastostatika,* 2nd part (Bihać-Sarajevo-Mostar, Tehnički fakultet)

[8] Veladžić E 2022 Truss deformation measurements using strain gauges. *Bachelor thesis* University of Bihać, Faculty of Technical Engineering

IOP Conf. Series: Materials Science and Engineering 1298 (2023) 012034 doi:10.1088/1757-899X/1298/1/012034

Analysis of foundation solution for the noise protection walls on the reinforced earth

Asmin Veladžić[1], **Dženana Gačo**[2] and **Aida Veladžić**[1]

[1] AVCON Ltd Sarajevo, Branislava Nušića bb, 71000 Sarajevo
[2] University of Bihac, Faculty of Technical Engineering, Dr. Irfana Ljubijankića bb, 77000 Bihać

Email: asmin@avcon.ba

Abstract. This paper brings the introduction of the analysis of foundation solution for the noise protection walls on the reinforced earth along the Highway Corridor Vc, a road and rail transport corridor that connects Hungary, Croatia and Bosnia and Herzegovina, included in the network of Pan-European transport corridors. A solution with pile foundations made of steel IPE profiles is presented in detail. During the design and modelling, as well as the calculation of the structural foundation, the modern software program PLAXIS was used, and the calculation was made in accordance with EN 1997 Eurocode 7, Geotechnical Design. Special attention is devoted to the analysis of the impact of driving IPE steel profiles, as foundation piles, on reinforced earth. The presented solution was then verified by an additional control calculation and a field test of the impact of the horizontal load on the bearing capacity of the IPE steel profile driven into the reinforced slope. Based on the conducted analyses and testing, it has been proven that it is possible to build noise protection walls on the reinforced earth without disturbing the internal stability and functionality of the reinforced earth.

Keywords: reinforced earth, noise protection wall, steel profile, foundation, analysis

1. Introduction

Noise protection walls (sound barriers) are long structures that prevent the direct transmission of road traffic noise through the air, and they are constructed of structural elements (foundations and load-bearing columns) and noise protection elements (sound panels). Noise protection walls are active protection measures used to reduce noise pollution in the environment. In addition to noise protection, sound barriers limit the spread of dust and smog caused by road traffic, which directly reduces the negative noise pollution impact on human health and increases the level of comfort, i.e. contributes to the improvement of the quality of life in populate areas near roads where the increased number and speed of motor vehicles are recorded, such as area next to highways, main roads, regional or local roads. [1]

Noise protection walls are built on the finished road, and their design and construction are conditioned by the existing elements on the road itself. Every now and then the noise protection walls collide with other installations or road equipment, and this is an issue that requires engineering decisions and the preparation of adequate technical solutions. During the construction of the Highway Corridor Vc, on the section "Zenica Bypass" and "Tarčin-Tunnel Ivan", a new technical challenge arose and that is the collision between the route of the noise protection wall and the reinforced earth.

Reinforced earth is a supporting structure formed by placing extendible or inextensible materials such as metallic strips or polymeric reinforcement within the soil. The reinforcement transfers the stresses that are created within the mass from soil to the reinforcement strips by friction. [2] In addition to the term "reinforced earth", the terms "reinforced/retained soil" or "mechanically stabilized earth" are also used. Simply put, this means that the earth is reinforced with tensile elements such as metal strips or geosynthetics. [3]

As per previous practice in Bosnia and Herzegovina, constructions on reinforced earth were not allowed by the Client, because it was assumed that the intervention would cause structural damage to the reinforced earth. This paper will provide an in-detail analysis of the case of driving single foundations into reinforced earth on Highway Corridor Vc, subsection Tarčin-Tunnel Ivan. The objective of the analysis is to determine whether driving IPE steel profiles into reinforced earth will negatively impact the stability and functionality of this type of supporting structure.

2. Reinforced earth

To provide further clarification on the current issues, the paper brings firstly the definition of the term "reinforced earth". Structures made of reinforced earth are composite constructions consisting of artificial backfill strengthened with linear laid reinforcement. A typical shape (cross-section) of a wall and slope with reinforced earth is shown in Figure 1. [4]

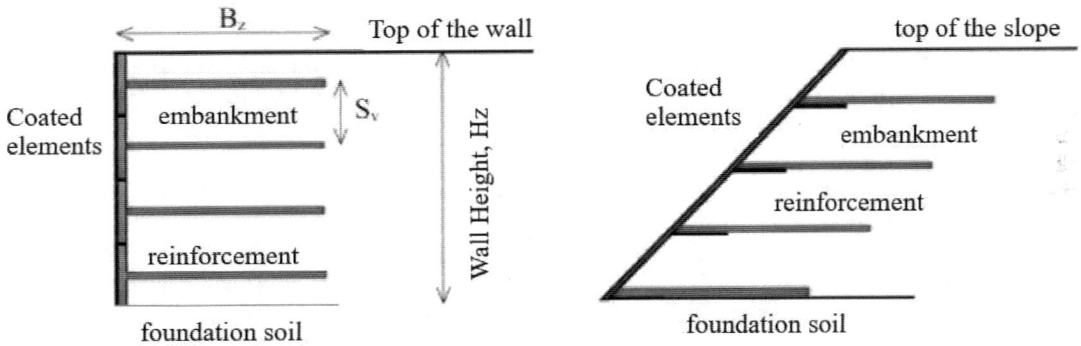

Figure 1. Typical cross-section of a wall and slope with reinforced earth [4]

Reinforced earth is widely used in many areas of civil engineering. One of the reasons for its wide application is the use of local materials, which significantly reduces construction costs. These types of structures can be constructed rather quickly and easily without the use of heavy machinery or intensive manual labour. [2] After the foundation soil is prepared, the wall is formed by combining earth and reinforcement material until the required height is reached. The typical ratio of width (Bz) and height (Hz) of reinforced earth is Bz/Hz=0.7. The length of the wall (Lz) is usually larger than the dimensions of a cross-sectional. Horizontally laid reinforcements that are placed at vertical distance (Sv) from 20 to 80 cm can be made of different materials. [4] As can be concluded from the technology used for producing reinforced earth, constructions built by reinforced earth are soil structures in which tensile elements or geosynthetics are placed during filling and compaction. There are many different types of geosynthetics available on the current market, including those shown in Figure 2. geotextiles, geogrids, geomembranes, geocomposites, etc. [3]

IOP Conf. Series: Materials Science and Engineering 1298 (2023) 012034 doi:10.1088/1757-899X/1298/1/012034

Figure 2. Most common geosynthetics [3]

Reinforced earth as a supporting structure has been used for over 50 years in modern construction. Later on, the paper will bring the analysis of solutions for noise on reinforced land in wall foundations on Highway Corridor Vc, subsection Tarčin-Tunnel Ivan.

3. Analysis of foundation solution for the noise protection walls on the reinforced earth on Corridor Vc, subsection Tarčin-Tunnel Ivan

Corridor Vc is a road and rail transport corridor that connects Hungary, Croatia and Bosnia and Herzegovina and is included in the network of Pan-European transport corridors, as shown on the Figure 3 [9]. Corridor Vc on the territory of Bosnia and Herzegovina is being built based on the "Feasibility Study of the Highway Corridor Vc through B&H" and it is estimated that the total length of the Corridor is about 320 km, of which about 275 km goes through the Federation of Bosnia and Herzegovina. The section of Corridor Vc Tarčin-Ivan is 6.9 km long and is divided into two subsections:

- subsection Tarčin-Tunnel Ivan,
- subsection Tunnel Ivan with two tunnel tubes. [5]

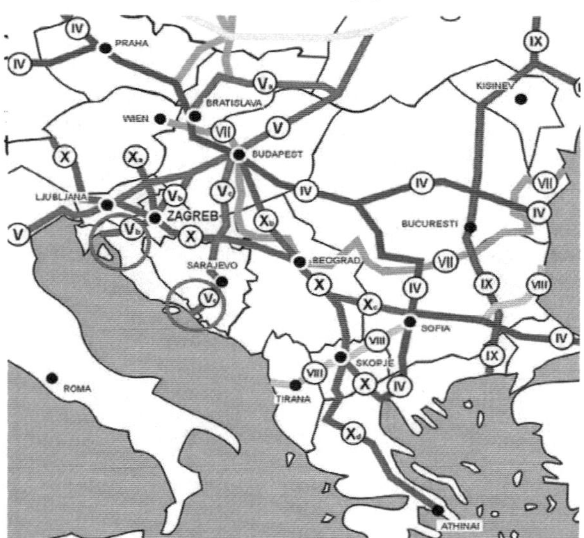

Figure 3. Highway Corridor Vc [9]

IOP Conf. Series: Materials Science and Engineering 1298 (2023) 012034 doi:10.1088/1757-899X/1298/1/012034

On the Tarčin-Tunnel Ivan subsection, a collision occurred where the noise protection walls, and reinforced earth collided over a total length of 229 m. The technical solution for the wall foundation on reinforced soil at this section of the highway route is subject to further analysis. The PLAXIS program for geotechnical calculations was used for the static calculation. The calculation was made according to EN 1997 Eurocode 7, Geotechnical Design. PLAXIS is a 3D geotechnical engineering software, primarily intended for geotechnical engineers. Keeping in mind that this is reinforced earth, it was not possible to make classic reinforced concrete single foundations for the noise protection walls, nor was it possible to make the foundation by driving steel pipes with a width circular profile. Therefore, the task was to find an effective solution which would not damage the stability of the reinforced earth, and which would enable the foundation of sound barriers. During the analysis, in accordance with DIN EN 1974, the load-bearing elements of noise protection walls are calculated for the following loads: aerodynamic wind load, aerodynamic pressure which was created from passing vehicles, self-weight of noise protection elements and dynamic load as a result of cleaning snow. [5]

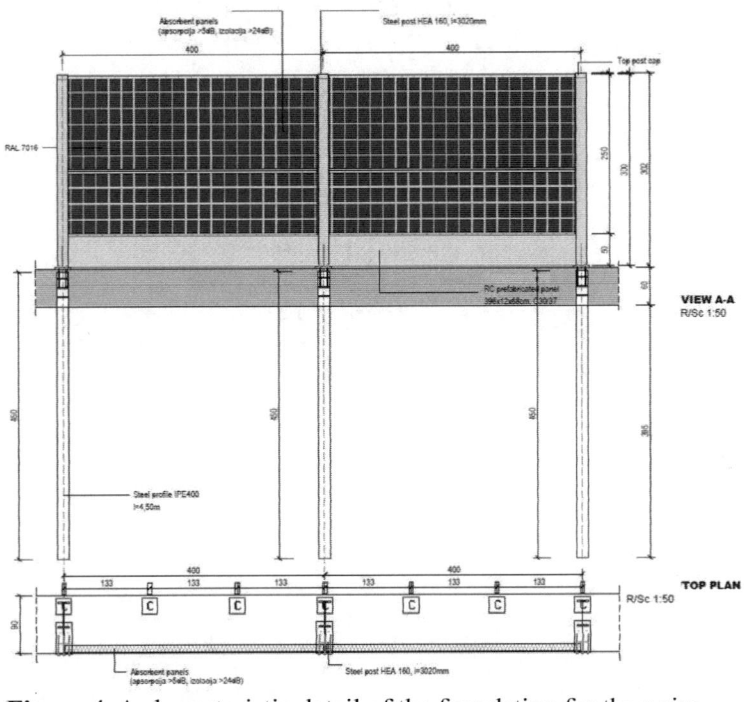

Figure 4. A characteristic detail of the foundation for the noise protection wall on the reinforced soil [5]

Furthermore, the calculation of the influence of shear force and bending moments was carried out for various wind zones, as well as the calculation of the horizontal displacement of the pile's top.
After a detailed analysis, it was determined that the optimal solution for the foundation of the wall, precisely because of the dimensions of the cross-section of the steel profile, is the IPE 400 profile, which is driven into the reinforced earth and takes over the function of a single foundation. Figure 4. shows a characteristic detail/enlarged view of the structure of the noise protection wall on reinforced earth. In the above-shown detail, it is evident that the load-bearing structure of the wall consists of HEA 160 steel columns, length l=3020 mm. The axial distance between the load-bearing columns is 400 cm. Between the columns, leaning against the ground, reinforced concrete prefabricated panels with dimensions of 396x12x68 cm, made of concrete quality C30/37, are placed. On these concrete panels, between the steel columns, the acoustic panels are installed, first row acoustic panel dim. 396x12x100cm, second row acoustic panel measuring dim. 396x12x150cm (total panel height 250cm, total wall height 302cm).

IOP Conf. Series: Materials Science and Engineering 1298 (2023) 012034 doi:10.1088/1757-899X/1298/1/012034

The color of the panel is anthracite gray - RAL 7016. The foundation of the wall is made by driving IPE 400 steel profiles, which are 450 cm long.

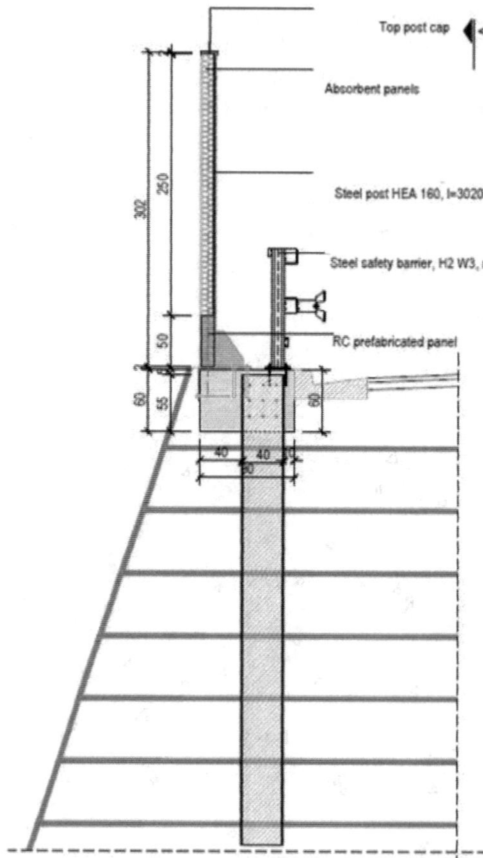

Figure 5. Characteristic cross-section of a noise protection wall on reinforced earth [5]

Figure 5. shows a characteristic cross-section of a noise protection wall on reinforced earth. In relation to the previous image, the section provides the following additional information:

- HEA 160 steel columns are welded to an anchor shoe made of sheet steel. The column is connected/anchored, through the sheet steel, to the foundation of the wall with anchor bolts;
- HEA 160 steel columns are additionally reinforced at the very bottom of the column. The reinforcement is made of a steel sheet in the shape of a triangle whose two tops are "cut off";
- steel profiles IPE 400, which serve as the single foundation of noise protection walls, are connected by a reinforced concrete beam, width 90 cm, height 60 cm. There is a distance of 40 cm from the edge of the panel to the beginning of the IPE 400 profile (the axis of the column and the axis of the foundation IPE profile are not in same line. The reason for this is to preserve the stability of the embankment / the IPE profile is distanced from the edge of the reinforced embankment - reinforced earth);
- the acoustic panels are finished with a cover element, while the steel supporting columns are finished with a cover cap made of an aluminium sheet;
- between the roadway and the panel, a safety road crash barrier (W beam guardrail) is installed. [5]

Considering that in Bosnia and Herzegovina, there is no recorded construction of any structure on reinforced earth, in order to confirm this innovative solution, control calculations and field testing of a sample - foundation pile IPE 400 driven into reinforced earth - was performed.

4. Control of the impact of installing steel IPE profiles on reinforced earth

In order to assess the impact of the steel IPE profiles, which serve as single foundations for noise protection walls, on the stability and functionality of the reinforced earth, an expertise was obtained by the Institute for Materials and Structures, which is the authorized public institution of Faculty of Civil Engineering of University of Sarajevo for this and similar issues. The expert's opinion was based on the examination of technical elements of the design solutions. For the purposes of testing the load-bearing capacity of driven foundation piles embedded in reinforced soil under horizontal loads, a testing program was prepared and conducted on a sample element on the section of the Highway Corridor Vc, subsection Tarčin-Ivan. The construction of a noise protection wall at the collision point between the wall and reinforced earth will depend on the results of these control calculations and field testing.

4.1. Control calculation

The primary objective of the control calculation is to assess the functionality and stability of the reinforced earth under these changed circumstances. The above-presented solution predicts damage to 7 upper geogrids in reinforced slopes with a face slope ratio of 3:1, and one (upper) geogrid is expected to be damaged in the zone of the reinforced slope with a face slope ratio of 1:1.25. The flange width of the IPE 400 profile is 18 cm, which defines the width at which local damage occurs to the geogrid. The control analysis for the characteristic profiles, shown in Table 1. was performed with the Plaxis 2D software package. The model of the reinforced slope was created, without simulating in detail the layers of the underlying earth that determine overall stability, but the stresses in the upper geogrids are controlled, which is also the subject of the analysis. [6]

Table 1. Designed elements and geometry of the reinforced earth structure and foundation profiles [6]

Sector	Cross-sectional profile	Gradient of the slope face [V:H]	Driving depth [m]	Strength of PET geogrid [kN/m]	Spacing of geogrid [m]
4	P124 to P131	3:1	4,0 to 4,5	80/30	0,6
6 & 7	P209 to P213	3:1	4,5	150,0	0,6
6 & 7	P218 to P226	1:1.25	<4,0	110,0	3,0

The obtained results for the characteristic profile P129 were compared with the results from the analysed solutions by comparing the maximum tensile forces in the zone reinforced with 80/30 geogrids. Results are almost the same (control calculation result: F=10,5 kN/m, monitored solution: F=10,63 kN/m), with deviations of approximately 1% which can be attributed to the use of a simpler Mohr-Coulomb model applied during the control. Figure 6. shows the tensile forces in the upper 7 geogrids that were damaged by the installation of steel profiles. The maximum force in these networks is 8,4 kN/m, which is approximately 20% less than the maximum force in the zone reinforced with 80/30 geogrids, shown in the given solution. The global safety factor against tensile failure of the most heavily loaded geogrid in the 80/3,0 grid zone is 80/10,5=7,6. Damaged nets with a maximum tensile strength of approximately 0.8*10,5=8,4 kN/m have a global tensile failure factor of (1-0,05)*80/8,4=9,0 (where 0,05 represents the width ratio of the damage, conservatively assumed as 0,20 m, and the profile spacing is 4,00 m). This means that damaged geonets have a higher over-dimensioning factor despite the damage

IOP Conf. Series: Materials Science and Engineering 1298 (2023) 012034 doi:10.1088/1757-899X/1298/1/012034

caused by drill piles with a width of 20 cm on every 4,00 m because they are less loaded compared to geonets at greater depths for which the calculation is done [6].

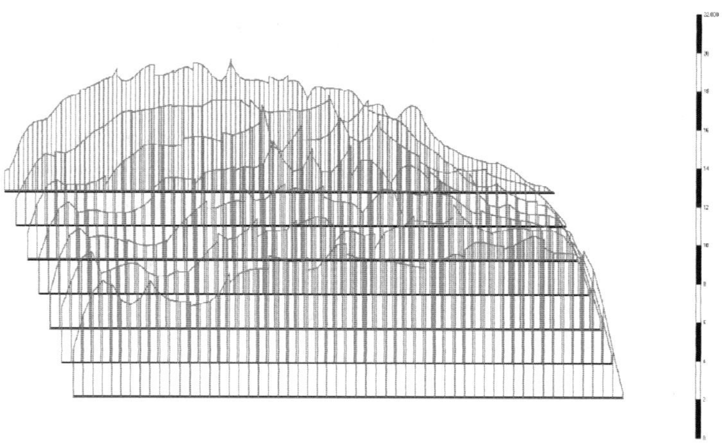

Figure 6. The tensile forces in the upper 7 geogrids for P129, F_{max}=8.4 kN/m < 10.5 kN/m (relevant for dimensioning) [6]

In addition, an analysis of the impact of load transfer from the foundation profile to the geogrids was carried out, which showed that the increase in tensile force is negligible and does not exceed 10% on one meter width. Based on the revision of the collected data and the results of the comprehensive analyses carried out as part of the expert opinion, the following conclusions were reached:

- performed simplified control calculation analyses generally give results consistent with those from the monitored project solution;
- the tensile forces in the upper geogrids that were damaged by the installation of steel profiles are at least 20% lower than the maximum tensile forces used for the purpose of the design calculation;
- damage to the geogrid in a width of approximately 20 cm (the width of the profile flange is 18 cm) at a distance of 4,00 m causes a reduction in tensile load capacity of 5%which is less than a 20% reduction in the tensile force that is relevant in the damaged zone compared to the force used during design calculation;
- the additional load from the sound barriers does not result in a significant increase in the maximum tensile forces in the geogrids because the basic supports of the sound barriers are in the zone closer to the face of the wall, which does not collide with the zone of maximum forces from the embankments own weight, which are further from the face of the slope.

Therefore, the installation of steel profiles and the related load will have no impact on the stability and functionality of the embankments made of reinforced earth designed for the specific conditions of the highway section in question. [6]

4.2. Horizontal Static Loads Testing

For the purposes of testing static bearing capacity of the sound barrier columns against horizontal loads, on the section of Highway Corridor Vc, subsection Tarčin-Ivan, the load-bearing capacity of the foundation steel IPE 400 profile was tested with the load applied at a height of 1,00 m over the temporarily welded HEA 160 profile. The IPE 400 profile is driven into the reinforced earth. According to the monitored design solution, the required pile driving depth is 4,00 m.

The columns were tested for the effects of static load applied at different heights above the top of the pile. The height is chosen in order to adjust to the position of the reactive system and generally, results in shear forces (bending moment and transverse force) are in accordance with the limit loads from the

IOP Conf. Series: Materials Science and Engineering 1298 (2023) 012034 doi:10.1088/1757-899X/1298/1/012034

analysed solution. The load was applied using a hydraulic press, in stages, with continuous monitoring of load and potential displacement (Figure 7.). [7]

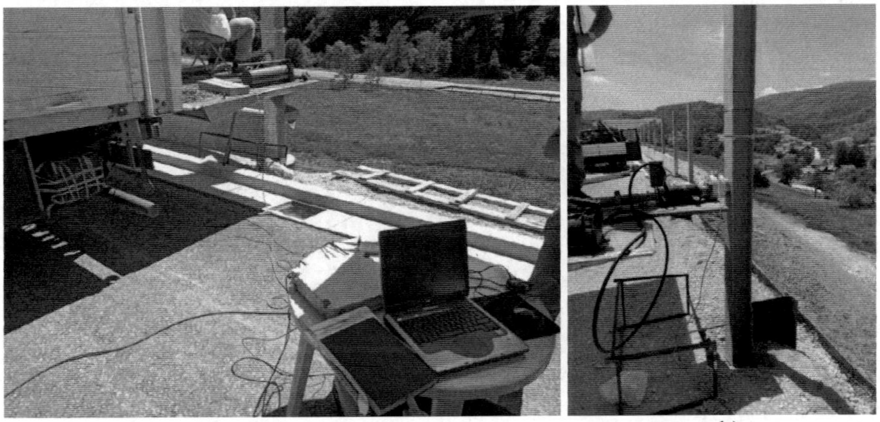

a) b)

Figure 7. Column testing: a) test setup, b) position of force application instrument, comparators, and reference points on the column [7]

The aim of the testing was to achieve stresses equivalent to those caused by wind. The displacements at the bottom of the column are continuously monitored by measuring comparators fixed to the stand which were connected to the Data collection unit (Spider). The displacements of the column at different heights above the top of the foundation were monitored using a geodetic levelling instrument. The test results are shown in Table 2. [7]

Table 2. Measured deflections for the tested columns [7]

Force [kN]	Displacement at the top of the pile [mm]	Column displacement – Geodetic levelling instrument [mm]
0.0	0.0	0.0
5.04	0.76	4.0
10.08	1.6	6.0
15.12	2.38	9.0
20.04	3.4	13.0
25.08	4.6	20.0
29.94	5.8	29.0
0.0	0.78	3.0

According to the measurement results after unloading, relatively small irreversible movements were determined. This shows that the stress around the horizontally loaded pile applied during the test did not cause deformation on the reinforced earth. The measured displacements at the top of the pile for the limit loads from the project are significantly less than 10% of the flange width of the IPE profile, which confirms that the external bearing capacity of the earth around the pile is not impaired.

Based on the previously performed load test and the analysis of the activity of the tested structure of the noise protection wall with foundation steel profiles (piles) intended for use on the subsection Tarčin-Tunnel Ivan on Highway Corridor Vc, the following conclusions were reached:

- the tested structure shows better parameters and characteristics compared to design assumptions;

- the tested structure meets the conditions specified in the manufacturer's technical sheet and has the necessary load capacity to take over the load foreseen by the design.

Based on the obtained expert's opinion and report of the above-mentioned Institute for Materials and Structures on the conducted tests [6] [7] [8], it is proven that foundation for noise protection walls with steel IPE profiles on reinforced earth meets all the required parameters with better-than-expected results.

5. Conclusion

As discussed in Introduction, in previous practice in Bosnia and Herzegovina, constructions on reinforced earth were not allowed by the Client, because it was assumed that the intervention would cause structural damage to the reinforced earth. Having this in mind, the purpose of the conducted analysis was to find a solution for construction on reinforced earth. Namely, due to the assumption that possible damage to the geogrid would lead to a negative impact on the reinforced earth structure, the possibility of construction on reinforced earth was not previously analysed in detail or thoroughly considered. With the use of the PLAXIS program for structural calculations, a solution was found for the foundation of noise protection walls with IPE 400 steel piles. IPE profiles have high strength and stability and are frequently used in the construction of structures that are often exposed to heavy loads. The IPE 400 profile was chosen precisely because of its advantages, which arise from the dimensions of the profile, i.e. due to its cross-sectional height, the steel profile IPE 400 has a high resistance to bending. In addition to the issue of load-carrying capacity, the objective was to check whether the pile driving will damage the stability and functionality of the reinforced earth embankment.

This was revised through a control calculation, and the load-bearing capacity was confirmed through on-site testing of the structure (example) on horizontal loads.

The results of testing and control numerical analysis confirmed that the tensile forces in the upper geogrids, which would be damaged by the driving of steel piles, are significantly less than the maximum tensile forces based on which the calculations were made, and at the same time, the driving of piles does not affect the face of the slope of the reinforced earth embankment. This is also confirmed by the fact that the planned driving of sound barrier piles is carried out in the zone closer to the face of the wall. Moreover, the testing proved that in the area of the damaged geogrids, their bearing capacity was reduced by only 5%, which is significantly less than the calculated assumption of 20%. Lastly, it can be concluded that the results of all performed tests and control of the analysed solution confirm that the applied design methods, the used materials, and the work technology fully met the expectations and set requirements.

6. References

[1] University of Ljubljana - Faculty of Civil and Geodetic Engineering and DDC Consulting & Engineering Ltd 2005 *Guidelines for the design, construction, maintenance and supervision of roads, Book I: Design, Chapter 6: Road and Environment,* Directorate of Roads of the Federation of Bosnia and Herzegovina and JP "Roads of Republika Srpska" p **5-42**

[2] Roje-Bonacci T 2005 *Support Structures and Construction Pits*, Faculty of Civil Engineering and Architecture 9 p **121**

[3] Bogdan A 2022 Reinforced *earth as a supporting structure*, Journal Građevinar 74, Zagreb 5 p **433-441**

[4] Skejić A 2017 *Numerical modelling of static bearing capacity test of bored pile built in stiff marly clay,* Faculty of Civil Engineering, University of Zagreb, Zagreb p **2**

[5] Veladžić A 2023 *Analysis of steel structural elements of sound barriers in traffic*, Technical Faculty of the University of Bihać p **42-71**

[6] Skejić A and Balić A 2022 *Expert opinion on the possible impact of the installation of steel profiles for the foundation of the sound barrier columns on the stability and functionality of the reinforced embankment,* Institute for Materials and Structures, Faculty of Civil Engineering, University of Sarajevo

[7] Medić S and Skejić A 2022 *Elaboration on testing the bearing capacity of columns for*

horizontal loads on the wall for noise protection on the section of the Tarčin-Ivan highway, Institute for Materials and Structures, Faculty of Civil Engineering, University of Sarajevo

[8] Medić S and Skejić A 2022 *Elaboration on testing the bearing capacity of columns for horizontal loads on the wall for noise protection on the section of the Tarčin-Ivan highway* Institute for Materials and Structures, Faculty of Civil Engineering, University of Sarajevo

[9] Bogović L, Jović M, Tijan E, Hadžić A P 2020 *The importance of the port of Rijeka within the - Pan-European corridor V* International Virtual Academic Conference - Education and Social Sciences Business and Economics, Republic of N. Macedonia p **69**

AUTHOR INDEX

Institute of Physics
Dirac House, Temple Back
Bristol BS1 6BE UK

ISSN: 1757-8981
ISBN 978-1-7138-8735-5
9 781713 887355